Honeybee Neurobiology and Behavior

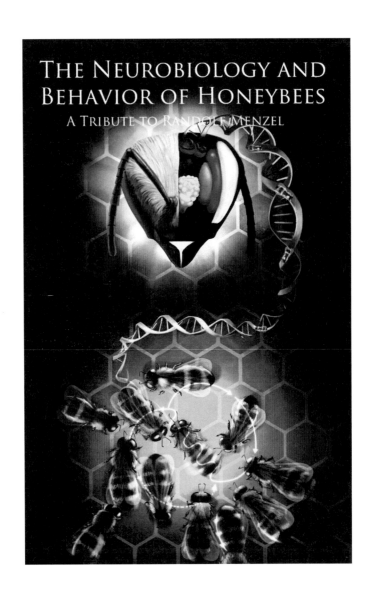

THE NEUROBIOLOGY AND BEHAVIOR OF HONEYBEES

A TRIBUTE TO RANDOLF MENZEL

C. Giovanni Galizia · Dorothea Eisenhardt
Martin Giurfa

Editors

Honeybee Neurobiology and Behavior

A Tribute to Randolf Menzel

 Springer

Editors
C. Giovanni Galizia
Department of Neurobiology
Universität Konstanz
Universitätsstrasse 10
78457 Konstanz
Germany
giovanni.galizia@uni-konstanz.de

Dorothea Eisenhardt
Department of Neurobiology
Freie Universität Berlin
Königin-Luise-Str. 28-30
14195 Berlin
Germany
theodora@neurobiologie.fu-berlin.de

Martin Giurfa
Centre de Recherches sur la
Cognition Animale
CNRS - Université Paul Sabatier
Route de Narbonne 118
31062 Toulouse Cedex 9
France
giurfa@cict.fr

ISBN 978-94-007-2098-5   e-ISBN 978-94-007-2099-2
DOI 10.1007/978-94-007-2099-2
Springer Dordrecht Heidelberg London New York

Library of Congress Control Number: 2011939201

Printed on acid-free paper

Springer is part of Springer Science+Business Media (www.springer.com)

Prologue

Twenty-five years ago, Randolf Menzel and Alison Mercer edited a book that marked several generations of researchers and students who had chosen, for different reasons, the honey bee as a model for their studies. The book, whose title was Neurobiology and Behavior of Honeybees, was published in 1987, at a time in which studies on honeybee sensory physiology and learning and memory – which occupied an important place in the book contents – were intensively developed. Some of us used to read this book as students, attracted by its content and scientific contributions, but more than that, marveled by the apparently inexhaustible potential of the honey bee as a model to understand basic questions in biology. In a sense, the book was the best demonstration that Karl von Frisch (1886–1982), the father of studies on the behavior of honeybees, was right when he described honeybees as a "magic well" for discoveries in biology because "the more is drawn from it, the more is to draw".

In these last two decades, researchers have continued to draw from this magic well. We have seen the adoption of new ideas and concepts in the study of bee behavior and neurobiology, we have benefited from new tools for the analyses of the bee brain that were unimaginable some 20 years ago, and we have opened new research pathways with the advent of the genomic era, which resulted in the sequencing of the entire honeybee genome. Thus, to the question of "why producing a new version of the book Neurobiology and Behavior of Honeybees in 2011?" it seemed to us that the answer could be multiple.

We wanted to underline the potential of the honey bee as a model system to tackle different fundamental scientific questions, particularly at the behavioral, neural and molecular levels. In that sense, appreciating how research has evolved in the last two decades is important to appreciate where research is heading and which are the essential questions that we need to answer in the immediate and not so immediate future. At the behavioral level, for instance, fundamental changes have occurred since the 1990s mostly based on a conceptual switch that changed the way honey bee researchers view their study object. Despite the fact that the bee, like many other invertebrates, was traditionally a powerful model for the study of learning and memory, the questions and behavioral analyses that had been undertaken were concerned by simple associative learning forms, usually the establishment of a simple link between a visual stimulus or an odorant and sucrose reward. Yet, understanding the behavioral plasticity of an organism implies acknowledging that such elemental links are just part of its plastic repertoire and that higher-order learning forms, not amenable to simple associative analyses underlie sometimes problem resolution in a natural context. This conceptual twist adopted by several bee researchers in the last two decades has made of the honey bee one of the most tractable models for the study of cognitive performances. Indeed, studies on categorization, rule extraction, transitivity, top-down modulation of perception, and numerosity (to cite just some examples) have been performed in the honey bee, which are still out of reach in other powerful invertebrate models which despite disposing of fantastic tools for neuronal control do not exhibit the behavioral richness expressed by bees in these recent studies. Studies on navigation benefited from novel radar tracking technologies that allowed, for the first time, reconstructing entire flight pathways in the field, and thereby yielding the view that bees build spatial representations akin to a cognitive map. It is to note that despite presenting several chapters on bee learning and memory, the former book by Menzel and Mercer did not mention a single time the word 'cognition'. It thus seems that the cognitive revolution has reached the bee as a model system, thus providing a novel glint to the already solid architecture of behavioral studies existing in this insect.

At a neural level, new approaches using imaging technology have emerged in these last two decades that allowed for simultaneous recording of neuronal assembles, and that revealed spatial functional architectures in the brain. In this way both olfactory coding and more recently specific aspects of visual coding have been uncovered. In electrophysiology, although single-unit recordings are still fundamental to understand the properties of network components, new approaches based on multielectrodes have been developed that allow the simultaneous recording of many neurons at the same time with very high temporal resolution. Both approaches could be coupled with behavioral analysis *in toto* thus starting to reveal, in the case of olfactory learning, how and where olfactory memories are located in the bee brain.

Molecular biology has helped us to understand many new aspects of honeybee neuroscience. The honey bee genome has been sequenced so that genetic and molecular architectures underlying several aspects of bee behavior are now known. Yet, the advent of the genomic era with its panoply of sophisticated molecular techniques that are being adapted to the honey bee has yielded new challenges both for the molecular biologist, the neuroscientist and the behavioral biologist. For instance, olfactory and gustatory receptor genes have been identified in the honey bees but their specific ligands are still unknown. We also know that a bee has fewer genes involved in its immune system as compared to other insects, but which consequence should be derived from this fact is unclear.

All in all, a first level of response as to why this book would be necessary in 2011 is given by the fact that honey bee studies have made tremendous progresses in the last two decades and that it was therefore timely to cover this evolution in order to provide an integrative view on our current state of knowledge.

Finally, there is the personal tribute that this book wants to convey. The book is conceived as homage to Randolf Menzel and his long and productive career devoted to understanding multiple aspects of honey bee behavior and neurobiology.

Since Karl von Frisch, no researcher has covered with such a success so many aspects of the biology of the honey bee as Randolf Menzel did. From color vision and its cellular bases, to olfactory learning and the formation of olfactory memories and their cellular and molecular bases, from large-field navigation to floral ecology, the profusion of topics covered in Randolf Menzel's career is truly unique and impressively rich. What is unique in him is the combination between the laboratory neurobiologist and the field researcher, between the molecular biologist interested in the intracellular cascades underlying memory and the ecologist who travels around the world studying floral coloration in order to understand the neural bases of color vision. The original combination of behavioral, neurobiological, pharmacological, molecular, psychological and ecological approaches defines the richness of his multidisciplinary work. In his institute, he established long ago a rule that had to be respected: any student willing to work on neural or molecular levels in the laboratory had to perform at least once a field experiment on bee navigation in order to appreciate the complexity, plasticity and richness of honeybee behavior in a natural context. Copying from the richness of discussions in his group, we have designed a new format for this book: after each section, Randolf Menzel contributes with a dedicated commentary, putting the various aspects into a timeline joining past and not yet foreseen future, and making this book not only a collection of the state-of-the-art, but a real workbook for future research.

All three of us, former disciples of Randolf Menzel, wanted to celebrate through this book the life's work of our former mentor. We believed that such esteem was extensive to many other colleagues that did not work directly with him but who nevertheless appreciate the richness of his contributions. We were right in this conclusion as many coauthors joined us in this initiative to produce this book. We all coincided in acknowledging that Randolf Menzel is a passionate researcher, a researcher having devoted his life to understanding a mini brain which has uncovered some of his mysteries thanks to his work. Despite of this, we – including him – all know that there are still many, too many, questions to answer.

We really wish that this book will enlighten the pathways to take in order to find such answers. If there is a lesson to be taken from this book and from the research work contributed by Randolf Menzel is that Karl von Frisch was right when he called the honeybee a magic well – no matter how much is discovered, more remains to be uncovered.

Let us end with thanking all those who made this book possible: our friends and colleagues who joined us to celebrate the 70th birthday of Randolf Menzel in June 2010, the authors of the chapters, who took the care of writing these excellent texts, Ignacio (Nacho) Malter Terrada for the cover art (we are sorry that the publisher could not print it on the external cover for commercial reasons), and especially Mihaela Mihaylova for the impressive amount of work in the production of the book, assuring the highest standards of quality.

Martin Giurfa
Dorothea Eisenhardt
Giovanni Galizia

Contents

Part I
Mechanisms of Social Organization

Chapter 1.1
The Spirit of the Hive and How a Superorganism Evolves

Robert E. Page, Jr.

Abstract Social insects presented Darwin (1859) with major difficulties for his fledgling theory of evolution by natural selection. How could differential survival and reproduction result in sterility, differential anatomy and behavior between sterile workers and queens, and differentiation among the sterile individuals of a colony? Maurice Maeterlink, Belgian author and Nobel Laureate, wrote (in 1901) about the "inverted city" of the honey bee noting that there is no central authority, that order and organization is achieved mysteriously through what he called the "spirit of the hive" [Maeterlink M, The life of the bee. Dodd, Mead, and Company, New York, 1913]. William Morton Wheeler [J Morphol 22:307–325, 1911; The social insects. Harcourt, Brace and company, New York, 1928], Harvard entomologist and philosopher, proposed that insect societies are true "superorganisms" because they are organized for nutrition, reproduction and defense, a view that was initially supported by biologists but lost favor by the early 1970s. Hölldobler and Wilson resurrected the superorganism in their book *The Superorganism: the Beauty, Elegance, and Strangeness of Insect Societies* [W. W. Norton, New York, 2008]. However, fundamental questions remain about the evolution of insect societies as superorganisms. Not only is there order without central control (the spirit of the hive), there is also no central genome on which natural selection can operate to sculpt a social system. Here, I will locate and define the honey bee "spirit of the hive" and show how selection operating on social traits involved in colony nutrition, a superorganismal trait of Wheeler, changes the genome, development, physiology, and behavior of individual workers that affect the "spirit of the hive" and, therefore, social organization.

R.E. Page, Jr. (✉)
School of Life Sciences, Arizona State University, Tempe, AZ, USA
e-mail: repage@asu.edu

C.G. Galizia et al. (eds.), *Honeybee Neurobiology and Behavior: A Tribute to Randolf Menzel*, DOI 10.1007/978-94-007-2099-2_1,
© Springer Science+Business Media B.V. 2012

Abbreviations

dsRNA   double stranded RNA
QTL   Quantitative trait loci

1.1.1 Introduction

Social insects have fascinated natural historians and philosophers since Aristotle and continue to fascinate us today with their self-sacrificing altruism, complex nest architecture, untiring industry, and division of labor. However, they presented Charles Darwin with special difficulties for his fledgling theory of evolution by natural selection. How can sterile castes, such as worker honey bees and ants, evolve when they don't normally reproduce? The existence of sterile castes seems to be in direct opposition to a theory that requires differential survival and reproductive success. However, Darwin considered a bigger difficulty to be the observation that the reproductive individuals in colonies are often anatomically differentiated from the sterile workers, showing adaptation of a sterile caste. He considered the biggest difficulty to be the anatomical differentiation within the worker caste that is dramatically demonstrated in many species of ants. Darwin waved his arms and invoked selection on families as an explanation, an explanation later shown not to be that simple.

Social insects provided additional difficulties for Darwin when he considered the architecture of the honey bee nest. Darwin had a Cambridge mathematician study the comb of the bee from an engineering perspective of strength and economy and concluded "for the comb of the hive bee, as far as we can see, is absolutely perfect in economizing labour and wax." How could the wax combs be built with such precision to maximize the strength of the comb and at the same time save costly building materials? And, as he pointed out, "this is effected by a crowd of bees working in a dark hive" ([8], p. 339). How could they achieve this architectural feat with instincts alone, working without any central control of construction tasks? Darwin experimented with honey bees and demonstrated to his satisfaction that bees could construct combs using just their instincts and local information regarding cell construction, thereby, solving his dilemma of perfection and instincts.

The Nobel Laureate poet, playwright, and author, Maurice Maeterlink also was fascinated by social insects. In his book *The Life of the Bee* ([20], pp. 38–39) he noted that there was no central control of cooperative behavior, thought by many to be the domain of the queen, and noted "She is not the queen in the sense in which men use the word. She issues no orders; she obeys, as meekly as the humblest of her subjects, the masked power, sovereignly wise, that for the present, and till we attempt to locate it, we will term the 'spirit of the hive'." Here he resorted to a mystical vitalism to explain how colonies full of individuals working in the dark organize into a cooperative whole, and left it for later for someone to identify the "spirit of the hive" and where it resides.

   William Morton Wheeler, the early twentieth century entomologist and philosopher rejected the vitalism of Maeterlink but also the strict interpretation of Darwin as an explanation for the existence of social insects, "... the 'struggle for existence,'survival of the fittest.' 'Nature red in tooth and claw,' ... depicts not more than half the whole truth" ([49], p. 3). He believed that Darwinian selection based on competition for survival and reproduction could not build the kind of cooperation he observed in colonies of his beloved ants. In 1911 he wrote an essay "The Ant Colony as an Organism". He defined an organism as a "complex, definitely coordinated and therefore individualized system of activities, which are primarily directed to obtaining and assimilating substances from an environment, to producing other similar systems, known as offspring, and to protecting the system itself and usually also its offspring from disturbances emanating from the environment" ([48], p. 308). In other words, they are organized for nutrition, reproduction, and defense. He later proposed [50] the term superorganism in an apparent attempt to set aside social insect evolution from, or expand on, the individual-based Darwinian struggle for existence. However, he did not provide an alternative to natural selection nor a mechanism for the evolution of the superorganism.

   The concept of a "Superorganism" lost favor among biologist by the early 1970s because it failed to provide rigorous experimental paradigms for studying insect sociality. However, in 2008, the superorganism was re-revived with publication of "The Superorganism: the Beauty, Elegance and Strangeness of Insect Societies" by Hölldobler and Wilson [13]. They present their view of how multi-level selection can shape social structure and address the three major themes presented by Darwin, Maeterlink, and Wheeler: (1) the evolution of sterility, (2) the evolution of insect castes, and (3) how colonies composed of large numbers of individuals organize themselves into a cooperative social unit without central control (the "spirit of the hive"). From their book it is apparent that in the 150 years since publication of the *Origin of Species* we have solved part of the problem of the evolution of sterility. Though the debate continues [22], we have worked out some of the developmental mechanisms of caste determination, and we have a better understanding of the ways in which colonies self-organize into social units with a division of labor. But, we still know little about how selection on colony (superorganismal) traits derived from complex social interactions, such as food storage, effects heritable changes that are reflected through different levels of organization such as development, physiology, and behavior of workers – the Darwinian explanation.

   How does a superorganism evolve a complex social organization? There is no centralized control of behavior, no social genome on which natural selection can act, and a hierarchy of organizational levels from genes to the society. In the following sections I am going to provide a brief overview of the biology and natural history of honey bees, define the mechanisms behind the mystical "spirit of the hive" of Maeterlink, and then discuss a 20 year selection program designed to map the effects of colony level selection on a single social trait, a characteristic of the superorganism, across different levels of biological organization.

1.1.2   Natural History of the Honey Bee

A honey bee colony typically consists of 10–40,000 worker bees who are all female, and depending on the time of year, zero to several 100 males (drones), and a single queen – the mother of the colony (see [51] for a description of the behavior and life history of honey bees). The nest is usually constructed within a dark cavity and is composed of vertically oriented, parallel combs made of wax secreted by the workers. Each comb can contain thousands of individual hexagonal cells on each of the vertical surfaces. The individual cells of the combs serve as vessels for the storage of honey (the carbohydrate food source for bees), pollen (the source of protein), and as individual nurseries for developing eggs, larvae, and pupae. In addition, the comb serves as the social substrate for the colony. The nest has an organizational structure that is similar to concentric hemispheres, only expressed in vertical planes, where the innermost hemisphere contains the larvae and pupae (the brood), the next hemisphere above and to the sides of the brood contains the stored pollen, and the upper and outer regions contain honey that is derived from the nectar of flowers. If you remove a comb that is near, but to the side of, the center of the nest it will contain three bands covering both sides: the outer band will be honey, the center band pollen, and the lower central part of the comb will contain the brood. The amount of surplus pollen is regulated by colonies, first shown by Jennifer Fewell and Mark Winston [12]. They added pollen to colonies and then looked at the effects on pollen foraging and pollen intake. Colonies reduced the intake of pollen until they consumed the "surplus". When pollen was removed from colonies, pollen intake increased until the surplus pollen was restored.

In addition to the social and nest structures, there is also a structured division of labor [51]. When workers first emerge from their cells as adults they engage in cleaning cells in the brood nest. When they are about a week old they feed and care for larvae, followed by tasks associated with nest construction and maintenance, food processing, receiving nectar from foragers, guarding the entrance, etc. Then in about their third or fourth week of life they initiate foraging. As foragers they tend to specialize on collecting pollen or nectar, demonstrated by a bias in the amount of each when they return to the nest. Once they initiate foraging they seldom perform any within-nest tasks for the duration of their short lives of 5–6 weeks.

1.1.3   What Is the Spirit of the Hive?

One cannot observe a hive of honey bees without getting the feeling that they are engaged in highly coordinated and cooperative behavior. As discussed above, both Darwin and Maeterlink struggled with how this can occur. It seems as if there must be some kind of central control, but on careful examination none can be found. This led Maeterlink to call upon the "spirit of the hive". But what is it? I will show you here that the coordinated behavior long observed and admired emerges from algorithms of self organization and requires only that worker honey bees respond to stimuli that they encounter; when they respond they change the amount of stimulus at that location and thereby affect the behavioral probabilities of their nestmates.

Stored pollen inhibits foragers from collecting pollen while young larvae stimulate pollen foraging. Young larvae produce a mixture of chemicals, called brood pheromone, that is secreted onto the surface of their bodies [18] (see also Chap. 1.3). It is the brood pheromone that releases pollen foraging behavior [33]. Pollen foragers returning from a foraging trip seek out combs with brood and pollen and walk along the margin where they have the opportunity to have contact with the pheromone produced by larvae, and contact stored pollen [10, 43]. Stored pollen that is located in the comb is consumed by nurse bees, so it is likely that returning foragers can assess the need for pollen by contacting empty cells along this margin. Empty cells would indicate that pollen had been consumed and fed to developing larvae. Behavior of a pollen forager is affected by direct contact with the brood and stored pollen. They apparently use information obtained from direct contact to assess "colony need" [9, 45].

Imagine that pollen foragers have response thresholds to empty cells encountered along the brood/pollen boundary. If a forager encounters more empty cells than some value representative of her response threshold, she will leave the hive and collect another load of pollen. If, however, she encounters fewer empty cells, she does not continue to forage for pollen, perhaps she is recruited to nectar or water foraging. This is a very simple view, but not unsupported. Seeley [43], reporting unpublished results of Scott Camazine, showed that the number of cells inspected before unloading increased with more stored pollen. In addition, the probability that a pollen forager performed a recruitment dance decreased with more stored pollen, showing that pollen foragers are able to make local pollen stores assessments. Figure 1.1.1 shows a cartoon of a returning pollen forager assessing empty cells on a comb. She has a response threshold function of 20 empty cells (f_{20}). If she encounters more than 20 empty cells, she will unload and make another foraging trip. If she encounters fewer, she will stop foraging for pollen. The other individual has a threshold function of 21 cells. The pollen forager unloads her pollen then makes another trip. By unloading her pollen, she changes the pollen stores stimulus from 22 empty cells to 21, which now is below the pollen foraging response threshold of the other individual. Thus by responding to the stimulus, the number of empty cells, the forager decreases the stimulus by depositing pollen, and affects the probability that other individuals will engage in that task. Though this is an over-simplified example, and I'm not implying bees can count, this example demonstrates the fundamental basis of self-organized division of labor in the nest.

1.1.4   How Does Complex Social Behavior Evolve?

Response to a stimulus and the correlated change in the stimulus as a result of the behavioral response is the fundamental mechanism of social organization and, therefore, the "spirit of the hive" of Maeterlink. But how does complex social

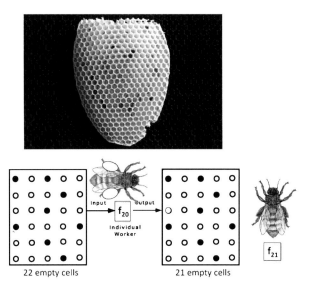

Fig. 1.1.1 Cartoon showing the response relationship between the stimulus level (empty cells) and behavior, and the correlation between behavior and the stimulus. At the *top* is a piece of comb used for pollen storage. *Below left* is a diagramatic representation of a comb with 22 empty cells. The worker honey bee between the combs has a response threshold function of 20 (f_{20}). The diagram of a comb on the *right* shows 21 empty cells. The bee on the *right* has a response threshold function of f_{21}. See text for full description

behavior evolve? To answer this question, in 1990 Kim Fondrk and I initiated a large scale breeding program. We selected for a single trait, the amount of surplus pollen stored in the comb, also known as pollen hoarding. Stored pollen is regulated by colonies of bees [12], therefore, it makes a good social phenotype for a study of colony-level selection. It is the consequence of the activities of thousands of individual workers. Nurse bees consume the protein rich pollen and convert it into glandular secretions that are fed to developing larvae. The larvae are the end point consumers. Thousands of workers engage in collecting and storing the pollen, and in recruiting new foragers.

Our selection program was successful in producing two "strains" of honey bees that differed dramatically in the social trait we selected, pollen hoarding [25]. Colonies of the high pollen hoarding strain ("high strain") store on average more than 10 times more pollen than colonies of the low strain. We now have completed more than 30 generations of selection. While we conducted the selection experiment we repeatedly asked the question, "what changes have occurred at different levels of organization?" To address this question, we conducted common garden experiments where we placed high and low strain workers into hives soon after they emerged as adults and observed their behavior. High and low strain bees were marked so we could determine their origins. We collected them as they returned from foraging

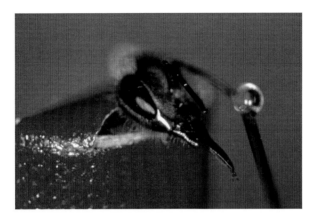

Fig. 1.1.2 Proboscis extension reflex (PER) of the honey bee. A droplet of sugar water is touched to the antenna of the bee. If the sugar concentration is sufficiently high the bee will reflexively extend her proboscis as shown with this bee. Using this technique we determine the sucrose response threshold of a bee by presenting her with a series of increasing concentrations. The concentration at which she first responds is her threshold (Photo by Joachim Erber)

trips and determined the age at which they initiated foraging, the weight of nectar collected, the concentration of the nectar, and the weight of their pollen loads [24, 30, 32, 34].

High strain bees foraged earlier in life (about 12 days in some experiments), collected relatively more pollen, less nectar, nectar of a lower average concentration of sugar, and were more likely to collect water than were the low strain bees [32]. In addition, high strain bees were less likely to return empty from a foraging trip. We expected colony level selection for pollen hoarding to affect pollen and nectar load sizes, but we didn't expect it to affect foraging age, the concentration of nectar collected, water foraging, or the likelihood of being an unsuccessful forager. I had no explanation until Joachim Erber from the Technical University of Berlin joined the effort (see also Chap. 6.4). Together we asked if pollen and nectar foragers differed in their reflex to sugar solutions of different concentrations. We used the proboscis extension reflex (PER) assay (Fig. 1.1.2) to measure the reflex thresholds of returning foragers. We found that pollen foragers responded more readily to water and to sucrose solutions of lower concentration than did nectar foragers. This was a surprising result that motivated us to test workers from the high and low pollen hoarding strains [24].

The sucrose responses of returning foragers could have been the result of their foraging activity. However, it could also have been a result of fundamental differences in pollen and nectar foragers that occur prior to initiating foraging that influence their foraging decisions. High and low strain bees differ in their foraging behavior, as discussed above. We tested the sucrose response of high and low strain workers soon (hours to days) after they emerged from their brood cells as adults [30].

This test was independent of foraging behavioral differences because bees don't normally initiate foraging until after their second week of adult life. We found that high-strain bees were significantly more responsive to water and to low concentrations of sugar solution than were low-strain bees. Therefore, sucrose responsiveness could be an indicator of fundamental neurological differences between bees that exist already early in adult life and that affect foraging decisions much later. These differences may be at least partially responsible for the division of labor and be selectable components of the "spirit of the hive".

If the sensory response system differences we observe are fundamental and affect foraging behavior, then we should be able to test very young bees and predict their foraging behavior 2 or 3 weeks later. Tanya Pankiw tested this hypothesis by taking very young and newly emerged bees from "wild-type" colonies (commercial colonies not derived from the high or low pollen hoarding strains). She tested them for their response thresholds to sucrose, marked them, put them in a commercial hive, and collected them when they returned from foraging trips weeks later [29, 31]. The results showed that bees that collected water were, on average, the most responsive to water and low concentrations of sucrose followed by bees that collected only pollen, those that collected both, and then those that collected only nectar. Bees that returned empty were those that were the least responsive to sucrose solutions when tested soon after emerging as adults. In addition, she found a significant negative correlation between the concentration of nectar collected by bees and the responsiveness of bees to sucrose. Bees that were the most sensitive to low concentrations of sucrose collected nectar that was more dilute than those that were less sensitive (see also [28]).

We selected for a single trait, the amount of surplus pollen stored in combs. We looked for differences between our strains at different levels of biological organization such as individual foraging behavior and sensory responses [11, 24, 30, 32]. We also looked at the differences in learning and memory and neurobiochemistry [14, 39, 40, 42]. We compared high and low strain workers and we looked for correlations of these traits in wild-type bees to determine if the relationships were specific to our selected strains or represented general properties of the behavioral organization of honey bees [23, 27, 29, 31, 37, 38, 41]. It is interesting how the response to sucrose correlates with such a broad set of behavioral and physiological traits, thus defining a phenotypic architecture associated with foraging behavior that can be changed by colony level selection on stored pollen (Fig. 1.1.3).

At the individual behavioral level, we have defined a pollen hoarding behavioral syndrome that consists of an early onset of foraging, a bias to collect more pollen and less nectar, and the tendency to collect nectar with lower concentrations of sugar, and water. This syndrome of correlated traits is linked together with a pleiotropic network of genes. Greg Hunt constructed the first genetic map of any social insect and mapped quantitative trait loci (QTL) that affected differences in quantities of stored pollen between the high and low pollen hoarding strains [15]. We initially identified three quantitative trait loci (mapped regions on chromosomes that contain genes) that explained significant amounts of the variation in stored pollen that we observed between the high- and low-pollen-hoarding strains (Fig. 1.1.4).

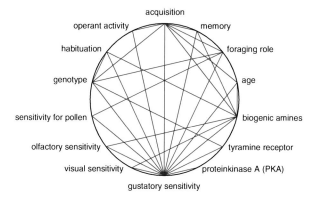

Fig. 1.1.3 The phenotypic architecture of the pollen hoarding syndrome. Phenotypic traits span levels of biological organization from the genotype to foraging behavior. Lines connect traits that have been demonstrated to be significantly correlated. Studies were performed on high and low strain workers as well as wild-type bees. (See [23, 27] for reviews of the studies that were used for this diagram). The connection between genotype and habituation is based on unpublished data (The original figure was drawn by Joachim Erber)

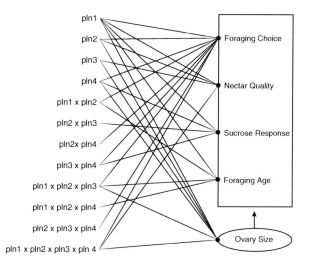

Fig. 1.1.4 The genetic architecture of the traits associated with the pollen hoarding syndrome, and ovary size (number of ovarioles). Ovary size is believed to be the cause of much of the pleiotropy observed. On the *left* are the QTL pln1–pln4 and their interactions. On the *right* are the mapped traits know to associate with the QTL

Next, with Olav Rueppell [35, 36], we mapped the individual foraging and sucrose response traits and found a fourth QTL. We also found that all QTL affected all traits (pleiotropy) and all QTL affected each other (epistasis). The honey bee genome sequence allowed us to look into these regions and seek candidate genes that we are currently testing for effects on foraging behavior. However, our attention has refocused on something that we believe gives us a broad explanation for the complex phenotypic and genetic architectures.

1.1.5   The Reproductive Ground Plan

Gro Amdam and I suggested that the explanation for the correlations of traits in the pollen hoarding syndrome may be found in the use of a reproductive regulatory network as a mechanism to establish a foraging division of labor in honey bees ([1–3, 5, 26, 27], see also Chap. 1.2). The network consists of many parts, some with known functions, others parts remain unknown. However, key players that we have focused on, so far, include the ovaries that make ecdysteriod hormones that along with juvenile hormone are thought to act on the fat body of insects and result in the production of vitellogenin, an egg yolk precursor protein [5]. There are certainly many other expanding effects of this network that we are trying to understand, but these give us points of insertion for manipulation of the network so we can study the effects.

We believe that the reproductive network involving these components is ancient in the insects, part of a fundamental reproductive ground plan, operating on the activation and maturation of ovaries, production of egg yolk proteins, and maternal behavior including making a nest, provisioning the nest with protein for their offspring, and laying eggs. It has been shown in many insects that behavior changes with the states of the ovaries. We believe that in honey bees, the ancient relationship between ovary and behavior has been co-opted, and is used now as a mechanism for canalizing workers into performing different tasks, and can explain differences in the age of onset of foraging and foraging behavior.

The ovaries are certainly key players. Honey bees have paired ovaries that consist of ovariole filaments in which eggs are made. Queens have on average more than 150 ovarioles per ovary [17] while workers usually have less than 30, depending on the population [19]. The number of ovarioles is determined during the 5th instar of larval development (about 5–7 days after hatching). At this time workers and queens have the same number of developing ovarioles, however, ovarioles are lost in workers through a process of apoptosis, or programmed cell death [7]. Ovarioles are rescued from apoptosis by juvenile hormone circulating at just the right time. Queens have a bigger spike of juvenile hormone compared to workers, and end up with more ovarioles. When we compare bees from the high and low pollen hoarding strains, we find that high strain workers have more ovarioles, newly emerged adult bees already have ovaries that are activated, ready to absorb vitellogenin, and have higher titers of circulating vitellogenin compared with low strain bees [1, 3]. In other words, they seem to be in more advanced stages of reproductive readiness. When we look at the amount of circulating juvenile hormone during the 5th larval instar we find what we would expect: high strain bees have higher titers, which may explain why they have more ovarioles [6]. We think of this difference in the titer of juvenile hormone as a developmental signature of colony-level selection from our breeding program. Natural selection should leave similar signatures, as it has with respect to queen and worker ovary development.

Are the results of our comparisons of high- and low-strain bees general results? We studied wild-type bees and found that workers with more ovarioles are more sensitive to sucrose solutions, forage earlier in life, show a bias for collecting pollen, collect nectar with less sugar, are less likely to return empty from a foraging trip, and have higher titers of vitellogenin when they are young [4, 44]. These traits fit exactly with those found for the high strain bees and independently verify the relationships between ovariole number and behavior. We also mapped QTL for ovariole number to the same QTL locations as for the behavioral traits suggesting that the behavioral effects are derived from the effects of these QTL on the ovary [46]. Gene expression for two candidate genes for two of the mapped QTLs correlate with the ovary and behavioral traits in crosses specifically designed to test for their direct effects [46].

We can remove the ovaries from one bee and put them into another [47]. When we do this, we can show that the transplanted (grafted) ovaries live, develop, and respond like the resident ovaries. For experiments, we inject glass beads as a control. They are immunologically inert but require the same surgical procedure. We conducted a study where we grafted ovaries into a test group, beads into the controls, placed the bees into an observation hive, then watched their behavioral transitions. Ovary grafted bees made the transitions through within nest behavior and into foraging faster than the bead controls. This is the same pattern we see in high strain bees versus lows (high strain bees have more ovarioles), and that we have shown for wild-type bees with more ovarioles versus those with fewer.

We can disrupt the reproductive regulatory network by eliminating or greatly reducing the presence of vitellogenin. We inject double stranded RNA (dsRNA), the template used by cells to make proteins from genes. The dsRNA is then taken up by the fat body cells where vitellogenin is normally made, but it blocks the production of the protein. Normal RNA is in a single strand. When we block vitellogenin we affect the behavioral traits associated with the pollen hoarding syndrome: bees are more responsive to sucrose, forage earlier in life, and show a bias for collecting nectar [4, 16, 21].

1.1.6 Conclusion

How does social organization evolve? At least in part, selection on the superorganismal trait of stored pollen changes frequencies of alternative alleles (forms) of genes or gene regulatory networks with broad pleiotropic effects including effects on reproductive signaling networks. Changes in signaling networks affect changes in development that affect the ovaries of workers, that in turn affect sensory physiology and response systems, and thus the behavior and interactions of thousands of individuals. And where does the "spirit of the hive" reside? It resides in the stimulus response interactions derived at least in part from the ovaries of "a crowd of bees working in a dark hive".

1.1.7   Outlook

Over the last 20 years my colleagues and I applied artificial selection as a surrogate for natural selection to produce two populations that vary in behavior, and a forward genetic approach (phenotype to gene) to understand the origins of division of labor and the genetic and environmental basis of variation in individual behavior and colony organization. Most recently my lab along with the labs of Olav Rueppell and Gro Amdam applied the breeding techniques and gene mapping strategies to crosses of natural populations of Africanized honey bees and commercial European honey bees that have not been selected for pollen hoarding (A. Graham, M. Munday, O. Kaftanoglu, R. Page, G. Amdam, and A. Siegel, unpublished data). We found that the same QTL, and presumably the same candidate genes, explain the natural varia-tion between these populations for both ovariole number and foraging behavior. Gro Amdam and her colleagues (see Chap. 1.2) have been pioneering gene silencing in social insects and exploring specific, important signaling pathways that are involved in reproduction and foraging behavior. The continuing development of silencing techniques coupled with the standard breeding and quantitative genetic techniques used in forward genetics, and studies of natural populations of social insects will continue to provide a powerful combination of strategy and technique to unravel the many remaining mysteries of complex social organization.

References

1. Amdam GV, Csondes A, Fondrk MK, Page RE Jr (2006) Complex social behaviour derived from maternal reproductive traits. Nature 439(7072):76–78
2. Amdam GV, Ihle KE, Page RE (2009) Regulation of honeybee worker (*Apis mellifera*) life histories by Vitellogenin. In: Donald WP, Arthur PA, Anne ME, Susan EF, Robert TR (eds) Hormones, brain and behavior, vol 2, 2nd edn. Academic, San Diego, pp 1003–1025
3. Amdam GV, Norberg K, Fondrk MK, Page RE Jr (2004) Reproductive ground plan may mediate colony-level selection effects on individual foraging behavior in honey bees. Proc Natl Acad Sci USA 101(31):11350–11355
4. Amdam GV, Norberg K, Page RE Jr, Erber J, Scheiner R (2006) Downregulation of vitellogenin gene activity increases the gustatory responsiveness of honey bee workers (*Apis mellifera*). Behav Brain Res 169(2):201–205
5. Amdam GV, Page RE (2010) The developmental genetics and physiology of honeybee societies. Anim Behav 79(5):973–980
6. Amdam GV, Page RE Jr, Fondrk MK, Brent CS (2010) Hormone response to bidirectional selection on social behavior. Evol Dev 12(5):428–436
7. Capella ICS, Hartfelder K (1998) Juvenile hormone effect on DNA synthesis and apoptosis in caste-specific differentiation of the larval honey bee (*Apis mellifera* L.) ovary. J Insect Physiol 44(5–6):385–391
8. Darwin C (1998) The origin of species by means of natural selection, or, the preservation of favored races in the struggle for life. 1998 Modern Library edn. Modern Library, New York

9. Dreller C, Page RE, Fondrk MK (1999) Regulation of pollen foraging in honeybee colonies: effects of young brood, stored pollen, and empty space. Behav Ecol Sociobiol 45(3–4):227–233
10. Dreller C, Tarpy DR (2000) Perception of the pollen need by foragers in a honeybee colony. Anim Behav 59(1):91–96
11. Fewell JH, Page RE (2000) Colony-level selection effects on individual and colony foraging task performance in honeybees, *Apis mellifera* L. Behav Ecol Sociobiol 48(3):173–181
12. Fewell JH, Winston ML (1992) Colony state and regulation of pollen foraging in the honey-bee, *Apis mellifera* L. Behav Ecol Sociobiol 30(6):387–393
13. Hölldobler B, Wilson EO (2008) The superorganism: the beauty, elegance and strangeness of insect societies. W. W. Norton, New York
14. Humphries MA, Müller U, Fondrk MK, Page RE Jr (2003) PKA and PKC content in the honey bee central brain differs in genotypic strains with distinct foraging behavior. J Comp Physiol A 189(7):555–562
15. Hunt GJ, Page RE, Fondrk MK, Dullum CJ (1995) Major quantitative trait loci affecting honey-bee foraging behavior. Genetics 141(4):1537–1545
16. Ihle KE, Page RE, Frederick K, Fondrk MK, Amdam GV (2010) Genotype effect on regulation of behaviour by vitellogenin supports reproductive origin of honeybee foraging bias. Anim Behav 79(5):1001–1006
17. Laidlaw HH, Page RE (1997) Queen rearing and bee breeding, 1st edn. Wicwas Press, Cheshire
18. Le Conte Y, Mohammedi A, Robinson GE (2001) Primer effects of a brood pheromone on honeybee behavioural development. Proc R Soc B 268(1463):163–168
19. Linksvayer TA, Rueppell O, Siegel A, Kaftanoglu O, Page RE Jr et al (2009) The genetic basis of transgressive ovary size in honeybee workers. Genetics 183(2):693–707, 691SI-613SI
20. Maeterlink M (1913) The life of the bee. Dodd, Mead, and Company, New York
21. Nelson CM, Ihle KE, Fondrk MK, Page RE, Amdam GV (2007) The gene vitellogenin has multiple coordinating effects on social organization. PLoS Biol 5(3):e62
22. Nowak MA, Tarnita CE, Wilson EO (2010) The evolution of eusociality. Nature 466(7310): 1057–1062
23. Page RE, Erber J (2002) Levels of behavioral organization and the evolution of division of labor. Naturwissenschaften 89(3):91–106
24. Page RE, Erber J, Fondrk MK (1998) The effect of genotype on response thresholds to sucrose and foraging behavior of honey bees (*Apis mellifera* L.). J Comp Physiol A 182(4):489–500
25. Page RE, Fondrk MK (1995) The effects of colony level selection on the social-organization of honey-bee (*Apis mellifera* L) colonies – colony level components of pollen hoarding. Behav Ecol Sociobiol 36(2):135–144
26. Page RE Jr, Amdam GV (2007) The making of a social insect: developmental architectures of social design. Bioessays 29(4):334–343
27. Page RE, Scheiner R, Erber J, Amdam GV (2006) The development and evolution of division of labor and foraging specialization in a social insect (*Apis mellifera* L.). Curr Top Dev Biol 74:253–286
28. Pankiw T (2003) Directional change in a suite of foraging behaviors in tropical and temperate evolved honey bees (*Apis mellifera* L.). Behav Ecol Sociobiol 54(5):458–464
29. Pankiw T, Nelson M, Page RE, Fondrk MK (2004) The communal crop: modulation of sucrose response thresholds of pre-foraging honey bees with incoming nectar quality. Behav Ecol Sociobiol 55(3):286–292
30. Pankiw T, Page RE (1999) The effect of genotype, age, sex, and caste on response thresholds to sucrose and foraging behavior of honey bees (*Apis mellifera* L.). J Comp Physiol A 185(2):207–213
31. Pankiw T, Page RE (2000) Response thresholds to sucrose predict foraging division of labor in honeybees. Behav Ecol Sociobiol 47(4):265–267
32. Pankiw T, Page RE (2001) Genotype and colony environment affect honeybee (*Apis mellifera* L.) development and foraging behavior. Behav Ecol Sociobiol 51(1):87–94

33. Pankiw T, Page RE, Fondrk MK (1998) Brood pheromone stimulates pollen foraging in honey bees (*Apis mellifera*). Behav Ecol Sociobiol 44(3):193–198
34. Pankiw T, Waddington KD, Page RE (2001) Modulation of sucrose response thresholds in honey bees (*Apis mellifera* L.): influence of genotype, feeding, and foraging experience. J Comp Physiol A 187(4):293–301
35. Rueppell O, Chandra SBC, Pankiw T, Fondrk MK, Beye M et al (2006) The genetic architecture of sucrose responsiveness in the honeybee (*Apis mellifera* L.). Genetics 172(1):243–251
36. Rueppell O, Pankiw T, Nielsen DI, Fondrk MK, Beye M et al (2004) The genetic architecture of the behavioral ontogeny of foraging in honeybee workers. Genetics 167(4):1767–1779
37. Scheiner R, Erber J, Page RE (1999) Tactile learning and the individual evaluation of the reward in honey bees (*Apis mellifera* L.). J Comp Physiol A 185(1):1–10
38. Scheiner R, Kuritz-Kaiser A, Menzel R, Erber J (2005) Sensory responsiveness and the effects of equal subjective rewards on tactile learning and memory of honeybees. Learn Mem 12(6):626–635
39. Scheiner R, Page RE, Erber J (2001) The effects of genotype, foraging role, and sucrose responsiveness on the tactile learning performance of honey bees (*Apis mellifera* L.). Neurobiol Learn Mem 76(2):138–150
40. Scheiner R, Page RE, Erber J (2001) Responsiveness to sucrose affects tactile and olfactory learning in preforaging honey bees of two genetic strains. Behav Brain Res 120(1):67–73
41. Scheiner R, Page RE, Erber J (2004) Sucrose responsiveness and behavioral plasticity in honey bees (*Apis mellifera*). Apidologie 35(2):133–142
42. Schulz DJ, Pankiw T, Fondrk MK, Robinson GE, Page RE (2004) Comparisons of juvenile hormone hemolymph and octopamine brain Titers in honey bees (Hymenoptera: Apidae) selected for high and low pollen hoarding. Ann Entomol Soc Am 97(6):1313–1319
43. Seeley TD (1995) The wisdom of the hive. Harvard University Press, Cambridge
44. Tsuruda JM, Amdam GV, Page RE Jr (2008) Sensory response system of social behavior tied to female reproductive traits. PLoS One 3(10):e3397
45. Vaughan DM, Calderone NW (2002) Assessment of pollen stores by foragers in colonies of the honey bee, *Apis mellifera* L. Insect Soc 49(1):23–27
46. Wang Y, Amdam GV, Rueppell O, Wallrichs MA, Fondrk MK et al (2009) PDK1 and HR46 gene homologs tie social behavior to ovary signals. PLoS One 4(4):e4899
47. Wang Y, Kaftanoglu O, Siegel AJ, Page RE, Amdam GV (2010) Surgically increased ovarian mass in the honey bee confirms link between reproductive physiology and worker behavior. J Insect Physiol 56(12):1816–1824
48. Wheeler WM (1911) The ant colony as an organism. J Morphol 22:307–325
49. Wheeler WM (1926) Social life among the insects. Constable and Company Limited, London
50. Wheeler WM (1928) The social insects. Harcourt, Brace and Company, New York
51. Winston ML (1987) The biology of the honey bee. Harvard University Press, Cambridge

Chapter 1.2
Vitellogenin in Honey Bee Behavior and Lifespan

Gro V. Amdam, Erin Fennern, and Heli Havukainen

Abstract Vitellogenin is a phospholipoglycoprotein that affects multiple aspects of honey bee life-history. Across the vast majority of oviparous taxa, vitellogenins are female-specific egg yolk proteins, with their essential function tied to oogenesis. In honey bees, however, vitellogenin is also expressed by female helpers, called workers, which are largely sterile. Here, vitellogenin influences behavior and stress resilience, and is believed to be important to honey bee social organization. Together with longtime collaborators, we have discovered roles of vitellogenin in worker behavioral traits such as nursing, foraging onset and foraging bias, and in survival traits such as oxidative stress resilience, cell-based immunity, and longevity. We have also identified a mutually inhibitory interaction between vitellogenin and the systemic endocrine factor juvenile hormone (JH), which is central to insect reproduction and stress response. This regulatory feedback loop has spurred hypotheses on how vitellogenin and JH together have become key life-history regulators in honey bees. A current research focus is on how this feedback loop is tied to nutrient-sensing insulin/insulin-like signaling that can govern expression of phenotypic plasticity. Here, we summarize this body of work in the context of new structural speculations that can lead to a modern understanding of vitellogenin function.

G.V. Amdam (✉)
School of Life Sciences, Arizona State University, PO Box 874501, Tempe, AZ 85287, USA

Department of Chemistry, Biotechnology and Food Science, Norwegian University of Life Sciences, PO Box 5002, N-1432 Aas, Norway
e-mail: Gro.Amdam@asu.edu

E. Fennern
School of Life Sciences, Arizona State University, PO Box 874501, Tempe, AZ 85287, USA

H. Havukainen
Department of Chemistry, Biotechnology and Food Science, Norwegian University of Life Sciences, PO Box 5002, N-1432, Aas, Norway

C.G. Galizia et al. (eds.), *Honeybee Neurobiology and Behavior: A Tribute to Randolf Menzel*, DOI 10.1007/978-94-007-2099-2_2,
© Springer Science+Business Media B.V. 2012

Abbreviations

DRH   Double repressor hypothesis
JH   Juvenile hormone
RGPH   Reproductive ground plan hypothesis
Vg   Vitellogenin
vWFD   von Willebrand factor D-type

1.2.1   Introduction

In colonies of social insects, groups of individuals perform distinct tasks. What factors determine this differentiation of behavior? How can the regulation of social behavior be studied and understood? Here we look at the influence of vitellogenin, a multifunctional yolk precursor protein, on honey bee (*Apis mellifera*) social organization. The honey bee is one of the most important and well-researched models for the study of social behavior [31, 45]. Honey bees allow us to ask how complex social organization emerges without centralized control – through the summation of the behavioral interactions between individuals. Within each colony of insects, these interactions produce a structured division of labor that is correlated with age and associated with changes in physiology and lifespan.

Most honey bee eggs develop into essentially sterile helper females called 'workers'. Colonies have about 10,000–40,000 workers that show complex division of labor between several behavioral groups, the most important being nurse bees that take care of brood (eggs, larvae, and pupae) and foragers that collect resources from the external environment: nectar, pollen, propolis (resin) or water [37]. Worker division of labor emerges as an association between age and behavior: young bees are most often nurses while older workers forage. However, ontogeny is flexible. Foragers can revert to within-nest tasks while nurses can accelerate behavioral development and forage precociously [26]. Once foragers, workers usually die within 7–10 days, but foraging efforts decline if no brood is present requiring care such as feeding. During periods without brood rearing, workers develop into stress-resistant '*diutinus* bees' with lifespans of 280 days or more. This phenotypic plasticity produces different life-histories with worker lifespans ranging from a few weeks to nearly a year (review: [6]).

Variation in worker behavior and lifespan correlates with vitellogenin (review: [6]). In a highly regulated manner, this protein is expressed during development and adult life by both sexes and all behavioral groups of honey bees. The primary site of synthesis is the abdominal fat body (functional homologue of vertebrate liver and adipose tissue), which lines the body-wall as a single cell-layer composed of trophocytes and oenocytes. From fat body, vitellogenin is secreted into the hemolymph (blood). This circulating vitellogenin titer is highest in queens, and lowest in males [21]. In workers, vitellogenin synthesis and hemolymph levels vary with social role and longevity: the short-lived foragers produce substantially less vitellogenin and have significantly lower titers than nurse bees, while the stress resilient *diutinus* bees have the highest levels of all the worker groups (see [3] for a recent review).

The expression of the *vitellogenin* gene is nutrient sensitive, and responds positively to the level of amino acids in hemolymph. Along with many genes and endocrine factors, such as protein-kinase G [13], *malvolio* that encodes a manganese transporter [12], and the systemic JH (review: [16]), *vitellogenin* has been linked to food-related worker behavior – specifically to the onset of foraging. In addition, vitellogenin takes part in brood-food synthesis in nurse bees, and it also affects the bees' propensity to collect pollen rather than nectar during foraging trips [32] (see also Chap. 1.1).

Added to the food-related effects, vitellogenin has an independent and positive influence on worker lifespan [32]. This effect may be explained by a nutritive role of the vitellogenin protein [21] and its positive influence on oxidative stress resilience and cell-based immunity [36]. Thereby, vitellogenin is among the most multifunctional life-history regulators known in honey bees, and is likely to be instrumental for colony social organization.

With this chapter, we summarize what is currently known about honey bee vitellogenin and its effects on worker life-histories. We outline mechanisms that may allow vitellogenin to influence worker phenotypic outcomes, and discuss how this protein has become such a central regulatory element during honey bee social evolution.

1.2.2   Vitellogenin Properties

Some effects of vitellogenin in honey bees are surprising (Fig. 1.2.1). Vitellogenins are female-specific yolk proteins that are central to egg development in most oviparous invertebrate and vertebrate animals. Worker honey bees, however, do not normally lay eggs. They have lost the ability to mate, their reproductive organs are greatly reduced, and further oocyte development is blocked by pheromonal inhibition (see [38] for a review).

1.2.2.1   Gene Sequence and Protein Structure

The honey bee *vitellogenin* gene (GenBank accession number: CAD56944.1) is 5,440 bp long and yields a polypeptide of 1,770 amino acids [33]. Many taxa, including *Xenopus* and *Caenorhabditis*, have multiple *vitellogenin* genes [35, 43], but only one is found in the honey bee genome. The vitellogenin amino acid sequence, as a whole, is only weakly conserved between species, e.g., the sequence similarity between honey bee and bumblebee (*Bombus ignitus*) vitellogenin is only 51%. However, vitellogenins contain conserved elements, including a GL/

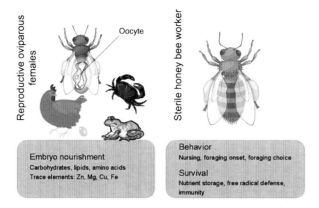

Fig. 1.2.1 The general roles of vitellogenin proteins in oviparous females of a variety of taxa versus honey bee vitellogenin functions in workers, which are largely sterile. In most oviparous taxa (some examples given in the *left panel*), vitellogenins are yolk precursors that are synthesized in liver (vertebrates) or fat body (invertebrates) for secretion into the blood. From circulation, vitellogenin protein is taken up by the ovary for transfer to the developing oocytes where it nourishes the embryo with macronutrients as well as zinc (Zn), magnesium (Mg), copper (Cu) or iron (Fe, as noted in the *gray box, lower left*). A bee (other than honey bee) is indicated with transparent abdomen to point out vitellogenic oocytes in the ovary. This process of vitellogenesis occurs in honey bee queens and in worker bees that are not inhibited from egg-laying. In worker bees, furthermore, vitellogenin influences behavior and survival (*right panel*, details on affected behavioral traits and mechanism of somatic maintenance in the corresponding *gray box, lower right*). Behavioral effects have been traced to workers, but the effect on longevity and survival is shared with queens [16, 36]

ICG (Glysine-Leucine/Isoleucine-Cysteine-Glycine) amino acid motif followed by nine cysteine residues near the C-terminus in all available insects sequences so far [33].

Honey bee vitellogenin is synthesized as a monomeric 180 kDa phospholipo-glycoprotein [46]. The protein consists of several parts: N-sheet, polyserine linker and a lipid cavity that includes an α-helical region and a vWFD (von Willebrand factor, type D) domain (Fig. 1.2.2a). A secondary structure can be tentatively estimated based on *in silico* prediction (PSIPRED; [27]). For the N-sheet and α-helical domain, furthermore, the crystal structure of lamprey lipovitellin can be used as a guide [34]. The lamprey lipovitellin crystal structure (Fig. 1.2.2b) is currently the only solved structure from the vitellogenin protein family.

A region of the N-sheet domain may confer binding between vitellogenin and its receptor in *Oreochromis aureus* [29], but this domain-association is not confirmed in other species. The N-sheet is connected to the rest of vitellogenin by a putative phosphorylated linker. This polyserine linker is not present in vertebrates and shows diversity in insects for location and length [14]. The role of the α-helical domain is unidentified, but may bind zinc in lamprey [10]. The vWFD domain has cysteine-residues conserved in insects [42] but its structure and function is unknown.

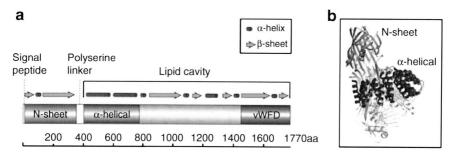

Fig. 1.2.2 Approximation of honey bee vitellogenin structure. (**a**) The amino terminal N-sheet domain starts after a 16 amino acid long signal sequence. This sheet is connected to the rest of vitellogenin by a polyserine linker. Based on the primary structure, the serine-rich linker is a large and flexible loop. The C-terminal region subsequent to the polyserine region forms a big lipid-carrying cavity, which can be approximated from a representation of the lamprey lipovitellin crystal structure. The C-terminal region of honey bee vitellogenin has two clear domains: an α-helical region (estimated to consist of total 16 helixes) and the vWFD-type domain. (**b**) Lamprey lipovitellin, shown as a monomer, is prepared with Pymol [19] based on its Protein Data Bank entry (ID: 1lsh). The β-barrel-like N-sheet (*green*) is estimated to resemble the honey bee N-sheet. The α-helical domain (*red*) forms an arch on the large β-sheets of the rest of the funnel-like lipid cavity. Lipids found in the crystal structure are depicted as yellow sticks. The solved structure lacks the vWFD domain, and the polyserine linker is missing

1.2.2.2 Vitellogenin Expression, Accumulation, Behavioral Correlation

In honey bees, most *vitellogenin*-expressing fat body cells are localized to the abdomen with some cells dispersed in the thorax and head [16, 39]. Vitellogenin synthesis is largely specific to trophocytes, which constitute one of two major cell types in fat body. Some transcript is also localized to queen ovarian tissue [24]. In most species, the vitellogenin pre-protein is cleaved, typically into a small N-terminal and a large C-terminal piece [42]. Honey bee vitellogenin lacks the conserved cleavage site found near the polyserine region of other insects, but appears as at least three forms in patterns that are tissue/compartment specific. The protein is detected as 180, 150 and 40 kDa in worker fatbody [25] as 180 kDa in hemolymph; and as 180 and 150 kDa in hypopharyngeal glands, which are the food-producing head glands of workers [5]. The 40 kDa fragment corresponds to the N-sheet of vitellogenin, and the 150 kDa fragment corresponds to the lipid cavity (Fig. 1.2.2a, b; [25]). The site of cleavage and the biological significance of these fragment patterns are unidentified.

The levels of mature vitellogenin protein in fat body and hemolymph change as a function of worker life-history progression. The rate of synthesis is enhanced soon after adult emergence from the pupal stage, and blood levels increase. During ambient conditions with active foraging by colonies, the vitellogenin level peaks in 7–10 day old nurse bees, and subsequently drops at foraging onset (see [3] and references therein). In nurse bees, vitellogenin provides amino acids and perhaps other nutrient building blocks for food synthesis or 'jelly production' by the hypopharyngeal head glands [5]. Jelly is central to the nourishment of larvae and the queen, and is also fed to other adult workers including foragers [17]. During unfavorable periods, larval rearing and foraging decline, and nurse bees accumulate very large amounts of vitellogenin in the fat body and hemolymph. This physiology is the hallmark of the long-lived and stress-resilient *diutinus* worker bees [36].

1.2.2.3 Experimental Manipulation, Behavior, and Frailty

The *vitellogenin* gene can be experimentally suppressed by RNA interference (RNAi) [8]. This *vitellogenin* gene silencing causes a drop in the circulating level of vitellogenin protein, resulting in cessation of nursing behavior followed by precocious foraging onset relative to controls [30, 32]. These results demonstrate that vitellogenin can slow worker behavioral progression, and may explain the functional significance of reduced vitellogenin levels that co-occur with the natural nurse-to-forager transition of worker bees [6, 32]. Moreover, knockdowns bias their foraging efforts toward nectar (carbohydrate source) rather than pollen (protein and lipid source) [32]. In other words, vitellogenin also influences worker food-choice behavior and presumably biases a bee's foraging effort toward pollen (see Chap. 1.1).

In addition to these effects on behavior, the RNAi-mediated repression of vitellogenin shortens worker lifespan [32, 36]. This outcome is in part conferred by the precocious foraging onset of knockdowns, since foragers rapidly perish [32]. Longevity, however, is also shortened independent of behavior. This influence is attributed to a positive effect of vitellogenin on resilience to stressors such as oxidative metabolic damage, starvation, and immune challenge [32, 36].

In summary, the drop in vitellogenin levels that generally occurs during worker adult ontogeny is associated with increased frailty as well as behavioral change. These relationships are directly supported by experimental repression of *vitellogenin* activity.

1.2.3 Vitellogenin Functions – Hypotheses and Molecular Mechanisms

How can vitellogenin influence behavior and lifespan? The traditional view of vitellogenins as yolk precursors contrasts with the modern understanding of honey bee vitellogenin as a protein that broadly affects social life-histories [3].

1.2.3.1 Proximate and Ultimate Explanations

To explain the effect of vitellogenin on foraging onset, Amdam and Omholt [6] proposed a double repressor model. This model entails a mutually inhibitory feedback loop between vitellogenin and the life-shortening systemic endocrine

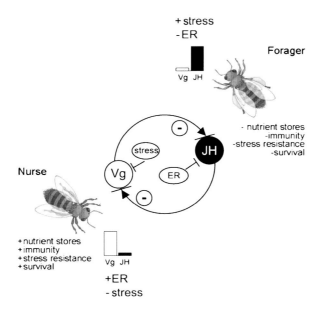

Fig. 1.2.3 The double repressor model (DRH). In nurse bees (*left*), vitellogenin (Vg, level indicated with white bars) and the presence (+) of an external repressor pheromone (+ER) (⊢ ER) inhibit juvenile hormone (JH, level indicated with black bars) and foraging activity in the absence of stress (show as minus (−) stress). If the ER signal is reduced (minus ER, −ER), workers with low Vg titers are at greater risk of JH signaling and foraging. Stress (+Stress) reduces Vg (⊢ Stress) and elicits foraging through JH signaling that is a conserved element of the insect stress response. Vg synthesis is suppressed by JH, closing a feedback loop between Vg and JH that locks the worker into the forager state (*right*). The feedback between Vg and JH is indicated with mutual repressor arrows (*gray*) and correspondence with increased vs. reduced levels of stress and ER is shown alongside these *arrows*

factor JH, which affects physiology and behavior in many insects via regulatory roles in reproduction and stress response. According to this hypothesis, honey bee nurse stage is governed by vitellogenin while JH is high in foragers and has pleiotropic effects on behavioral physiology that confers the depletion of nutrient stores and reduced somatic maintenances (Fig. 1.2.3). The model proposed that foraging onset initiates when a decline in vitellogenin (reducing nutrient concentration and stress resilience) allows JH to increase. JH feeds back to suppress vitellogenin further – in effect locking the bee into the forager state [6].

In forming the double repressor hypothesis (DRH), Amdam and Omholt proposed vitellogenin as an *internal* repressor that could slow the nurse bee-to-forager transition of bees. In reference to 'double' repression, the effect of vitellogenin joined an existing explanation where foraging onset was delayed by an *external* repressor contact pheromone. This repressor contains ethyl oleate, and appears to be produced by foragers to inhibit recruitment of nurses to foraging tasks [28]. The DRH also provided reasoning for adaptive division of labor between nurse bees and foragers. An internal repressor like vitellogenin conferred a selective advantage for honey bee societies, because it retained nutrient rich and healthy nurse bees in the nest while

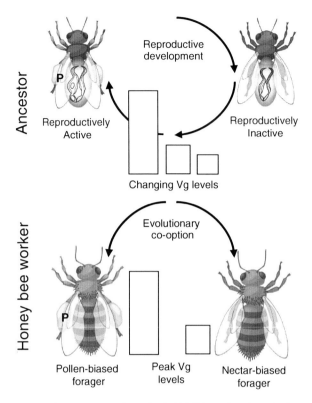

Fig. 1.2.4 The reproductive ground plan hypothesis (RGPH) outlines how the female reproductive biology could have been co-opted in the division of labour between foraging worker honey bees. From the *top*: In ancestors of honey bees, the sequential progression of the reproductive cycle was linked to changes in female food-related behavior. During periods of no active reproduction (*top, right*), the ovary was undeveloped (not enlarged with yolk) and the Vg titre (*white bars*) was low. During reproduction (*top left*), *vg* gene expression increased prior to yolk deposition, and pollen (P in bee, *upper left*) was required for nest provisioning. *Below*: Vg and foraging behavior is similarly associated in worker honey bees, where the Vg peak titer has a positive and seemingly direct influence on pollen collection [32]

vitellogenin-poor or strained workers (less useful in nursing) would be first to respond with a shift to foraging, should levels of external repressor drop [6].

The role of vitellogenin in foragers' preference for pollen or nectar was addressed by Amdam and Page's reproductive ground plan hypothesis, or RGPH [4]. While the DRH describes the regulation of foraging onset, the RGPH is a broader hypothesis that seeks to explain what gene networks natural selection might have acted upon during honey bee social evolution (Fig. 1.2.4). This framework views the relationship between vitellogenin and pollen foraging preference in honey bees as a 'footprint' of ancestral trait-associations. The RGPH was inspired by correlative relationships between yolk production, egg development, and division of labor in several species of wasps, ants, and bees (see [4] and references therein). Most solitary bees, moreover, feed on nectar and pollen while pollen-collecting (hoarding) behavior is specific to the stage of larval nest-provisioning that is evident during vitellogenesis (references in [4]). Similarly, high levels of vitellogenin are associated with pollen foraging in worker honey bees that – like solitary females – collect pollen for storage in the brood nest. The RGPH suggests that natural selection exploited the reproductive genetics of female bees to facilitate advanced division of labour between workers with different foraging biases.

The frailty of foragers compared to nurse worker bees, at least in part resulting from declining vitellogenin levels at foraging onset, could be advantageous at a colony-level [7]. By reducing vitellogenin, workers attain a physiology of high JH, reduced cell-based immunity, and increased oxidative stress susceptibility [32, 36]. Foragers can be exposed to a heavy load of pathogens in the field, and will return to the same source of food over the course of days. By being susceptible to substances that could harm the colony, foragers die before bringing much contamination to the nest. Likewise, stress by handling, injury, and disease causes JH to increase and bees to initiate foraging (references in [2]). The regulatory connection between vitellogenin, JH, and stress thereby ensures that bees in poor condition transition to a state where frailty and extrinsic mortality pretty much guarantee rapid death at a safe distance from the colony [7].

1.2.3.2   Putative Molecular Action

The pleiotropic effects of vitellogenin on behavior and stress resilience make sense in theory; however, the structural and regulatory aspects of the underlying processes are not well understood. To date, one structural mechanism has been proposed for the role of vitellogenin in oxidative stress resistance and cell-based immunity – effects that might emerge from zinc-binding properties [9, 36]. Also, a regulatory association between vitellogenin and insulin/insulin-like signaling has been hypothesized to explain the pleiotropic effect of vitellogenin on JH, behavior and lifespan [16, 36]. Insulin/insulin-like signaling governs responses to food-intake in eukaryotes, including eating behavior and metabolic physiology related to lifespan [11, 18]. In insect model systems, inhibition of insulin/insulin-like signaling (reducing nutrient sensing) will thereby lower JH and boost survival capacity [22, 41]. In honey bees, on the other hand, this relationship can be reversed, since increased nutrient status (vitellogenin) can confer low JH and increased survival [6, 32, 36]. It is envisioned that this inversion results from an ability of vitellogenin to suppress insulin/insulin like signals (review: [2]).This inhibitory function, however, is not yet validated experimentally.

Current knowledge on the basic sequence and structure of vitellogenin (Fig. 1.2.2) suggests that phosphorylation, cleavage, or ligand binding can format vitellogenin for different functions. Building on this information and exploiting the crystal structure of lipovitellin [34], we can identify motifs with potential to serve as regulatory sites (Fig. 1.2.5).

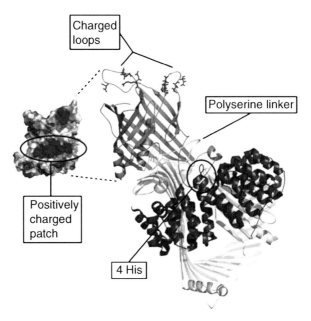

Fig. 1.2.5 A preliminary view of structures of honey bee vitellogenin. The model combines the lamprey crystal structure with a homology model of the vitellogenin N-sheet (*green*) and by adding a sketch of the polyserine linker (*yellow*) between the N-sheet and the lipid cavity. The polyserine loop was oriented randomly. Unlike the lamprey N-sheet, honey bee vitellogenin has two loops (Charged loops) whose charged residues are visualized as *red* (negative) and *blue* (positive) sticks. An electrostatic surface potential map (*left side* of the main structure) of the honey bee vitellogenin N-sheet model was prepared with Pymol [19]. The map visualizes a positively charged patch not present in lamprey lipovitellin, formed by residues of the β-sheets. In the middle of the α-helical region, four residues of histidine – a zinc coordinator – are found close to each other in sequence (approximation location *circled*)

One of the most striking structural elements in honey bee vitellogenin is the polyserine linker that offers multiple potential phosphoryl-acceptor sites (Fig. 1.2.2a and 1.2.5). The location of the polyserine linker is ideal for structural speculation: the linker separates the loosely attached N-sheet from the rest of the molecule. It is unknown how cleavage of the N-sheet is achieved, but this can be a phosphorylation-regulated process, such as in the case of presenilin-2 (cleavage inhibited by phosphorylation [44]) or cohesin (cleavage triggered by phosphorylation, [1]). It is well-documented that phosphorylation-induced structural changes can have effects on signaling, activation or deactivation of enzymes, and degradation, localization and binding (review: [15]).

The N-sheet contains several structurally interesting features. There are two insect-specific loops with a considerable number of conserved charged residues (Fig. 1.2.5; 'Charged loops'). A loop is not necessarily a passive linker, but can play an intrinsic structural role, e.g. [20]. In some cases, even active site residues are located in a loop, e.g. [47]. Also, there is a positively charged patch on the β-sheets of the N-sheet of honey bee vitellogenin (Fig. 1.2.5). Charged patches typically attract negatively charged binding partners (for recent data, see [23]). The N-sheet and the polyserine linker of honey bee vitellogenin offer possibilities for regulation: phosphorylation, cleavage, flexible regions for conformational changes, and an attractive charge for binding. The vWFD and α-helical domains provide opportunity for metal binding. A proximate understanding of vitellogenin function will require targeted studies of structure and dynamic action at these sites.

1.2.4 Outlook

It is an open question how vitellogenin, and its associated molecular processes, can impact the development or function of the central nervous system to influence honey bee behavior. Vitellogenin and JH are intimately connected, and there is considerable evidence linking JH to insect brain development and more recently to the circadian cycle of insects (see [40] and references therein). Yet it is unclear, for example, whether vitellogenin levels vary during the circadian cycle of the bee, as well as whether and how receptor molecules are required for vitellogenin function.

In future, we need to ask how a modern understanding of vitellogenin can apply to processes of the honey bee brain. Perhaps molecular developmental queues associated with JH and vitellogenin, that are essential for normal holometabolic and reproductive development, were fine-tuned to accommodate more subtle development shifts required by the adult brain. Part of the answer may be that cells/neurons require development queues for a burst of neurite outgrowth during the first 2 weeks of adult brain development, and may be part of a program of worker behavioral progression.

References

1. Alexandru G, Uhlmann F, Mechtler K, Poupart MA, Nasmyth K (2001) Phosphorylation of the cohesin subunit Scc1 by Polo/Cdc5 kinase regulates sister chromatid separation in yeast. Cell 105(4):459–472
2. Amdam GV (2011) Social context, stress, and plasticity of aging. Aging Cell 10(1):18–27
3. Amdam GV, Ihle KE, Page RE (2009) Regulation of honey bee (*Apis mellifera*) life histories by vitellogenin. In: Pfaff D, Arnold A, Etgen A, Fahrbach S, Rubin R (eds) Hormones, brain and behavior, 2nd edn. Elsevier Academic Press, San Diego
4. Amdam GV, Norberg K, Fondrk MK, Page RE (2004) Reproductive ground plan may mediate colony-level selection effects on individual foraging behavior in honey bees. Proc Natl Acad Sci USA 101:11350–11355
5. Amdam GV, Norberg K, Hagen A, Omholt SW (2003) Social exploitation of vitellogenin. Proc Natl Acad Sci USA 100:1799–1802
6. Amdam GV, Omholt SW (2003) The hive bee to forager transition in honeybee colonies: the double repressor hypothesis. J Theor Biol 223:451–464
7. Amdam GV, Seehuus SC (2006) Order, disorder, death: lessons from a superorganism. Adv Cancer Res 95:31–60

 8. Amdam GV, Simões ZLP, Guidugli KR, Norberg K, Omholt SW (2003) Disruption of vitel-logenin gene function in adult honeybees by intra-abdominal injection of double-stranded RNA. BMC Biotechnol 3:1–8
 9. Amdam GV, Simões ZLP, Hagen A, Norberg K, Schrøder K et al (2004) Hormonal control of the yolk precursor vitellogenin regulates immune function and longevity in honeybees. Exp Gerontol 39:767–773
10. Anderson TA, Levitt DG, Banaszak LJ (1998) The structural basis of lipid interactions in lipovitellin, a soluble lipoprotein. Struct Fold Des 6(7):895–909
11. Bartke A (2005) Minireview: role of the growth hormone/insulin-like growth factor system in mammalian aging. Endocrinology 146(9):3718–3723
12. Ben-Shahar Y, Dudek NL, Robinson GE (2004) Phenotypic deconstruction reveals involve-ment of manganese transporter malvolio in honey bee division of labor. J Exp Biol 207:3281–3288
13. Ben-Shahar Y, Robichon A, Sokolowski MB, Robinson GE (2002) Influence of gene action across different time scales on behavior. Science 296:741–744
14. Chen JS, Sappington TW, Raikhel AS (1997) Extensive sequence conservation among insect, nem-atode, and vertebrate vitellogenins reveals ancient common ancestry. J Mol Evol 44(4):440–451
15. Cohen P (2000) The regulation of protein function by multisite phosphorylation – a 25 year update. Trends Biochem Sci 25(12):596–601
16. Corona M, Velarde RA, Remolina S, Moran-Lauter A, Wang Y et al (2007) Vitellogenin, juvenile hormone, insulin signaling, and queen honey bee longevity. Proc Natl Acad Sci USA 104:7128–7133
17. Crailsheim K (1990) The protein balance of the honey bee worker. Apidologie 21:417–429
18. Dallman MF, Warne JP, Foster MT, Pecoraro NC (2007) Glucocorticoids and insulin both modulate caloric intake through actions on the brain. J Physiol 583:431–436
19. DeLano WL (2002) The PyMOL molecular graphics system. De Lano Scientific, Palo Alto
20. Doucet N, Watt ED, Loria JP (2009) The flexibility of a distant loop modulates active site motion and product release in ribonuclease A. Biochemistry 48(30):7160–7168
21. Engels W, Kaatz H, Zillikens A, Simões ZLP, Truve A et al (1990) Honey bee reproduction: vitellogenin and caste-specific regulation of fertility. In: Hoshi M, Yamashita O (eds) Advances in invertebrate reproduction, vol 5. Elsevier Science Publishers B.V., Amsterdam, pp 495–502
22. Flatt T, Min KJ, D'Alterio C, Villa-Cuesta E, Cumbers J et al (2008) *Drosophila* germ-line modulation of insulin signaling and lifespan. Proc Natl Acad Sci USA 105(17):6368–6373
23. Grigg JC, Cooper JD, Cheung J, Heinrichs DE, Murphy ME (2010) The *Staphylococcus aureus* siderophore receptor HtsA undergoes localized conformational changes to enclose staphyloferrin A in an arginine-rich binding pocket. J Biol Chem 285(15):11162–11171
24. Guidugli KR, Piulachs MD, Belles X, Lourenco AP, Simões ZLP (2005) Vitellogenin expres-sion in queen ovaries and in larvae of both sexes of *Apis mellifera*. Arch Insect Biochem Physiol 59:211–218
25. Havukainen H, Halskau Ø, Sjærven L, Amdam GV (2011) Deconstructing honeybee vitello-genin: novel 40 kDa fragment assigned to its N-terminus. J Exp Biol 214:582–592
26. Huang Z-Y, Robinson GE (1996) Regulation of honey bee division of labor by colony age demography. Behav Ecol Sociobiol 39:147–158
27. Jones DT (1999) Protein secondary structure prediction based on position-specific scoring matrices. J Mol Biol 292(2):195–202
28. Leoncini I, Le Conte Y, Costagliola G, Plettner E, Toth AL et al (2004) Regulation of behav-ioral maturation by a primer pheromone produced by adult worker honey bees. Proc Natl Acad Sci USA 101(50):17559–17564
29. Li A, Sadasivam M, Ding JL (2003) Receptor-ligand interaction between vitellogenin receptor (VtgR) and vitellogenin (Vtg), implications on low density lipoprotein receptor and apolipoprotein B/E. The first three ligand-binding repeats of VtgR interact with the amino-terminal region of Vtg. J Biol Chem 278(5):2799–2806

30. Marco Antonio DS, Guidugli-Lazzarini KR, Nascimento AM, Simões ZLP, Hartfelder K (2008) RNAi-mediated silencing of *vitellogenin* gene function turns honeybee (*Apis mellifera*) workers into extremely precocious foragers. Naturwissenschaften 95:953–961
31. Menzel R, Leboulle G, Eisenhardt D (2006) Small brains, bright minds. Cell 124:237–239
32. Nelson CM, Ihle K, Amdam GV, Fondrk MK, Page RE (2007) The gene *vitellogenin* has multiple coordinating effects on social organization. PLoS Biol 5:673–677
33. Piulachs MD, Guidugli KR, Barchuk AR, Cruz J, Simões ZLP et al (2003) The vitellogenin of the honey bee, *Apis mellifera*: structural analysis of the cDNA and expression studies. Insect Biochem Mol Biol 33:459–465
34. Raag R, Appelt K, Xuong NH, Banaszak L (1988) Structure of the lamprey yolk lipid-protein complex lipovitellin-phosvitin at 2.8 A resolution. J Mol Biol 200(3):553–569
35. Rina M, Savakis C (1991) A cluster of vitellogenin genes in the Mediterranean fruit fly *Ceratitis capitata*: sequence and structural conservation in dipteran yolk proteins and their genes. Genetics 127(4):769–780
36. Seehuus SC, Norberg K, Gimsa U, Krekling T, Amdam GV (2006) Reproductive protein protects sterile honey bee workers from oxidative stress. Proc Natl Acad Sci USA 103:962–967
37. Seeley TD (1995) The wisdom of the hive. Harvard University Press, Cambridge
38. Slessor KN, Winston ML, Le Conte Y (2005) Pheromone communication in the honeybee (*Apis mellifera* L.). J Chem Ecol 31(11):2731–2745
39. Snodgrass RE (1956) Anatomy of the honey bee. Comstock, New York
40. Stay B, Zera AJ (2010) Morph-specific diurnal variation in allatostatin immunostaining in the corpora allata of Gryllus firmus: implications for the regulation of a morph-specific circadian rhythm for JH biosynthetic rate. J Insect Physiol 56(3):266–270
41. Tatar M, Bartke A, Antebi A (2003) The endocrine regulation of aging by insulin-like signals. Science 299:1346–1350
42. Tufail M, Takeda M (2008) Molecular characteristics of insect vitellogenins. J Insect Physiol 54(12):1447–1458
43. Wahli W, Dawid IB (1980) Isolation of two closely related vitellogenin genes, including their flanking regions, from a *Xenopus laevis* gene library. Proc Natl Acad Sci USA 77(3):1437–1441
44. Walter J, Schindzielorz A, Grünberg J, Haass C (1999) Phosphorylation of presenilin-2 regulates its cleavage by caspases and retards progression of apoptosis. Proc Natl Acad Sci USA 96(4):1391–1396
45. Weinstock GM, Robinson GE, Gibbs RA, Weinstock GM, Weinstock GM et al (2006) Insights into social insects from the genome of the honeybee *Apis mellifera*. Nature 443(7114):931–949
46. Wheeler DE, Kawooya JK (1990) Purification and characterization of honey bee vitellogenin. Arch Insect Biochem Physiol 14:253–267
47. Williams SL, Essex JW (2009) Study of the conformational dynamics of the catalytic loop of WT and G140A/G149A HIV-1 Integrase core domain using reversible digitally filtered molecular dynamics. J Chem Theor Comp 5:411–421

Chapter 1.3
Circadian Rhythms and Sleep in Honey Bees

Ada Eban-Rothschild and Guy Bloch

Abstract The circadian clock of the honey bee is involved in complex behaviors and is socially regulated. Initial molecular characterization suggests that in many ways the clock of the bee is more similar to mammals than to *Drosophila*. Foragers rely on the circadian clock to anticipate day–night fluctuations in their environment, time visits to flowers, and for time compensation when referring to the sun in sun-compass orientation and dance language communication. Both workers and queens show plasticity in circadian rhythms. In workers, circadian rhythms are influenced by task specialization and regulated by direct contact with the brood; nurse bees tend the brood around the clock with no circadian rhythms in behavior or clock gene expression. An important function of the circadian clock is the regulation of sleep. Bees show a clear sleep state with a characteristic posture, reduced muscle tonus, and elevated response threshold. Honey bee sleep is a dynamic process with common transitions between stages of deep and light sleep. The sleep stages of workers active around-the-clock are overall similar to foragers. Sleep deprivation leads to an increase in the expression of sleep characteristics the following day, and may interfere with some learning paradigms. This review shows that the honey bee is an excellent model with which to study circadian rhythms and sleep in an ecologically and socially relevant context. Future research needs to deepen our understanding of these fascinating behaviors, reveal their neuronal and molecular bases, and explore their interactions with other physiological processes.

A. Eban-Rothschild • G. Bloch (✉)
Department of Ecology, Evolution and Behavior, The Alexander Silberman Institute of Life Sciences, The Hebrew University of Jerusalem, Jerusalem 91904, Israel
e-mail: bloch@vms.huji.ac.il

C.G. Galizia et al. (eds.), *Honeybee Neurobiology and Behavior: A Tribute to Randolf Menzel*, DOI 10.1007/978-94-007-2099-2_3,
© Springer Science+Business Media B.V. 2012

Abbreviations (Excluding Gene and Protein Names)

FS	First sleep stage
MB	Mushroom body
mRNA	messenger RNA
OL	Optic lobe
SS	Second sleep stage
TS	Third sleep stage

1.3.1   Circadian Rhythms

1.3.1.1   What Are Circadian Rhythms?

Circadian rhythms are defined as biological rhythms that meet the following three criteria: (1) they persist, or "*free-run*", with a period of about 24 h in the absence of external time cues, (2) they are reset, or *entrained*, by environmental cues, in particular light and temperature, and (3) they have a stable period length in a wide range of physiologically relevant temperatures. This phenomenon, commonly termed '*temperature compensation*', is thought to require specific mechanisms because most biological reactions accelerate with rising temperature. The circadian clock influences many physiological and behavioral processes. These include activity, sleep-wake cycles, feeding, mating, oviposition, egg hatching, and pupal eclosion. The circadian clock is also involved in measuring day length, and influences photoperiodism and annual rhythms such as diapause and seasonal reproduction [12].

The circadian system is commonly described as having three functional components. The *core* of the clock is composed of *pacemakers*, cell autonomous rhythm generators that cycle approximately, but not exactly with a 24-h period. The central pacemaker is entrained by *input pathways* in which environmental signals are detected, converted to sensory information, and transmitted to the central pacemakers. *Output pathways* carry temporal signals away from pacemaker cells to various biochemical, physiological, and behavioral processes [2, 12].

The molecular bases of rhythm generation in organisms as diverse as fungi, plants, fruit flies and mammals consists of interlocked autoregulatory transcriptional/translational feedback loops with positive and negative elements [12]. The pacemaker cells are interconnected in a circadian network that couples their activities and orchestrates normal rhythms in physiology and behavior [2, 12].

The molecular clockwork of the fruit fly *Drosophila melanogaster* has been well characterized and provides a model for studies on animals, and insects in particular. The positive elements *Clock (Clk) and Cycle (Cyc)* activate the transcription of the negative elements, the transcription factors *Period* (*Per*) and *Timeless* (*Tim1*). *Per* and *Tim1* are translated into proteins that enter the nucleus where they interfere with the transcriptional activity of the CLK: CYC complex, and by that shut down their own expression. *Par Domain Protein 1 (Pdp1), Vrille (Vri)*, and *Clockwork Orange (Cwo)* are thought to act together with *Clk* in an interlocked feedback loop that is thought to stabilize the *Per/Tim1* loop. Several kinases including *Double-time (Dbt), Shaggy (Sgg), Casein Kinase II (CKII)* and *Protein Phosphatase 2A (PP2A)* fine tune this cell-autonomous rhythm generation machinery [2, 12]. Drosophila-type *Cryptochrome (Cry-d,* also known as *insect Cry1)* has a photic input function. Although the genes and the organization principles of the molecular clockwork are similar in *Drosophila* and mammals, there are some important differences. For example, mammals do not have orthologs to *Tim1* and *Cry-d,* but have three paralogs for *Per.* They also have two paralogs for *Cry* (mammalian-type *Cry*) that act together with the *Per* genes in the negative loop of the clock.

1.3.1.2   Circadian Rhythms in Honey Bees

Circadian rhythms in the honey bee have been recently reviewed in Bloch [3] and [4] and are therefore only briefly discussed below. The first behavior with a rhythm of about a day described for bees was the flying of foragers outside the hive at a specific time of day [46]. The observation that foragers can learn to associate a food reward with a specific time of day led to the discovery that the circadian clock is involved in *time memory* ("Zeitgedächtnis"). Bees have excellent time memory and can learn to arrive at a specified location at any time of the day; they can learn as many as 9 time points with intervals of only 45 min between food availability (reviews: [3, 27]). Foragers also rely on their circadian clock to compensate for the sun's movement with time (*time-compensated sun-compass orientation*), since they orient themselves by maintaining a fixed angle to the sun, and the sun moves during the day. Foragers staying for long periods inside the hive use the clock to correct their waggle-dance in accordance with the shift in the sun's azimuth [47]. *Locomotor activity*, the best studied behavioral circadian rhythm in animals, has been well-characterized in honey bees (reviews: [3, 27]).

1.3.1.3   The Molecular and Neuronal Organization of the Honey Bee Circadian Clock

The honey bee genome does not encode orthologs to *Cry-d* and *Tim1* genes, but does have orthologs to the mammalian-type paralog *Cry-m* (also known as insect *Cry2*) [35]. The CRY-m proteins of bees and other insects, like mammalian CRY proteins, are effective transcriptional repressors and are not sensitive to light. Thus, *amCry-m* is not likely to fill the photic input function of *Drosophila Cry* [50]. The absence of orthologs to *Cry-d* and *Tim1* and the evidence that *Cry-m* is not sensitive to light suggest that honey bees use a novel light input pathway. The honey bee genome also encodes a single ortholog for the clock genes *Per, Cyc, Clk, Cwo*, and *Tim2*. Furthermore, there are highly conserved otologs to *Vri* and *Pdp1*, but no true orthologs to the orphan nuclear receptors REV-ERB (α and β) and ROR (α, β, γ) that are thought to orchestrate the expression of BMAL1 (the vertebrate ortholog of *Cycle*) in the mammalian clock [35]. These findings suggest that amVRI and amPDP1 are involved in an interlocked loop regulating *amCyc* expression, reminiscent of their function in the Drosophila clockwork (in which they regulate Clk expression). This hypothesis, however, has not yet been explicitly tested.

In foragers, and other bees with strong circadian rhythms, brain mRNA levels of both *Cry-m* and *Per* oscillate with strong amplitude and with a similar phase under both light–dark and constant darkness illumination regimes. In contrast to *Drosophila*, the predicted honey bee CYC protein contains a transactivation domain and its brain transcript levels oscillate virtually in anti-phase to *Per*, as in the mouse [35, 42]. Based on the known organization principles of the molecular clockwork, and studies on clock genes in the honey bee, a working model for the honey bee circadian clock can be proposed (Fig. 1.3.1). In this model *amPer* and *amCry-m* act together as the negative elements of the interlocked feedback loop and *amCyc* is the oscillating factor in the positive limb that probably also includes *Vri* and *Pdp1*.

The anatomical organization of the circadian clock has not been described in detail for the honey bee or for any other bee. The current picture of the anatomical organization of the brain clock is based largely on immunocytochemical studies with antibodies against PER and the neuropeptide Pigment Dispersing Factor (PDF) [7, 49]. Both the PER-ir and PDF-ir clusters are located in brain areas that are implicated in the regulation of circadian rhythms in *Drosophila* and other insects. The most consistent PER immunoreactivity (PER-ir) was detected in the cytoplasm of about eight large cells in the area between the calyces and the alpha and beta lobes of the mushroom bodies. Additional neurons in the optic lobes (OLs) and other parts of the brain showed nuclear staining.

1.3.1.4   Plasticity in Circadian Rhythms and Its Social Regulation

By contrast to most insects, newly-emerged honey bees typically have no circadian rhythms in locomotor activity or metabolism (Fig. 1.3.2a; reviews: [3, 27]). The ontogeny of circadian rhythms is endogenous because it occurs under constant conditions and rhythms free-run with a period of about, but not exactly 24 h. The development of overt circadian rhythms is associated with age-related changes in brain *Per* expression (reviews: [3, 4]).

In colonies foraging in the field the expression of behavioral rhythms is associated with worker age and task specialization. Young workers typically care for the brood around-the-clock inside the constantly dark and homeostatically regulated hive and sleep in irregular intervals [28, 42, 13, 20] (Fig. 1.3.2b). Foragers have strong circadian rhythms with a consolidated period of sleep during the night [5, 18–20, 28, 42]. Honey bee larvae are frequently attended by nurse bees. Around-the-clock activity

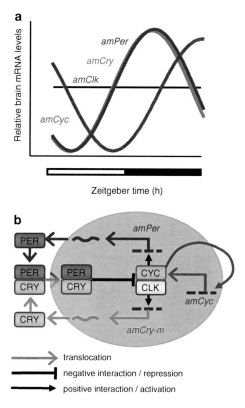

Fig. 1.3.1 A model for the honey bee molecular clockwork (**a**) Schematic representation of the oscillations of clock genes in the honey bee brain. The phase of mRNA cycling is shown for *Period* (*amPer*), *Cryptochrome-m* (*amCry*) and *Cycle* (*amCyc*). The phase of *amCyc* transcript is almost in anti-phase to that of *amPer* and *amCry*. The plots were generated by fitting a cosine model with about a 24 h cycle to brain mRNA levels measured in [35]. A straight line is depicted for *Clock* (*amClk*) that appears to have similar transcript levels throughout the day. A model is not shown for *Timeout (amTim2)* for which the pattern of mRNA variation over time was not consistent across experiments. Relative amplitudes for the various genes are not to scale. The bar at the *bottom* of the plot shows the illumination regime. Filled box – night or subjective night; *open box* – day or subjective day. (**b**) A schematic working model of the molecular clockwork in the honey bee brain. Gene name abbreviations in capital letters and italic lower case letters refer to proteins and DNA locus, respectively. The mRNA and protein for each gene is illustrated by similarly *colored wavy lines* and geometric shapes, respectively. The orange oval shape illustrates the nucleus, *purple dashed lines* depict DNA sequences, *arrows* with open wings depict translocation, *arrows* with closed wings depict positive interactions/activation, lines with a T end depict negative interactions/repression. The putative positive loop for *Cyc* is shown with no details (Reproduced by permission from [4])

may therefore enable nurses to provide better care for the brood, whereas foraging is limited to day time and relies on the circadian clock. The hypothesis that plasticity in circadian rhythms is functionally significant is supported by the strong link between division of labor and circadian rhythmicity [5, 6, 8], and by comparative studies.

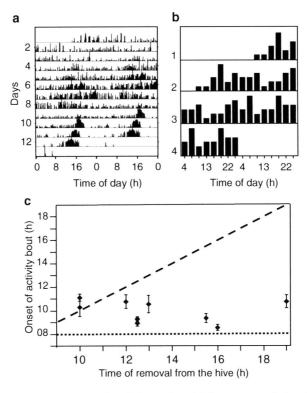

Fig. 1.3.2 Plasticity in circadian rhythms in honey bees (**a**) the ontogeny of circadian rhythms in locomotor activity. A double-plot actogram showing two consecutive days in each row. The height of the black columns in each row corresponds to levels of locomotor activity. A newly-emerged bee was placed individually in a cage in a constantly dark laboratory environment and locomotor activity was monitored automatically (for more details on the data acquisition system see [48]). This bee showed circadian rhythms for the first time on day 9. (**b**) Reversion from activity with, to activity without, circadian rhythms. The double-plot actogram depicts the observed brood care activity of a forager that was induced to revert to nursing behavior (based on data from [5]). (**c**) The onset of the morning bout of locomotor activity for nurse bees removed from the hive. Nurse bees that cared for the brood around the clock in a light–dark illuminated hive were transferred to individual monitoring cages in a constant laboratory environment. Each point shows the onset of the morning bout of activity (mean ± SE) for nurse bees from one experiment. The *dashed line* depicts a perfect correlation between the time of removal and the onset of morning locomotor activity. The horizontal dotted line depicts the onset of the subjective morning (08:00; based on data from [43])

There is a similar task-related plasticity in the bumblebee *B. terrestris* in which division of labor is based primarily on size rather than age as in honey bees and in ants whose division of labor evolved independently of that in honey bees [3, 4].

Plasticity in circadian rhythms is a social behavior and therefore it is important to identify the social signals inducing a bee to switch between activity with and without circadian rhythms. The most straightforward hypothesis is that the brood regulates activity rhythms because brood care is the main activity of nurse bees and the brood may benefit from being attended around-the-clock. Recent studies in which nurse-age bees were caged on broodless combs inside or outside the hive indicate that plasticity in circadian rhythms is modulated by direct contact with the brood [43]. The identity of the brood signal(s) and the sensory modality by which the signal is detected have yet to be identified.

Another important line of research addresses the mechanisms underlying plasticity in the circadian system. Variation in the environment of nurses and foragers (e.g., light and temperature) cannot account for task-related plasticity in circadian rhythms because nurses are active around-the-clock even when experiencing a light–dark illumination regime, and foragers continue to show strong circadian rhythms under constant conditions [27, 35, 42, 43]. Nurses are typically younger than foragers but their attenuated rhythms are apparently not because their circadian system is undeveloped or underdeveloped. Nurses switch to activity with strong circadian rhythms shortly after transfer to the laboratory, suggesting that their circadian system was capable of generating robust rhythms when they were in the hive [42, 43]. In addition, in colonies with a severe shortage in nurses, some old foragers with strong circadian rhythms revert to care for the brood and are active around-the-clock like nurses in normal colonies [5, 6]. An additional hypothesis is that the molecular feedback loop in brain pacemaker cells in the nurse brain is fixed at a certain state. The molecular and behavioral cycling would take up again from this point when the bee is released from the hive environment. Therefore, if the nurse is removed from the hive, the phase of the oscillations outside the hive would be predicted to be determined by the time of removal. However, when this hypothesis was tested the onset of activity was correlated with the subjective morning in the hive from which the nurses were collected, and not with the time of removal from the hive (Fig. 1.3.2c; [43]). Thus, a more likely explanation is that plasticity in circadian rhythms is mediated by modifications in the functioning or organization of the circadian network. For example, it is possible that some oscillators in the brain of around-the-clock active nurses generate circadian rhythms but these are not synchronized with each other. The oscillators are synchronized again when the nurse is removed from the hive or switches to activities with little or no direct contact with the brood. It is also possible that oscillators in the nurse brain do in fact cycle, but with a low amplitude relative to foragers.

1.3.1.5 Mating-Related Plasticity in Circadian Rhythms of Queens

There is also plasticity in the circadian behavior of queens. Virgin gynes rely on their circadian clock for timing their nuptial flights to a species-specific time of day [22]. Egg-laying queens on the other hand have no diurnal periodicity in behavior [16]. A similar plasticity has been more thoroughly investigated in ants in which the switch to arrhythmicity was shown to be associated with mating; virgin queens that were kept for similar periods, with or without wings, continued to exhibit robust circadian rhythms [24, 40]. Although plasticity in circadian rhythms of queens is reminiscent of that described above for nurses and foragers, it is probably regulated differently because queens of ants and honey bees do not care for their brood. The functional significance of this behavior may be related to increasing their fecundity, which is critical for the growth and maintenance of their colonies.

1.3.2   Sleep

The circadian clock influences many essential physiological processes, one of which is sleep. In honey bees, it is important to study sleep not only because it is significant for health and functions but also for their behavioral plasticity, remarkable learning capacities, and natural plasticity in circadian rhythms. Surprisingly however, relatively little is known about sleep in honey bees.

1.3.2.1   What Is Sleep?

Three main characteristics are commonly used to define sleep: (1) a period of *quiescence* associated with a specific posture and/or resting place, which is typically accompanied by reduced motor activity, (2) an increased *arousal threshold* (i.e. a higher intensity stimulus is needed to produce a response) and (3) a *homeostatic regulation* mechanism, which is manifested in a sleep rebound after periods of sleep deprivation [45]. A sleep state is distinguished from quiet wakefulness by a decrease in the ability to react to stimuli, whereas the reversibility to an awake state distinguishes sleep from coma [44]. Sleep is regulated by circadian and homeostatic mechanisms which are partly independent. The circadian system plays a crucial role in the timing and consolidation of sleep to an ecologically appropriate period; diurnal animals typically sleep during the night and nocturnal animals during the day [9]. The homeostatic mechanism reflects the need for sleep that accumulates during prolonged periods of wakefulness and dissipates during sustained sleep.

Sleep research has traditionally focused on humans and other mammals. It was commonly thought that true sleep is not found in lower taxa. Over the past three decades studies on diverse non-mammalian species, including fish [31], insects [14, 19, 41, 45], and even nematode worms [32], have shown that rest in these animals meets many criteria of sleep. The molecular pathways associated with sleep in mammals, flies, fish and worms show much conservation, suggesting an ancient and common origin for sleep [1, 9]. Three main areas of molecular conservation in the pathways controlling sleep are the involvement of circadian clock genes (such as *Per*), signaling pathways (such as EGF receptor) and genes involved in neurotransmission (such as GABA receptors) [1].

Fig. 1.3.3 Body posture of honey bee workers in various arousal states. Each photograph is a single frame taken from a continuous 24-h video recording. (**a**) Immobile-active state [IA] – an awake bee stays in the same place, the thorax, abdomen, and head are clearly raised above the substrate. (**b**) First sleep stage [FS] – the abdomen and thorax are clearly raised above the substrate, and the antennae are extended at an angle of 90–180°, between the pedicle and the scape. (**c**) Second sleep stage [SS] – body is typically more adjacent to the substrate, and the antennae are extended at an angle of ~90° between the pedicle and the scape. (**d**) Third sleep stage [TS] – the muscle tonus is reduced, and the body is adjacent to the substrate. The angle between the pedicle and scape < 90°, with the antennae tips typically touching the substrate. The three sleep stages also differ in bout duration, antenna movements, and response threshold (From [13])

Although sleep is ubiquitous in the animal kingdom its adaptive value remains an ongoing enigma. Many explanations have been proposed for sleep function, including energy conservation, restoration at the cellular and network levels, maintenance of synaptic homeostasis and memory consolidation (e.g., [26, 34, 44]). Sleep seems to be particularly important for the brain, since the most immediate effect of sleep deprivation is cognitive impairment [10].

1.3.2.2   Do Honey Bees Sleep?

Honey bees are among the first invertebrates for which a sleep-like state was described. In a set of seminal studies, Walter Kaiser and his colleagues characterized the nightly rest behavior of honey bee foragers and proposed that this state shares many behavioral and physiological characteristics of sleep with mammals and birds [18, 19, 36, 37]. Foragers exhibit all three behavioral characteristics of sleep: a period of quiescence, an increased response threshold, and a homeostatic regulation mechanism [18, 23, 36, 37]. Foragers sleep in a ***characteristic posture*** with relaxed thorax, head and antennae, and with little antennae movements (Fig. 1.3.3). In the hive foragers typically sleep at the periphery of the nest [18, 20]. Both in the hive and in the lab, foragers prefer to rest in locations with an ambient temperature around 28°C. This preference for a relatively low temperature may allow them to conserve energy since during sleep they are ectothermic and their body temperature is similar to the ambient temperature [38].

Several studies have shown that sleep in honey bees is associated with an increase in **response threshold**. Long-term, extracellular, single-unit recordings from optomotor interneurons in the OLs of honey bee foragers revealed a diurnal oscillation in their sensitivity to moving visual stimuli; the response threshold was higher during the subjective night than during the subjective day [19]. Elevated response thresholds were also found for heat and light stimuli ([13, 18], respectively). For example, the light intensity needed to elicit a response (moving the head and antennae) from a bee in sleep stage three (TS, deep sleep, see Fig. 1.3.3) was about 10,000 times higher than that needed to obtain a similar response from an immobile awake bee [13].

Antennae movement was commonly used as a proxy for honey bee sleep. Kaiser [18] defined sleep as a state of antenna immobility or small amplitude antennal movements. Sauer et al. [37] further showed that this sleep state is dynamic and is correlated with additional characteristics such as a typical head inclination and abdominal ventilatory cycles. Eban-Rothschild and Bloch [13] suggested that honey bee sleep is not uniform and described three different sleep stages (that they termed "First", "Second" and "Third"; abbreviated FS, SS, and TS, respectively, see Fig. 1.3.3) that differ in body and antennae posture, sleep bout duration, antenna movements, and response threshold. Reduced antennal motility and more pronounced downward tilting of the head, which probably corresponds to deep sleep (TS in [13]) is also associated with an increase in ventilatory cycle duration [37], reduced body temperature, and the low sensitivity (high response threshold) for heat stimuli [18]. In honey bees as in mammals, the transitions from arousal to deep sleep and from deep sleep to awake states are gradual; bees typically enter sleep through the first sleep stage and progress to the second and third stages. These behavioral analyses of sleep dynamics, however, did not find relatively regular sleep cycles with a consistent period as seen in humans and other mammals [13]. There is also one preliminary study suggesting that deep sleep is correlated with rhythmic electrophysiological activity in the mushroom bodies (MBs) [39]. This is an interesting suggestion because the MBs are implicated in sleep in *Drosophila* [17, 30]. In mammals and birds, electroencephalogram (EEG) records, which correspond to neuronal activity in the cerebral cortex, are used for defining sleep and sleep stages. There is also evidence that arousal state is correlated with characteristic electrical brain activities in invertebrates (e.g., [29, 33]). Additional studies are needed to confirm these observations and establish electrophysiological correlates for sleep in bees.

There is evidence that sleep in insects is **homeostatically regulated**, similar to mammals and birds (e.g., [23]). For example, fruit flies exhibit a proportional increase in sleep duration (the index for sleep was continuous bouts of >5 min. with no locomotor activity) following sleep deprivation [14, 41]. Sleep rebound following sleep deprivation in honey bees differed across studies. Hussaini et al. [15] reported that bees that were sleep-deprived for 15 h (during all of the dark phase and some of the light phase), increased sleep duration during the following light phase, but not during the next dark phase (the index for sleep was the amount of flagella immobility lasting 5 min or more). By contrast, Sauer et al. [36] sleep-deprived foragers for 12 h during the dark phase, and found an increase in sleep (antennae immobility) only during the following dark phase. The latency from the beginning of the dark period to the first episode of antennal immobility ("sleep latency") tended to decrease following sleep deprivation. These authors also showed that disturbing the bees during the light period (day), did not result in a similar rebound, suggesting that the response was due to sleep lost and not the stress associated with the sleep deprivation procedure [36]. Although these studies differ in the time of sleep rebound (day vs. night), both suggest that bees compensate for sleep lost by intensifying their sleep the following day.

Taken together, the studies reviewed above show that the consolidated nightly rest of honey bee foragers meet the major behavioral and physiological criteria for defining it as sleep.

1.3.2.3   Do Bees That Are Active Around the Clock Sleep?

Because sleep is typically associated with a consolidated period of inactivity, it was not clear whether bees that are active around-the-clock sleep, and if they do, whether their sleep is similar to that of foragers. Two recent studies addressed this question and suggest that around-the-clock active bees do sleep. Young bees that were placed in individual cages in the lab were active around-the-clock but still showed the same three sleep stages as seen in foragers (Fig. 1.3.4). Moreover, the body and antenna postures, antenna movements, and response thresholds for each sleep stage was similar to that in foragers from the same colonies [13]. A precise determination of arousal state is much more challenging in the hive, which is densely populated and where motionless bees may be awake but busy in heat production or brood incubation [21]. Nevertheless, there is evidence suggesting that worker bees, including around-the-clock active cell cleaners and nurses, sleep both inside and outside the comb cells ([20]; sleep was defined as a quiescent state, with no antennae movements for ≥ 3 s). Young honey bees spent more time inside the comb cells, and as they grew older and changed task they tended to sleep more outside the cells. These observations are consistent with previous studies in which the amount of 'standing motionless' was recorded for bees performing various tasks in observation hives [28].

Young bees appear to differ from foragers in their sleep dynamics as well. Young bees have fewer sleep bouts during the whole day; however these bouts tend to be longer in comparison to foragers. Foragers tend to progress mainly from light to deep sleep, and from deep sleep they pass directly to awake states, switching less often between sleep stages. Young bees tend to pass more often between the three sleep stages without switching to being awake. It is still unclear whether these differences relate to age or to differences in circadian rhythms between these two groups of bees [13] (Fig. 1.3.4).

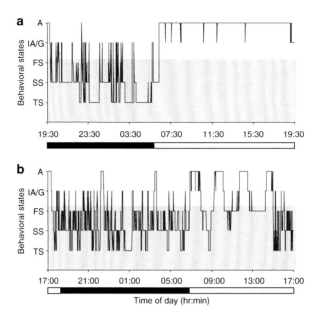

Fig. 1.3.4 Sleep dynamics in honey bees with and without circadian rhythms (**a**) a forager with consolidated activity during the subjective day. (**b**) A young bee that was active around-the-clock. The vertical (Y) axis depicts arousal state (A = active, G = grooming, for other abbreviations see legend to Fig. 1.3.3). *Gray* background indicates that the bee is asleep; white background indicates that the bee is awake. The *horizontal bars* at the *bottom* of the plots depict the subjective time. Filled bar = subjective night; open bar = subjective day (From [13])

1.3.2.4   Memory Consolidation During Sleep

In many animals sleep is associated with memory consolidation, the process that transforms new memories to more stable representations that become integrated into the network of pre-existing long-term memories. In both mammals and birds, brain structures implicated in specific learning tasks appear to show the same electrical activity pattern during sleep as during learning (e.g., [11]). Honey bees can provide an excellent model to study the relationships between memory consolidation and sleep, since the functional significance of learning and memory in foraging behavior is well established (see Chaps. 2.5, 6.2, and 6.6) and there has been much progress in understanding the molecular, biochemical and neuronal mechanisms of learning and memory in bees [25]. Surprisingly, this relationship was not explored until recently.

Hussaini et al. [15] tested the influence of sleep deprivation on the acquisition and extinction of new memories. They conditioned foragers to associate odors with a food reward and tested their memory for this association at various time intervals. Sleep deprivation had no effect on memory acquisition, but significantly reduced extinction learning. These findings are consistent with studies in mammals in which sleep deprivation impaired performance in some learning paradigms but not in others.

1.3.3 Outlook

The circadian clock of the honey bee is involved in complex behaviors such as sun-compass orientation, time-memory, division of labor, and social coordination of worker activities, all of which can be studied in a relatively natural context. Recent studies on time memory have shown that there is still much to learn even about circadian behaviors, which were described years ago in bees (reviews: [3, 4]).

The circadian clock is emerging as an important element in the temporal coordination of honey bee society. The circadian system of honey bees is very sensitive to social influences and shows remarkable plasticity. These characteristics may have been shaped by social evolution [3, 4]. The hope is that future studies will discover the specific social signals and the sensory modalities by which the social environment modulates the circadian behavior of bees. Unveiling the molecular and neuronal bases of plasticity in circadian rhythms is another important line of future research, with possible implications that are far beyond the sociobiology of bees. Another line of socio-chronobiological research that should be pursued in the future is social entrainment [3, 4].

Their social behavior, chronobiological plasticity, and remarkable learning capacities also make honey bees an attractive model to study sleep in a natural context. For example, foragers orient themselves over long distances and rely more heavily than nurses on visual learning. This natural variability between nurses and foragers that develop in a similar hive environment and are genetically related creates a natural model system with which to study the adaptive value of sleep, which has been commonly hypothesized to be linked to memory consolidation and synaptic plasticity. In order to effectively study honey bee sleep, protocols for sleep deprivation need to be developed, including in the hive. It is also needed to develop methods for precisely recording sleep in bees in the complex social environment of the hive.

We need to know much more about the neuronal and molecular mechanisms governing circadian rhythms and sleep in honey bees. This is particularly true for sleep, which has not yet been correlated with specific genes, anatomical structures or neurophysiological processes. There are also significant gaps in our knowledge of the molecular biology of the circadian clock, and the neuroanatomical characterization of the circadian network is only at its very initial stages. Molecular and neurobiological studies on sleep and circadian rhythms will not only help enhance our understanding of these important systems, but will also set the stage for studies on the ways these systems interact with each other and affect the social and foraging behavior of bees.

References

1. Allada R, Siegel JM (2008) Unearthing the phylogenetic roots of sleep. Curr Biol 18:R670–R679
2. Bell-Pedersen D, Cassone VM, Earnest DJ, Golden SS, Hardin PE, Thomas TL, Zoran MJ (2005) Circadian rhythms from multiple oscillators: lessons from diverse organisms. Nat Rev Genet 6:544–556

3. Bloch G (2009) Plasticity in the circadian clock and the temporal organization of insect societies. In: Gadau J, Fewell J (eds) Organization of insect societies: from genome to sociocomplexity. Harvard University Press, Cambridge, pp 402–432
4. Bloch G (2010) The social clock of the honeybee. J Biol Rhythm 25:307–317
5. Bloch G, Robinson GE (2001) Reversal of honey bee behavioural rhythms. Nature 410:1048
6. Bloch G, Toma DP, Robinson GE (2001) Behavioral rhythmicity, age, division of labor and period expression in the honey bee brain. J Biol Rhythm 16:444–456
7. Bloch G, Solomon SM, Robinson GE, Fahrbach SE (2003) Patterns of PERIOD and pigment-dispersing hormone immunoreactivity in the brain of the European honey bee (*Apis mellifera*): age- and time-related plasticity. J Comp Neurol 464:269–284
8. Bloch G, Rubinstein CD, Robinson GE (2004) Period expression in the honey bee brain is developmentally regulated and not affected by light, flight experience, or colony type. Insect Biochem Mol Biol 34:879–891
9. Cirelli C (2009) The genetic and molecular regulation of sleep: from fruit flies to humans. Nat Rev Neurosci 10:549–560
10. Cirelli C, Tononi G (2008) Is sleep essential? PLoS Biol 6:1605–1611
11. Dave AS, Margoliash D (2000) Song replay during sleep and computational rules for sensorimotor vocal learning. Science 290:812–816
12. Dunlap. JC, Loros JJ, DeCoursey PJ (2004) Chronobiology: biological timekeeping. Sinauer, Sunderland, 382 p
13. Eban-Rothschild AD, Bloch G (2008) Differences in the sleep architecture of forager and young honey bees (*Apis mellifera*). J Exp Biol 211:2408–2416
14. Hendricks C, Finn SM, Panckeri KA, Chavkin J, Williams JA, Sehgal A, Pack AI (2000) Rest in *Drosophila* is a sleep-like state. Neuron 25:129–138
15. Hussaini SA, Bogusch L, Landgraf T, Menzel R (2009) Sleep deprivation affects extinction but not acquisition memory in honey bees. Learn Mem 16:698–705
16. Johnson JN, Hardgrave E, Gill C, Moore D (2010) Absence of consistent diel rhythmicity in mated honey bee queen behavior. J Insect Physiol 56:761–773
17. Joiner WJ, Crocker A, White BH, Sehgal A (2006) Sleep in *Drosophila* is regulated by adult mushroom bodies. Nature 441:757–760
18. Kaiser W (1988) Busy bees need rest, too – behavioral and electomyographical sleep signs in honey bees. J Comp Physiol A 163:565–584
19. Kaiser W, Steiner-Kaiser J (1983) Neuronal correlates of sleep, wakefulness and arousal in a diurnal insect. Nature 301:707–709
20. Klein BA, Olzsowy KM, Klein A, Saunders KM, Seeley TD (2008) Caste-dependent sleep of worker honey bees. J Exp Biol 211:3028–3040
21. Kleinhenz M, Bujok B, Fuchs S, Tautz H (2003) Hot bees in empty broodnest cells: heating from within. J Exp Biol 206:4217–4231
22. Koeniger N, Koeniger G (2000) Reproductive isolation among species of the genus *Apis*. Apidologie 31:313–339
23. Martinez-Gonzalez D, Lesku JA, Rattenborg NC (2008) Increased EEG spectral power density during sleep following short-term sleep deprivation in pigeons (*Columba livia*): evidence for avian sleep homeostasis. J Sleep Res 17:140–153
24. McCluskey ES (1992) Periodicity and diversity in ant mating flights. Comp Biochem Physiol 103:241–243
25. Menzel R, Leboulle G, Eisenhardt D (2006) Small brains, bright minds. Cell 124:237–239
26. Mignot E (2008) Why we sleep: the temporal organization of recovery. PLoS Biol 6:661–669
27. Moore D (2001) Honey bee circadian clocks: behavioral control from individual workers to whole-colony rhythms. J Insect Physiol 47:843–857
28. Moore D, Angel JE, Cheeseman IM, Fahrbach SE, Robinson GE (1998) Timekeeping in the honey bee colony: integration of circadian rhythms and division of labor. Behav Ecol Sociobiol 43:147–160
29. Nitz DA, van Swinderen B, Tononi G, Greenspan RJ (2002) Electrophysiological correlates of rest and activity in *Drosophila melanogaster*. Curr Biol 12:1934–1940

30. Pitman JL, McGill JJ, Keegan KP, Allada R (2006) A dynamic role for the mushroom bodies in promoting sleep in *Drosophila*. Nature 441:753–756
31. Prober DA, Rihel J, Onah AA, Sung RJ, Schier AF (2006) Hypocretin/orexin overexpression induces an insomnia-like phenotype in zebrafish. J Neurosci 26:13400–13410
32. Raizen DM, Zimmerman JE, Maycock MH, Ta UD, You YJ, Sundaram MV, Pack AI (2008) Lethargus is a *Caenorhabditis elegans* sleep-like state. Nature 451:569–U6
33. Ramon F, Hernandez-Falcon J, Nguyen B, Bullock TH (2004) Slow wave sleep in crayfish. Proc Natl Acad Sci USA 101:11857–11861
34. Roth TC, Rattenborg NC, Pravosudov VV (2010) The ecological relevance of sleep: the trade-off between sleep, memory and energy conservation. Philos Trans R Soc B Biol Sci 365:945–959
35. Rubin EB, Shemesh Y, Cohen M, Elgavish S, Robertson HM, Bloch G (2006) Molecular and phylogenetic analyses reveal mammalian-like clockwork in the honey bee (*Apis mellifera*) and shed new light on the molecular evolution of the circadian clock. Genome Res 16:1352–1365
36. Sauer S, Herrmann E, Kaiser W (2004) Sleep deprivation in honey bees. J Sleep Res 13:145–152
37. Sauer S, Kinkelin M, Herrmann E, Kaiser W (2003) The dynamics of sleep-like behaviour in honey bees. J Comp Physiol A 189:599–607
38. Schmolz E, Hoffmeister D, Lamprecht I (2002) Calorimetric investigations on metabolic rates and thermoregulation of sleeping honey bee s (*Apis mellifera carnica*). Thermochim Acta 382:221–227
39. Schuppe H (1995) Rhythmic brain activity in sleeping bees. Wien Med Wochenschr 145:463–4
40. Sharma VK, Lone SR, Goel A (2004) Clocks for sex: loss of circadian rhythms in ants after mating? Naturwissenschaften 91:334–337
41. Shaw PJ, Cirelli C, Greenspan RJ, Tononi G (2000) Correlates of sleep and waking in *Drosophila melanogaster*. Science 287:1834–1837
42. Shemesh Y, Cohen M, Bloch G (2007) Natural plasticity in circadian rhythms is mediated by reorganization in the molecular clockwork in honey bees. FASEB J 21:2304–2311
43. Shemesh Y, Eban-Rothschild AD, Cohen M, Bloch G (2010) Molecular dynamics and social regulation of context-dependent plasticity in the circadian clockwork of the honey bee. J Neurosci 30:12517–12525
44. Siegel JM (2005) Clues to the functions of mammalian sleep. Nature 437:1264–1271
45. Tobler I (1983) Effect of forced locomotor on the rest activity cycle of the cockroach. Behav Brain Res 8:351–360
46. von Buttel-Reepen HB (1900) Sind die bienen reflexmaschinen? Arthur Georgi, Leipzig
47. von Frisch K (1967) The dance language and orientation of bees. Harvard University Press, Cambridge
48. Yerushalmi S, Bodenhaimer S, Bloch G (2006) Developmentally determined attenuation in circadian rhythms links chronobiology to social organization in bees. J Exp Biol 209:1044–1051
49. Zavodska R, Sauman I, Sehnal F (2003) Distribution of PER protein, pigment-dispersing hormone, prothoracicotropic hormone, and eclosion hormone in the cephalic nervous system of insects. J Biol Rhythm 18:106–122
50. Zhu HS, Sauman I, Yuan Q, Casselman A, Emery-Le M, Emery P, Reppert SM (2008) Cryptochromes define a novel circadian clock mechanism in monarch butterflies that may underlie sun compass navigation. PLoS Biol 6:138–155

Chapter 1.4
Mechanisms of Social Organization: Commentary

Randolf Menzel

The social life of honeybees harbors many more secrets than those already discovered since more than 100 years of extensive work. How shall we proceed uncovering these secrets? This will depend strongly on how we think about the mental capacities of the individual as a member of the community. It appears to me that we acknowledge a rather high level of neural functions when considering the honeybee's individual behavior, e.g. when it is foraging, navigating, handling flowers, or deciding between options outside the hive. Yet, when looking into the hive and analyzing the bees' behavior in a collective context, a single bee suddenly becomes a rather stereotyped element whose behavior is predominantly controlled by innate sensory-motor links. It is even argued that in this social context, in which self-organization phenomena would determine the emergence of collective behaviors, individuals are all identical and interchangeable. We are asked to believe that the impressive functions of a colony as a whole result from emergent properties based on rather simple rules and interactions followed reflexively and stereotypically by the individual bee. Are bees switching off their central brain when entering the hive? Are they just rather stupid members in a network of community functions, e.g. a reproductive network? I have the suspicion that the limits imposed on us when we look into a community of thousands of animals, which all look more or less the same, gives us little chance to include in our concepts more than the typical robot view of insect societies, and this may be inadequate.

Does indeed the "spirit of the hive" reside in the stimulus–response interactions derived in part from the ovaries? The data presented by Rob Page are interpreted as support of a rather linear relationship between genes-hormones-ovaries-response threshold-behavior. Although the reproductive regulatory network is considered as

R. Menzel (✉)
Institut für Biologie, Neurobiologie, Freie Universität Berlin, Berlin, Germany
e-mail: menzel@neurobiologie.fu-berlin.de

C.G. Galizia et al. (eds.), *Honeybee Neurobiology and Behavior: A Tribute to Randolf Menzel*, DOI 10.1007/978-94-007-2099-2_4,
© Springer Science+Business Media B.V. 2012

involving many components (most of them yet unknown) I ask myself whether the logic of the arguments needs to be critically scrutinized when we try to uncover the governing of rules within the society.

Establishing correlations between gene functions, hormonal control and behavior in a social context is an extremely important first step, and the data presented are impressive. However, as mentioned in the Introduction of Rob Pages' chapter, individual workers regulate their socially-relevant behavior (e.g. pollen foraging) according to their sequential experience with empty or filled pollen cells and the amount of brood. Thus, individuals learn continuously about the reproductive state of the whole colony and adapt their behavior accordingly. The regulatory network, therefore, needs to include experience-dependent feedback loops at the level of the individual. Such feedback loops determine that the information flow is not linear anymore, and as a consequence correlations are harder to interpret. Furthermore, when it comes to mechanisms the level of the individual will also be important. The wonderful experiment with the grafted ovaries presented in Rob Page's chapter may also be interpreted as documenting a homeostatic effect of ovaries on many other functions including those that regulate neural-control systems. In short, I believe that more emphasis is required to include feedback loops reaching central neural circuits related to the individual adaptation by learning.

One of the many behaviors of individuals embedded in a social context is circadian rhythm and sleep. The beauty in the study on circadian rhythm and sleep as presented in Ada Eban-Rothschild's and Guy Bloch's chapter is the finding that sleep does not need to be coupled to circadian rhythm. Young bees sleep but not in a circadian rhythm, whereas foraging bees do. Furthermore, the requirements for sleep, whatever they may be, appear to be fulfilled also under arrhythmic distributions of rather short sleep phases. Old bees may revert to non-rhythmic behavior (likely including arrhythmic sleep). Thus, conditions for sleep appear not to depend on age but are a function of social status. These unique characters combined with the social regulation will hopefully provide us with tools allowing to unravel the mysteries of sleep. What can honeybee studies contribute to this general problem?

Input and output of the brain are disconnected during sleep, a condition that may allow the brain to listen to itself and use this for its reorganization. This is the general idea behind the notion that sleep supports the consolidation of memory. Guy Bloch is right in stating that surprisingly little is known about sleep in bees and memory consolidation although so much has been discovered about molecular, neural and behavioral phenomena related to memory consolidation. It was indeed an unexpected result that consolidation of appetitive olfactory memory is not com-promised by deprivation of sleep, but extinction memory is [2]. This is a particularly relevant finding in the context of the argument that sleep deprivation may simple reduce sensory and motor performance. Retention scores are in fact enhanced after sleep deprivation indicating impaired memory consolidation for extinction learning. Thus, as in other animals, sleep in bees is involved in the consolidation of specific forms of memory [1]. Which memories could these be? Extinction memory provides a clue. In extinction learning an animal learns that a stimulus that was initially learned as being associated with a specific outcome is no longer associated with that outcome. It is now both predictive and non-predictive for reward. The task of the learning and memory systems will be to extract those conditions (e.g. context, the animal's own behavior) in which one (stimulus-outcome) or the other contingency (stimulus-no outcome) applies. We may speculate that self-organizing processes during sleep could search for those additional conditions by replay mechanisms searching for co-existing neural excitation patterns associated with the contradictory stimulus-outcome conditions. In any case, consolidation of the new memory (the extinction memory) needs to be related to other memories (innate or required) already existing in the brain, and these other memories may involve multimodal and operant forms of learning. These speculations lead to a working hypothesis. Sleep dependent memory consolidation may be characterized by their cross reference to other already existing memories. A hint in this direction comes from yet unpublished data from our lab which indicate sleep dependent memory consolidation for novel navigation tasks. Navigation memory is characterized by the association of a novel flight path to an already existing spatial reference frame, the spatial memory developed during exploratory orientation flights of young foragers. In more general terms I would expect that learning requiring the association between already existing and novel forms of related memory contents and thus the extraction of rules via multiple learning trials may require sleep-promoted consolidation.

References

1. Diekelmann S, Born J (2010) The memory function of sleep. Nat Rev Neurosci 11(2):114–126
2. Hussaini SA, Bogusch L, Landgraf T, Menzel R (2009) Sleep deprivation affects extinction but not acquisition memory in honeybees. Learn Mem 16(11):698–705

Part II
Communication and Navigation

Chapter 2.1
Foraging Honey Bees: How Foragers Determine and Transmit Information About Feeding Site Locations

Harald Esch

Abstract Experiments by Karl von Frisch in the 1940s revealed that dances of successful foragers in honeybee colonies describe the locations of feeding sites by timing and direction of patterns in their dances. v. Frisch and his students believed that energy expanded on the way to a feeder is used as a measure of distance ("The Energy Hypothesis"). More recent experiments showed that the optic flow is used to determine distance ("The Optic Flow Hypothesis"): Bees flying through a narrow tunnel, that generates a large amount of optic flow while moving forward, indicate a much larger distance than bees flying in the field. The dances of tunnel bees can be used to deceive hive mates to search at much larger distances when they leave their hives. "Robot Bee Experiments", intended to reveal the *nature* of bee communication, did not lead to completely satisfying results: Robots are not as efficient as real dancers. Four chemical compounds, detected with gas chromatograph/mass spectrometry on the abdomen of dancers, might be used as "pheromones". The following contribution does not discuss the *efficiency* of dances, which is covered in other chapters of this book, but it rather tries to explain the *physiology* of this remarkable behavior.

2.1.1 Introduction

Von Frisch first described the dances of honey bees in a summary of previous work [30]. He believed that he understood their significance. Taking up experimentation 20 years later, he discovered that he had overlooked the most important facts: dances report distance and direction of feeding sites visited by foragers, at least to a human observer [31] (for details see the next two paragraphs). These new observations led

H. Esch (✉)
Department of Biological Sciences, University of Notre Dame, Notre Dame, IN 46556, USA
e-mail: Harald.E.Esch.1@nd.edu

C.G. Galizia et al. (eds.), *Honeybee Neurobiology and Behavior: A Tribute to Randolf Menzel*, DOI 10.1007/978-94-007-2099-2_5,
© Springer Science+Business Media B.V. 2012

to intensive experimentation by himself and his coworkers with the goal to establish how foragers *perceive* and *use* the dance information. Von Frisch's work on dancing honey bees has been recapitulated in his book *The Dance Language and Orientation of Honey bees* [32]. Over the years new results explained many of the sensory tasks underlying the dance behavior and initiated studies of communication in other species of bees.

2.1.2   The Elements of Communication

2.1.2.1   Communicating Direction

For a human observer it seems obvious that the direction of wagging runs reflects the feeding site direction when they occur on a horizontal platform at the hive entrance. In this particular case, wagging runs point directly to the feeder. Direction depends on the location of the sun, represented by the point of maximal brightness, or if the sun is not directly visible, by patterns of polarized skylight [33]. However, dances occur mainly in the darkness of the hive and on vertical combs. Inside a hive, on a vertical surface, gravity serves as reference. In the absence of visual cues, direction "against gravity" represents the direction of the sun. The angle between sun and feeding site, as seen from the hive, is reproduced on the vertical comb. Perception of a cue to "feeder direction" on a vertical comb inside the hive *immediately* initiates a wagging run: When a gravity oriented wagging run is finished, and the sun is reflected through a mirror from the bottom of the comb, the dancer turns around and performs the next wagging run, this time with orientation to the sun's picture in the mirror. By removing and adding the mirror at appropriate times, one can stimulate wagging run after wagging run. These dances successfully recruit newcomers to appropriate feeding sites [32]. Dancers and followers obviously change reference cues without a problem. This, and other observations, suggests that the wagging run itself, with its direction, contains the complete information about feeding site location ("the vector information") (Fig. 2.1.1).

2.1.2.2   Communicating Distance

Von Frisch first believed that the dance rhythm (dance cycles/time) is the measure of feeding site distance. Basic experiments used the average of dance cycles per 15 s as a measure of distance. Typical functions relating distance and dance cycles per unit time show a rather exponential decay indicating that the closer the food source the faster bees dance and vice versa for distant food sources. Due to lack of suitable instrumentation available at the time, most of the early experiments were done with only a stopwatch. Work using this method, and published by von Frisch and

Fig. 2.1.1 The first robot [32]. The wooden dummy was kept inside the hive for a day before an experiment and it assumed the hive odor. It waggled to the *left* and *right* driven by step motor

coworkers, might be deceiving to a naive reader. Single data points represent *averaged* observations. The data in the early literature are the result of *averaging the averages* of different dances of one or several individuals. This gives a wrong picture of the *accuracy of single dances*. During development of a first bee robot in the 1950s, it became possible to measure duration of single dance cycles directly with a microphone, or by an electromagnetic method [3–5, 7, 34]. The accuracy of wagging times within and between individuals of the same hive varies significantly. This is also true for dances of individuals from different hives [8]. The accuracy is much less than what von Frisch type graphs suggest. The *"between individuals error"* is often ±30% (standard deviation) of distance signaled in a given hive.

2.1.3 The Nature of the Dance

2.1.3.1 Robot Bees

A critical line of experimentation attempted to develop a "robot bee", a machine capable of performing all important behaviors of dancing foragers. It was hoped that changing various parts of complete dances would enable us to identify the behavior patterns required for successful recruitment. Such an attempt was repeated several times, namely in the 1950s by Esch and coworkers [26, 32], in the 1980s by Michelsen and coworkers [19–21] and more recently by the Menzel group in Berlin (personal communication). In all cases, the recruiting success of robot bees is consistently lower than that of real dancers.

In many of *our* experiments performed in the 1950s, robots had repeatedly touched real dancers. Subsequently dance followers became more attentive to robots, and more recruits appeared at advertised feeding sites (Esch, personal observation). Newer experiments showed an increase in foraging activity in hives into which air from another hive with intensive dancing was injected [28]. This suggested that real dancers might produce a *pheromone* that stimulates nest mates to *pay attention* to dancers, and possibly *extract information* from their dances. Solid-phase micro extraction, in conjunction with gas chromatography and mass spectrometry, revealed that *only dancing honey bees* release four specific substances: Two alkanes, tricosene and pentacosene, and two alkenes, Z-(9)-tricosene and Z-(9)-pentacosene. They appear on the abdomen of *active dancers* and cannot be found on individuals performing other tasks within the hive, regardless of their ages. Non dancing foragers show only barely detectable traces [29]. In behavioral experiments injection of these substances into a hive significantly increased the number of foragers leaving the hive, very similar to injection of scent from real dancers [28, 29]. The use of specific olfactory clues ("pheromones") in recruitment of hive mates is very common in the family of bees. Bumblebees, close relatives of the *honey bee*, use pheromones from a tergal gland inside their nests to initiate foraging activity [2]. Stingless bees stimulate their hive mates to forage and guide to their feeders using pheromones [18]. Even other social insects, as ants, use trail pheromones to guide nest mates to food [16].

2.1.3.2   The Role of Odors

The odor of a feeding site can affect the spatial information component of a waggle dance. If recruits are sent into the field with *vector information*, they know only the *general area* where to go. Without odor identifying a *specific target*, they would not know what to look for. The necessary information can be obtained from odor clinging to the dancer [32]. The importance of odors for stimulating dance-following by recruited bees (see above) and for close-up recognition of the appropriate food source to be exploited has led to misleading arguments denying the fact that bees use dance information to localize food sources. There have been repeated attempts to question or minimize the importance of *vector information* [35, 36]. Wenner completely denies the existence of navigational information in dances. He claims that odor alone provides adequate information to find a food source location by flying up an odor plume ("The Odor Search Hypothesis") [36]. This claim has been repeatedly proved to be wrong (see Chap. 2.5). A recent study shows that previous experience during food collection ("private navigational information") can be more important than the information transmitted by dances ("social vector information") [13]. Unlike Wenner, the latter authors do not deny the existence of vector information. However, an uncritical reading of their theses might lead one to accept Wenner's ideas, but this is not their intention ([14], see Chap. 2.4). They argue that recruits, that had visited feeders of the same odor as they find in present dances, prefer to go to the sites they remember from the past ("private navigational information"). The question whether navigational information is used or not depends on the adaptive significance of the situation. Recruits who obtained samples of food that point to places with the same type of food they had found before, return to the old sites and spend a minimum of energy. The use of vector information obtained by attending dances increases the cost of finding food, but since foragers dance only for high quality food sources, this cost increase might be compensated by the high quality of the food advertized by dances [24]. The real importance of "social vector information" shows when a recruit begins her foraging life, or when specific resources are exhausted and new food sources have to be found ([23], see also Chap. 2.3).

2.1.3.3 The Energy Hypothesis

Wind direction affects distance perception. Foragers flying into the wind on their way to a feeder indicate greater distances than ground distance. Individuals flying with wind indicate shorter distances. This observation lead von Frisch to suspect that energy spent on the way to a feeder is used as a measure of distance ("The Energy Hypothesis"). Final confirmation for the energy hypothesis seemed to come from foragers that had to fly up a steep hill. They needed more energy to reach their feeders, and signaled distances larger than ground distance ("The Mountain Experiments"). Heran stated that the results of his mountain experiments were not completely convincing [15]. This called for a repetition. Mountain experiments require a special terrain, and they are physically very demanding. Further experiments were planned in the 1960s but were never done. Many years later we determined the actual energy expenditure of foragers on the way to a feeder by measuring a forager's metabolic expenditures directly [6, 12]. It became evident that the energy hypothesis could not be correct. That led to a repetition of the mountain experiments under better controlled and less demanding physical conditions [9].

2.1.3.4 The Optic Flow Hypothesis

To eliminate the physical stress of walking up and down a steep mountain, we attached the feeder to a weather balloon that could be raised to different altitudes. A group of foragers was trained to this feeder at ground level 70 m from the hive [9]. The foragers performed waggle dances. The balloon was then slowly raised to higher altitudes. Most foragers kept visiting the feeder. As the balloon ascended, distances signaled by dancers became *shorter and shorter* despite the fact that the bees had to fly *greater and greater distances*. In addition, they had to raise their bodies to *higher and higher altitudes*. This showed clearly that the energy hypothesis was not acceptable. It suggested that *optic flow* (*i.e. the flow of retinal images experienced by the bee through its motion*) *in feeder direction* might be necessary to assess feeder distance ("The Optical Flow Hypothesis"). During the climb to the balloon the optic flow provided by the structured ground was progressively reduced, thus leading to a decrease in dance duration. To check this hypothesis, foragers were trained from a 50 m high building to a feeder atop of another building of similar height at a distance of 228 m. As a control, bees had to fly the same distance to a feeder on the ground. With optic flow diminished (i.e. flying at high altitude from one building to the next), dancers signaled a distance of only 128 m. This result is in line with the optic flow hypothesis [10].

Use of optic flow in sensing distances is not unique to *honey bees*. Stingless bees can use optic flow to measure flight distances [17].

2.1.4   Tunnel Experiments

2.1.4.1   Dancing After Tunnel Flights

Waggle duration can be used as measure of the optic flow sensed [9, 10]. Srinivasan manipulated the optic flow of bees flying to an artificial food site in an elegant experiment: foragers visited a feeder at the end of a narrow tunnel (6.4 m long, 20 cm high, and 11 cm wide) whose walls and floor were marked with randomly distributed black and white *vertical* patterns. The tunnel was located 35 m from the hive. The fact that tunnel walls were very close to the trajectory of the bees flying in the middle of the narrow tunnel generated a large amount of optic flow while they moved forward. Clearly, the longer the tunnel, the higher the optic flow perceived. With *increased optic flow* in the narrow tunnel, foragers flying 35 m to the entrance, and 6.4 m through the tunnel, signaled a feeder distance of 186 m. However, with tunnel walls covered by *axial* stripes, *parallel* to the bees' flight trajectory, foragers performed round dances, indicating a feeder location of 35 m. This makes sense from the perspective of the optic flow hypothesis: Axial stripes do not allow perception of a change in the visual panorama as the bee flies forwards and thus suppresses optic flow [25]. Srinivasan determined the exact calibration of the "odometer" by measuring optic flow through *image motion and the corresponding increase in waggle duration*. One millisecond of wagging required 17.7° of *image motion* from front to back in the tunnel [25].

We used a similar setup to test how recruits *respond* to tunnel dances (Fig. 2.1.2). The tunnel was located 3 m from a hive, pointing into the southern direction. We obtained a calibration of dance duration for the southern direction. After that a group of marked foragers was trained to a feeder at the end of the tunnel. Their waggle duration pointed to a distance of 70 m. Human observers were stationed at clearly visible control stations at various distances to the south. They had to count arrivals and put them into alcohol when they sat down at a feeder. The tunnel was shut down at the beginning of each experiment and the observers made sure that no bees arrived during a period of at least 30 min. Recruits arrived at control sites 10 min after the opening of the tunnel and the beginning of dances. The distribution of individuals searching at different distances showed that most recruits searched near 70 m. Recruits must have had information about the distance

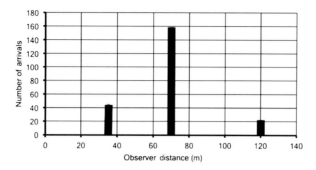

Fig. 2.1.2 Number of recruits arriving at control stations after following "tunnel dances" [11]

signaled by tunnel dancers. This result was extremely important as it dismissed in a categorical way Wenner's hypothesis about finding the food source following odor plumes. *Dancers* received vector information about distances by flying through the tunnel, and *recruits* read that information and followed it in the field after leaving the hive [11].

The experiments had an unexpected result: When the tunnel was turned into the northwestern direction for a control experiment, most recruits searched at 140 m. We obtained a new distance calibration for the sites in this direction. The same tunnel dances as in the southern direction now pointed to 140 m in northwestern direction. The optical environment in both directions was very different: To the south bushes and trees were located on both sides of the path to the control stations. In the northwestern direction control sites were placed on a meadow with no striking landmarks. Even for the same hive, at the same location, the distance perception can be different for different directions. This should not create problems for recruits. Dancers and recruits usually fly through the same environment. However, this observation might be important for an interpretation of the "dialects" of various races of bees [1, 27]. Experimenters have to pay close attention to the appearance of the environment otherwise some of their results could be questionable.

2.1.4.2 Measuring Image Motion in the Field

Dancing *honey bees* do not measure distances flown in *distance units*, but by the *amount of image motion* of the environment moving from front to back while flying to the feeder. The integral of image motion is the measure of feeder distance. It depends only on distance flown, not on flight velocity [10, 11, 25]. It is possible to derive the amount of optic flow bees experienced during foraging flights directly from the duration of their waggle runs [25].

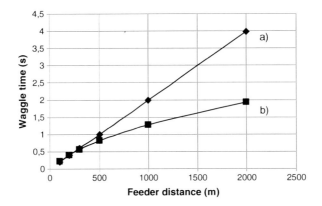

Fig. 2.1.3 Waggle data used to calculate image motion [4]. (**a**) Linear relationship if a given unit of waggle time is added for each meter flown. (**b**) Real waggle times per meter. There is a decrease in time added for each meter with increasing feeder distance

Srinivasan measured the optic flow and calibrated the bee's odometer by using *image motion* in the confined environment of a tunnel. We used the same methods for foragers in the field (for methods see [25]). It is important to realize that *angles* represent *distances* in these measurements. The image motion is defined by the angle through which a given stretch of tunnel wall or ground moves, while seen from a certain *distance* or *altitude*. In Srinivasan's case 1 cm on the tunnel wall, seen from a distance of 5.5 cm, was represented by 10.3°. Flying through 6,000 cm of tunnel produced an *image motion of* (6,000*10.3°)=6180°. The 6,000 cm tunnel flight was signaled by 350 ms of wagging. Thus the calibration of the tunnel odometer showed that 1 ms encoded (6,180/350)=17.7° of image motion.

We assumed in the field calculations that foragers obtain optical flow mainly from the ground. We used a flight altitude of 2.5 m for a 200 m feeding site ([32], p. 188). In the field environment 1 m on the ground, seen from an altitude of 2.5 m, is defined by 21.8°. Thus when a forager flies forwards for 1 m at an altitude of 2.5 m, it sees the ground *move backwards* for an angle of 21.8°. The flight to a 200 m feeder moves the image backwards for (200*21.8°)=4,360°. In dances upon return to the hive from 200 m waggle runs last for an average of 398 ms (data from an experiment in an open field: [4]). One millisecond of wagging thus encodes an image motion of (4,360°/398 ms)=10.95°. This is our calibration for the forager's odometer in the field.

With this calibration we can calculate the "*image motion*" that dancers would experience during foraging flights to feeding sites at different distances (Fig. 2.1.3). If we assume that foragers fly at an average altitude of 2.5 m to feeders *at all distances*, we would find a linear increase of image motion with distance (distance*21.8°), "*the calculated image motion*". We can derive the *real image motion* experienced by dancers for all distances from the waggle times of our data set (image motion=waggle

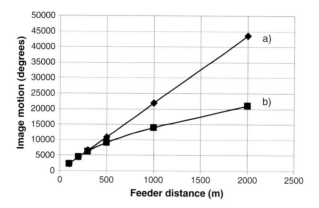

Fig. 2.1.4   (a) *Calculated image motion* assumes that foragers fly at an average altitude of 2.5 m at all distances. **(b)** *Real image* motion derived from waggle times and the calibration of the odometer (1 ms = 10.95°)

time × 10.95°). A comparison of *calculated* and *real image motion* shows that the *real image motion/meter decreases with distance*. The relationship between waggle time and feeder distance is shown in Fig. 2.1.4. The increase with distance is not linear. This was already noticed by von Frisch. He tried to find a reason for this relationship ([32] pp. 121–126). He suspected that the non-linear increase with distance was caused by foragers "forgetting" part of the distance flown. Our calculations suggest a better explanation: The shape of the distance curve is caused by a decrease in optic flow with increasing distance. *This decrease in optic flow is probably caused by an increase in flight altitude with increasing feeder distance.*

2.1.5   Conclusions

Karl von Frisch suspected that bee dances have similarities with a real language. He believed that recruits that had attended dances could reach the advertized feeding sites by themselves without further guidance. This idea led to heated discussions. Frisch's experiments, the efforts of his students, and investigations by many other members of the scientific community, supported his assumption. His critics completely denied the existence of navigational information in dances. They claimed that recruits could find a feeder location with the locale odor they obtained during dances. With this information they had just to fly up odor plumes in the field ("The Odor Search Hypothesis") [22, 35, 36]. Tunnel experiments clearly dismissed Wenner's objections: Tunnel dancers can acquire vector information about feeding site locations *in a tunnel*. They can transmit this information to recruits who search for these feeding sites *in the field* [11]. Further experiments using radar technology and measuring large-scale bee displacements further dismissed Wenner's arguments (see Chap. 2.5).

The acquisition and use of the location's coordinates ("*vector information*"), as expressed in dances, is generally understood and accepted. The direction of feeding sites is encoded in the direction of the waggle runs. Sun or gravity is used as a directional cue [32]. How foragers sense feeder distance has been answered in different ways over time. von Frisch believed that the energy required to reach a feeder is used as a measure distance ("The Energy Hypothesis"). By measuring the metabolic efforts of feeding flights, and by manipulating the flights energy requirements, we found that optic flow over the retina *while flying in feeder direction* must be used to assess distance ("The Optic Flow Hypothesis"). The optic flow can actually be measured trough *image motion*. Srinivasan determined image motion of bees flying in a tunnel. *He managed to calibrate the odometer of bees in terms of waggle time* [25]. For this contribution we used Srinivasan's methods and calibrated the odometer of foragers flying in the field: One millisecond of wagging requires 10.95° of image motion. We could use the image motion, as derived from waggle times, to estimate flight altitude on foraging flights. Flight altitude might increase with feeder distance. This explains the shape of the graphs that depict the relationships between dance parameters and feeder distance that bothered von Frisch [32]. Since flight velocity increases with altitude [10], foragers must speed up on more distant flights.

2.1.6 Outlook

Some questions raised in our contribution require more experimentation. The hypothesis that pheromones motivate nest mates to pay attention to dances and to extract their information has to be tested in more detail. The nature of suspected pheromones is of great interest. Use of pheromones to initiate and guide foraging behavior is common in many relatives of the honey bees ([2, 18]). Knowledge of their nature and origin could be very important for an understanding of the *evolution* of bee dances.

The role of image motion during foraging and dancing can help to investigate the "nature" of bee dances. The waggle dance might be a "symbolic replay" of a foraging flight: Waggle movements could exactly represent the meandering flight patterns on the way to a feeder ("one waggle, one flight oscillation"). We suspect that the whole waggle dance is an act of *conditioning*: A recruit "learns" the location of a feeding site during attendance of a symbolic replay of a foraging flight inside the hive. A food sample delivered by the dancer through trophallaxis serves as a reward. We know that bees can perform most of the behaviors that are required for this task.

References

1. Boch R (1957) Rassenmäßige Unterschiede bei den Tänzen der Honigbiene (*Apis mellifica L.*). Z vergl Physiol 40:289–320
2. Dornhaus A, Brockmann A, Chittka L (2003) Bumble bees alert to food with pheromone from tergal gland. J Comp Physiol A 189(1):47–51
3. Esch H (1956) Analyse der Schwänzelphase im Tanz der Bienen. Naturwissenschaften 43(9):207
4. Esch H (1956) Die Elemente der Entfernungsmitteilung im Tanz der Bienen (*Apis mellifica*). Experientia 12(11):439–441
5. Esch H (1961) Über die Schallerzeugung beim Werbetanz der Honigbiene. Z vergl Physiol 45(1):1–11
6. Esch H, Goller F, Burns JE (1994) Honeybee waggle dances – the energy hypothesis and thermoregulatory behavior of foragers. J Comp Physiol B 163(8):621–625
7. Esch HE (1963) Über die Auswirkung der Futterplatzqualität auf die Schallerzeugung im Werbetanz der Honigbiene. Verh Dtsch Zool Ges 26:302–309
8. Esch HE (1978) On the accuracy of the distance message in the dances of honey bees. J Comp Physiol 128:339–347
9. Esch HE, Burns JE (1995) Honeybees use optic flow to measure the distance of a food source. Naturwissenschaften 82(1):38–40
10. Esch HE, Burns JE (1996) Distance estimation by foraging honeybees. J Exp Biol 199(1):155–162
11. Esch HE, Zhang SW, Srinivasan MV, Tautz J (2001) Honeybee dances communicate distances measured by optic flow. Nature 411(6837):581–583
12. Goller F, Esch HE (1990) Waggle dances of honeybees: is distance measured through energy expenditure on outward flight. Naturwissenschaften 77:594–595
13. Grüter C, Balbuena MS, Farina WM (2008) Informational conflicts created by the waggle dance. Proc R Soc B 275:1321–1327
14. Grüter C, Farina WM (2009) The honeybee waggle dance: can we follow the steps? Trends Ecol Evol 24(5):242–247
15. Heran H (1956) Ein Beitrag zur Frage der Wahrnehmungsgrundlage der Entfernungsweisung der Bienen. Z vergl Physiol 38:168–218
16. Hölldobler B, Wilson EO (1990) The ants. Belknap, Cambridge
17. Hrncir M, Jarau S, Zucchi R, Barth FG (2003) A stingless bee (*Melipona seminigra*) uses optic flow to estimate flight distances. J Comp Physiol A 189(10):761–768
18. Lindauer M, Kerr WE (1958) Die gegenseitige Verständigung bei den stachellosen Bienen. Z vergl Physiol 41(4):405–434
19. Michelsen A (2003) Signals and flexibility in the dance communication of honeybees. J Comp Physiol A 189(3):165–174
20. Michelsen A, Andersen BB, Kirchner WH, Lindauer M (1989) Honeybees can be recruited by a mechanical model of a dancing bee. Naturwissenschaften 76(6):277–280
21. Michelsen A, Andersen BB, Storm J, Kirchner WH, Lindauer M (1992) How honeybees perceive communication dances, studied by means of a mechanical model. Behav Ecol Sociobiol 30(3–4):143–150
22. Rosin R (1980) The honey-bee "dance language" and the foundations of biology and behavior. J Theor Biol 87(3):457–481
23. Seeley TD (1995) The wisdom of the hive. Harvard University Press, Cambridge
24. Seeley TD, Mikheyev S, Pagano GJ (2000) Dancing bees tune both duration and rate of waggle-run production in relation to nectar-source profitability. J Comp Physiol A 186:813–819
25. Srinivasan MV, Zhang S, Altwein M, Tautz J (2000) Honeybee navigation: nature and calibration of the "odometer". Science 287(5454):851–853
26. Steche W (1957) Gelenkter Bienenflug durch Attrappentänze. Naturwissenschaften 44(22):598
27. Su S, Cai F, Si A, Zhang S, Tautz J et al (2008) East learns from West: asiatic honeybees can understand dance language of European honeybees. PLoS One 3(6):e2365

28. Thom C, Dornhaus A (2007) Preliminary report on the use of volatile compounds by foraging honey bees in the hive (Hymenoptera: Apidae: *Apis mellifera*). Entomol Gen 29(2–4):299–304
29. Thom C, Gilley DC, Hooper J, Esch HE (2007) The scent of the waggle dance. PLoS Biol 5(9):e228
30. von Frisch K (1923) Über die Sprache der Bienen, eine tierphysiologische Untersuchung. Zool Jb (Physiol) 40:1–186
31. von Frisch K (1946) Die Tänze der Bienen. Österreich Zool Z 1:1–48
32. von Frisch K (1967) The dance language and orientation of bees. Belknap Press of Harvard University Press, Cambridge
33. Wehner R, Rossel S (1985) The bee's celestial compass – a case-study in behavioral neurobiology. Fortschr Zool 31:11–53
34. Wenner AM (1962) Sound production during the waggle dance of the honeybee. Anim Behav 10:79–95
35. Wenner AM (1971) The bee language controversy: an experience in science. Educational Program Improvement Corp, Boulder
36. Wenner AM (2002) The elusive honey bee dance "language" hypothesis. J Insect Behav 15(6):859–878

Chapter 2.2
How Do Honey Bees Obtain Information About Direction by Following Dances?

Axel Michelsen

Abstract Several strategies (touch, vision, hearing, substrate vibrations, and air flows) have been proposed for how follower bees obtain information about the distance and direction announced in waggle dances. This review deals with the sounds and air flows generated by dancing bees. The vibrating wings of the dancer act as dipoles, and the surprisingly large sound pressures and air flows decrease rapidly with distance, thus restricting the possible range of communication. The movements and air flows have been mimicked in a robot dancer, which was able to direct follower bees to positions in the field, but caused less recruitment than live dancers. Subsequent measurements with a laser technique showed that the oscillating air flows caused by the wing vibration and wagging movements are probably too complicated to transmit the information about the direction to the target. However, the laser studies showed that the vibrating wings could cause a jet air flow behind the dancer's abdomen. The jet is generated by the vibrating wings, and it is so narrow in a plane parallel to the comb that it may provide information about direction to follower bees located behind the dancer. Measurements with hot wire anemometers confirmed the existence of the narrow jet behind live dancers. In addition, it was found that the narrow jets may exist together with a broad flow of air, which seems ideally suited for transporting dance pheromones. Both the narrow and broad flows can be switched on and off by the dancer, apparently by adjustments of the positions of the wings.

Abbreviations

Pa   Pascal $= 94$ dB SPL
PIV   Particle image velocimetry

A. Michelsen (✉)
Institute of Biology, University of Southern Denmark, DK-5230 Odense M, Denmark
e-mail: A.Michelsen@biology.sdu.dk

C.G. Galizia et al. (eds.), *Honeybee Neurobiology and Behavior: A Tribute to Randolf Menzel*, DOI 10.1007/978-94-007-2099-2_6,
© Springer Science+Business Media B.V. 2012

2.2.1   Introduction

More than 60 years ago, Karl von Frisch interpreted the dances of honey bees, and
since then the missing link in our understanding of this communication system has
been how the follower bees obtain information about direction and distance from
the movements of the dancer. Von Frisch [13] suggested that the followers might
touch the dancer, and that vibrations in the wax comb caused by the 12–15 Hz
wagging motion and/or the 250–300 Hz wing vibrations might be perceived by the
follower bees. He also argued that vision was not involved, since the follower bees
can obtain the specific information in the darkness of the hive (this view has recently
been questioned, see [3]). Other strategies may be possible, since oscillatory motions
like wagging and wing vibrations may generate sounds as well as oscillatory air
flows and jet air flows.

   Honey bees are probably deaf to sound pressures, but they can sense oscillatory
air flows and vibrations of the substrate [5, 6]. However, the vibrations seem only to
attract followers rather than to transmit information about the direction and distance
to food [11], and a transfer of specific information is possible from a robot dancer,
which does not touch the comb [8]. Studies using high speed video have shown that
follower bees have a chance of touching the dancer with one or both antennae during
each waggle movement [10], but the most obvious information to be obtained from
touching is whether the follower bee is facing the dancer's abdomen laterally or
from behind. This information is probably not sufficient for allowing the follower
bee to determine the direction of the waggle run.

2.2.2   The Dance Sounds

Fifty years ago, Wenner [14] and Esch [1] reported that waggle dances were
associated with brief (about 4 ms) sound pulses repeated ca. 30 times per second.
Esch found that the sounds had a frequency of ca. 250 Hz and pressures up to 1.3
microbar (ca. 76 dB SPL) at a distance of 1 cm dorsal to the dancer. Lower values
were measured behind the dancer and lateral to the dancer. Esch suggested that the
follower bees might sense the vibrations with their antennae, but Wenner referred
to a general agreement in the literature that "there is a lack of response by bees to
airborne sounds of normal intensity, but evidence for response to substrate sounds".
Shortly after these studies, Esch [2] reported that silent dances occurred, and that
the silent dancers were not successful in the recruitment of nest mates.

   After these pioneering studies, the case rested for 20 years until Donald R. Griffin
investigated the effect of the dancer's wagging on the amplitude of the sounds received
by a pair of mechanically coupled microphones positioned lateral and symmetric to
the dancer. Griffin (in [9]) confirmed the observation by Esch that the sound pressure
was much smaller behind and lateral to the dancer than above the dancer, but
he also observed that, although the sounds measured by the two microphones were

generally in phase, they were often out of phase. The bee sounds thus appeared to be quite directional, which is not what one would expect from a simple sound emitter at low frequencies. In addition, the sounds seemed to be inherently variable.

The reason for this confusion was that the sound source (the vibrating wings) is not a monopole source, but a dipole source. At a certain time, the entire sound radiating surface of a monopole "agrees" on producing a surplus pressure and a little later a rarefaction. In contrast, in dipoles and higher order radiators some parts of the surface produce a surplus pressure while other parts produce a rarefaction, and the two components tend to cancel. In the case of a vibrating wing, there will be an almost perfect cancellation in the plane of the wing (thus the small sound pressures lateral to and behind the dancer). In contrast, the sound pressures will be at maximum above and below the wings. Because most of the generated sound is cancelled by sound of opposite phase, the radiation of sound energy from the dancer is small.

In order to investigate the sound field of dancing bees, we used a pair of mechanically coupled probe microphones with the tips 1.5 or 2 mm apart [9]. The results were used to calculate the air flows driven by the differences in pressure. The up and down vibrations of the wings were found to cause pressures above and below the wings of ca. 1 Pa (Pascal = 94 dB SPL), that is almost 10 times larger than the value measured by Esch [1]. The pressures at the two surfaces of the wings are totally out of phase, and a pressure gradient of ca. 2 Pa thus exists at the edge of a wing. This causes the air to oscillate with a peak velocity of ca. 0.5 m/s. The pressure gradient and air velocity decrease with the third power of the distance from the dancer. (In contrast, the sound pressure from a monopole source decreases with only the second power of distance). At a distance of one bee length, the amplitude of the 250–300 Hz oscillatory air flows has decreased to less than 1% of the value close to the dancer (Fig. 2.2.1). The 12–15 Hz wagging motion also causes air flows, but the much smaller sound emission means that neither the driving force nor the air flows can be estimated with the microphone technique.

2.2.3   Experiments with a Robot Dancer

The measurements of sound pressures generated by dancing bees led to the hypothesis that the follower bees could obtain the specific information about direction and distance by perceiving the oscillating air flows around the dancer that are caused by the wing vibrations and the wagging motion. In order to test this hypothesis, we decided to build a robot dancer, which could perform the figure-of-eight dance and was surrounded by air flows similar to those of live dancers [8].

In brief, the robot was made of brass covered with a thin layer of beeswax. It was the same length (13 mm) as a worker honey bee, but somewhat broader (5 mm). The wings (a single piece of razorblade) was vibrated by an electromagnetic driver and caused a 280 Hz acoustic near-field around the robot similar to that around live dancers.

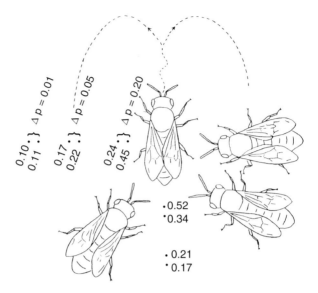

Fig. 2.2.1 Average maximum pressure amplitudes (given in Pa) of the dance sounds measured in two directions radially away from the dancer. Note that the pressure difference (Δ p), which is proportional to the velocity of the air, decreases rapidly with distance (From [9])

A step motor rotated the robot during the figure-of-eight path and also caused the robot to waggle during the waggle run. The robot was also moved in a figure-of-eight path. All motors were controlled by a computer. The software made it possible to vary the individual dance components independently and to create dances other than the normal waggle dance. The robot was located on the dance floor in an observation hive, and observers in the field noted the number of bees visiting baits at various locations.

In experiments with normal dances the majority of the bees were noted in the direction indicated by the dance, but some bees were noted in other directions – very similar to the result of experiments with live dancers. The main difference is that the number of bees recruited by a live dancer is 5–10 times larger. That live dancers are more efficient than metal models was not a surprise. The results of some manipulated dances were more informative. For example, the waggle run could be displaced from the centre of the figure eight to one of the return runs, so that the bees received conflicting information about direction. The bees followed the instructions given by the waggle run and ignored the information given by the dance figure. That the wagging run is the "master parameter" for the transfer of information was also found in experiments on the coding of the distance to the target, where the robot was running fast during the waggle run, but slowly during the return run (or vice versa). The bees followed the information about distance given by the waggle run and ignored the information given by the return run. The follower bees thus obtain information about both distance and direction to the food mainly by "observing" the waggle run.

2.2.4   Doubts

The robot dancer was made in such a manner that it simulated both the movements and the oscillating air flows that were known at the time of its construction. The experiments showed that its signals were understood. It was therefore logical to conclude that the oscillating air flows were playing a major role in the transfer of information, but after some time I began to doubt whether this explanation really was true. Although the air flows perpendicular to the wings are well above the perception thresholds determined by Kirchner et al. [6], these air flows do not contain any information about the direction of the waggle run. In contrast, the air flows parallel or normal to the dancer's body, which are the likely carriers of directional information, had not received much attention.

In the middle 1990s we investigated the air flows with two new laser instruments. Neither of them could be used for measurements in the hive, but they allowed us to map the air flows around models of bees that were caused to waggle and perform wing vibrations by various drivers. A laser anemometer allowed us to study the flows in a small volume of air as a function of time. Very accurate maps of air flows close to models of dancing bees were made with Particle Image Velocimetry (PIV). This method exploits a standard trick in fluid mechanics: a thin sheet of laser light is photographed twice with a very brief time interval by a digital camera mounted normal to the sheet. The changes in the positions of a large number of smoke particles from the first to the second digital photograph are then used for computing a map of the flows.

When the PIV technique was used for mapping the air flows around a wagging bee (a metal model or a real bee made robust by being stuffed with plastic) it became obvious that our ideas about the air flows during wagging had been very naïve. We had imagined that the air was just moved to and from by the wagging body. In reality, a thin (1–2 mm) boundary layer follows the movement of the body exactly, whereas other flows are lagging somewhat behind the body motion. The other flows are therefore not sufficient to fill the volume left by the body or remove the air from the space to be occupied by the body, and they have to be supplemented by short-circuit air flows opposite to the body motion (Fig. 2.2.2b). The flows often collide and lead to the formation of short-lived eddies (Fig. 2.2.2a). The life times of eddies are below one half of an oscillation period, so the air flows are basically laminar. Nevertheless, the complicated flow pattern must make it difficult for the follower bees to monitor the air flows due to the wagging motion with their antennae.

The most prominent component of the air flows oscillating with the frequency of the wing vibration is the short-circuiting flow around the edges of the wings. The measured air velocity (between 400 and 500 mm/s, peak) is very close to that predicted from the values obtained with the probe microphones. As predicted, the magnitude decreases rapidly with distance. Another component, which had not been predicted from the probe measurements, consists of flows in and out of the space between the wing and the abdomen. Again, it must be difficult for follower bees to obtain useful information about the direction of the waggle run by monitoring these flows with their antennae.

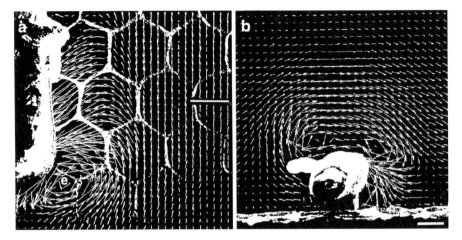

Fig. 2.2.2 Air flows caused by the body wagging with an amplitude of ±15° (rotating around a point about 1 mm in front of the head). Bars for velocity vectors (*middle right* for (**a**), *lower right* for (**b**)) 100 mm/s. Measurements taken using particle image velocimetry. (**a**) The bee seen from above when moving towards the *left* and illuminated by a 0.6 mm thick sheet of laser light parallel to the comb and 1–2 mm *below* the dorsal surface of the bee. Simple laminar flows occur in the air next to the thorax and most of the abdomen, but a collision with displaced air from the *left* side of the abdomen creates an eddy (e) close to the abdominal tip. (**b**) The bee seen from behind. A thin sheet of laser light (normal to the comb) hits the *middle* of the abdomen and a part of the *left* wing. Displaced air flows to the *left* close to the *left part* of the abdomen, then upwards and to the *right* over the back of the bee. Above the back, the air close to the body flows towards the *left*, whereas the air further away flows towards the *right*

2.2.5   The Jet Air Flow

To our big surprise, the up and down movements of the wing of our bee models generated not only oscillating air flows, but also a non-oscillating jet air flow behind the tip of the wings. In contrast to the oscillatory air flows, where the masses of air are flowing to and from, the air flows in only one direction in the jet (away from the dancer). The generation and propagation of such jet flows cannot be detected by measuring local sound pressures.

The jet shown in Fig. 2.2.3 was generated by continuous wing vibration in a bee model that did not waggle. The jet is broad in the dancer's dorso-ventral direction, but quite narrow in the direction normal to the plane of the photograph. The width of the jet increases linearly with distance from about 1 mm close to the wing tip to 11 mm at a distance of 5 cm. When the jet propagates away from the wing tip, it recruits air from its surroundings, increases in width, and slows down.

A more complicated pattern was seen with discontinuous wing vibration similar to that of live dancers. The velocity profiles in the PIV recordings now varied with the phases of the cycle of vibrations and pause. That air puffs produced during

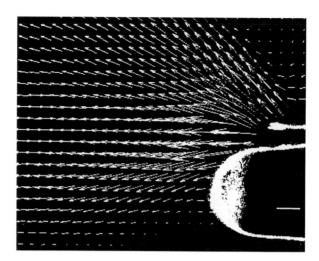

Fig. 2.2.3 Jet air flow caused by continuous wing vibration in a non-wagging metal model. Lateral view of a part of the abdomen when the wing was moving up. Bar for velocity vectors (*lower right*): 100 mm/s

each series of wing vibrations travel away was verified with laser anemometry of the air flows in small volumes of air in the direction away from the wing tip. The build-up and decay of the air puffs take some time, however, and the flow amplitude does not decrease to zero during pauses in the wing vibration similar to those of live dancers.

The PIV apparatus is very large and certainly not suited for measurements in a bee hive. In addition, the bees do not like smoke. Obviously, a more suitable instrument is needed for measurements in the hive. The hot wire anemometer is the classical method for measuring air flows, but in standard hot wire anemometers the noise in the output signal is too large to allow measurements of the bees' air flows. In addition, the bees are hostile towards the very hot wires (typically 300°C). By reducing the wire temperature to about 30°C above the ambient temperature we reduced the noise and made the wire acceptable to the bees. The disadvantage is that the output voltage now varies with the ambient temperature, so that a careful calibration of the nonlinear relationship between air flow velocity and output voltage is necessary. During our initial measurements, the hot wire probe was held by hand close to the dancing bees in an open observation hive. The direction of the hot wire was perpendicular to the surface of the wax comb. Much to our surprise, the 3 mm long hot wire is surprisingly robust and generally survives being hit by dancing bees.

Measurements with the hot wire anemometer showed that jet air flows are also generated by live dancers. Figure 2.2.4 shows the output signal from an anemometer

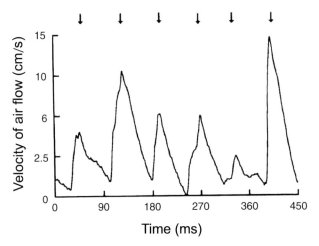

Fig. 2.2.4 Air flows measured by a hot wire anemometer at an extreme lateral position within the angle of wagging behind a dancing bee. The *arrows* indicate the times when the dancer's abdomen pointed towards the hot wire (From [7])

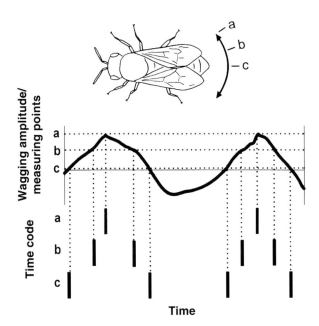

Fig. 2.2.5 Predicted time patterns for reception of a jet emitted by a wagging dancer at positions a, b, and c behind a bee. The predicted time patterns have been observed in recordings with hot wire anemometers behind live dancers. With a 13 Hz wagging, the time between consecutive receptions of the jet would be about 77 ms in **a** and 38 ms in **b** (From [7])

when the hot wire is held stationary behind the dancer at an extreme lateral position within the angle of wagging (position **a** in Fig. 2.2.5). The velocity signal recorded by the probe reflects the 14 Hz wagging frequency. The maximum values occur when the dancer's abdomen points directly at the probe (indicated by arrows). The shape of the signal shows that the wire has been hit very briefly by the maximum air velocity. In other words, the jet must have been quite narrow. As already discussed, consecutive peaks may show much variation in amplitude, because the wing vibration occurs in bursts of 3–4 vibrations about 30 times per second.

2.2.6 A Hypothesis for the Transfer of Directional Information

Assuming that the antennae are the air flow receivers of honey bees, we can predict the time patterns with which an antenna of a follower bee is maximally activated (Fig. 2.2.5). The time pattern is a simple function of the angular position of the follower relative to the axis of the waggle run. I suggest that the follower bees exploit this to obtain fairly precise information about the direction to the target reported in the dance. I further suggest that the brain of the bee compares the timing of the two antennal signals in order to judge whether she is located to the left or to the right of the axis of the waggle run.

One can now imagine a strategy for the transfer of information about direction: when bees are attracted to a dancer, they align themselves perpendicularly to the body contours of the dancer [10]. Perhaps the bees perceive the oscillating wing air flows with their antennae and exploit them for the alignment. From their positions in this typical alignment, the followers may then be able to use the antennal contact pattern to determine their approximate position relative to the dancer and finally reach a position behind the dancer within the angle of wagging.

One can imagine two possible strategies for the followers located behind the dancer within the angle of wagging. The followers may try to find the middle position (in line with the waggle run, position **c** in Fig. 2.2.5) by searching for the corresponding temporal pattern, or they may use the temporal pattern for estimating their angular position relative to the axis of the waggle run. In both cases, touching and/or perception of the direction of the air flow may serve to align the body of the follower with the body of the wagging dancer during the periods of contact. Judd [4] found two examples of followers, which were aligned with the axis of the waggle run (and which found the target after having followed just two waggle runs each) and four examples of followers that had located the target after having followed waggle runs within the angle of wagging, but far away from the middle position. This may suggest that the bees are able to use both strategies.

In three independent studies ([4, 10], a study by Martin Lindauer and our group) it was found that the specific information about the location of the food is available only to the follower bees that have spent some time behind the dancer within the angle of the wagging motion.

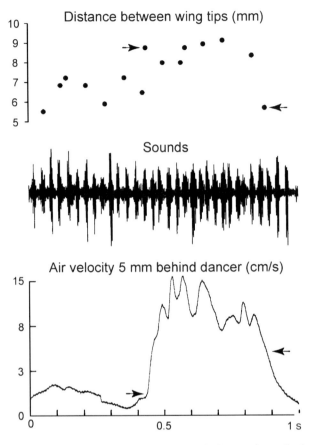

Fig. 2.2.6 Simultaneous recordings (during the last second of a waggle run lasting 2.2 s) of the distance between the tips of the wings (in mm, above), of sound generated by the vibrating wings (*middle*), and of the velocity in cm/s of air flowing away from the dancer about 5 mm behind the tip of the abdomen. The sound recording shows that the dancer vibrated its wings during the entire record, but it only produced a jet air flow during the 0.5 s when the distance between the tips of the wings had increased by 2–3 mm. *Arrows* indicate the times of wing opening and closure (From [7])

2.2.7 The Variable Jets

During the laboratory studies with the PIV technique and model bees we always observed a jet air flow of the shape already described, and we saw no difference between the jets made by a metal wing and real bee wings. The generation of jets by live dancers is more variable. Most of the jets start from a velocity of ~1 cm/s, which seems to be the background level of air flows in the hive (Fig. 2.2.4). In other recordings, the jets were accompanied by a broad flow of air with a velocity of about

8–15 cm/s (Fig. 2.2.6). With this broad flow the dancer displaces at least 10 ml of air per second.

A further complication is the observation that live dancers may switch the jet air flows and the broad air flow on and off (Fig. 2.2.6). From the high speed films it can be seen that the distance between the tips of the wings (which are held over the abdomen during the dances) increases when the jets and broad flow are switched on (and decreases when they are switched off). In the high speed films it is not possible to see whether the wings are rotating around their length axis (like during flight). One may speculate that the broad jet air flow serves to mark the zone behind the dancer, where the follower bees can obtain specific information, and/or to mark periods of particularly stable dancing (perhaps by means of an odor?). It remains to be learned how and why the dancers control the generation of the two kinds of air flow.

The honey bee waggle dances have, until recently, been considered a very stereotype behavior, where the main variation was the duration and direction of the waggle run, although the duration of dancing and the tendency to vibrate the wings were known to depend on the motivation of the dancer [2]. We now know that the duration of the return run (between consecutive waggle runs) varies with the profitability of the food source. A similar flexibility is found in the dance following. The patterns of following dances are related to the foraging experiences of the follower bees and to the presence, on the dance floor, of dancers announcing other targets. Finally, the patterns of moving with the dancer also show much flexibility.

2.2.8   Outlook

These findings raise a number of questions, which call for additional research. Do bees have the neural circuits necessary for handling the temporal information obtained when the jets hit the antennae? Are the waggle dance scents [12] released through the broad flow of air pumped by the dancer? If yes, why are the dancers not always making the broad air flows? Are the jets the reliable channel of information for the follower bees? If yes, is it possible to provide the bees with the necessary information for finding the target by means of a new robot, which only produces jets with a suitable time pattern, but does not dance? And so on. It seems fair to conclude that the dance language will keep the investigators busy for many years to come.

References

1. Esch H (1961) Über die Schallerzeugung beim Werbetanz der Honigbiene. Z vergl Physiol 45(1):1–11
2. Esch HE (1963) Über die Auswirkung der Futterplatzqualität auf die Schallerzeugung im Werbetanz der Honigbiene. Verh Dtsch Zool Ges:302–309

3. Horridge A (2009) What does the honeybee see? And how do we know?: A critique of scientific reason, ANU E Press, Canberra
4. Judd TM (1995) The waggle dance of the honey bee – which bees following a dancer successfully acquire the information. J Insect Behav 8(3):343–354
5. Kilpinen O, Storm J (1997) Biophysics of the subgenual organ of the honeybee, *Apis mellifera*. J Comp Physiol A 181(4):309–318
6. Kirchner WH, Dreller C, Towne WF (1991) Hearing in honeybees – operant-conditioning and spontaneous reactions to airborne sound. J Comp Physiol A 168(1):85–89
7. Michelsen A (2003) Signals and flexibility in the dance communication of honeybees. J Comp Physiol A 189(3):165–174
8. Michelsen A, Andersen BB, Storm J, Kirchner WH, Lindauer M (1992) How honeybees perceive communication dances, studied by means of a mechanical model. Behav Ecol Sociobiol 30(3–4):143–150
9. Michelsen A, Towne WF, Kirchner WH, Kryger P (1987) The acoustic near field of a dancing honeybee. J Comp Physiol A 161:633–643
10. Rohrseitz K, Tautz J (1999) Honey bee dance communication: waggle run direction coded in antennal contacts? J Comp Physiol A 184(4):463–470
11. Tautz J (1996) Honeybee waggle dance: recruitment success depends on the dance floor. J Exp Biol 199(6):1375–1381
12. Thom C, Gilley DC, Hooper J, Esch HE (2007) The scent of the waggle dance. PLoS Biol 5(9):1862–1867
13. von Frisch K (1967) The dance language and orientation of bees. Belknap Press of Harvard University Press, Cambridge
14. Wenner AM (1962) Sound production during the waggle dance of the honeybee. Anim Behav 10:134–164

Chapter 2.3
Progress in Understanding How the Waggle Dance Improves the Foraging Efficiency of Honey Bee Colonies

Thomas D. Seeley

Abstract The waggle dance of the honey bee is one of the most extensively studied forms of animal communication, but only recently have investigators closely examined its adaptive significance, that is, how it improves the foraging efficiency of a honey bee colony. Studies at the colony level, in which investigators have compared the effectiveness of food collection between colonies with normal and disoriented dances, have found that the waggle dance improves a colony's foraging performance when food sources are hard to find, variable in profitability, and ephemeral. Studies at the individual level, in which investigators compared the effectiveness of food collection between bees that do and do not use waggle dance information to find new food sources (recruits and scouts) have found that following a waggle dance to find a new food source raises the cost of doing so but that this cost is outweighed by a benefit in the quality of the food source that is found. The emerging picture of the adaptive significance of the honey bee's waggle dance is that it makes it possible for a colony to function as a collective decision-making unit that is skilled at deciding how to distribute its forager workforce over an array of widely scattered, highly variable, and ever changing patches of flowers.

2.3.1 Introduction

Ever since Karl von Frisch [31] deciphered the message encoded in the honey bee's famous waggle dance, numerous investigators have taken up the study of this remarkable form of animal communication. They have conducted hundreds of studies that have revealed many of the mechanisms of the dance, i.e. how bees acquire, encode,

T.D. Seeley (✉)
Department of Neurobiology and Behavior, Cornell University, Ithaca, NY 14853, USA
e-mail: tds5@cornell.edu

C.G. Galizia et al. (eds.), *Honeybee Neurobiology and Behavior: A Tribute to Randolf Menzel*, DOI 10.1007/978-94-007-2099-2_7,
© Springer Science+Business Media B.V. 2012

decode, and use the information in the dance about the direction and distance to a desirable food source (reviews: [7, 18, 19, 32], see Chap. 2.1). Moreover, the long-standing controversy between Karl von Frisch's claim that bees that have followed a waggle dance can use the location information expressed in the dance to find a desirable food source (the "dance-language hypothesis") and Adrian Wenner's claim that the bees cannot do so (the "odor-search hypothesis"; see [34] and Chap. 2.1) has been resolved through various experiments [9, 12, 22]. For example, when bees were caught upon leaving their hive after attending a waggle dance, released in a novel location, and then tracked using harmonic radar, it was found that they flew in the direction and distance indicated by the dances they had followed, just as predicted by the dance-language hypothesis but contrary to the odor-search hypothesis ([22]; see also Chaps. 2.1 and 2.5).

A second important, but less heavily traveled, avenue of investigation of the waggle dance concerns its adaptive significance. How exactly does a colony of honey bees benefit from the ability of its foragers to use the waggle dance communication system to share information about the direction, distance, and desirability of food sources? This is a question that has begun to attract careful work and in this chapter I will present the answer to it that is gradually taking shape. We will see that it is a question that requires probing both at the level of whole colonies, because natural selection in honey bees operates mainly through fitness differences between colonies, and at the level of individual bees, because the costs and benefits of this communication system depend on the details of how a colony's foragers make use of waggle dance information.

2.3.2   Colony-Level Studies

In principle, the ideal way to determine the fitness benefits gained by possessing a behavioral trait is to compare the fitnesses of individuals that are identical except that some do and some do not possess the trait. Of course, nature rarely provides the variation needed for this approach (i.e., individuals with and without the behavior of interest) but sometimes it is possible to create it using experimental manipulations. Two recent studies used the experimental approach to create honey bee colonies with and without oriented waggle dances, hence with and without the ability to signal the directions of food sources [6, 27]. Both studies took advantage of the fact that honey bees (*Apis mellifera*) usually perform their dances on a vertical comb in the dark, using upward as a sensory reference for orienting the waggle run in each circuit of the dance. The angle of the waggle run relative to upward represents the angle of the heading to the food source relative to the sun's azimuth. Thus the angle of the waggle run indicates the direction to the food source. When forced to live on horizontal combs in the dark or under diffused light, honey bees will still perform dances but the directions of their waggle runs will be random. The indication of direction is thereby disrupted. If, however, bees on a horizontal comb are given a direct view of the sun (or polarized skylight) or of a bright artificial light, they will again produce properly oriented and thus fully informative waggle dances [32].

In both studies, pairs of matched colonies living on horizontal combs were set up and in each pair one colony's bees were treated with a view of a bright light [27] or of the sun [6]. Thus, in each pair one colony's bees performed well-oriented dances while the other colony's bees performed completely disoriented dances. The two treatments were reversed every few days to control for any extraneous differences between the colonies in a pair. The foraging success of each colony was assayed by measuring its gain in weight, which is due mainly to nectar collection.

Sherman and Visscher [27] conducted their study in a suburban habitat in southern California, where bees can fly throughout the year. They found that colonies with oriented dances gained more weight than the colonies with disoriented dances in the winter, but not in the summer (when bee forage is plentiful) or in the autumn (when bee forage is sparse). Sherman and Visscher attribute these seasonal differences in their results to seasonal changes in the spatial distribution of food resources. Oriented dances only made a difference in the winter, a time when rich floral resources were available but probably were more dispersed and more ephemeral than in the summer.

Dornhaus and Chittka [6] conducted their study in three habitats (temperate shrubland, temperate agricultural land, and tropical forest) and they too found that only in one condition, the tropical setting, did colonies with oriented dances gain more weight than colonies with disoriented dances. By analyzing videorecordings of the dances performed in their study colonies with properly oriented dances, Dornhaus and Chittka constructed maps of the foraging locations of their bees in each habitat, and these revealed that the foraging sites were distributed much more patchily in their tropical habitat than in their two temperate habitats. The message from both studies is that the waggle dance system of recruitment communication can increase the food collection of honey bee colonies, but that it does not do so under all conditions.

What exactly are the conditions under which the ability to signal food-source location increases a colony's ability to collect food and thereby boost its fitness? The Dornhaus and Chittka [6] study points to patchiness in food sources as one of the conditions in which the waggle dance benefits a colony. This is not surprising. After all, if there is just one highly profitable patch of flowers located somewhere within a colony's flight range, then the benefit of communicating its location is obvious, since only one individual would need to find the lone bonanza for the whole colony to benefit. At the opposite extreme, if flowers brimming with nectar are plentiful and uniformly distributed in all directions, then a colony won't benefit by its foragers sharing information about where they've found good forage. Each forager can easily find by herself flowers laden with food. In nature, the spatial distribution of rich sources of nectar and pollen is generally somewhere between these two extremes [2, 30], which raises the question of how patchy (or how ephemeral, see below) food resources must be for the waggle dance to provide an advantage.

Beekman and Lew [1] have recently analyzed the specific foraging conditions in which signaling food-source location benefits a colony. They did so by developing a sophisticated individual-based simulation of a foraging honey bee colony. The individual bees in their model mimicked all the known behavioral rules employed by foraging bees (reviews: [24]), except that in one version of the model the bees did not perform waggle dances, so that a colony represented by this version of the model would not experience the costs and benefits of foragers sharing information about where food is to be found. In their simulations, the environment contained four flower patches. To vary the difficulty faced by the bees in finding these patches by independent search (scouting), Beekman and Lew varied the distances of the patches (2, 4, 6, or 8 km from the hive) and the size of the patches (each one a sector 5°, 15°, 30°, 45° or 60° wide). When they performed runs of their model so that all four patches had the same profitability, they found a clear result: endowing a colony with waggle dance communication enhanced its nectar collection as soon as the average success of scouts (bees that search for food independently) in locating flower patches fell below the average success of recruits. Thus, if flower patches were easily discovered by independent search, then the colony that was without dance communication did better than the colony that had the waggle dance. In other words, when flowers were easy to find, it did not pay to spend time and energy producing and following dances. But if flower patches were hard to find, either because they were small (less than 15° wide) or far away (6–8 km), then being able to signal flower patch location was clearly beneficial. And when flower patches were neither easy nor hard to find (e.g., 15° wide and 4 km away), the two types of colonies did about the same.

Beekman and Lew [1] further explored the conditions that favor colonies with waggle dance communication by adding variation in profitability among the four flower patches to the situation where patches are neither easy nor hard to find (patches 15° wide and 4 km away). When they did so, they found that possessing the waggle dance was extremely beneficial because foragers from a colony with the waggle dance focused rapidly on exploiting the richest patch whereas foragers from a colony without the waggle dance remained dispersed over the four available patches. This shows that having the waggle dance increases a colony's ability to collect food not only because it helps a colony's foragers exploit food sources that are *hard to find*, but also because it helps them focus their efforts on the food sources that are *best to exploit*. In other words, the ability of a colony's foragers to share information about food source location and quality endows the colony with a collective decision-making ability, or "swarm intelligence" [16], regarding where its foragers should work. The modeling analysis of Beekman and Lew [1] indicates that under conditions that are likely to be common in nature (widely dispersed flower patches that vary in profitability) this collective cognitive ability yields a better, more productive deployment of the colony's foragers than what would be achieved if each forager worked independently.

The dance-based enhancement of a colony's ability to keep its forager force focused on superior flower patches is probably especially important in situations where the best foraging opportunities are not only scattered widely over the land-scape (large spatial variation) but are also changing rapidly over time (large temporal variation). In nature, honey bee colonies are confronted with a kaleidoscopic array of patchy food sources. Studies that have tracked the recruitment foci of colonies over several days, by monitoring the dances performed inside colonies, have found that honey bee colonies make daily changes in the sites advertised by waggle dancing bees, presumably in response to changes in the locations of the best foraging opportunities ([2, 30], review: [24]).

I suggest that what we need next in the colony-level analysis of the adaptive significance of the waggle dance are experimental field studies to clarify the specific ecological conditions in which having the waggle dance increases the foraging success of a colony. Ideally, we would compare the foraging performances of colonies with and without oriented waggle dances when they are living in a location where an investigator can manipulate several properties of the food sources: their spatial distribution, their variance in profitability, and their turnover rate (the rate of change in the locations of the richest ones). This is doable. In North America, there exist vast, heavily forested parks that are essentially devoid of honey bee colonies and natural food sources (see [24]), so by introducing hives of bees and sugar water feeders to one of these locations it should be possible to measure how well colonies with and without waggle dance communication perform in different foraging scenarios.

2.3.3   Individual-Level Studies

A second approach to investigating how honey bees benefit from possessing the waggle dance is to look at the details of the behavior of individual foragers within a colony. As in the colony-level studies, the important thing is to compare the foraging success of bees for which waggle dances do and do not serve as a source of information about food sources. One important difference between the individual-level and the colony-level approaches, however, is that at the individual level these two types of bees do not need to be created experimentally because they exist naturally.

Consider the two contexts in which the foragers in a honey bee colony follow waggle dances (Fig. 2.3.1). First, there is the situation in which a forager seeks information to help her find a *new* food source. This forager can be a novice forager looking for her first food source or an experienced forager looking for a replacement food source (her previous source having become depleted). Second, there is the situation in which a forager seeks information about a *known* food source. This forager can be one whose exploitation of a rich food source has been interrupted (say, by nightfall or a rainstorm) and who seeks to learn whether this food source is again worth visiting. (In principle, it can also be a forager who seeks to confirm that the food source she has been working steadily is still worth visiting.) As discussed by Biesmeijer and de Vries [4], it is important to distinguish between these two contexts and to do so it helps to use distinct names for the bees which do and do not use dance information in each context. In the first context, they are called *recruits* and *scouts*, and in the second context they are called *reactivated foragers* and *inspectors*.

Recently, Biesmeijer and Seeley [3] examined the extent to which worker honey bees follow waggle dances throughout their careers as foragers. They found that only about 20% of their dance followings were conducted to start work at new,

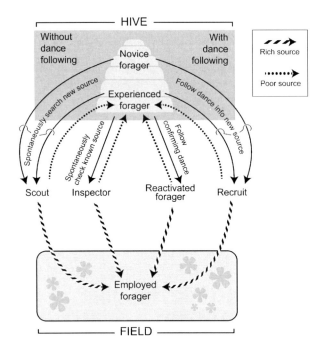

Fig. 2.3.1 The possible behavioral transitions in the career of a honey bee forager. Transitions (*arrows*) on the *right half* of the diagram involve following waggle dances: transitions on the *left half* do not involve following waggle dances (After [4])

unfamiliar food sources (i.e., recruitment) and that the other 80% or so were done to resume work at old, familiar food sources (i.e., reactivation). At first glance, the fact that following waggle dances to find new food sources accounts for only a small minority of the total instances of dance following suggests that recruitment is not the main way that bees benefit from having the waggle dance. This is, however, probably incorrect. Recruitment helps a bee solve the difficult problem of finding a new, high-quality food source somewhere within her colony's vast ($100 + km^2$) foraging area whereas reactivation helps a bee solve the relatively simple problem of knowing when to resume revisiting a known food source.

Several studies of the reactivation process have provided evidence that recruitment (not reactivation) is the main way that bees benefit from having the waggle dance, even though recruitment is not the most common context of dance following. These studies have shown that when a forager follows a dance to determine whether she should resume foraging at a familiar food source, she can be stimulated to resume foraging by just the floral scents that she detects on the body of the dancing bee or in a nectar sample that she receives from the dancing bee ([11, 15, 31], see also Chap. 2.1). Indeed, the mere injection of floral scents into a hive will induce many experienced foragers to resume foraging at a familiar food source ([32], p. 23, [20, 21]), though contacts with a dancer may strengthen the reactivation process [33].

A recent experiment [13] further demonstrates how in the context of reactivation bees tend not to use the location information in waggle dances. In this experiment inactive foragers who were familiar with a feeder that had been shut off for 4 h encountered dances produced by bees who carried the familiar ("right") odor but did not advertise the familiar location. The inactive foragers followed these dances closely but evidently ignored the unfamiliar ("wrong") location information in the dances; they always flew to the feeder with which they were familiar. In short, they were reactivated to their known feeder, not recruited to a new feeder. It is important to note too that the inactive foragers were reactivated to their familiar feeder just as effectively when they encountered an odor-right/location-wrong dance as when they encountered an odor-right/location-right dance [13]. It appears, therefore, that when a forager follows dances for reactivation to a familiar food source, she acquires mainly odor information, and little or no location information, from the dancing bee (see also Chap. 2.1).

A bee uses the information in waggle dances very differently when she follows dances for recruitment to an unfamiliar food source. Here the dance follower must acquire information about both location and odor, for both are needed to find the new food source. A recruited bee uses the location information to guide her to the general vicinity of the rich flowers advertised by the dancer, and upon reaching this location she uses the odor information to find flowers of the specific type that the dancer had found to be rewarding [9, 12, 22, 32].

Given that bees use the location information in waggle dances primarily when they need to find new food sources, it makes sense to focus on comparing bees that do (recruits) and do not (scouts) follow waggle dances to find new food sources in order to understand how the waggle dance affects foraging efficiency. Two studies [23, 26] have looked closely at how recruits and scouts compare in terms of foraging behavior and foraging success. In both studies a colony living in an observation hive was moved to a heavily forested area in northeast Connecticut. A group of foragers labeled for individual identification was allowed to gather sugar syrup at a feeder for 3 days. On the fourth day the feeder was not refilled so all the bees were forced to find new food sources, and they did so either by following dances or by searching on their own. By working with just 15 bees at a time, it was possible to make detailed records of each bee's behavior inside the observation hive. This work revealed that using the waggle dance actually increased the time and energy that a bee spends in finding a new food source. For example, recruits spent more time outside the hive searching for a new food source than did the scouts (121 min vs. 82 min). The details of the behavioral records of the recruits explain this counter-intuitive finding (see Fig. 2.3.2). On average, a recruit needed to make more than four attempts at being recruited (= following a waggle dance inside the hive and then searching for the indicated flowers outside the hive) before she achieved success. This result is consistent with previous studies of the recruitment of bees to feeders, which have reported that most recruits need to make multiple dance-guided search trips to find a feeder [8, 17]. We should not necessarily conclude from these findings, however, that recruitment by the waggle dance is inefficient [14]. Instead, it may be that the need to make multiple dance-guided search trips reflects the consid-

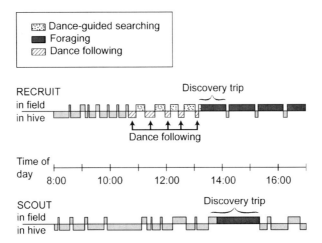

Fig. 2.3.2 Time records across a day for the behaviors of a typical recruit and scout as each one located a new food source after the feeder they had been working was shut off. Both bees made mostly short (<10 min) trips outside the hive at first, probably to inspect the empty feeder. The recruit required five episodes of dance following coupled with outside searching before she found a new food source. Note that the recruit made a steady series of successful foraging trips once she found her new food source, but that the scout did not bring back forage more than once from her new food source. The new food sources of these two bees were natural patches of flowers in the surrounding countryside (After [26])

erable difficulty of finding the particular patch of flowers represented by a dance. If the patch advertised by a dance is a meadow filled with bright blossoms a few 100 m from the hive, then recruits probably can find it easily. But if the patch is a clump of flowering shrubs bordering a swamp several 1,000 m from the hive, then recruits may need to make several dance-guided search trips to find the recruitment target. There is imprecision in the *encoding* of location information in dances for *nearby* targets [10, 28, 29], and it is likely that there is also imprecision in the *decoding* of location information in dances for *distant* targets.

Although on average recruits spent more time to find new food sources than did scouts, the food sources that recruits eventually found were generally much higher in quality than those found by scouts [23, 26]. The loads of nectar and pollen brought back by recruits were markedly larger than those brought back by scouts. Recruits bearing nectar, for example, needed on average 35 s to unload their nectar to receiver bees inside the hive whereas scouts with nectar needed on average only 17 s for the transfer. Recruits and scouts also differed markedly in their probabilities of returning with forage on the next trip outside the hive following the discovery trip (the first trip back from a new food source). Recruits almost always (93%) came back from their next excursion with more food, whereas scouts did so less than half the time (43%). Figure 2.3.2 shows how a typical recruit had a string of successful foraging trips once she found a new food source, but a typical scout did not.

It is not surprising that on average the new food sources found by recruits were substantially superior to those found by scouts. We now know that even though every successful forager brings home information about her food source's location and profitability, only bees returning from highly profitable sources perform waggle dances and so *share* their information with their nestmates ([24], p. 88). Furthermore, each forager that performs a dance adjusts the strength of her dance in accordance with the profitability of her food source ([24], p. 90, [5, 25]). The pool of shared information within the hive consists, therefore, almost exclusively of information about highly profitable food sources. This means that when recruits tap into this pool of information to help them find a new work site, they are directed to the most desirable of all the flower patches that their thousands of fellow foragers are currently visiting. Scouts, however, find their new work sites by searching on their own. In doing so, they have the potential to make fresh discoveries, but they generally encounter only mediocre food sources because by working alone they do not have access to broad information about the current foraging opportunities.

2.3.4 Outlook

The waggle dance of the honey bee is perhaps the most extensively studied form of animal communication, but we still have only a rather fuzzy understanding of how it improves the foraging efficiency of a honey bee colony. As described above, a few studies have been reported recently that address this mystery. Studies at the colony level, in which investigators compared the effectiveness of food collection between actual colonies with normal and disoriented dances, or between virtual colonies with and without waggle dances, have found that the waggle dance is important when food sources are hard to find, variable in profitability, and ephemeral. Studies at the individual level, in which investigators compared the effectiveness of food collection between bees that do and do not use waggle dance information to find new food sources (recruits and scouts) have shown that using waggle dance information raises the cost of finding a new food source but that this cost is outweighed by a benefit in the quality of the food source that is found. The emerging picture of the adaptive significance of the honey bee's waggle dance is that it makes it possible for a colony to function as a collective decision-making unit that is able to optimally distribute its forager workforce over an array of widely scattered, highly variable, and ever changing patches of flowers. The next step is to bring this emerging picture into sharper focus. One way to do so will be to present a colony with a controlled foraging environment (one in which the spatial distribution of the food sources, the variance in their profitabilities, and their turnover rate can be manipulated) and then to measure the colony's foraging performance in various scenarios, when its foragers are with vs. without oriented waggle dances. Although challenging, this approach holds the promise of providing a crystal clear picture of how a honey bee colony benefits from possessing the amazing waggle dance.

References

1. Beekman M, Bin Lew J (2008) Foraging in honeybees – when does it pay to dance? Behav Ecol 19(2):255–262
2. Beekman M, Ratnieks FLW (2000) Long-range foraging by the honey-bee, *Apis mellifera L.* Funct Ecol 14(4):490–496
3. Biesmeijer JC, Seeley TD (2005) The use of waggle dance information by honey bees throughout their foraging careers. Behav Ecol Sociobiol 59(1):133–142
4. Biesmeijer JC, de Vries H (2001) Exploration and exploitation of food sources by social insect colonies: a revision of the scout-recruit concept. Behav Ecol Sociobiol 49(2–3):89–99
5. De Marco RJ, Farina WM (2001) Changes in food source profitability affect the trophallactic and dance behavior of forager honeybees (*Apis mellifera L.*). Behav Ecol Sociobiol 50(5):441–449
6. Dornhaus A, Chittka L (2004) Why do honey bees dance? Behav Ecol Sociobiol 55(4):395–401
7. Dyer FC (2002) The biology of the dance language. Annu Rev Entomol 47:917–949
8. Esch HE, Bastian JA (1970) How do newly recruited honeybees approach a food site. Z vergl Physiol 68:175–181
9. Esch HE, Zhang SW, Srinivasan MV, Tautz J (2001) Honeybee dances communicate distances measured by optic flow. Nature 411(6837):581–583
10. Gardner KE, Seeley TD, Calderone NW (2007) Hypotheses on the adaptiveness or non-adaptiveness of the directional imprecision in the honey bee's waggle dance (Hymenoptera: Apidae: *Apis mellifera*). Entomol Gen 29(2–4):285–298
11. Gil M, Farina WM (2002) Foraging reactivation in the honeybee *Apis mellifera* L.: factors affecting the return to known nectar sources. Naturwissenschaften 89(7):322–325
12. Gould JL (1975) Honey bee recruitment: the dance-language controversy. Science 189(4204):685–693
13. Grüter C, Balbuena MS, Farina WM (2008) Informational conflicts created by the waggle dance. Proc R Soc B 275(1640):1321–1327
14. Grüter C, Farina WM (2009) The honeybee waggle dance: can we follow the steps? Trends Ecol Evol 24(5):242–247
15. Johnson DL, Wenner AM (1966) A relationship between conditioning and communication in honey bees. Anim Behav 14(2):261–265
16. Krause J, Ruxton GD, Krause S (2010) Swarm intelligence in animals and humans. Trends Ecol Evol 25(1):28–34
17. Mautz D (1971) Communication effect of wag-tail dances of *Apis mellifera carnica* (pollm.). Z vergl Physiol 72(2):197–220
18. Menzel R, Giurfa M (2006) Dimensions of cognition in an insect, the honeybee. Behav Cogn Neurosci Rev 5(1):24–40
19. Michelsen A (2003) Karl von Frisch lecture. Signals and flexibility in the dance communication of honeybees. J Comp Physiol A 189(3):165–174
20. Reinhard J, Srinivasan MV, Zhang S (2004) Scent-triggered navigation in honeybees. Nature 427(6973):411
21. Ribbands CR (1954) Communication between honeybees. I: The response of crop-attached bees to the scent of their crop. Proc R Entomol Soc Lond A 30:1–3
22. Riley JR, Greggers U, Smith AD, Reynolds DR, Menzel R (2005) The flight paths of honeybees recruited by the waggle dance. Nature 435(7039):205–207
23. Seeley TD (1983) Division of labor between scouts and recruits in honeybee foraging. Behav Ecol Sociobiol 12:253–259
24. Seeley TD (1995) The wisdom of the hive. Harvard University Press, Cambridge
25. Seeley TD, Mikheyev AS, Pagano GJ (2000) Dancing bees tune both duration and rate of waggle-run production in relation to nectar-source profitability. J Comp Physiol A 186(9):813–819

26. Seeley TD, Visscher PK (1988) Assessing the benefits of cooperation in honeybee foraging – search costs, forage quality, and competitive ability. Behav Ecol Sociobiol 22(4):229–237
27. Sherman G, Visscher PK (2002) Honeybee colonies achieve fitness through dancing. Nature 419(6910):920–922
28. Tanner DA, Visscher PK (2010) Adaptation or constraint? Reference-dependent scatter in honey bee dances. Behav Ecol Sociobiol 64(7):1081–1086
29. Towne WF, Gould JL (1988) The spatial precision of the dance communication of honey bees. J Insect Behav 1:129–155
30. Visscher PK, Seeley TD (1982) Foraging strategy of honeybee colonies in a temperate deciduous forest. Ecology 63(6):1790–1801
31. von Frisch K (1946) Die Tänze der Bienen. Österreich Zool Z 1:1–48
32. von Frisch K (1967) The dance language and orientation of bees. Harvard University Press, Cambridge
33. von Frisch K (1968) The role of dances in recruiting bees to familiar sites. Anim Behav 16:531–533
34. Wenner AM, Wells PH (1990) Anatomy of a controversy: the question of a "language" among bees. Columbia University Press, New York

Chapter 2.4
Olfactory Information Transfer During Recruitment in Honey Bees

Walter M. Farina, Christoph Grüter, and Andrés Arenas

Abstract Honey bee colonies use a number of signals and information cues to coordinate collective foraging. The best known signal is the waggle dance by which dancers provide nest-mates with information about the location of a foraging or nest site. The efficiency of this nest-based recruitment strategy partly depends on olfactory information about food sources that is transferred from dancer to receivers in parallel to spatial information. Here we will address how the waggle dance facilitates the acquisition and the retrieval of food odor information and how olfactory memory affects the interaction patterns among nest-mates within the dancing and the food-unloading context. We further discuss how olfactory information affects the food preferences of foragers acquired directly from scented-food offered inside the hive. The discussed results show that odor learning in this context is an important component of the honey bee recruitment system that has long-term consequences for foraging decisions.

Abbreviations

LIO	Linalool
PER	Proboscis extension reflex
PHE	Phenylacetaldehyde
US	Unconditioned stimulus

W.M. Farina (✉) • A. Arenas
Grupo de Estudio de Insectos Sociales, IFIBYNE-CONICET, Departamento de Biodiversidad y Biología Experimental, Facultad de Ciencias Exactas y Naturales, Universidad de Buenos Aires, Pabellón II, Ciudad Universitaria, C1428EHA Buenos Aires, Argentina
e-mail: walter@fbmc.fcen.uba.ar

C. Grüter
Laboratory of Apiculture & Social Insects, School of Life Sciences, John Maynard-Smith Building, University of Sussex, Falmer BN1 9QG, UK

C.G. Galizia et al. (eds.), *Honeybee Neurobiology and Behavior: A Tribute to Randolf Menzel*, DOI 10.1007/978-94-007-2099-2_8,
© Springer Science+Business Media B.V. 2012

2.4.1 Background

Honey bees (*Apis mellifera*) are excellent models to study the formation and development of information networks in decentralized biological systems [39]. The different worker groups in honey bee colonies show a strong operational cohesion that emerges from the continuous interactions amongst nest-mates. In this chapter we discuss olfactory information transfer in this network, its role for the organization within worker groups and how olfactory learning leads to cohesion between worker groups in the context of food collection.

In honey bees, like in many other social insects, collective tasks such as foraging, nest-climate regulation, nest-building/repair or brood care constantly need to be adjusted in response to changes in the environment or within the nest. Food source locations, for example, are often stable only during some days or a few weeks [40]. Responses to these changes at a colony level are often the result of individuals responding to simple local information [39]. This local information is provided either by nest-mates, environmental cues or by modifications of the environment by nest-mates. Information from nest-mates is transferred via different sensory modalities (e.g. chemosensory, tactile), either inadvertently (cues) or based on traits specifically designed by selection to convey information (signals).

An example for inadvertent information transfer is the propagation of gustatory and olfactory information about the exploited resources during the distribution of liquid food within the hive ([12, 19, 21, 33, 34], see below). The frequent food sharing among workers connects different worker groups. It allows workers not directly involved in foraging, like nurse bees, to obtain information from bees that actively participate in resource exploitation and processing, i.e. foragers and food processor bees [19, 21, 33, 34].

The best known behavior in honey bees, the waggle dance, involves signal transmission [41] and provides nest-mates with information about the location of a foraging or nest site [9, 37, 41]. When foragers find highly-quality food sources, they perform waggle dances inside the nest to recruit other bees to the same location. The occurrence, duration and the rate of waggle-run production are tuned to the profitability of the feeding site, thereby allowing for an adaptive distribution of recruits among the various food sources ([26, 39, 41], see also Chaps. 2.1 and 2.3).

Honey bees are able to use floral scents as guiding cues for long-distance flights [42]. However, there are also some problems with using olfaction for long distance orientation because (1) floral bouquets can change in the air due to differences in volatility of the different compounds [24, 43] and (2) floral bouquets of some species change during the day [43]. Therefore, honey bees also strongly rely on visual information to find food sources, such as celestial and terrestrial references or familiar landmarks [7, 41]. Once a bee returns from collecting resources to the closed and dark environment of the hive cavity, other sensory modalities that are less relevant in the field become more important, e.g. acoustical signals and vibrations [8, 31].

When a successful forager returns with nectar, she brings back the food scent in the honey crop or clinging on her body. These odor cues act as attractant or orientation guide for nest mates while the forager walks or dances [5, 41]. Usually, several bees simultaneously follow a dancer for a few dance circuits. Most of these followers are foragers but also food-processing bees are attracted by dances [10]. Recruits are rarely able to find the precise location of the food source for landing with the vector information alone, but need additional information sources [37, 42]. These can be olfactory and visual cues from the flowers themselves or cues and signals provided by other bees on the food source [41]. Given the importance of olfactory cues for locating a food source, in-hive recruitment strategies involving not only the transfer of vector information of the waggle dance but also the transfer of olfactory cues are likely to have a positive effect on the foraging performance of honey bee colonies (review: [22]).

In this chapter we will discuss how the waggle dance facilitates the acquisition and the retrieval of olfactory information about food sources and consider how this information affects the food preferences of foragers. Finally, we show that olfactory learning also has more subtle effects on honey bee foraging through its effects on perceived interaction patterns.

2.4.2 Floral Scents as Guiding Cues for Hive Mates

Apart from the transmission of spatial information mentioned in the previous section, the dance serves at least two other informational purposes: first, it increases the attention and activity of bees in the vicinity by communicating the presence of an attractive food source ([41], review: [22]). If a forager is performing a waggle dance, the increased attention of unemployed foragers facilitates the perception of the acoustic–vibratory signals emitted by the intensive movements of the wings that form the acoustic near field of the dancer ([8, 31], see also Chap. 2.2). Second, the honey bee dance is important to transfer food odor information [41, 42]. The molecules of floral odors clinging on the foragers' body as well as the pollen loads carried on hind legs can be perceived by other foragers. Dancing bees often shortly interrupt dancing and offer samples of food to surrounding bees ([41], review: [11]). These interactions are often brief (<2 s) which suggests that in these cases followers do not actually receive nectar but just probe it. There is good evidence, that these interactions play an important role in olfactory learning (see below).

Thus, a returning forager provides different kinds of information during dancing, which functions as a compound or multicomponent signal [22]. Information cues are transmitted in parallel or complementary to the signal. The transfer of floral odor information linked to the display of a stereotypic behavior can be seen in other social insects as well [27]. However, as far as we know only honey bees perform a nest-based signal that transmits location information [22, 41].

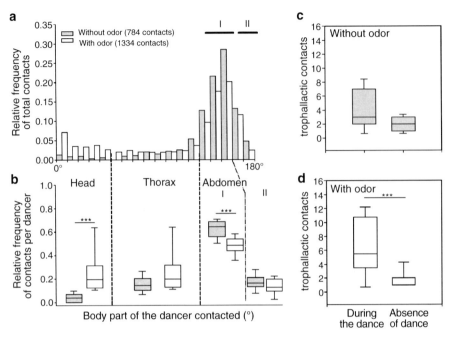

Fig. 2.4.1 Floral scents as guiding cues for dance followers. Distribution of the head contacts of hive bees onto the incoming foragers displaying waggles and the number of trophallactic contacts offered by the dancer inside the hive. (**a**) Head contacts onto the body of dancers. Comparisons were done between dancers that collected unscented 1.8 M sucrose solution (*gray bars*, without odor, N = 14) and dancers that foraged a scented (Linalool, LIO) sucrose solution (*white bars*, with odor, N = 14). 0° corresponds to the frontal part of the dancer's head and 180° to the posterior extreme of its abdomen. (**b**) Total head contacts (in relative frequencies) grouped according to the part of the dancer's body that was contacted. (**c**) and (**d**) The number of trophallactic contacts offered by dancers carrying either unscented (**c**, without odor, N = 15) or scented sucrose solution (**d**, with odor, N = 16). Trophallaxis events were grouped according to the period they occurred: during the dancing display, before the first waggle-run and/or after the last waggle-run phase observed. Medians, quartiles, and the 5 and 95° percentiles are shown. Asterisks indicate statistical differences (Mann–Whitney *U*-test in (**c**), Wilcoxon-test in (**d**), ***p < 0.001) (After [5]. With permission)

2.4.2.1 Interactions During Scented Dances

The highest proportion of head contacts of hive bees around the dancers during the waggle-run phases is observed around the abdomen of the dancers (more than 60% of all cases, Fig. 2.4.1a, b). This can be expected if this position improves the acquisition of the transmitted signal [31] (see Chap. 2.2). However, the distribution of head contacts around dancers differs significantly for the different types of collected resources. Head contacts around the hind legs are more frequent when hive bees follow pollen dancers (with pollen loads on their hind legs) than when they follow non-pollen dancers [5]. Dancers that forage at natural nectar sources are contacted in a higher proportion at their thorax and head (for details see [5]). These quantitative descriptions suggest that the general pattern of contacts depends on the type of the collected food source and the presence of odor in the collected food. Accordingly, a higher proportion of head-to-head contacts was found when dancers collected scented sugar solution at artificial feeders (Fig. 2.4.1a). Food odor effects seem to be stronger when odors are located on the dancers' mouthparts, probably due their persistency within the honey sac [41]. In the mentioned experiment (Fig. 2.4.1), the hive entrance was scented with a different odor in order to reduce the effect of the food odors clinging onto the foragers' body surface. The presence of odor in the collected solution does not seem to modify the number of trophallaxis events during hive stays [5], but it increases the proportion of these interactions during dancing (Fig. 2.4.1c, d). Hence, the dance acts as a congregating mechanism while the crop scent concentrated on the mouthparts helps other bees to orientate and taste the liquid food. Moreover, the higher proportion of the head-to-head contacts between "scented nectar" dancers and hive bees compared to the unscented situation might lead to a higher number of mouth-to-mouth contacts during the return phases of the waggle dance (Fig. 2.4.1c, d).

2.4.2.2   Associative Learning Within the Dancing Context

As mentioned above dance followers often receive or taste samples of the collected food during short interruptions of dancing. Dirschedl [6] showed that most of the recruits arriving at the food source (ca. 95%) received food samples collected by the recruiting foragers inside the hive. This might explain why von Frisch, during his pioneering study published in 1923, found that recruits had strong preferences for food containing odors that were brought back by the recruiting bee. These findings suggested that recruits learn the contingency odor-reward through these food offerings during the short interruptions of the dance. Only recently this idea has been tested using the proboscis extension reflex (PER) assay on recruited foragers [12].

The proboscis extension reflex (PER) assay offers a powerful method to test associations established between odor and sugar within a variety of behavioral contexts in honey bees (review: [29]). Bees reflexively extend their proboscis to drink solution when the antennae are touched with sucrose solution (unconditioned stimulus, US). In classical conditioning in the laboratory, an odor (conditioned stimulus, CS) is paired with the US, which causes the odor itself to become capable of eliciting the proboscis extension as a conditioned response [4]. The solution transferred or probed during trophallaxis events functions as an US, like the small samples of sugar solution applied in the laboratory during olfactory conditioning. The food odor functions as conditioned stimulus [17]. Farina et al. [12] used the PER assay to test whether foragers that were recruited by dancers to a food source 160 m from the hive learned the odor of the food source during interactions with a dancing bee. Once recruited bees arrived at a feeding platform they were captured before drinking the solution offered by the feeder. In the laboratory, the PER of recruits for the food odor and a novel odor was evaluated. The proportion of bees that showed the proboscis extension after presenting the food odor linearly increased as increasing amounts of the scented food were carried into the hive by trained foragers. Four days after offering scented food to the foragers, the PER to the conditioned odor was still elevated but weaker than on previous days (see Fig. 2.4.1 in [12]).

This study suggests that associative learning does not seem to depend strongly on the duration of the oral contact. An excitatory motor display like the honey bee dance might be a particularly efficient context for olfactory learning where brief trophallactic interactions taking place within a context that increases arousal might increase the probability to establish olfactory memories [30].

2.4.2.3   The Extent of Information Propagation About Floral Odors

While information on distance and direction transferred during dance maneuvers is perceived only by the dance followers, olfactory and gustatory information about the discovered nectar source can be acquired by a much broader audience. It has been demonstrated that once the fresh nectar enters the hive its distribution can be rapid and extensive amongst colony members [19, 32–34]. Nixon and Ribbands [32], for example, found that 62% of all sampled foragers of a colony were in contact with sucrose solution collected by only six bees after 4 h. In this sense, chemosensory cues of nectars such as floral odors and food quality may provide the colony with global information about the available resources [33], which means that the information has the potential to affect most colony members [19, 21, 34].

Grüter et al. [19], for example, showed that the sharing of scented food within the hive leads to a propagation of olfactory information. Bees of different age/sub-caste groups learned the odor of relatively small amounts of liquid food that had been collected during 5 days by five to nine foragers. Furthermore, information propagation is more extensive, i.e. reaches more bees, as the food profitability (in this case sugar concentration) of the collected nectar increases (Fig. 2.4.2). The figure shows the proportion of nurse-aged bees (hive bees of 4–9 days old, N), receiver-aged bees (hive bees of 12–16 days old, R) and foragers (F) extending the proboscis after presenting a treatment odor. The PER levels for the solution odor were higher for all the age categories when the foragers of a colony collected 2 M sucrose solution versus 0.5 M sucrose solution. Apart from the positive relationship between learning and US strength [4], a higher level dancing in the 2 M treatment might have contributed to the increased propagation (e.g. due to arousal or a higher number of trophallaxis events). Hence, more bees learn the odor of high quality food sources than of food sources producing nectar of low sugar concentration.

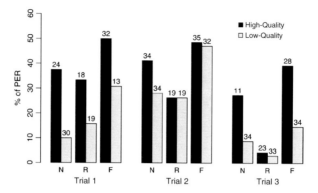

Fig. 2.4.2 Effect of food quality on information propagation. Proportion of nurse-aged bees (4/9-day-old bees, N), receiver-aged bees (12/16-day-old bees, R) and foragers (F) extending the proboscis (PER) after presenting a treatment odor to the antennae. Prior to the PER test, entire colonies were treated by feeding 8 foragers per colony with high-quality food (2 M sucrose solution) or with low-quality food (0.5 M sucrose solution) at an artificial feeding station. During 5 days, these eight foragers collected 14 ml of scented solution per day (totally 70 ml per colony). The scents were PHE (Trial 1, 3) and LIO (Trial 2). Two colonies were used in Trial 1, two other colonies in Trial 2. The two colonies used in Trial 2 were used again in Trial 3, but with reversed treatments. Overall, a higher proportion of bees responded to the treatment odor after treating colonies with 2 M sucrose solution compared to 0.5 M solution (For experimental procedure see [19]) (Grüter, Balbuena and Farina, unpublished results)

2.4.2.4 Recalling Olfactory Memories While Following Dancers

While bees forage they learn the odor of the food source and other characteristics like its location, color or shape [29, 30]. Active foragers rely on these memories to find particular locations when visiting food patches repeatedly [35, 36, 41]. Inside the hive the crop scent can cause a conditioned response in inactive foragers with knowledge about this scent from earlier foraging trips. After perceiving a familiar odor, these foragers often leave the nest and fly to the food sites where the odor was learned in the first place. Thus, inactive foragers use the odor of the floral type inside the nest as a prediction for the reappearance of their food source [35, 36, 41]. Most dance followers are experienced foragers [3] and the dance is the most frequent social interaction during this reactivation process to a profitable foraging site. They are attracted to dancers carrying familiar scents [20, 41]. However, the dancer to which an experienced forager is attracted to due to the familiarity with the odor does not necessarily advertise the foraging location where the dance follower learned the odor. As a consequence this follower has to decide whether to fly to the memorized location or whether to decode the vector information of the dance and follow its instructions. Hence, the waggle dance can create a conflict between the self-acquired navigational memory and the vector information of the waggle dance. Grüter et al. [20] found that in these situations of informational conflict, followers with field experience mostly ignored the spatial information encoded by the waggle dance even if they followed a dance for several waggle runs. They relied on their own memory about food source locations in 93% of all cases (see Fig. 2.4.2 in [20]). This strategy is likely to be adaptive if nectar production of a plant species is synchronized so that the flowers of a species provide nectar at the same time of day at different locations and if finding familiar food patches is easier than finding novel patches advertised by dances ([22], see also Chap. 2.3). However, more research is needed to explore the role of flight distances, food quality or the amount of foraging experience on the use of self-acquired information against location information.

2.4.3 Outside Behavior After in-Hive Olfactory Learning

How does olfactory learning inside the hive affect food choice during foraging? Olfactory cues can be learned from the stores of the nest. Free [15] counted the number of visits (i.e. hovers and landings) either at a feeder scented with the odor of a currently collected food source or at a feeder scented with the odor of their stored food. He observed more visits for the currently collected food odor at the beginning of the test phase (during the first 10–30 min). After this period the bees' preferences shifted to the odor of the food stores. Hence, preferences are shaped by both the transfer of scented food from successful foragers and by the scent presented in the food stores but not communicated by mates.

Foraging preferences after offering scented sugar solution were evaluated recently in foragers that had to choose between two feeders either scented with a previously experienced odor or with a novel odor [1]. Scented food, obtained by mixing a pure odorant with a sugar solution was offered in an in-hive feeder that was left in the hive for a 4 day period. Honey bee foragers, trained to visit an unscented training feeder, were evaluated for their first landing choice when the feeder was removed and two similar feeders (testing feeders) were placed 6 m to the hive and 1.3 m from each other. A higher number of landings was recorded at the feeder scented with the solution odor compared to a novel odor. Thus, food odors learned within the hive were used to guide searching during short-range foraging flights [1]. Preferences for the solution odor were found to last for at least 4 days after all the scented stores of the hive were removed and replaced by combs that contained non-scented food (Fig. 2.4.3a). This implies that bees are able to retain olfactory memories established within the hive for several days and use this information for foraging decisions, though the time periods are slightly shorter than when associations are established outside the colony (13 days: [28], 10 days: [2]). In a study that allowed foragers to collect the scented solution rather than treating the colonies with in-hive feeders, in-hive olfactory memories analyzed via the PER paradigm showed high levels of responses to the food odor for up to 10–11 days after the end of offering food [21].

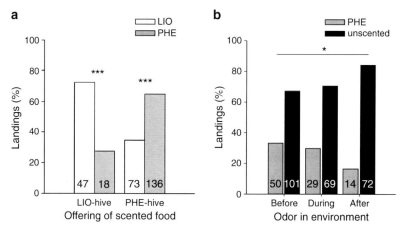

Fig. 2.4.3 Outside behavior after in-hive olfactory learning (**a**) Landings on a dual-choice device (PHE-feeder vs. LIO-feeder) 4 days after a scented food store was removed from within the hive. LIO-hive received sugar solution scented with LIO while PHE-hive received sugar solution scented with PHE. (**b**) Percentages of landings on a dual-choice device (PHE-feeder vs. unscented feeder) after PHE was presented in the hive environment as volatile for different times during the experiment: before (2 and 1 days before the test), during (2 and 4 days after initiating the odor experience), and after the odor experience (3 h and 2 days after removing the odor source and replacing all honeycombs). The number of bees landing on each feeder is shown at the *bottom* of each bar. *Asterisks* indicate statistical differences: *p < 0.05, *** p < 0.001, G-test (After [1]. With permission)

The offering of scented food inside the hive as a method to increase the yield of honey or the production of seeds of a particular crop has been studied extensively during the 1940s and 1950s [41]. For instance, von Frisch [41] found positive effects of artificial feeding of scented food (increased yields of seed) in several floral species (e.g. red-clover, Swedish clover, rape). The fields to which bees were guided by the scent were visited by about 3–4 times more bees than control fields where hives were fed with unscented food. He concluded that round dances performed by foragers after finding the scented sugar solution recruited other bees to search for the food odor in the surrounding area. However, another study using a similar method did not succeed in guiding bees to the particular flower species [14].

Floral odors can also have inhibitory effects depending on how odors are exposed inside the hive. So far only a few works tested the effect of volatile odor exposure, i.e. not associated with a reward, in the hive environment on appetitive behaviors, both in the classical context of the PER [12, 13, 16, 38] or in an operant context in the field [15]. A recent study indeed found that honey bees avoided the odor while searching for food outside (Fig. 2.4.3b) after a 5-day-exposure period [1]. A reduced landing motivation was observed towards the previously experienced odor (PHE-scented feeder) against an unscented feeder. This bias might be related to forms of non-associative learning causing an avoidance response which might prevent the nectar foraging of a particular floral type in a context of searching for food (Fig. 2.4.3b). Note that differences between PHE and unscented choice before the odor exposure (Fig 2.4.3b) might reside in the amount of information available in both testing feeders and their similarities with the immediately prior experience gained at this site. Because experimental bees were previously trained to an unscented feeder, to ensure a number of individuals approaching to the site, the unscented options of the choice device perfectly matched with the searching image obtained before.

2.4.4 Social Feedbacks During in-Hive Recruitment

Martin Lindauer observed that the dances of a forager did not only depend on the profitability of the food source itself but also on the amount of food collected by other foragers, i.e. the overall nectar (or water) influx [25]. He noted that on days of good foraging conditions when most foragers of the colony are active, returning foragers had to wait a long time until they found a receiver for their load. When foraging conditions were poor returning foragers usually quickly found several unloading receivers. This social feedback, the time a forager had to wait for a receiver bee, had a strong effect on the occurrence of dances [26, 39]. If the delay is short, a forager performs a waggle dance to recruit more foragers to work. If the delay is long, the forager does not perform waggle dances and her own foraging motivation decreases. We now know that also the number of unloading bees positively affects dancing [10]. The ability of foragers to adjust their recruitment behavior according to the availability of receiver bees enables the colony to keep a healthy balance between food collection and food processing [39].

Receiver bees that have experienced a certain odor during unloading contacts show a preference to receive food containing the same odor from foragers during subsequent unloading [18]. In this experiment, receiver bees had a 78% chance of unloading a forager returning with a particular odor if the receiver bee experienced the same odor during unloading contacts in the past. Only 12% of all receivers unloaded an unfamiliar odor (10% unloaded both odors). This finding challenges the assumption of random unloading made in many theoretical studies on the informational value of transfer delays and the causes of multiple unloading contacts in honey bees [39]. Rather, foragers returning with odors that are well known to most receiver bees of a colony experience a stronger social feedback than foragers returning with a new food odor. A food odor could be well-known to receiver bees if a particular plant species has been exploited intensively in the past. This idea was tested by treating entire colonies with scented food. A group of ca. 30–100 foragers collected 200 ml scented food at an artificial feeder (Fig. 2.4.4, see details in [23]). The authors tested how foragers returning with a familiar food scent were unloaded. Interestingly, no effect on the unloading delay was found (Fig. 2.4.4a). As discussed by [23], latent inhibition effects on foragers caused by the treatment might have made the experimental design unsuitable to test this particular question. On the other hand, foragers returning with the familiar odor were unloaded by more receivers than when returning with a novel odor (Fig. 2.4.4b). During dancing they were also attended by more follower bees (Fig. 2.4.4c). Hence, the social feedback experienced by foragers that collect from apple might depend on the availability of receivers that

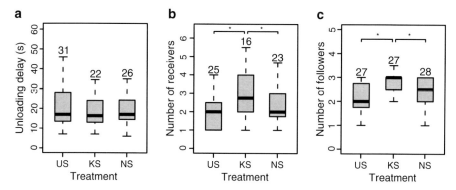

Fig. 2.4.4 Social feedbacks during in-hive recruitment. (**a**) Unloading delays for foragers coming back with nectar solution US, no scent in solution; KS, a known scent in solution; NS, a novel scent in solution. (**b**) Number of receivers during unloading. (**c**) Number of followers in case incoming foragers danced. The boxplots show medians, quartiles, 5th and 95th percentiles. The number of bees are presented above the bars. Asterisks indicate statistical differences (*$P<0.05$, permutation test) (After [23]. With permission)

unload apple, and not on the total number of available receivers. Having these associations between foragers and processor bees based on odors could be seen as a form of team formation and one might speculate that it increases the efficiency of nectar collection compared to completely random interactions. However, more research is needed to confirm or refute this idea.

2.4.5   Concluding Remarks

Food odors learned inside the colony lead to a preference for these odors and help foraging workers to find a particular food source in the surrounding area. Odor information can have an effect on foraging decisions even if it is not acquired within a recruitment context. This effect persists for several days which suggests that olfactory learning in young bees not directly involved in foraging-related tasks might affect their food preferences at forager age. Moreover, previously acquired olfactory information seems to play a significant role within the hive in that it leads to non-random interactions within the dancing and the food-unloading context. Known odors affect interaction patterns between foragers and followers and between foragers and receivers. This might have consequences for the operational balance between food collection and processing capacities. However, more empirical and theoretical research is needed to explore the consequences of non-random unloading for the efficiency of food collection and processing. The discussed results show that odor learning inside the hive is an important component of the recruitment system of honey bees (and other insects) with long-term consequences for foraging decisions. Future directions in this research field should consider precocious olfactory experiences and their role related to resource exploitation.

References

1. Arenas A, Fernandez VM, Farina WM (2008) Floral scents experienced within the colony affect long-term foraging preferences in honeybees. Apidologie 39(6):714–722
2. Beekman M (2005) How long will honey bees (*Apis mellifera* L.) be stimulated by scent to revisit past-profitable forage sites? J Comp Physiol A 191(12):1115–1120
3. Biesmeijer J, Seeley T (2005) The use of waggle dance information by honey bees throughout their foraging careers. Behav Ecol Sociobiol 59(1):133–142
4. Bitterman ME, Menzel R, Fietz A, Schäfer S (1983) Classical conditioning of proboscis extension in honeybees (*Apis mellifera*). J Comp Psychol 97(2):107–119
5. Diaz PC, Grüter C, Farina WM (2007) Floral scents affect the distribution of hive bees around dancers. Behav Ecol Sociobiol 61(10):1589–1597
6. Dirschedl H (1960) Die Vermittlung des Blütenduftes bei der Verständigung im Bienenstock. Doctoral thesis, University of Munich, Germany
7. Dyer FC (2002) The biology of the dance language. Annu Rev Entomol 47:917–949
8. Esch H (1961) Über die Schallerzeugung beim Werbetanz der Honigbiene. Z vergl Physiol 45(1):1–11
9. Esch HE, Zhang SW, Srinivasan MV, Tautz J (2001) Honeybee dances communicate distances measured by optic flow. Nature 411(6837):581–583
10. Farina WM (2000) The interplay between dancing and trophallactic behavior in the honey bee *Apis mellifera*. J Comp Physiol A 186(3):239–245
11. Farina WM, Grüter C (2009) Trophallaxis: a mechanism of information transfer. In: Jarau S, Hrncir M (eds) Food exploitation by social insects: ecological, behavioral, and theoretical approaches. CRC Press, Boca Raton, pp 173–187
12. Farina WM, Grüter C, Diaz PC (2005) Social learning of floral odours inside the honeybee hive. Proc. R. Soc. B 272(1575):1923–1928
13. Fernandez VM, Arenas A, Farina WM (2009) Volatile exposure within the honeybee hive and its effect on olfactory discrimination. J Comp Physiol A 195(8):759–768
14. Free JB (1958) Attempts to condition bees to visit selected crops. Bee World 39:221–230
15. Free JB (1969) Influence of odour of a honeybee colonys food stores on behaviour of its foragers. Nature 222(5195):778
16. Gerber B, Geberzahn N, Hellstern F, Klein J, Kowalksy O et al (1996) Honey bees transfer olfactory memories established during flower visits to a proboscis extension paradigm in the laboratory. Anim Behav 52:1079–1085
17. Gil M, De Marco RJ (2005) Olfactory learning by means of trophallaxis in *Apis mellifera*. J Exp Biol 208(Pt 4):671–680
18. Goyret J, Farina WM (2005) Non-random nectar unloading interactions between foragers and their receivers in the honeybee hive. Naturwissenschaften 92(9):440–443
19. Grüter C, Acosta LE, Farina WM (2006) Propagation of olfactory information within the honeybee hive. Behav Ecol Sociobiol 60(5):707–715
20. Grüter C, Balbuena MS, Farina WM (2008) Informational conflicts created by the waggle dance. Proc. R. Soc. B 275(1640):1321–1327
21. Grüter C, Balbuena MS, Farina WM (2009) Retention of long-term memories in different age groups of honeybee (*Apis mellifera*) workers. Insect Soc 56(4):385–387
22. Grüter C, Farina WM (2009) The honeybee waggle dance: can we follow the steps? Trends Ecol Evol 24(5):242–247
23. Grüter C, Farina WM (2009) Past experiences affect interaction patterns among foragers and hive-mates in honeybees. Ethology 115(8):790–797
24. Knudsen JT, Tollsten L, Bergstrom LG (1993) Floral scents – a checklist of volatile compounds isolated by headspace techniques. Phytochemistry 33(2):253–280
25. Lindauer M (1949) Über die Einwirkung von Duftmacksstoffen und Geschmacksstoffen sowie anderer Faktoren auf die Tanze der Bienen. Z vergl Physiol 31(3):348–412

26. Lindauer M (1954) Temperaturregulierung und Wasserhaushalt im Bienenstaat. Z vergl Physiol 36 (4):391–432
27. Lindauer M, Kerr WE (1960) Communication between the workers of stingless bees. Bee World 41:29–71
28. Menzel R (1969) Das Gedächtnis der Honigbiene für Spektralfarben. II. Umlernen und Mehrfachlernen. Z vergl Physiol 63:290–309
29. Menzel R (1999) Memory dynamics in the honeybee. J Comp Physiol A 185(4):323–340
30. Menzel R, Müller U (1996) Learning and memory in honeybees: from behavior to neural substrates. Annu Rev Neurosci 19:379–404
31. Michelsen A (2003) Signals and flexibility in the dance communication of honeybees. J Comp Physiol A 189(3):165–174
32. Nixon HL, Ribbands CR (1952) Food transmission within the honeybee community. Proc. R. Soc. B 140(898):43–50
33. Pankiw T, Nelson M, Page RE, Fondrk MK (2004) The communal crop: modulation of sucrose response thresholds of pre-foraging honey bees with incoming nectar quality. Behav Ecol Sociobiol 55(3):286–292
34. Ramirez GP, Martinez AS, Fernandez VM, Bielsa GC, Farina WM (2010) The influence of gustatory and olfactory experiences on responsiveness to reward in the honeybee. Plos One 5(10):1–12, e134898
35. Reinhard J, Srinivasan MV, Guez D, Zhang SW (2004) Floral scents induce recall of navigational and visual memories in honeybees. J Exp Biol 207(Pt 25):4371–4381
36. Ribbands CR (1954) Communication between honeybees. I: the response of crop-attached bees to the scent of their crop. Proc R Entomol Soc Lond A 30:1–3
37. Riley JR, Greggers U, Smith AD, Reynolds DR, Menzel R (2005) The flight paths of honeybees recruited by the waggle dance. Nature 435(7039):205–207
38. Sandoz JC, Laloi D, Odoux JF, Pham-Delegue MH (2000) Olfactory information transfer in the honeybee: compared efficiency of classical conditioning and early exposure. Anim Behav 59(5):1025–1034
39. Seeley TD (1995) The wisdom of the hive. Harvard University Press, Cambridge
40. Vogel S (1983) Ecophysiology of zoophilic pollination. In: Lange OL, Nobel PS, Osmond CB, Ziegler H (eds) Physiological plant ecology III, (encyclopedia of plant physiology). Springer, Berlin/Heidelberg/New York, pp 559–624
41. von Frisch K (1967) The dance language and orientation of bees. Belknap Press of Harvard University Press, Cambridge
42. Wenner AM, Wells PH, Johnson DL (1969) Honey bee recruitment to food sources: olfaction or language? Science 164(3875):84–86
43. Wright GA, Schiestl FP (2009) The evolution of floral scent: the influence of olfactory learning by insect pollinators on the honest signalling of floral rewards. Funct Ecol 23(5):841–851

Chapter 2.5
Navigation and Communication in Honey Bees

Randolf Menzel, Jacqueline Fuchs, Andreas Kirbach,
Konstantin Lehmann, and Uwe Greggers

Abstract Honey bees navigate and communicate in the context of foraging and nest selection. A novel technique (harmonic radar tracking) has been applied to foraging behavior. On the basis of the data collected, a concept that assumes an integrated map-like structure of spatial memory was developed. Characteristic features (long ranging landmarks) and local characteristics are learned during exploratory flights. Route flights and information about target destinations transferred during the waggle dance are integrated into the map-like memory, enabling bees to make decisions about their flight routes. Cognitive terminology is applied to describe these implicit knowledge properties in bee navigation.

Abbreviations

FD   Dance indicated food site
FT   Trained food site

2.5.1 Do Bees Navigate According to an Egocentric Path Integration Mechanism?

Honey bees are central place foragers. They start their exploratory orientation flights as young bees at their hive, they begin and end their foraging flights at their hive and they report vectors linking the hive and a profitable food source. In all these cases they relate their flights to a common reference frame comprising the directional

R. Menzel (✉) • J. Fuchs • A. Kirbach • K. Lehmann • U. Greggers
Institut für Biologie, Neurobiologie, Freie Universität Berlin, Berlin, Germany
e-mail: menzel@neurobiologie.fu-berlin.de

C.G. Galizia et al. (eds.), *Honeybee Neurobiology and Behavior: A Tribute
to Randolf Menzel*, DOI 10.1007/978-94-007-2099-2_9,
© Springer Science+Business Media B.V. 2012

component (i.e. the direction towards the food) with respect to the sun compass or learned landmarks [42] and the distance travelled to the food as measured via a visual odometer [13, 37] (see also Chap. 2.1). The traditional view of honey bee communication and navigation is that an experienced bee could well perform its navigational and communication tasks by relying solely on egocentric vector information. However, as we shall see, the bee appears to know much more about its environment and uses this information for novel flight paths.

An animal learns about the egocentric vector information by a process called path integration [44]. Egocentric is self-referred, whereas allocentric (see below) is referenced to the external world. Egocentric frames of reference follow the animal around, as it were, relating its movements to many reference frames. These can be anchored to the eye or body, and need to be integrated in order for actions to be co-ordinated. In egocentric path integration the directional component is weighted by segments of the distance components, and these measures are continuously integrated and stored in a kind of autopilot working memory. The information the animal uses is bound to its own body movement (thus the term egocentric), and therefore it can only return to its starting point by applying the integrating segments subtracting 180° for the return path. If the animal is transported to a remote release site it may apply the current status of the path integration memory, but then should be lost. Initially this concept appeared to be supported by observations in which foraging bees were caught after arrival or departure at either the hive entrance or a feeder, transported to a release site a few 100 m away and observing in which direction the bees departed from the release site. Their initial flight path as detected by the vanishing bearings showed that they flew as if they had not been displaced [10, 22, 45]. Since the distance component of the vector could not be determined with this method it was assumed that bees fly according to their vector memory, including the distance measure. Until the early '90s, vanishing bearings of displaced bees were never seen to point directly toward the intended goal. It was therefore concluded that bees refer only to an egocentric reference frame while navigating in their environment. The only data not concurring with this conclusion were reported by Gould [15] who indeed observed displaced bees vanishing from the release site to the intended goal. He therefore suggested that bees may possess a memory structure equivalent to a cognitive map, i.e. an allocentric representation of space which would allow them to travel novel routes between two locations. However, Gould's results could not be repeated despite multiple attempts to do so, therefore they were assessed with reservation. It was suspected that bees in his experiments could have used special landscape features at the test site for their novel shortcutting goal directed behavior [11].

If path integration is the only mechanism bees use in navigation they should be lost after release at an unexpected site. Results reported in Menzel and others [21] indicated that bees are not lost after displacement and thus must apply additional navigational strategies. In a different experiment we trained the same bees to two different locations, one in the morning and another in the afternoon. Then we collected bees at the moment of departure from the hive and transported them to the incorrect feeding site (the afternoon site in the morning and vice versa). In both cases, they vanished predominantly toward the hive, indicating that they used the local landmarks to identify the location, switched motivation (they were collected when motivated to fly out to the feeder) and retrieved the correct vector memory to return to the hive [24]. Next we asked which direction hive-departing bees take when released halfway between the morning and afternoon feeding sites, a site they had not visited before. Half of these bees behaved according to their current status of path integration memory: they flew into the direction they would have taken from the hive if they had not been displaced. The other half flew toward the hive. These latter bees must have changed their motivation (return to the hive) and applied a novel shortcutting flight.

Novel shortcuts are taken as evidence for a form of spatial memory that cannot be explained by an egocentric reference but rather must include a memory of the spatial relations between landmarks. Such a form of navigation is called allocentric. Multiple allocentric navigation strategies exist one of which could be based on a spatial memory that has the structure of a geocentric map, often also called cognitive or mental map [40]. Do bees perform novel shortcuts indicative of a map-like memory structure? To address this question it is necessary to monitor the full flight path of displaced bees because the memory from path integration dominates initial behavior, and bees may switch to an allocentric reference later when they are out of sight. It is also necessary to rule out the possibility that bees could pilot toward a beacon at the goal or perform sequential visual matching with distant cues (panorama) they might have learned at the goal.

2.5.2   Proving Allocentric Navigation

The method of choice for monitoring the full flight path of released bees is harmonic radar tracking [29]. Figure 2.5.1 presents three examples in which the vector flight component is given in red, the search component in blue and the novel shortcutting return flight to the hive in green. Figure 2.5.1a shows the flight path of a foraging bee that was collected at the feeder in the moment it departed from the feeder. Figure 2.5.1b gives the flight path of a bee that was recruited by a dancing foraging bee that danced for a feeder 200 m to the east from the hive. Bees first perform a vector flight if they had learned one either by experience with a distant feeding place (Fig. 2.5.1a) or from dance communication (Fig. 2.5.1b), then they switched to a search phase and initiated a straight return flight to the hive over distances that do not allow them to aim toward the hive with respect to a beacon close to the hive or the panorama of the horizon [26]. Figure 2.5.1c shows the flight path of a foraging bee that collected sucrose solution at a feeder very close to the hive (10 m distance). This feeder circled around the hive. Therefore, these foraging bees did not learn a vector component, and indeed the bee motivated to fly back to the feeder lacks a vector component and searches around the release site first before it switched to the direct home flight.

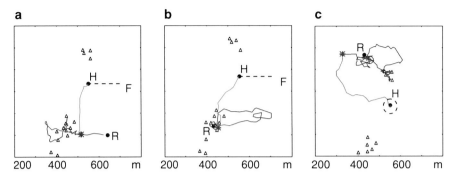

Fig. 2.5.1 (**a**) Flight path of a bee that was trained from the hive (H) to a feeder (F) 200 m to the east of the hive. The bee was captured at departure at F, transported to the release site R, equipped with a radar transponder and released. It first flew according to the vector it would have taken from F if it were not transported (*red line*), then searched (*blue line*) and then flew straight back to the hive (*green line*). (**b**) Radar track of a bee that was recruited by a dancer indicating a feeding place 200 m to the east. The recruited bee was caught when leaving the hive, transported to a release site 300 m south of the hive, equipped with a transponder for harmonic radar tracking and released. It flew first 200 m to the east (*red line*), searched there briefly (*blue line*), returned to the release site (*upper blue line*), and after some searching returned home along a straight flight (*green line*). (**c**) Flight path of a bee that was trained to a feeder close to the hive (10 m, *dotted line*) that rotated around the hive. Thus the bee did not learn a vector between the hive and the feeder. When departing from the feeder the bee was treated the same way as the bee shown in (**a**). (After [26]).The *red star marks* the beginning of the direct return flight according to a definition based on the directedness of the flight path [26]. The *triangles* indicate the locations of colored tents (height 3.7 m) as landmarks

Most interestingly, bees trained to a distant feeder returned home not only by direct flights to the hive but also via the feeder. Figure 2.5.2 shows not only the 10 (out of 29) animals that were published as examples of those displaced foragers that took the route via the feeder [26], but also three more flights that appeared to follow the same strategy, although with less accuracy. The ability to decide between the hive and the feeder as the destination for a homing flight requires some form of relational representation of the two locations. Given that neither of these two locations could be approached with the help of a beacon or the help of the panorama of the horizon as seen from the location where these flights started, it is tempting to conclude that bees indeed perform novel shortcuts and make decisions between potential goals in reference to a map-like structure of their spatial memory. However, one can also argue they may have learned to associate home directed vectors with local landmarks. Such a concept would explain the direct home flights, but an additional process would be required to explain the results shown in Fig. 2.5.2. This additional process may be based on the integration of memory about far ranging vectors, one that leads to the hive from a particular location and one pointing from the hive to the feeder. All these vector operations would have to be made at the level of a working memory in which representations of these vectors are available for integration. It may be argued that such operations at the level of working memory are basically not different from a map-like memory structure. Indeed,

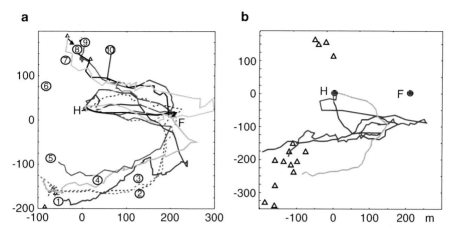

Fig. 2.5.2 Final part of 13 flight tracks of bees that flew back to the hive via an area close to the feeder (*F*). (**a**) 10 (out of 29) flights. One of these bees landed at the feeder. The numbers in (**a**) mark the flight tracks of the 10 animals. (**b**) Three tracks not shown in Menzel et al. [26] that did not quite reach the area of the feeder but performed somewhat similar flights. The tracks in (**a**) and (**b**) give only the final flight part of the respective animals via the feeder back to the hive. The animals were released at different locations either north or south of the hive (see [26])

bees trained to two feeders simultaneously are able to fly novel short cuts from an unexpected site to three locations, each of the two feeders and the hive. It is arguable whether multiple vector operations or a map-like representation are the more parsimonious explanation.

2.5.2.1   Search Flights

Bees perform search flights before heading straight to the far distant goal. What are they searching for? They might either localize their current position by identifying the spatial relationship of close landmarks and retrieving the appropriate trajectory to the distant goal from a map-like memory, or try to recognize a mismatch between the visual appearance of far distant landmarks as seen from the goal and reduce this mismatch gradually. In the latter case no map-like structure of spatial memory is needed, and a rather simple picture matching procedure would suffice. Although the environment in which our radar tracking experiments were performed did not provide such far distant landmarks we further examined the possibility that they still might be able to identify their location by matching distant landmarks. We argued that in this case there should be a tendency to approach local landmarks from a particular direction because a stepwise reduction of visual mismatch may lead to stereotypical flight performances in an attempt to recapitulate the visual appearance of initial learning. We selected the landmarks which we put up in the study area, which were colored tents with the height of 3.5 m. Figure 2.5.3 shows that the tents were approached from all directions. These results are consistent with the interpretation

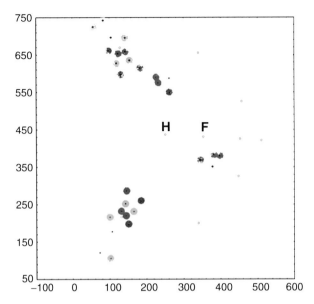

Fig. 2.5.3 Directions of approaches to the tents in our study area. Each tent is marked by a *dot*, and the directions of approaches are indicated by *colored ticks* (different colors for different tents). The angular orientation of each *tick* indicates the direction from which a bee approached the tent. If there is a close colored ring around the *black dot* it means that bees have approached the respective tent from all directions. H marks the position of the hive, F that of the feeder. The other small *grey circles* indicate different release sites

that bees have not learned the tents relative to the appearance of far distant cues. Furthermore, we observed that also the hive is approached over long distances from practically all directions.

2.5.2.2 What Is Learned During Orientation Flights?

Young honey bees on their first exploratory orientations flights need to return safely to their colony. As pointed out above it was assumed so far that path integration mechanisms provide them with the necessary information for safe return. We asked what else young bees learn during their first flights.

Before initiating foraging flights bees also need to learn a range of properties of their environment. These properties relate to the sun compass, the time of day and the local ephemeris function, and possibly they also have to calibrate their visual odometer. In one of the most fascinating series of experiments, von Frisch and Lindauer [43] showed that bees use long-stretching landmarks as guides for sun compass orientation. Later Dyer and Gould [12] called the same phenomenon a backup system for cloudy days and related the connection between sun compass orientation and landmark orientation to a safety system. However, it is more likely that the tight connections between long-stretching landmarks and sun compass need to be seen in the context of calibrating the properties of the sun compass. In either case, long-stretching landmarks are of special importance for the bee.

Local landmark and picture memories are formed during scanning behavior, an elaborate behavior of young bees when leaving the hive for the first time. These first excursions from the hive are followed by orientation flights, bringing the bee in long stretched loops from the hive into the surrounding environment [2, 3]. If only path integration would be applied a young bee captured after return and transported into the explored area should not be able to return home, but we found that they were able to return home along fast and directed flights. In contrast if such a young bee was released over the same distance but into an area it had not explored it needed a long time of searching, or failed. Thus bees learn about the spatial relations of landmarks during their first orientation flights indicating an allocentric form of spatial reference. To test whether this allocentric reference forms a map-like structure of spatial memory we performed an experiment in which bees that had made their orientation flights in different landscapes were tested for their search flight patterns in the same test area. We found that the search flight patterns differed between animals from different landscapes, and that the search flights partially resembled prominent long ranging landmarks that they had learned in their home landscapes. Thus, after experiencing that the test area did not provide the learned landmarks these bees retrieved from their memory the learned prominent landmarks and flew as if they existed in their relation to the sun compass and their geometric layout. Taken together these findings support the conclusion that the allocentric, map-like structure of spatial memory is established during exploratory orientation flights in young bees.

2.5.3   What Do Dancers Report?

Honey bees use various kinds of stereotypical motion patterns for social communication [32, 42]. A bee may shake its body back and forth, rotating it while walking slowly across the comb [31, 35]. This type of motion pattern has been called the 'tremble dance' [41], and appears to help the colony members to coordinate their activities while handling the collected food. In the 'grooming invitation dance', the bee remains stationary and briefly vibrates its body laterally at a frequency of 4–9 Hz, sometimes with alternating brief periods of self-grooming. As a consequence, grooming by nest mates increases [18]. These two dance forms do not convey spatial information. But the round and waggle dances do. Round dances signal food sources in the close vicinity, and may give some indication of the flight direction toward these sources.

In the waggle dance, a dancing bee executes fast and short forward movements in a straight direction on the vertical comb surface, returning in a semicircle in the opposite direction and repeating the cycle in regular. The straight portion of this course, called the waggle run, is emphasized by a combination of lateral waggling of the abdomen and sound pulses (see Chaps. 2.1–2.3). The length of each waggle run and the number of sound pulses emitted increases with the distance flown to reach the source, and their angles relative to gravity correlate with the direction of the outbound foraging flights relative to the sun's azimuth in the field and/or the sun-linked patterns of polarized skylight. Thus, by encoding the visually measured distance and the direction toward the goal, the waggle dance provides vector information toward a desirable goal. But what does the dancer really indicate? This will depend on both the sender (dancer) and receiver (recruit).

Early detour experiments by von Frisch and colleagues (review: [42]) indicated that the bees' distance estimation is decoupled from the actual flight performance, indicating that no global flight vector is reported. These early findings were recently confirmed by manipulating the navigational information experienced by the dancing bee [6] taking advantage of the fact that bees flying in a narrow tunnel with b/w stripes evaluate the distance longer by a factor of >5 (see Chap. 2.1). Thus one might ask whether the waggle dance encodes spatial information provided only by the actual flight path. So far, the role of landmarks and ground structure has been considered only in the context of resetting and calibration [38, 39]. Distances are measured not only by the visual odometer but also by the learned sequential appearance of landmarks [4, 23]. The idea that knowing the landscape influences dance communication is not without precedent. Early experiments showed that with increasing experience of the terrain, directional information available during the inbound flight may also be computed for the waggle dance besides the dominant influence of the outbound flight component [28]. If the waggle dance depends not only on the current state of the animal's path integrator, but also on information that the animal has associated with landmarks, the dance communication process would rely not only on egocentric vector measures but also on some form of allocentric reference system, a proposal that needs to be tested in future experiments.

The waggle dance provides much more than just information about distance and direction. The number of dancing events performed by the dancer varies across dances, possibly reflecting regulatory phenomena that operate between sender and receiver. The strength of the dance depends on: the flow rate [27] and sugar content [42] of the nectar that the dancers bring into the colony; the nectar influx of the whole colony [7, 19, 27, 32] (see also Chap. 2.4); the dancer's past foraging experience [8, 9]; and the nature of the indicated goal, i.e. either a nest site or a food, water or resin source [33]. The distance measure is coded independently of the profitability of the food source [36]. It has even been reported that "danger" at the food source somehow degrades the probability of initiating dancing for a food source [1]. Honey bees also adjust the rate of waggle runs by modifying the duration of the return phase based on specific properties of the indicated goal [34], and most importantly by means of signals derived from their interactions with their fellow mates [7, 19, 20, 27, 30]. These relations allow the dance communication system to be tuned according to the dancer's experience, the particular properties of the indicated goal, the demands of the colony and the availability of resource opportunities. Additional cues, i.e. floral odors, are learned by the recruits, remind them about odors they had learned, and are used to pinpoint the location of the targeted goal [14, 42] (see Chap. 2.4). Thus there is a rich semantics involved in this form of symbolic-like communication.

Taking all these components of dance communication together one may ask whether the waggle dance just transmits information about a motor performance isolated from the knowledge the communicating animals have about the environment. Such a conclusion is suggested by the observation that recruited bees flew according to the vector information after being transported to a release site but did not fly from the release site to the location indicated by the dance [26]. However, it could well be that under such rather unnatural conditions the recruits apply just the vector information stored in working memory as do feeder departing bees when transported to an unexpected release site. Indeed, vector information in both foraging and recruits appears to be of high salience, and dominate behavior initially (see above), but animals may be able to refer to other reference systems after the vector memory did not lead them to the intended goal. Therefore we performed experiments in which the recruited bees were asked whether they compare the dance indicated vector information with their own experience about flight routes to a food source.

2.5.4   Integration of Experienced Flight Routes and Communicated Vectors

In these experiments a group of bees foraged at a feeding site (the trained food site FT) and experienced that FT did not provide any food anymore. As a consequence they gave up foraging at FT and became recruits to two bees performing dances for a food site (the dance indicated food site FD) at the same distance as FT but at either 30° or 60° to FT. As in all other experiments with the harmonic radar we did not use any odor at the food site and the two locations could not be seen by the animals over distances of >50 m and not with the help of the panorama. We found that recruits performed depending on their own foraging experience and on the information transmitted in dance communication. The number of outbound flights to either FT or FD depended on the angular difference between FT and FD. Furthermore, recruits performed a range of novel flight behaviors. In the 30° arrangement some of them deviated from the course toward FD during their outbound flights and crossed over to FT. Most importantly, after arriving at either FD or FT some of them performed cross flights to the respective other location (Fig. 2.5.4). From these observations we conclude that locations FD and FT are both stored in spatial memory in such a way that bees are able to fly directly from one location to the other following a novel shortcut.

We asked whether the decision for FD or FT depended on the number of waggle runs followed by the recruited bee and found that more information is needed by recruits to fly to FD, the dance-indicated location. Bees that followed fewer waggle runs either flew to their experienced feeding site, returned to the hive after a short excursion, or did not leave the hive. Following more waggle runs (in our experiment on average 25 runs) resulted in FD flights indicating that the motivation to apply the information collected about FD is enhanced after longer dance following. However, the information about FD has been learned also during shorter dance following since animals that flew first to FT performed short cut flights from FT to FD (Fig. 2.5.4).

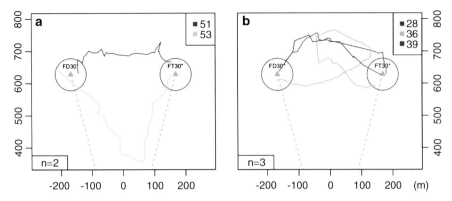

Fig. 2.5.4 (**a, b**) Radar paints of recruited bees in an area close to the locations, FT and FD. The scale on the *left side* gives the distance to the hive. The recruited bees had been foraging for a few days at the feeding site FT. Then they experienced for a day that their feeding site FT did not provide any more food. A day later two other bees danced for feeder FD. The vectors toward FD and FT appeared under an angle of 30° as seen from the hive and had about the same distance (650 m). The FT foraging bees attending the dance were equipped with a transponder when leaving the hive. The radar tracks show two cross flights of recruits after they had arrived at FD (**a**) three cross flights after arriving at FT (**b**)

Obviously dance communication involves two separate components, a motivational and an instructive component, the former requiring less information transfer. The motivational component appears to remind a recruit about its own foraging experience. The signals included in this form of communication are certainly manifold (olfactory, gustatory, acoustic and vibratory, in addition to the specific signals from the waggle runs). It is well documented that floral odors carried by the dancer stimulate recruits to leave the hive, and if the odor reminds them about their own foraging goals they return to these feeding sites ([16, 42], see also Chaps. 2.1, 2.3, and 2.4). Thus floral odors may have a particularly high potential to motivate recruits to take up their own foraging behavior again. Since our experiments did not include any artificial odor marks recruits flying to either FD or FT were guided only by their memory, the recent memory from dance communication or the old memory from their foraging experience. Thus the motivational component can also be triggered by the particular motor components of the waggle runs and the sensory stimuli emanating from a dancing bee which are not specific to the indicated goal.

Given the bees' rich navigational memory one may ask what exactly is communicated by the waggle dance: just the outbound vector or the location of the goal? In the first case the amount of vector information accumulated by the recruit may have to pass a certain threshold before the new vector information can be applied. In the latter case the recruit would compare the expected properties of the indicated location with its own knowledge of this location and other potential foraging options from its own experience before reaching a decision about where to fly. Since we interpret our radar tracking data to document an allocentric navigational memory it is tempting to conclude that vector information from the waggle dance is incorporated into it, and thus it too has an allocentric structure.

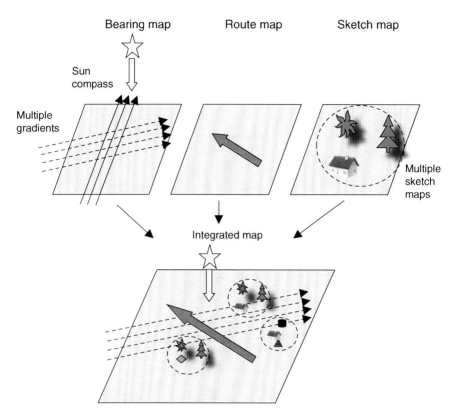

Fig. 2.5.5 Concept of spatial memory structure in the honey bee adopted from Jacobs and Schenk [17]. A coarse map (bearing map) is established by relating far-ranging landscape features (gradients) to the sun compass. In our experiments gradients were irrigation channels, far-ranging ground structures, a tree line and roads. Bees learned these gradients and generalized them when translocated to a different landscape. Isolated islands of sketch maps are thought to be placed into the bearing map and provide snapshot memories of topographic arrangements of landmarks, possibly from a vantage point. Since bees travel along routes multiple times when foraging they also establish a route memory. Route memories combine both gradient and sketch map properties. They are embedded in the sun compass and gradients, replacing one of the other feature if one is missing (e.g. when the sky is fully overcast), they contain an uninterrupted sequence of snapshot memories, and the vector components are both tightly stored in working memory and communicated in the waggle dance

2.5.5 Conclusion and Outlook

The map-like structure of navigation memory in bees may still be incomplete and partial. Jacobs and Schenk [17] developed a concept that may be helpful (Fig. 2.5.5). A coarse map (bearing map) is established by relating far-ranging landscape features (gradients) to the sun compass. Isolated islands of sketch maps are thought to be placed into the bearing map. Such snapshot memories have been studied in great detail at the nest site and the feeding place (e.g. [5]). This concept assumes that the integrated map does not provide the same information at different locations, in fact, it can be full of "white regions" that lack sketch map memories and are characterized only by the coarse gradient map. Evidence for this property is lacking so far and requires additional experiments.

The kinds of questions to be asked in future studies on navigation and communication in honey bees differ from those addressed so far. The sensory-motor routines involved are well understood and they have been analyzed by asking "What can the animal do?". Now we need to ask what is stored in the bee's working memory, how is this information processed, and how are decisions made? To find out we will have to analyze the structure of internal representations. Dance communication provides us with a window into these processes, and carefully designed experiments will allow access to processes beyond behavioral acts. These operations are far from simple and transcend elemental forms of association [25]. The richness of these operations is accessible only in animals acting in their natural environment, and the methods are now available for collecting the relevant data. Ultimately, we want to know how and where the bee's small brain performs these operations – the answer lies in the future.

References

1. Abbott KR, Dukas R (2009) Honeybees consider flower danger in their waggle dance. Anim Behav 78(3):633–635
2. Capaldi EA, Dyer FC (1999) The role of orientation flights on homing performance in honeybees. J Exp Biol 202(Pt 12):1655–1666
3. Capaldi EA, Smith AD, Osborne JL, Fahrbach SE, Farris SM et al (2000) Ontogeny of orientation flight in the honeybee revealed by harmonic radar. Nature 403(6769):537–540
4. Chittka L, Geiger K (1995) Can honeybees count landmarks? Anim Behav 49:159–164
5. Collett TS, Graham P (2004) Animal navigation: path integration, visual landmarks and cognitive maps. Curr Biol 14(12):R475–R477
6. De Marco R, Menzel R (2005) Encoding spatial information in the waggle dance. J Exp Biol 208(Pt 20):3885–3894
7. De Marco RJ (2006) How bees tune their dancing according to their colony's nectar influx: re-examining the role of the food-receivers' 'eagerness'. J Exp Biol 209:421–432
8. De Marco RJ, Farina WM (2001) Changes in food source profitability affect the trophallactic and dance behavior of forager honeybees (Apis mellifera L.). Behav Ecol Sociobiol 50(5):441–449
9. De Marco RJ, Gil M, Farina WM (2005) Does an increase in reward affect the precision of the encoding of directional information in the honeybee waggle dance? J Comp Physiol A 191(5): 413–419
10. Dyer FC (1991) Bees acquire route-based memories but not cognitive maps in a familiar landscape. Anim Behav 41:239–246
11. Dyer FC (1998) Spatial cognition: lessons from central-place foraging insects. In: Balda R, Pepperburg I, Kamil A (eds) Animal cognition in nature. Academic Press, London, pp 119–154
12. Dyer FC, Gould JL (1981) Honey bee orientation: a backup system for cloudy days. Science 214(4524):1041–1042
13. Esch H, Burns J (1996) Distance estimation by foraging honeybees. J Exp Biol 199(Pt 1):155–162

14. Farina WM, Grüter C, Acosta L, Mc Cabe S (2007) Honeybees learn floral odors while receiving nectar from foragers within the hive. Naturwissenschaften 94(1):55–60
15. Gould JL (1986) The local map of honey bees: do insects have cognitive maps? Science 232:861–863
16. Grüter C, Farina WM (2009) The honeybee waggle dance: can we follow the steps? Trends Ecol Evol 24(5):242–247
17. Jacobs LF, Schenk F (2003) Unpacking the cognitive map: the parallel map theory of hippocampal function. Psychol Rev 110(2):285–315
18. Land BB, Seeley TD (2004) The grooming invitation dance of the honey bee. Ethology 110(1):1–10
19. Lindauer M (1949) Über die Einwirkung von Duft- und Geschmacksstoffen sowie anderer Faktoren auf die Tanze der Bienen. Z vergl Physiol 31(3):348–412
20. Lindauer M (1954) Dauertänze im Bienenstock und ihre Beziehung zur Sonnenbahn. Naturwissenschaften 41(21):506–507
21. Menzel R, Brandt R, Gumbert A, Komischke B, Kunze J (2000) Two spatial memories for honeybee navigation. Proc R Soc B 267(1447):961–968
22. Menzel R, Chittka L, Eichmüller S, Geiger K, Peitsch D et al (1990) Dominance of celestial cues over landmarks disproves map-like orientation in honey bees. Z Naturforsch C 45(6):723–726
23. Menzel R, Fuchs J, Nadler L, Weiss B, Kumbischinski N et al (2010) Dominance of the odometer over serial landmark learning in honeybee navigation. Naturwissenschaften 97(8):763–767
24. Menzel R, Geiger K, Chittka L, Joerges J, Kunze J et al (1996) The knowledge base of bee navigation. J Exp Biol 199(1):141–146
25. Menzel R, Giurfa M (2001) Cognitive architecture of a mini-brain: the honeybee. Trends Cogn Sci 5(2):62–71
26. Menzel R, Greggers U, Smith A, Berger S, Brandt R et al (2005) Honey bees navigate according to a map-like spatial memory. Proc Natl Acad Sci USA 102(8):3040–3045
27. Núñez JA (1970) The relationship between sugar flow and foraging and recruiting behaviour of honey bees (Apis mellifera L.). Anim Behav 18:527–538
28. Otto F (1959) Die Bedeutung des Rückfluges für die Richtungs- und Entfernungsangabe der Bienen. Z vergl Physiol 42(4):303–333
29. Riley JR, Smith AD, Reynolds DR, Edwards AS, Osborne JL et al (1996) Tracking bees with harmonic radar. Nature 379(6560):29–30
30. Seeley TD (1986) Social foraging by honeybees – how colonies allocate foragers among patches of flowers. Behav Ecol Sociobiol 19(5):343–354
31. Seeley TD (1992) The tremble dance of the honey bee – message and meanings. Behav Ecol Sociobiol 31(6):375–383
32. Seeley TD (1995) The wisdom of the hive: the social physiology of honey bee colonies. Harvard University Press, Cambridge
33. Seeley TD, Buhrman SC (2001) Nest-site selection in honey bees: how well do swarms implement the "best-of-N" decision rule? Behav Ecol Sociobiol 49(5):416–427
34. Seeley TD, Mikheyev AS, Pagano GJ (2000) Dancing bees tune both duration and rate of waggle-run production in relation to nectar-source profitability. J Comp Physiol A 186(9):813–819
35. Seeley TD, Weidenmüller A, Kuhnholz S (1998) The shaking signal of the honey bee informs workers to prepare for greater activity. Ethology 104(1):10–26
36. Shafir S, Barron AB (2010) Optic flow informs distance but not profitability for honeybees. Proc R Soc B 277(1685):1241–1245
37. Srinivasan MV, Zhang S (2004) Visual motor computations in insects. Annu Rev Neurosci 27:679–696
38. Srinivasan MV, Zhang S, Altwein M, Tautz J (2000) Honeybee navigation: nature and calibration of the "odometer". Science 287(5454):851–853
39. Tautz J, Zhang S, Spaethe J, Brockmann A, Si A et al (2004) Honeybee odometry: performance in varying natural terrain. PLoS Biol 2(7):e211
40. Tolman EC (1948) Cognitive maps in rats and men. Psychol Rev 55(4):189–208

41. von Frisch K (1923) Über die Sprache der Bienen, eine tierphysiologische Untersuchung. Zool
    Jb (Physiol) 40:1–186
42. von Frisch K (1965) Tanzsprache und Orientierung der Bienen. Springer, Heidelberg
43. von Frisch K, Lindauer M (1954) Himmel und Erde in Konkurrenz bei der Orientierung der
    Bienen. Naturwissenschaften 41:245–253
44. Wehner R (1992) Arthropods. In: Papi F (ed) Animal homing. Chapman & Hall, London,
    pp 45–144
45. Wehner R, Menzel R (1990) Do insects have cognitive maps? Annu Rev Neurosci 13:403–414

Chapter 2.6
Communication and Navigation: Commentary

Randolf Menzel

Karl von Frisch's discovery of dance communication in honeybees has fascinated researchers for many decades and will prolong this fascination for many years to come. What will be in the focus of future research on honeybee dances and communication? My suspicion is that priority will be given to determine on the one side what is actually communicated and on the other side which neural mechanisms are involved in integrating the signals that are communicated. The first question relates to the cognitive dimensions of this formidable communication process while the second question refers to their neural underpinning.

Karl von Frisch used the term "dance language", and may have understood this in an allegorical or metaphorical sense, but the term "language" appears frequently in the literature on bee dance communication: "The honeybee dance language, in which foragers perform dances containing information about the distance and direction to food sources, is the quintessential example of symbolic communication in non-primates." (first sentence of the summary in [13]). Premack and Premack [10] stated that the honeybee dances should not be called a language, based on the argument that there is no evidence that the bees can judge whether their dances conform to anything in their surroundings. They also stated that there is also no evidence yet for chain communication whereby an animal picks up on the received information without experiencing itself the primary signals inducing the dance. In his studies of dance communication within a swarm Lindauer [6] did not observe a bee changing its dance pattern until it had actually visited the second cavity, and these observations were verified more recently by Visscher and Camazine [14] who observed no higher attraction of bees to dances which indicated the same location as the one for which they had previously been dancing. However, the authors found that it takes a swarm longer to get started if the decision must be made between alternative nest sites, and

R. Menzel (✉)
Institut für Biologie, Neurobiologie, Freie Universität Berlin, Berlin, Germany
e-mail: menzel@neurobiologie.fu-berlin.de

C.G. Galizia et al. (eds.), *Honeybee Neurobiology and Behavior: A Tribute to Randolf Menzel*, DOI 10.1007/978-94-007-2099-2_10,
© Springer Science+Business Media B.V. 2012

they present arguments for some form of collective "quorum sensing" [12] indicating that some form of "evaluation" of the incoming information is performed by those individuals in the swarm that guide the whole swarm.

Communication codes can be of three kinds, indexical, iconic and symbolic [1, 9]. Signals of an indexical code are directly connected to the object they refer to, such as the odor referring to a flower. Iconic codes rely on similarities between the signal and the object to be communicated, as in the case of pointing with a finger towards a location. Symbolic codes are based on conventions relating signals to objects with no recourse to causal relations or similarity between signal and object. For humans conventions can be fixed explicitly, as for instance in traffic signs, or develop implicitly, as in human language. The bee dance contains undoubtedly indexical components, a property particularly strongly expressed by the odor and taste of the nectar fed to recruits during the dances via trophallactic contacts. Certainly the bee dance contains also iconical components particularly well recognizable in the dance of *Apis florea*, the small Indian bee that dances on the upper horizontal level of the combs hanging on tree branches in the open. In this case, foragers' dances replay both the outbound flight direction relative to the sun as directly experienced during the flight and the distance travelled to the food through the vigor of dance movements [7]. Cave dwelling bees and also *Apis dorsata*, the giant Indian bee which builds combs that hang from tree branches in the open, dance on vertical combs and transfer thereby the directional component derived from the sun-compass in the horizontal to the gravity field in the vertical. Is it justified to assign a symbolic component to this transfer between two sensory systems and dimensions? As said above, symbolic codes emerge from the relation between the rules to read the code, and do not require any similarity between the content of information and its code. Indeed, the transposition between navigation according to celestial cues and waggle runs relative to gravity lacks similarity between the content of information (direction) and its code. However, this transposition may be facilitated because insect menotactic behavior switches quickly between guidance by light and gravity keeping the directional component constant relative to either stimulus component [5]. The conditions in the bee dance are, however, more complex, because the direction of flight can also be extracted from the pattern of polarized light in the sky [11] and the arrangement of extended landmarks experienced during the flight [2, 15]. In both cases, flight direction can be determined without the necessity of directly viewing the sun. Thus the bee dance contains symbolic components though at a low level of complexity.

Given the combined indexical, iconical and symbolic components in dance communication, the essential question refers to the kind of mental states that the communicating bees access when sending and receiving information about a feeding or a nest site. Does the dancer transmit only the motor performances to be applied by the recruit, or does she express her memory about the location of the site in the same geometric reference frame as the recruit? Does the recruit evaluate the information in the context of its own experience? Does the dancer read out the memory of its experience made with the site or does she just convert a stereotypical measure of direction, distance and quality into separate dance parameters? We look at one of these questions from the perspective of the recruit in our chapter on navigation and come to the conclusion that recruits incorporate the message of the dance into their spatial memory in such a way that experienced and communicated spatial information are integrated. A conservative interpretation of these results presumes a common reference frame for locations that are experienced and indicated by the dance [8]. However, we may not exclude at this stage a higher level of cognitive processing in dance communication including some form of evaluation of the received information by the recruit, e.g. deciding between places already insight the hive with respect to the expected outcome of this decision and the economical travel between these two places. The ultimate experimental approach to these and many other questions around the dance communication process will be the dancing robot. The rather mixed experience with attempts in this direction (see the chapters by Esch and Michelsen) tells us already that much more goes on between dancer and recruits than a "simple learning act" as proposed by Harald Esch. So far emphasis has been on perfecting the dancing robot with respect to the stimuli it emits. Possibly it may be equally important to consider the "cognitive" dimension of the communication process as a whole including the motivational and instructive conditions of the receiving bees.

Harald Esch provides us with a wealth of ideas about the dance communication process born from his rich experience with the topic ever since the early 1960s of the last century. These ideas can now be critically tested using the existing harmonic radar and further improvements of this in-flight measuring device, which allows reconstructing entire paths of flying bees in an open field. The improvements require higher temporal resolution leading also to higher spatial resolution and measurements of flight height as well as measurements in more structured and complex landscapes. Spatial resolution of the existing harmonic radar ranges around 10 m, and the height of flight cannot be measured. Improving these aspects will allow addressing the following questions: (1) Do indeed bees fly at higher altitude when flying over longer distances, and does this explain the non–linear distance code in the dance? Observations by eye under optimal conditions do not support such an assumption making it rather unlikely that the shape of the distance function is caused by the decrease in optic flow. (2) Do bees in flight perform regular oscillations (saccades) possibly leading to or reflected in the wagging movements during dancing? Higher spatial/temporal resolutions of the radar system may allow to address this question. (3) Harald Esch suggests that a mixture of four odor components emitted by the dancing bee possibly enhances dance and recruiting performance. Our preliminary experiments applying such a mixture locally to a dancing robot did not suggest that these odor components act as a dance pheromone (Landgraf and Kirbach, personal communication). (4) Is dance communication a "simple learning act" "not requiring complex cognitive functions"? Does the dance just transmit procedural information by presenting a kind of a "symbolic flight"? As pointed out above, the answer to these questions will come from a better understanding of the navigation memory referred to by both the dancer and the recruit. We need to ask whether just flight vectors are communicated or qualified locations that are evaluated both by the dancer and the recruit on the basis of their respective memories. Attempts in this direction have been unsatisfactory because of methodological limitations (e.g. the bees where not tracked during their flights, odor needed to be used to induce landing at test sites, the experience of recruits before dance following was not known in most cases or not quantified), [4, 16]. At least simple reward learning of procedural information based on reinforcing particular dance components through throphallactic sucrose delivery to recruits is unlikely because we found (see our chapter on navigation) that not all recruits receive regurgitated food, and that "unrewarded" recruits perform similarly as "rewarded" ones with respect to novel short cuttings. Furthermore, recruits perform usually trophallaxis with the dancer before attending the dance.

The still unresolved question is whether Axel Michelson's robot recruited bees successfully. The experience of the bees landing at the test stations after attending robot dances was unknown, odor needed to be used as baits at the test sites, and only bees landing at these baited sites were counted. Ten percent efficiency (as compared to real bees dancing) is a very low number indeed, and a typical following behavior of bees around the robot could not be detected. It is quite possible that the robot provided some form of motivational signal but lacked or provided only partial instructive components informing the bee where to go. In my view the only way of clarifying these issues is to follow the flight path of individual recruited bees under conditions in which their experience is known and odors are avoided.

Dance communication occurs between individuals. The decisions of the individuals have global consequences as exemplified so instructively by Tom Seeley. These global consequences lead to community effects, a phenomenon well known for any communication process in social groups. Although in evolutionary terms the colony represents the relevant unit, in functional terms the individuals are the units of information transfer. In my view there is no collective decision making in dance communication at least not in the context of food foraging. What counts here are the decisions of the individuals and the group phenomena emerge from the coordinated performance of individuals.

Multiple questions arise with respect to the decision making process of the individual bee. For example, why are some recruits motivated to fly back to a formerly visited feeding site and others to follow the instruction about distance and direction and choose to fly to the dance indicated place? Biesmeijer and de Vries (2001) (cited in the chapter by Seeley) distinguish between reactivated foragers and inspectors, recruits, and scouts (the latter an unfortunate term since it was introduced by Karl von Frisch in just the opposite meaning: scouts in von Frisch's terminology are those that collect novel information and not food), and give the impression that there may be subgroups of foragers like e.g. pollen and nectar foragers. We found in our study that the switch between these behaviors depends on the amount of information received during dance communication. Recruits with the same experience at their exhausted feeding site were motivated to visit their old place if they attended less dance rounds than recruits that attended more dance rounds. Furthermore, and in our view most importantly, the difference between the own experience and the spatial information communicated via the dance influenced the decision process of the recruits. Thus the balance between the motivational and instructive components of the waggle dance depends on several parameters, and practically all of them relate to the former experience of the individual bee receiving the information. Even in the case of a young bee without foraging experience innate search images will guide their choices in the field as nicely shown for color choices [3]. Odors are learned insight the hive by young bees (see chapter by Farina et al.) and add to the search image that is likely to control the balance between motivational and instructive component already in the "novice" forager. One could postulate "a knowledge of the colony", but that is a construct without a corresponding mechanism because the knowledge houses in the brain of the individual bee. Whether a food source is poor or rich, whether the odor emitting from the dancer indicates an attractive food source, whether the indicated location is close to an already visited place, all of these and many more qualifications depend on the information stored in the brain of the receiving individual bee. Therefore, what we need to know is how decisions are made by a single bee during the communication process. My suspicion is that the community effects will simply emerge from these rules.

From a learning perspective odors are highly salient cues, they are learned particularly fast, shifted to long term memory quickly and generalized across different contexts. Odors are reliable cues in the close vicinity of the odor source but become less reliable as far distant cues because of changing environmental conditions (wind speed and direction). Therefore, spatial information can be reliably connected to odor distribution over short distances in constant or zero air flow conditions. The hive is such an environment. It would be interesting to know how much of the within-hive locations depend on local odor cues and whether bees take advantage of such distributions for the orientation within the 3D environment of the hive. Furthermore, how quickly do bees learn changes in the odor distribution?

Odors are very hard to characterize and measure. This is particularly true for long-chain hydrocarbons of contact odors embedded in the cuticle of the body and the wax of the comb. How well do bees discriminate such odors, and what are the thresholds for the detection and spatial separation? These questions are of particular relevance for social organization including discrimination between subgroups or even individuals, and the detection of infected brood. The olfactory conditioning paradigm of the proboscis extension response offers a useful tool to study such questions but comes to its limits when contact chemicals have to be tested. Often just the mechanical contact with the odor carrying substrate releases proboscis extension in a hungry bee making it impossible to train and test odor detection and discrimination. Also Ca^{2+} imaging has not yet been helpful to determine possible regions of the antennal lobe devoted to the processing of contact chemicals, possibly because contact chemoreception does not involve the glomeruli of the lACT accessible from the front part of the bee brain. It will be important to include imaging of the mushroom body lip region in such studies because there it should be possible to reach also the projections of the mACT.

Given the situation of odors being an unreliable cue for far distance orientation in the environment one may ask how unreliable cues are learned and used in decision making. Since odor detection in the open is related to wind direction and wind direction will be measured relative to the sun compass and landmarks it is possible that an odor cue combined with a particular wind direction forms a compound that needs to be experienced as a whole. Therefore, the least one needs to do in any training experiment using odors as a cue is to measure continuously wind direction and wind speed. Tracking the flight path will allow in addition identifying special behavioral routines to cope with the changing wind and olfactory conditions.

References

1. Bierwisch M (2008) Der Tanz der Symbole – Wie die Sprache die Welt berechnet. In: Fink H (ed) Neuronen im Gespräch, Sprache und Gehirn. Mentis, Paderborn, pp 17–45
2. Dyer FC, Gould JL (1981) Honey bee orientation: a backup system for cloudy days. Science 214(4524):1041–1042
3. Giurfa M, Núñez J, Chittka L, Menzel R (1995) Color preferences of flower-naive honeybees. J Comp Physiol A 177(3):247–259
4. Gould JL, Gould CG (1982) The insect mind: physics or metaphysics? In: Griffin DR (ed) Animal mind-human mind. Springer, New York, pp 269–298
5. Jander R, Jander U (1970) Über die Phylogenie der Geotaxis innerhalb der Bienen (Apoidea). Z vergl Physiol 66:355–368
6. Lindauer M (1955) Schwarmbienen auf Wohnungssuche. Z vergl Physiol 37(4):263–324
7. Lindauer M (1956) Über die Verstandigung bei indischen Bienen. Z vergl Physiol 38(6): 521–557
8. Menzel R, Kirbach A, Haass WD, Fischer B, Fuchs J et al (2011) A common frame of reference for learned and communicated vectors in honeybee navigation. Curr Biol 21(8):645–650
9. Peirce CS (1931) Collected papers. Harvard University Press, Cambridge
10. Premack D, Premack AJ (1983) The mind of the ape. Norton, New York
11. Rossel S, Wehner R (1982) The bee's map of the e-vector pattern in the sky. Proc Natl Acad Sci USA 79:4451–4455
12. Seeley TD, Visscher PK (2004) Quorum sensing during nest-site selection by honeybee swarms. Behav Ecol Sociobiol 56(6):594–601
13. Sherman G, Visscher PK (2002) Honeybee colonies achieve fitness through dancing. Nature 419(6910):920–922
14. Visscher PK, Camazine S (1999) Collective decisions and cognition in bees. Nature 397:400
15. von Frisch K (1967) The dance language and orientation of bees. Harvard University Press, Cambridge
16. Wray MK, Klein BA, Mattila HR, Seeley TD (2008) Honeybees do not reject dances for 'implausible' locations: reconsidering the evidence for cognitive maps in insects. Anim Behav 76:261–269

Part III
Brain Anatomy and Physiology

Chapter 3.1
The Digital Honey Bee Brain Atlas

Jürgen Rybak

Abstract For a comprehensive understanding of brain function, compiling data from a range of experiments is necessary. Digital brain atlases provide useful reference systems at the interface of neuroanatomy, neurophysiology, behavioral biology and neuroinformatics. Insect brains are particularly useful because they constitute complete three-dimensional references for the integration of morphological and functional data. Image acquisition is favored by small sized brains permitting whole brain scans using confocal microscopy. Insect brain atlases thus serve different purposes, e.g. quantitative volume analyses of brain neuropils for studying closely related species, developmental processes and neuronal plasticity; documenting and storing the Gestalt and spatial relations of neurons, neural networks and neuropils; structuring large amounts of anatomical and physiological data, thus providing a repository for data sharing among researchers. This chapter focuses on the spatial relations of neurons in the honey bee brain using the Honey bee Standard Brain (HSB). The integration of neurons into the HSB requires standardized image processing, computer algorithms and protocols that aid reconstruction and visualization. A statistical shape model has been developed in order to facilitate the segmentation process. Examples from the olfactory and mechanosensory pathways in the bee brain and the organization of the mushroom bodies (MBs) are used to illustrate the implementation and strength of the HSB. An outline will be given for the use of the brain atlas to link semantic information (e.g. from physiology, biochemistry, genetics) and neuronal morphology.

J. Rybak (✉)
Department of Evolutionary Neuroethology, Max-Planck-Institute for Chemical Ecology,
Hans-Knöll Strasse 8, D-07745 Jena, Germany
e-mail: jrybak@ice.mpg.de

C.G. Galizia et al. (eds.), *Honeybee Neurobiology and Behavior: A Tribute to Randolf Menzel*, DOI 10.1007/978-94-007-2099-2_11,
© Springer Science+Business Media B.V. 2012

Abbreviations (Excluding Brain Areas)

GABA   Gamma-aminobutyric acid (neurotransmitter)
HSB   Honey bee Standard Brain
ISA   Iterative Shape Averaging
SSM   Statistical shape model
VIB   Virtual insect brain

3.1.1 Introduction

Since the pioneering work of Camillo Golgi and Ramon y Cajal on the fine structure of the nervous system numerous anatomical studies on the honeybee brain was performed [18, 31, 50]. These classical studies provide the basis for state-of-the-art anatomical, physiological and molecular studies. Such data have to be related to the brain structures, and to the morphology of single neurons and neuronal networks. At the same time progress in computer science and neuroinformatics made it possible to create brain atlases as digital databases aiming to organize and visualize experimental data from different sources [3, 6, 9].

Digital anatomical atlases provide a three-dimensional (3D) map, and a coordinate system that contains information about the relative locations of neurons (and networks) within the brain. They provide a scaffold, that eventually can tie semantic (e.g. bibliographical or experimental data from other sources than anatomy) with the spatial information about the brain compartments and its neural components [49]. A further advantage is the computer readability of digitized atlases [4], i.e. the possibility to index anatomical structures in a controlled vocabulary (nomenclature), designing an ontology and allowing an interactive search through the 3D database as well as cross references to other databases [5, 23, 26]. Requirements for digital atlases are spatial normalization procedures that align individual brains to a template and a common coordinate system (e.g. the Talairach stereotactic atlas for the human brain: [47]), and thus detect and account for structural variability among specimen or eliminate differences [6, 21, 38, 48].

In insects, confocal microscopy allows for whole-brain scans and fast data acquisition. Single neurons can be identified in 3D and provide the anatomical substrate for the identified neuron concept relating functional properties of the whole animal to single neurons [29, 30]. Studies that focus on the neuronal correlates underlying neuronal processing, learning and memory are reviewed in detail in other chapters of the book. Parts of the relevant networks have been integrated in to the Honey bee Standard Brain (HSB) providing a framework for further studies directed towards a more detailed map of the functional organization of the bee brain ([3, 6, 19, 20, 41], review: [30]).

3.1.2   Brain Atlases

The use of anatomical atlases has a long tradition in neuroscience. Historically, composed as paper-based atlases [14], they were used to describe the location of anatomical structures, bring them into a standardized reference coordinate system using anatomical landmarks and nominal conventions and are often used as stereotactic tools as well as a guidance to plan and interpret experimental results (e.g. insects: [45]; humans: [47]). However, these atlases were obtained through manual generalization of many brains by the researcher. 3D digitized, forms of these atlases were possible through fast, computer based data acquisition in neuroimaging (magnetic resonance imaging (MRI), Positron emission tomography (PET), optical imaging, confocal microscopy), new neuron tracer techniques, and computer algorithms in neuroinformatics that allow to handle and manipulate digital data in virtual space [37, 48]. Digital atlases are often web-based, rapidly searchable and can be used interactively for a 3D-visualization of data [4, 41]. For example, the Allen Brain Atlases are interactive, multimodal image databases of gene expression for the mouse and the human brain which relate genetic expression data mapped onto high-resolution histological scans of cross sections of the brain ([16], http://www.brain-map.org).

3.1.3   Standard Brain Atlases in Insects

Insect standard atlases were generated as population-based reference systems. They combine the features of multiple specimen in order to generate a representative atlas [6, 10, 38], (see also the special edition 'Digital Brain atlases' in Frontiers of System Neuroscience, 2010). In all these approaches a standardized methodology: synaptic staining of neuropil, confocal microscopy on whole-mount brain, and standardized protocols from neuroinformatics for handling the digital data, were used allowing the integration of neuronal structures with high accuracy [8, 41].

Standard atlases come in several forms, depending on the question or purpose pursued in using them. For insect brains, two standardization methods of a given species or sex have been employed: (1) The virtual insect brain (VIB) protocol allows a comparative volume analysis of brain neuropils, developmental studies, and studies on neuronal plasticity [9, 21, 38]. (2) The iterative Shape Averaging (ISA) procedure eliminates specimen shape variability [6, 24, 39] to accumulate structural data in the reference space of the Standard Brain Atlas. In comparison to the VIB protocol, which is based on selecting an individual representative brain (*Drosophila* Standard Brain, [38]), the ISA averaging procedure is best suited for the registration of neurons collected from different brains; [9]. This is the procedure applied for the HSB (Fig. 3.1.1)

Neuro-informatics provides computer algorithms for image processing, visualization and integration of digitized anatomical data (bioimage informatics: [37]). The workflow, or pipeline, for the incorporation of neuronal morphologies into brain

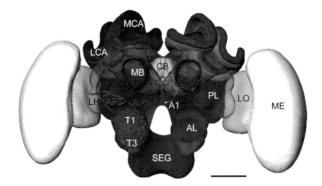

Fig. 3.1.1 Surface model of the Honey bee Standard Brain (*HSB*). Neuropiles of the midbrain, mushroom bodies (*MB*), protocerebral lobes (*PL*) and the subesophageal ganglion (*SEG*) are shown in transparency. Neuronal elements at different level of resolution are integrated to the HSB: antennal lobe (*AL*) glomeruli groups of sensory tracts T1, T3, the antennal lobe-protocerebral tracts (*APT*) and a single mushroom body extrinsic neuron type A1. Scale: 250 μm. *MCA* median calyx, *LCA* lateral calyx, *LH* lateral horn, *CB* central body, *LO* lobula, *ME* medulla

atlases has to be standardized to a high degree in order to make the representation of data accessible to investigators from different labs [4, 15, 28, 41], Fiji: http://pacific.mpi-cbg.de/wiki/index.php/Fiji.

3.1.3.1   The Honey Bee Standard Brain

The HSB was created as an average-shape atlas comprising 22 neuropils, calculated from 20 individual immunostained worker bee brains. The delineation of major neuropil borders by semi-automatic manual segmentation defined the brain compartments creating digitized 3D regions. The averaging method includes correction for the global size and positioning differences of the individual brains by repeatedly applying linear and nonlinear registration algorithm. This registration (or matching) process involves geometric transformations like translation and rotation (affine or linear transformation) as well as local deformations (non-linear or elastic transformation) and results in a stack of average label images [6, 39].

3.1.3.2   Registration of Single Neurons into the HSB
and the Statistical Shape Model

The spatial transformation of neurons into the HSB is a four-step procedure: (1) The manual reconstruction of the neuron, facilitated by an automatic extraction of the neuron's skeleton based on threshold segmentation (examples are given in Fig. 3.1.2),

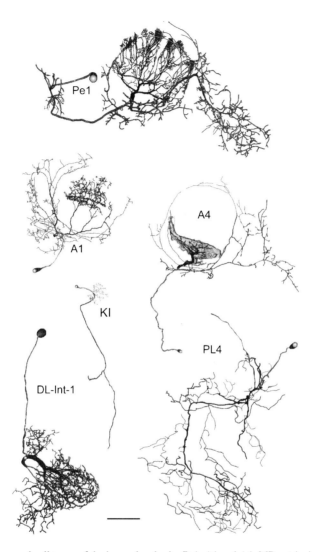

Fig. 3.1.2 Neuronal cell types of the honey bee brain. Pe1, A1 and A4: MB extrinsic neurons. *KI*
Kenyon cell type 1 of the collar region of the MB calyces; *DL-Int-1* mechanosensory dorsal lobe
interneuron 1; *PL4*: protocerebral interneuron 4. Note, that the colored subparts of the neurons
correspond to respective brain compartments (neuropils) of the honey bee brain. *green*: protocerebral lobe, *red*: pedunculus-lobe of the MB, *yellow*: calyx of the mushroom body, *blue*: antenno
mechanosensory motor center (AMMC). Scale 100 μm. Data courtesy of Alvar Prönneke (for
PL4) and Hiro Ai (for DL-Int-1)

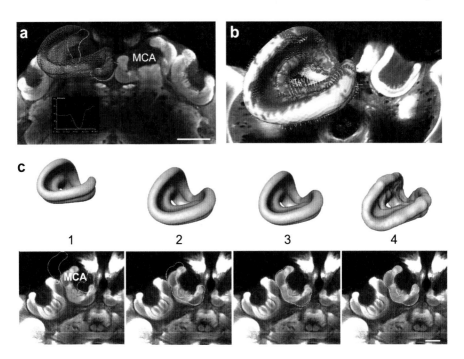

Fig. 3.1.3 Autosegmentation of the right median calyx of the mushroom bodies. (**a**) Confocal image with synaptic antibody staining and a surface view of the statistical shape model of the right median calyx (*MCA*) in transparent colors, *red* line: 1D *grey* value profile along a surface normal (see inset: corresponding intensity profile). (**b**) Displacement visualization of the calyx model: non-linear, elastic deformations that point outwards (in *blue*) and inwards (in *red*) and displacement vectors (*yellow arrows*) are shown. (**c**) Automatic segmentation process: (1) initial positioning of the calyx model. (2) transformation of the model by affine registration of the image to a reference dataset. (3) rigid transformation and (4) elastic deformation of the model by optimization of position and shape parameters using local 1D grey value profiles (as seen in **a**) and inset (**a**). Scale in (**a**) 200 μm, in (**c**) 100 μm (Adapted from [35].)

(2) the 3D segmentation of brain neuropils containing the neuron, (3) their registration with the HSB, (4) the integration of the neuron into the HSB by applying the transformation coordinates computed in step (3) to the reconstructed neuron of step (1). Thus, in order to integrate the structural data of neurons into the HSB the neuropil boundaries of the stained brain have to be registered into the three-dimensional space of the HSB (labelfield registration). Detailed protocols for the pipeline are given in [22, 28, 41].

The accurate and reliable localization of neuropil boundaries during the registration process is a prerequisite for integrating neurons with high accuracy into any standard atlas [9, 21, 41]. Therefore, in order to facilitate and standardize the segmentation process a statistical shape model (SSM) was developed based on a method for automatic segmentation of medical imaging data (e.g. [25]). The model-based auto-segmentation of neuropil boundaries utilizes *a priori* knowledge about the 3D shape of an object, here, the bee brain neuropils, and characteristic features of the confocal imaging data (grey-value intensity profiles of the confocal images) [35, 41]. The general ideas is to roughly position a SSM, or part of it, in the imaging data of the current study case and subsequently vary the shape parameters and the spatial location of the model until the SSM matches the object (i.e. the neuropil boundary) in the imaging data as closely as possible (exemplified for the median calyx of the mushroom bodies (MCA) in Fig. 3.1.3).

3.1.4   Landmark Registration of Other Structural Modalities

A further challenge is to map anatomical data obtained by different techniques, for example digitized histological sections containing morphological information of the overall brain organization and ultrastructural datasets into the HSB. In Fig. 3.1.4

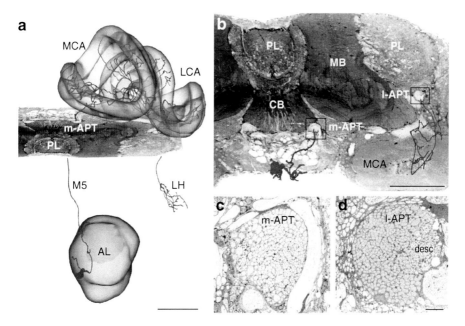

Fig. 3.1.4 Neural architecture of the central brain. (**a**) Shown are a HSB surface model of left mushroom body calyces (*MCA, LCA*) and the left antennal lobe (*AL*). A horizontal section of an ethyl gallate staining (landmark registration) is mapped together in the HSB space with a uniglomerular projection neuron (M5) connecting the AL to the mushroom body calyces (*MCA, LCA*) and lateral horn (*LH*) (labelfield registration) (**b**) antennal lobe-protocerebral tracts (m-APT, l-APT) are identified in the ethyl gallate section. The section corresponds to that seen in (**a**), viewed from below. The M5 axon runs in the m-APT and its descending axon in the l-APT (*rectangular boxes*). View of the M5 neuron in the MB calyx is concealed by the ethyl gallate section. The spatial locations of the two modalities (ethylgallate, digital reconstruction of the M5 neuron) demonstrate the spatial accuracy of fitting neuronal data from different sources registered to the HSB. (**c**), (**d**) An electron microscopic image (different preparation) depicts the number of axons in the m-APT (app. 410 axon profiles) and l-APT (app. 510 profiles), respectively, at the corresponding depth of the brain (see boxes in (**b**). desc: descending axon collaterals of the m-APT. Scale: A: 100 μm, (**b**): 200 μm, (**c**): 5 μm. Data courtesy of Sabine Krofzcik (for M5 neuron). (**b**), (**c**): Adapted from [41] (**b**), and [40] (**c**). *CB* central body, *m-APT, l-APT* median, lateral antenno-protocerebralis tract, *PL* protocerebral lobe

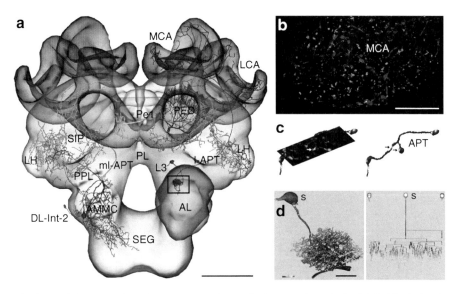

Fig. 3.1.5 Olfactory and mechanosensory pathways in the central bee brain. (**a**) *Left* brain hemisphere: The L3 projection neuron (*red*) projects to the lateral horn (*LH*) and mushroom body calyces (*MCA, LCA*). The mushroom body output neuron (Pe1) with dendritic input in the pedunculus (*PED*) (*red*) and output targets in the protocerebral lobe (*PL*) (*green*) overlaps with L3 in the LH. Right hemisphere: a ml-APT neuron (*red and green*) and the DL-Int-2 (*blue*) targeting disperse regions in the protocerebral lobe (*PL*) and subesophageal ganglion (*SEG*). (**b**) Confocal microscopy analysis allows the mapping of GABA-immunoreactive profiles (green laser channel) onto axonal terminals of APT projection neurons (red laser channel). (**c**) Bright spots (*arrows*) indicate a high probability of close attached putative GABA-immunoreactive synaptic contacts on a axonal terminal of an antennal lobe-protocerebral tract neuron (APT). (**d**) The glomerular arborization of the L3 neuron (inset in (**a**)) at high resolution (*left*) and the corresponding dendrogram (*right*). Neuronal distance measured from the L3 soma (s) is false-color coded. PPL: posterior protocerebral lobe, SIP: ring neuropil of the vertical lobe (superior intermediate protocerebrum), AMMC: antennal mechanosensory and motor center Scale: 200 μm in (**a**), 20 μm in (**b**) and (**d**). Data courtesy of Daniel Münch and Gisela Manz (for L3 and ml-APT neuron)

an olfactory projection neuron of the m-APT(M5) is visualized together with an histological section stained by ethyl gallate showing central brain regions (MBs, central body (CB), and protocerebral lobe (PL)). The ethyl gallate method provides detailed information about neuropils, somata, and fascicles, thus capturing spatial context information of brain structures. A single section was registered into the HSB by finding corresponding points or landmarks in the HSB and the histology section (landmark registration). Separately, the single stained and reconstructed M5 neuron was transformed to the HSB using a label field registration (see: last preceding paragraph). Figure 3.1.4a-d shows the composition in the HSB: the M5 axon runs through the corresponding ascending and descending parts of the median APT (m-APT, m-APT-desz). Ethyl gallate stained sections were used to identify the median and lateral antennal lobe-protocerebral tract (m- and l-APTs) in a light and electron microscopy study revealing app. 410 and 510 axons, respectively, for each tract [40] (Fig. 3.1.4c, d). A future goal will be to integrate data from histological procedures into the HSB by mapping whole image stacks onto the HSB.

Registration of neurons into the HSB allows the visualization and assembly of neuronal networks gained from different experiments and researchers [3, 6, 41]. Additionally, an iterative procedure allows composing networks of registered neurons at different levels of resolution, for example, the target areas of olfactory interneurons in subregions of the mushroom body calyces or their overlapping projection with a mushroom body extrinsic neuron, the Pe1, in the lateral horn (LH) (Fig. 3.1.5a). High-resolution scans deriving from the same study can thus be fitted with high accuracy into the HSB (Fig. 3.1.5a, d).

Fitting neurons into the HSB can be achieved with a certain degree of accuracy with regard to spatial relationships, but cannot replace ultrastructural studies on synaptic connectivity. Approximation using co-localization studies on the light microscopy level is an alternative (Fig. 3.1.5b). Combining high-resolution confocal microscopy with precise three-dimensional dendritic surface reconstruction allows for automated co-localization analysis in order to map the distribution of potential synaptic contacts on axon terminals (Fig. 3.1.5b, c). This procedure was also used to estimate the distribution of putative GABAergic synaptic contacts on the dendrites of the Pe1 [36]. Note that synaptic contacts mapped in this way can be registered to the HSB, and thus information is stored in the HSB repository.

3.1.5   Neural Organization of the Central Bee Brain

3.1.5.1   The Olfactory Pathway

Axons of antennal olfactory receptors form four tracts (T1 to T4) and converge in app. 160 glomeruli in the antennal lobe (AL), where they feed a network of local interneurons and projection neurons (PN) [1, 10, 19, 31]. Mechanosensory neurons running in tracts T5 and T6 bypass the AL and target the antenno-mechanosensory motor center (AMMC), the posterior protocerebral lobe (PPL), and the subesophageal ganglion (SEG) where they overlap with mechanosensory interneurons and fiber tracts originating in the optic lobes (OLs) [3, 27]. Multiglomerular projection neurons (mPNs) run via the mediolateral antennal lobe-protocerebral tracts (ml-APT 1–3) into the lateral, median and posterior protocerebral lobe. In contrast, uniglomerular projection neurons (uPN) constitute a dual pathway leaving the AL via the median- and lateral APT. They target subregions of MB calyces (CAs) and the LH (Fig. 3.1.5) ([33], reviews: [11, 43], see also Chap. 4.1).

The protocerebral lobes (PL) comprise the MBs, the central body (CB), and a large tangled neuropil characterized by interneuronal tracts and more densely packed subcompartments formed by overlapping arborizations of output neurons from the AL, the OLs, and mushroom body extrinsic neurons.

Olfactory information is predominantly processed ipsilaterally with two exceptions: (1) A class of projection neurons of the AL T4 region project via the m-APT

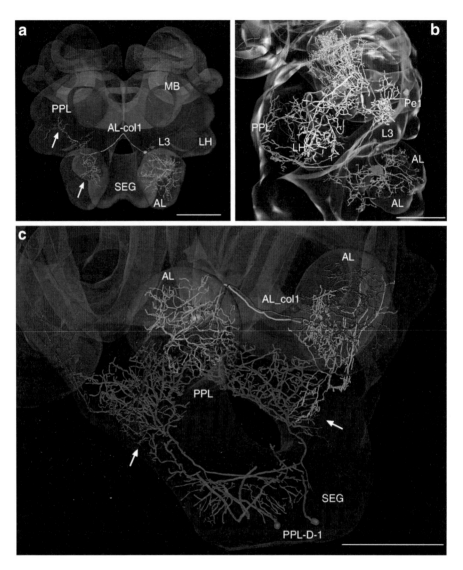

Fig. 3.1.6 Olfactory and mechanosensory neuronal circuitry of the protocerebral lobe. (**a**) The antennal lobe neuron, AL-col1, connects the glomeruli of the antennal lobe (*AL*) with the coarse neuropil of the contralateral AL and the contralateral posterior protocerebral lobe (*PPL*) (*arrows*). As a comparison the uniglomerular projection neuron L3 is depicted. (**b**) Comparison of target areas of mushroom body output neuron, Pe1 (*yellow*), projection neuron L3 (*blue*) and the AL-col1 (*green*) in the lateral protocerebral lobe (LH and PPL). (**c**) overlapping arborizations of the AL-col1 neuron and the mechanosensory posterior protocerebral interneuron (PPL-D-1) in the posterior PL (*PPL*) (*arrows*). Both neuron types are mirror imaged. (for the PPL-D1: see also Chap.4.3). Scales: 250 μm

towards the basal ring and lip region of the MB calyces on both brain sides, and send collaterals to the lateral protocerebral lobe [1]. (2) A multiglomerular neuron (AL-col1) connects both ALs and projects to the contralateral protocerebral lobe excluding the LH. Al-col1 exhibits arborizations in the cortex region of almost all glomeruli of the ipsilateral AL and projects to the central coarse neuropil, but not to the glomerular neuropil, of the contralateral AL (Fig. 3.1.6).

As shown in Fig. 3.1.6a-c the HSB can be used to analyze the topographic relationship between the AL projection neurons, mushroom body extrinsic neurons (Pe1) and protocerebral interneurones. The spatial analysis shows overlapping areas of mechanosensory neurons with the AL-col1 in the posterior protocerebral lobe (Fig. 3.1.6c). The mechanosensory interneuron PPL-D-1 was previously reported by Ai [2] (see also Chap.4.3). It responds to vibratory signals detected by the Johnston organ as well as to olfactory stimuli.

3.1.5.2 *Topography of the Mushroom Bodies*

The honey bee MBs consist of double calyces and a parallel system of axons forming the pedunculus (PED) and lobes [50]. The large volume of the calyces with numerous intrinsic Kenyon cells (KCs) [18] are particular pronounced in social hymenoptera and seem to be correlated with social behavior and learning (review: [30]). The calyces are supplied by tracts from sensory neuropils, and from the SEG. The pedunculi and lobes (alpha and beta or vertical and medial lobe) receive inputs and provide outputs via extrinsic neurons [42]. The calyces are divided into the lip, collar and basal ring [31, 43, 46].

The ordered arrangement of strictly parallel KCs and their respective neighborhood relations leads to so-called corresponding zones, i.e. the calycal dendritic arborizations of the KCs from type I, restricted to a certain calyx (CA) region (lip, collar, or basal ring) correspond to respective zones occupied by their axonal projections in the peduncle, the alpha and the beta-lobe. For example, lip K I cells are represented in a horizontal, medial layer of the alpha and beta-lobe respectively (Fig. 3.1.7a), whereas the basal ring KCs are represented in the dorsal layer of the alpha-lobe. In contrast, narrow-banded KII cells with claw-like arborizations in the CA and their somata located outside the CA neuropil occupy all regions of the CA and project to the ventral part of the alpha-lobe and the anterior beta-lobe, where modality specific input is not maintained [42, 43, 46]. Recurrent MB extrinsic neurons (type A3) with somata position, number of cells (app. 120) and morphology resembling the cluster of GABA-IR neurons described by Schäfer and Bicker [44] arborise at all levels of the lobes and peduncle and either constitute a feedback to the calyces, or they may interconnect peduncle-lobes without innervating the CA [13]. An example of using the HSB for allocation and documenting topographical relations of extrinsic and intrinsic elements in the MB is shown in Fig. 3.1.7a. A Kenyon cell, type I, with dendritic fields in the collar of the posterior lateral calyx (LCA) send axons through

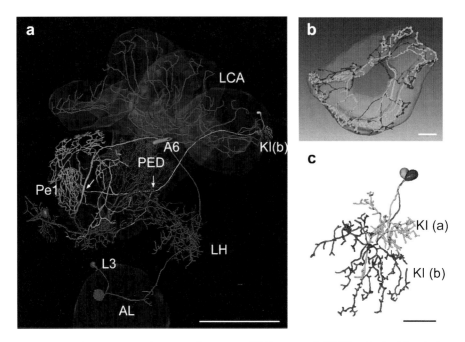

Fig. 3.1.7 MB topography. (**a**) an axon from a type KI Kenyon cell (KI(b)) with dendrites in the posterior lateral calyx (*LCA*) project into the pedunculus (*PED*) and lobes to corresponding zones of a Pe1 cell (*red*) and a A6 extrinsic neurons (*green*) (*arrows*). The L3 neuron projection neuron overlaps with the KI dendrites in the collar region of the LCA and the output terminals of the PE1 in the lateral horn (LH). The HSB is shown in transparency. Scale 200 μm. (**b**) The axonal terminals of single olfactory projection neuron of the m-APT (*red*) and l-APT (*green*) form micro domains in the lip region of the calyces. Scale: 50 μm, Modified after: [32]. (**c**) two distinct dendritic domains of Kenyon cells: 1 KI(a), KI(b) depicted at high resolution, Scale: 20 μm. Data courtesy of Ruth Bartels (for A6 neuron)

the peduncle to the medial alpha- and beta-lobe. They are overlapping at certain depths and zones within the pedunculus-lobe system with the Pe1 and the A6 extrinsic neurons [42].

3.1.5.3 Microcircuits of the Mushroom Bodies

Fibers from the sensory neuropils and SEG occupy distinct regions in the calyces. Modulatory neurons are widely distributed and immunoreactive to antibodies to octopamin (OA), serotonin, and dopamine (DA) (review: [43]). Olfactory projection neurons occupy distinct domains in the circumference of each CA in the lip region and zones in the basal ring depending on their type. Single m- or l-APT neurons differ in their width of zonal expansions within the CA [1, 19, 32]. As indicated in Fig. 3.1.7a–c assembling morphological data from extrinsic cell types and intrinsic

KCs integrated to the HSB from different experiments help to define microcircuit domains in the mushroom body. Ultrastructural studies revealed the organization of the microglomerular microcircuts indicating that the large diameter boutons of projection neurons are surrounded by small postsynaptic profiles of KCs and dendrites of A3 neurons, the latter being both pre- and postsynaptic to KCs ([12, 40], see also Chap. 3.2).

3.1.6   Outlook, Ontology's and Databases

The motivation for creating the HSB was to pool morphological data in a reference system which eventually will represent a biologically realistic model of the bee brain. Such a 3D model takes into account the whole entity of neurons and neural networks at different levels of neural organization. The HSB is primarily used for visual inspection of the neuroanatomical data and as a neuron knowledge base. The HSB also functions as data resource for sharing experimental data from different research groups. By means of a searchable web-based 3D data repository the user can explore and compare spatial relationships of neurons and relate this to bibliographical information of published results ([3, 6, 19, 41], http://www.neurobiologie.fu-berlin.de/beebrain).

Biological variability is an important issue to be considered when creating and using brain atlases. Specially designed atlases using the VIB approach have been proposed in order to detect and quantify structural differences in brain organization [9, 15, 21, 38]. The statistical shape model of the honey bee brain [41] developed for the automatic segmentation provides the following additional advantages in this respect: (1) a priori knowledge about the 3D shape of the average bee brain, (2) a measure of the variability of the structures allowing to analyze and quantify morphological volumetric changes in neuropils. These advantages will allow us to extend the HSB by using statistical models of the bee brain.

The exploration of subcompartments of the brain requires the development of sub-atlases which have to be integrated into the HSB, e.g. the AL atlas of the bee [10] (Fig. 3.1.1), or the high-resolution atlas of the central body (CB) developed for the locust brain [8]. A further challenge will be to relate structural data across all scales, from the whole brains to microcircuits or to single neurons at the LM and EM level (Fig. 3.1.4). Therefore, ultrastructural knowledge of synaptic connectivity has to be integrated requiring additional registration algorithms [7].

The structural framework of the HSB is necessary for modeling the functional properties of single neurons and microcircuits. Such a reverse engineering approach tries to integrate and map the physiological properties of identified neurons on to exact anatomical locations of the neuron or network. A first attempt was made for the AL in insects [20, 34].

The effective use of digital atlases in neuroscience requires steps of spatial and semantic normalization procedures [48]. Semantic information has to be added to the structural knowledge embodied in the atlas in conjunction with the information provided from data of different origin. Such data may range from the molecular to the macroscopic level resulting in a comprehensive database of the bee brain. Because of the heterogeneity of the data, an essential step is the organization of the data in a ontology of structured databases. Ontologies are formal descriptions of biological attributes that allow to organize the relations of data by using a vocabulary making it readable by computers [17, 23]. The HSB created so far contains geometric and topological information about neurons and neuropils. In its current form the ontology of the bee brain is implemented in the 3D atlas allowing grasping the neuroanatomical networks using a 3D graphical surface [23]. In the future a core ontology specifically designed for insects has to be developed.

References

1. Abel R, Rybak J, Menzel R (2001) Structure and response patterns of olfactory interneurons in the honeybee, *Apis mellifera*. J Comp Neurol 437(3):363–383
2. Ai H (2010) Vibration-processing interneurons in the honeybee brain. Front Syst Neurosci 3(19):1–10
3. Ai H, Rybak J, Menzel R, Itoh T (2009) Response characteristics of vibration-sensitive interneurons related to Johnston's organ in the honeybee, *Apis mellifera*. J Comp Neurol 515(2):145–160
4. Bertrand L, Nissanov J (2008) The neuroterrain 3D mouse brain atlas. Front Neuroinform 2(3):1–8
5. Boline J, Lee E-F, Toga AW (2008) Digital atlases as a framework for data sharing. Front Neurosci 2(1):100–6
6. Brandt R, Rohlfing T, Rybak J, Krofczik S, Maye A et al (2005) Three-dimensional average-shape atlas of the honeybee brain and its applications. J Comp Neurol 492(1):1–19
7. Cardona A, Saalfeld S, Preibisch S, Schmid B, Cheng A et al (2010) An integrated micro- and macroarchitectural analysis of the *Drosophila* brain by computer-assisted serial section electron microscopy. PLoS Biol 8(10):1–17
8. el Jundi B, Heinze S, Lenschow C, Kurylas A, Rohlfing T et al (2010) The locust standard brain: a 3D standard of the central complex as a platform for neural network analysis. Front Syst Neurosci 3(21):1–15
9. el Jundi B, Huetteroth W, Kurylas AE, Schachtner J (2009) Anisometric brain dimorphism revisited: implementation of a volumetric 3D standard brain in *Manduca sexta*. J Comp Neurol 517(2):210–225
10. Galizia CG, McIlwrath SL, Menzel R (1999) A digital three-dimensional atlas of the honeybee antennal lobe based on optical sections acquired by confocal microscopy. Cell Tissue Res 295(3):383–394
11. Galizia CG, Rössler W (2010) Parallel olfactory systems in insects: anatomy and function. Annu Rev Entomol 55(1):399–420
12. Ganeshina O, Menzel R (2001) GABA-immunoreactive neurons in the mushroom bodies of the honeybee: an electron microscopic study. J Comp Neurol 437(3):335–349
13. Grünewald B (1999) Morphology of feedback neurons in the mushroom body of the honeybee, *Apis mellifera*. J Comp Neurol 404(1):114–126
14. Ito K (2010) Technical and organizational considerations for the long-term maintenance and development of the digital brain atlases and web-based databases. Front Syst Neurosci 4(26):1–15
15. Jenett A, Schindelin J, Heisenberg M (2006) The Virtual Insect Brain protocol: creating and comparing standardized neuroanatomy. BMC Bioinforma 7(1):544
16. Jones AR, Overly CC, Sunkin SM (2009) The allen brain atlas: 5 years and beyond. Nat Rev Neurosci 10(11):821–828

17. Joshi SH, Van Horn J, Toga AW (2009) Interactive exploration of neuroanatomical meta-spaces. Front in Neuroinform 3(38):1–10
18. Kenyon CF (1896) The brain of the bee. A preliminary contribution to the morphology of the nervous system of the arthropoda. J Comp Neurol 6:133–210
19. Kirschner S, Kleineidam CJ, Zube C, Rybak J, Grünewald B et al (2006) Dual olfactory pathway in the honeybee, *Apis mellifera*. J Comp Neurol 499(6):933–952
20. Krofczik S, Menzel R, Nawrot MP (2009) Rapid odor processing in the honeybee antennal lobe network. Front Comput Neurosci 2(9):1–13
21. Kurylas AE, Rohlfing T, Krofczik S, Jenett A, Homberg U (2008) Standardized atlas of the brain of the desert locust, Schistocerca gregaria. Cell Tissue Res 333(1):125–145
22. Kuß A, Hege H-C, Krofczik S, Börner J (2007) Pipeline for the creation of surface-based averaged brain atlases. Proceedings of WSCG 2007 - the 15-th International Conference in Central Europe on Computer Graphics, Visualization and Computer Vision. Plzen, Czech Republic. 15:17–24.
23. Kuß A, Prohaska S, Meyer B, Rybak J, Hege H-C (2008) Ontology-based visualization of hierarchical neuroanatomical structures. In: Botha CP et al (eds) Proc Visual Computing for Biomedicine, Delft, pp 177–184
24. Kvello P, Lofaldli BB, Rybak J, Menzel R, Mustaparta H (2009) Digital, three-dimensional average shaped atlas of the Heliothis virescens brain with integrated gustatory and olfactory neurons. Front Syst Neurosci 3(14):1–14
25. Lamecker H, Lange T, Seebaß M, Eulenstein S, Westerhoff M et al (2003) Automatic segmentation of the liver for preoperative planning of resections. Ios Press, Newport Beach. 171–173
26. Larson SD, Martone ME (2009) Ontologies for neuroscience: what are they and what are they good for? Front Neurosci 3:1, 60–67
27. Maronde U (1991) Common projection areas of antennal and visual pathways in the honeybee brain, *Apis mellifera*. J Comp Neurol 309(3):328–340
28. Maye A, Wenckebach T, Hege H (2006) Visualization, reconstruction, and integration of neuronal structures in digital brain atlases. Int J Neurosci 116:431–459
29. Menzel R (2001) Searching for the memory trace in a mini-brain, the honeybee. Learn Mem 8(2):53–62
30. Menzel R (2009) Conditioning: simple neural circuits in the honeybee. In: Squire LR (ed) Encyclopedia of neuroscience, vol 3. Academic, Oxford, pp 43–47
31. Mobbs PG (1982) The brain of the honeybee *Apis mellifera*. I. The connections and spatial organization of the mushroom bodies. Phil Trans R Soc Lond B 298:309–354
32. Müller D, Abel R, Brandt R, Zockler M, Menzel R (2002) Differential parallel processing of olfactory information in the honeybee, *Apis mellifera* L. J Comp Physiol A 188(5):359–370
33. Münch D (2008) Intrazelluläre Färbungen und 3D Strukturanalyse von olfaktorischen Projektionsneuronen im Bienengehirn. MSc, FU Berlin, Berlin
34. Namiki S, Haupt SS, Kazawa T, Takashima A, Ikeno H et al (2009) Reconstruction of virtual neural circuits in an insect brain. Front Neurosci 3(2):206–213
35. Neubert K (2007) Model-based autosegmentation of brain structures in the honeybee, *Apis mellifera*. FU Berlin, Berlin
36. Okada R, Rybak J, Manz G, Menzel R (2007) Learning-related plasticity in PE1 and other mushroom body-extrinsic neurons in the honeybee brain. J Neurosci 27(43):11736–11747
37. Peng H (2008) Bioimage informatics: a new area of engineering biology. Bioinformatics 24(17):1827–1836
38. Rein K, Zockler M, Mader M, Grubel C, Heisenberg M (2002) The *Drosophila* standard brain. Curr Biol 12:227–231
39. Rohlfing T, Brandt R, Maurer Jr. CR, Menzel R (2001) Bee brains, B-splines and computational democracy: generating an average shape atlas. In: Proceedings of the IEEE workshop on mathematical methods in Biomedical Image Analysis, Kauai, Hawaii. 187–194
40. Rybak J (1994) Die strukturelle Organisation der Pilzkörper und synaptische Konnektivität protocerebraler Interneuronen im Gehirn der Honigbiene, *Apis mellifera*: eine licht- und elektronenmikroskopische Studie. Dissertation, FU Berlin, Berlin

41. Rybak J, Kuss A, Lamecker H, Zachow S, Hege HC et al (2010) The digital bee brain: integrating and managing neurons in a common 3D reference system. Front Syst Neurosci 4(30):1–14

42. Rybak J, Menzel R (1993) Anatomy of the mushroom bodies in the honey bee brain: the neuronal connections of the alpha-lobe. J Comp Neurol 334(3):444–465

43. Rybak J, Menzel R (2010) Mushroom body of the honeybee. In: Shepherd GM, Grillner S (eds) Handbook of brain microcircuits. Oxford University Press, New York, pp 433–440

44. Schäfer S, Bicker G (1986) Distribution of GABA-like immunoreactivity in the brain of the honeybee. J Comp Neurol 246(3):287–300

45. Strausfeld NJ (1976) Atlas of an insect brain. Springer, Berlin/Heidelberg/New York

46. Strausfeld NJ (2002) Organization of the honey bee mushroom body: representation of the calyx within the vertical and gamma lobes. J Comp Neurol 450(1):4–33

47. Talairach J, Tournoux P (1988) Co-planar stereotaxic atlas of the human brain: 3-dimensional proportional system – an approach to cerebral imaging. Thieme Medical Publishers, New York

48. Toga AW, Mazziotta JC (2002) Brain mapping: the methods, 2nd edn. Academic Press, San Diego

49. Toga AW, Thompson PM (2001) Maps of the brain. Anat Rec 265(2):37–53

50. Vowles DM (1955) The structure and connections of the corpora pedunculata in bees and ants. Q J Microsc Sci 96:239–255

Chapter 3.2
Plasticity of Synaptic Microcircuits in the Mushroom-Body Calyx of the Honey Bee

Wolfgang Rössler and Claudia Groh

Abstract Mushroom bodies (MBs) are prominent neuropils in the insect brain that have been implicated in higher order processing such as sensory integration, learning and memory, and spatial orientation. Hymenoptera, like the honey bee, possess particularly large MBs with doubled MB calyces (major sensory input structures of the MBs) that are divided into compartments. In this review we focus on characteristic modular synaptic complexes (microglomeruli, MG) in the honey bee MB calyx (CA). The main components of MG comprise a presynaptic bouton from projection neurons (PNs) (e.g. olfactory, visual), numerous dendritic spines from MB intrinsic neurons (Kenyon cells, KC), and processes from recurrent GABAergic neurons. Recent work has demonstrated a remarkable structural plasticity of MG associated with postembryonic brood care, age, sensory experience, and stable long-term memory. The mechanisms and functional significance of this neuronal plasticity are discussed and related to behavioral plasticity and social organization.

Abbreviations (Excluding Anatomical Structures)

ActD   Actinomycin D
CS   Conditioned stimulus
LTM   Long term memory
US   Unconditioned stimulus

W. Rössler (✉) • C. Groh
Behavioral Physiology and Sociobiology, Biozentrum, University of Würzburg,
Am Hubland, 97074 Würzburg, Germany
e-mail: roessler@biozentrum.uni-wuerzburg.de

C.G. Galizia et al. (eds.), *Honeybee Neurobiology and Behavior: A Tribute to Randolf Menzel*, DOI 10.1007/978-94-007-2099-2_12,
© Springer Science+Business Media B.V. 2012

3.2.1 The Honey Bee Mushroom Bodies

The mushroom bodies (MBs) are prominent paired structures on each side of the central brain of the honey bee (Fig. 3.2.1a). Anatomically, each MB is subdivided into the cup-shaped calyces – major sensory input regions of the MBs – the pedunculus (PED), and the medial (ML) and vertical lobes (VL) – the main output regions of the MBs (nomenclature after [43], Fig. 3.2.1), (see Chap. 3.1). In the honey bee and other Hymenoptera the MBs form a lateral (LCA) and medial (MCA) calyx in each brain hemisphere. Various studies in the honey bee and in other insects (e.g. *Drosophila*) have assigned the MBs important roles in higher sensory integration and in the organization of complex behaviors that involve learning and the formation of associative memories (e.g. [13, 20, 28]; see also Chaps. 6.1–6.5).

The total number of neurons in the brain of a worker bee was estimated with ~850,000, and the number of Kenyon cells (KCs), with ~184,000 on each side [30, 38, 43, 48]. This adds to ~368,000 KCs making up more than ~40% of the total number of neurons in the honey bee brain. For comparison, in *Drosophila* the total brain neuron population was estimated ~150,000 [2] and the KC population with ~2,000 in each brain hemisphere [3] representing less than ~4% of the total number of brain neurons.

What is the adaptive value of very large MBs with a high number of intrinsic KC neurons as found in the honey bee? The large population of KCs and associated neuronal microcircuits is highly suggestive for an elaborated computational potential and increased neuronal plasticity and associated storage capacities [40]. It is interesting in this context that a recent developmental study suggests that the invertebrate MBs and vertebrate pallium (including the cerebral cortex) have a common origin [45].

Fig. 3.2.1 Brain of the honey bee triple-labeled with an antibody to synapsin (*red*), f-actin-phalloidin (*green*), and Hoechst nuclear marker (*blue*). (**a**) Overview of a central plane in the brain with the major neuropil regions. (**b**) Higher magnification of one branch of the mushroom body (*MB*) calyx and anatomically distinct olfactory and visual subregions. (**c**) and (**d**) Higher magnification of double-labeled microglomeruli (*MG*) in the lip and collar region (a physical indentation separates the lip from the collar). (**e**) Individual double-labeled microglomerulus with synapsin immunoreactivity in a bouton of a projection neuron (*PN*) in *red* and f-actin-phalloidin staining of Kenyon cell (*KC*) dendritic profiles in *green*. (**f**) and (**g**) Electron micrographs of a section through a MG in the lip (**f**) and collar (**g**) of a young worker bee (Large presynaptic boutons with multiple dark-labeled active zones are surrounded by numerous small profiles, most of them presumably from KC dendrites). (**h**) Schematic diagram of one MG (modified after [14]). For clarity, only the presynaptic bouton of a projection neuron (*PN*) and profiles of KC dendritic spines are shown. (**i**) Schematic diagram indicating putative excitatory (+) and inhibitory (−) synaptic connections between projections neurons (*PN*), GABAergic recurrent neurons (GABA), and Kenyon cells (*KC*) based on immune-electron microscopic studies by Ganeshina and Menzel [11] (scheme modified after [11]). Further abbreviations: *AL* antennal lobe, *BR* basal ring, *CX* central complex, *DCO* dense collar region, *CO* collar region, *LA* lamina, *LCO* loosely arranged collar region, *LO* lobula, *LCA* lateral MB calyx, *LIP* lip region, *MCA* medial *MB* calyx, *ME* medulla, and *PED* pedunculus. Scale bar in A = 200 μm; B = 25 μm; C, D = 10 μm; E-G = 1 μm

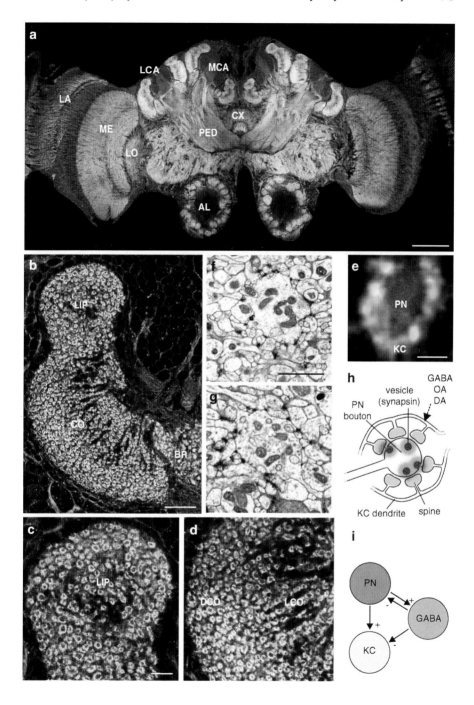

    The main focus of this chapter is to review recent work on the synaptic organization
of the major input structures of the MBs, the calyces, in particular their structural
and functional plasticity. Further neuroanatomical and functional aspects like
neurotransmitter systems and the role of the MBs in learning and memory are
summarized in other chapters of this book. Whereas in many insects the MB
calyces predominantly receive olfactory input, the MB calyces in the honey bee
(and in other Hymenoptera like ants) form prominent multimodal sensory input
regions (Fig. 3.2.1). Dendrites from different classes of KCs receive input from projection neurons (PNs) of primary olfactory centers and from the optic lobes (OL)
(e.g. [17, 30, 38, 43]), and they are innervated by PNs form the subesophageal calycal
tract – mediating most likely gustatory and/or tactile input [37]. This chapter
summarizes recent work on structural plasticity of synaptic complexes in the CA – the
MG, in particular plasticity associated with brood care and environmental influences,
sensory experience, age, and stable long-term memory. The resulting structural
and functional changes in MG are assumed to contribute to long-term changes and
adaptations in behavior associated with social organization.

3.2.1.1   Anatomical Subdivisions of the Mushroom-Body Calyces

In Hymenoptera, in particular in bees [14, 16, 17, 30, 43], social wasps [32], and
ants (e.g. [42]), the MBs are large in relation to other brain regions. The MB calyces
comprise three anatomically distinct and clearly delineated subdivisions: the lip
(LIP), the collar (CO) and the basal ring (BR) receiving olfactory, visual and both
sensory modalities, respectively (Fig. 3.2.1b). The lip region can further be subdivided into a cortical and central input zone innervated by olfactory PNs from two
antennal lobe-protocerebral tracts (the medial and the lateral APT) ([1, 23], tract
nomenclature after [10]). In the honey bee, a distinct region at the transition between
the lip and collar region was shown to receive input from PNs of the subesophageal
calycal tract (SCT), most likely transmitting information from gustatory and mechanosensory neurons of the proboscis or other mouthparts [37]. The basal ring (BR)
is organized in distinct layers receiving input from olfactory and bilateral visual
PNs [17, 23, 37]. The visual collar is further subdivided in distinct layers innervated
by visual PNs that transfer chromatic, temporal, and motion sensitive input from the
OLs (lobula (LO) and medulla (ME)) [34].

3.2.1.2   MG – Characteristic Synaptic Complexes
         in the Mushroom-Body Calyx

The different neuropil subregions of the MB calyces contain distinct synaptic
complexes that were termed MG (Fig. 3.2.1e–h). Anatomical details of MG were
pioneered by electron microscopic (EM) studies in ants [41]. Detailed EM studies

of the synaptic organization of MG in the honey bee were performed in Randolf Menzel's lab [11], showing synaptic microcircuits formed between olfactory PN boutons, KC dendrites and recurrent GABAergic extrinsic neurons. These investigations demonstrate the complexity of putative excitatory and inhibitory synaptic profiles in MG (Fig. 3.2.1h, i). EM studies by Yasuyama et al. [49] revealed a similar organization in the CA of *Drosophila* indicating that MG may represent evolutionary conserved functional units. Double staining of PN axons and KC dendritic profiles indicate that the majority of MG in the lip region comprise a central presynaptic bouton from olfactory PNs surrounded by a dense shell of numerous postsynaptic profiles (Fig. 3.2.1e, and schematic drawing in Fig. 3.2.1h), mainly formed by KC dendritic spines [9, 11, 12, 14–16, 39, 42, 49]. Double-immunolabeling with an antibody to the *Drosophila* synaptic-vesicle associated protein synapsin (from E. Buchner, Univ. Würzburg) in combination with f-actin-phalloidin staining and EM studies showed that KC dendritic tips are highly enriched with f-actin supporting their spine-like nature ([9, 14, 16, 42], Fig. 3.2.1e, h). In the honey bee, in addition to supply from PNs, MG microcircuits are innervated by a widely branched network of recurrent GABAergic neurons originating from the MB lobes (e.g. [18]). Furthermore, the MB calyces in the honey bee receive input from modulatory systems, in particular octopaminergic and dopaminergic neurons (e.g. [19]), which play an important role in associative learning (see also Chaps. 3.1 and 6.1–6.5). The precise synaptic connection of neuromodulatory and other extrinsic neurons into the MG microcircuits requires further investigation.

3.2.2 Structural Plasticity of MG at Different Life Stages of the Honey Bee

3.2.2.1 *Influences of Postembryonic Brood Care*

Cooperative brood care represents a most important feature in the organization of insect societies. Differential larval feeding and a remarkable phenotypic plasticity form the basis for the queen worker polymorphism and the determination of female castes [46]. Honey bee queens develop from fertilized eggs that are genetically not different from those that develop into workers, but they develop faster, are larger, live longer and differ markedly in their behavior. In addition to larval feeding the temperature of pupae is tightly controlled to 35 ± 0.5°C by thermoregulatory behavior of adult worker bees to ensure proper metamorphic brood development. Similarly in ants, pupae are exposed to controlled temperature ranges to regulate postembryonic development, and ant nurses respond to changes in the ambient temperature by carrying the brood to nest areas with the appropriate temperatures [47]. Recent studies in honey bees and ants have shown that pupae that developed at different temperatures within the range of naturally occurring temperatures in the brood area (33–36°C in the case of the honey bee) exhibit differences in adult behavior [4, 44, 47].

Groh et al. [14, 16] and Groh and Rössler [15] investigated effects of both larval feeding and naturally occurring variations in pupal temperature on the number of MG in freshly emerged adult bees. The results revealed that both larval feeding and pupal thermoregulation affect the adult number of MG. One day old queens, compared to workers, had significantly lower numbers of MG in both the lip and collar regions at all temperatures tested, and worker pupae raised at the lower range of naturally occurring brood temperatures (33°C compared to 34.5°C) had lower numbers of MG in the adult CA lip (but not in the collar region) compared to those raised at the natural temperature in the central brood region (34.5°C). Developmental studies showed that the temporal sequence of MG formation was substantially faster (by about 4 days) in queens compared to workers reared at the natural temperature [15]. Behavioral experiments indicate that adult bees reared at the lower range of natural occurring brood temperatures perform less well in associative memory tasks, forage later, and differ in dance-communication performance and undertaking behavior compared to bees raised at higher temperature [4, 44]. Similarly, ants that were raised at different temperatures during pupal development differed in their adult temperature thresholds that are needed to induce brood carrying behavior [47].

The results indicate that differential brood care may cause substantial changes in MG numbers in the adult CA. Whether or how these changes are causally linked to the changes demonstrated in complex behaviors including learning and memory remains to be shown. It is appealing to speculate that developmental plasticity of MG in the CA may also contribute to changes in sensory thresholds for the display of certain behaviors. This in turn may influence associative learning performance and division of labor according to a "threshold model" proposed by Pankiw and Page [33], (see Chap. 1.1). The most challenging task for future work will be to causally link differential brood-care to changes in behavior at both the individual and colony level.

3.2.2.2 Effects of Adult Behavioral Maturation, Sensory Experience and Age

Do adult sensory experience and age affect the structural plasticity of MG? A robust volume increase of the MB calyces at the onset of foraging was reported in several studies on social Hymenoptera – in bees (e.g. [5]), in ants (e.g. [42]), and in wasps, [32]. The cellular mechanism for these volume changes remained unclear. A Golgi study by [8] first indicated that in the visual subregions (collar) outgrowth of KC dendrites may contribute to volume increase in the CA during the transition to foraging. Volume studies by Ismail et al. [22] combined with pharmacological stimulation suggest that activity mediated by muscarinic cholinergic pathways trigger a volume increase comparable to the one induced by foraging experience.

What are the cellular mechanisms for these remarkable volume changes in the MBs – and how do sensory experience and age contribute to it? To address this question, effects of sensory experience and age (or an intrinsic program) need to be dissected. Due to the close contact between individuals in social-insect colonies, complete olfactory deprivation is impossible without removing the sensory organs (the antennae) or isolation. Kleineidam and Rössler [24] performed an olfactory deprivation study in *Camponotus* ants by unilateral removal of the antenna on the first day of adult life. This caused a substantial reduction of the volume of AL glomeruli after ~15 days. Interestingly, no obvious changes in MG numbers were found in the olfactory subregions of the ipsilateral MB-calyx lip, which is innervated only by input from the ipsilateral antenna. Similar observations were made in the honey bee (Rössler, unpublished results). This may indicate that input from the contralateral side at the level of the MBs may compensate for unilateral loss of sensory olfactory input.

Another study focused on manipulations of the environment (natural environment versus artificially reduced environment) and/or manipulations at the colony level (experimental induction of precocious or delayed foragers) [25]. This study revealed effects on the number of MG in olfactory and visual regions of the CA, but the problem with broad manipulations of the social environment is that many variables (olfactory, visual stimuli and physical contact), are changed at the same time.

Long living honey bee queens show a remarkable age-related decrease of MG in the collar and an increase in the olfactory lip [14]. A study by Stieb et al. [42] on *Cataglyphis* ants was able to dissect effects of age and sensory (in this case visual) experience. Ants from dark-reared *Cataglyphis fortis* colonies were exposed to light precociously or delayed. The results clearly demonstrate that precocious sensory exposure (on the first day of adult life) as well as delayed light exposure (even after 6 and 12 months) triggered a decrease in MG numbers associated with CA volume change similar to that observed at the transition from indoor activities to foraging. Artificially aged ants that had lived in constant darkness over a period of 6–12 months showed a very slow increase in MG numbers, even in the collar. This is in contrast to aged honey bee queens [14], which showed a decrease of MG in the collar, indicating that caste-specific (or species-specific) differences in aging phenomena have to be considered.

Anti-tubulin staining is restricted to the dendritic shafts of KCs [42]. Combined f-actin and tubulin staining of KC dendritic processes revealed that the most drastic effect of precocious light exposure and natural maturation appears to be a massive outgrowth of KC-dendritic shafts and an associated increase in tubulin positive profiles in between MG. At the same time, the number of PN boutons per area decreased, and both processes (decease in presynaptic bouton density and KC dendrite expansion) resulted in a net expansion of the total MB-calyx volume [42]. This indicates that pruning of MG was involved, confirming an EM study in *Cataglyphis* ants by Seid and Wehner [39]. Future experiments need to show whether the effects on MG reorganization are causally linked to the onset of foraging behavior. To be able to further dissect mechanisms and effects of age and sensory experience on MB synaptic plasticity in the honey bee, we need to learn more about the molecular pathways involved, to combine experiments with pharmacology, RNA interference, or neuroendocrine and hormonal manipulations.

Furthermore, interesting ageing models like winter bees or artificially aged dark reared (visually deprived) cohorts will offer very promising targets to dissect the mechanisms and consequences of age- and experience dependent structural synaptic plasticity in the MB calyces.

3.2.2.3 Structural Plasticity of MG Related to Stable Long-Term Memory

Transcription-dependent formation of a stable long-term memory after olfactory conditioning is associated with structural synaptic plasticity of MG in the olfactory lip of the honey bee CA [21]. In this study honey bees were trained to associate a sugar reward with odor using a five-trial conditioning paradigm and the proboscis extension response. Only the bees that had received paired stimulation of the conditioned (odor pulse) and unconditioned stimulus (sugar water) (CS and US), and that were not injected with Actinomycin D (ActD, a specific transcription inhibitor) 3 h after training, retained a stable long-term memory (LTM) after 3 days when they were tested with the CS. Most importantly, the formation of a stable LTM was associated with an increase in the density of MG in the olfactory lip. The increase in MG was modality specific as only the olfactory lip, but not the visual collar was affected. Naïve (unstimulated) bees, and bees that had received unpaired stimulation, as well as paired bees that had been injected with ActD did not show memory retrieval for the rewarded odor after 3 days (and the unpaired and naïve groups showed similar MG density values). The density of MG was sampled only in a central region of the CA, but there may be a wide distribution of structural plasticity in MG across the MB-calyx lip. This is supported by the fact that PNs have widely scattered boutons ([1, 31], see also Chap. 3.1). On the other hand, individual PNs may differ substantially regarding the contribution to MG-density changes depending on their odor specificity. Transcription-independent memories, such as early-long-term memory (eLTM) did not lead to any detectable stable changes as the unpaired and paired ActD groups displayed similar MG densities. Therefore, Hourcade et al. [21] conclude that the formation of a transcription-dependent LTM is accompanied by a stable increase in the density of MG in the MB-calyx lip. They further speculate that structural synaptic rearrangements and growth of new synapses may be a common feature involved in stable LTM in mammalian as well as in insect brains.

Previous calcium imaging experiments in Randolf Menzel's lab had shown that five spaced CS-US presentations lead to an increase in calcium activity in the lip in response to the rewarded odor [6] indicating that calcium may be correlated with structural plasticity after multi trial learning. In fact, Perisse et al. [35] showed that calcium is essential for LTM formation, which further adds to a potential role of calcium in mediating structural plasticity associated with stable LTM. The structural synaptic changes may be part of a memory trace, but whether they are actually required for memory storage remains to be determined.

3.2.3 Outlook

The results from studies over the past years provide increasing evidence that MG in the CA undergo structural reorganization related to brood care (food and thermo-regulation) [14–16], behavioral maturation [25, 42], sensory experience and age [14, 42], and are associated with the formation of stable LTM [21].

Due to the well known structural and functional properties of the olfactory pathway, the CA lip provides a particularly feasible neuronal substrate to study the mechanisms and consequences of long-term plasticity. MG in the lip region receive input from olfactory PNs via two antennal lobe-protocerebral tracts [10, 23, 31], and they receive putative modulatory reinforcement via the octopaminergic VUM_{mx1} neuron [11, 19, 28]. Future work is necessary to identify the physiological properties and functional role of different assemblies of MG within the CA lip, which may be diverse in their composition.

Whereas structural features of the CA input from the OLs to the collar region and from the subesophageal ganglion (SEG) to a region between lip and collar is well investigated, much less is known about the function [17, 34, 37], and almost nothing is known about the multimodal function of the BR. Furthermore, reverberant activity via GABAergic neurons and its potential influence on MG plasticity requires further investigations at both the structural and functional levels [18, 36].

The presence of f-actin rich KC dendritic spines [9, 14, 16] indicates that rearrangements of cytosceletal elements are likely to mediate structural plasticity of MG, similar to activity dependent structural plasticity of synaptic spines in hippocampal neurons (e.g. [50]). Interference with actin dynamics and associated molecular pathways will be very elusive to investigate the molecular mechanisms. Another interesting point is the potential role of epigenetic changes on neuronal networks in the MBs. Recent studies have shown a remarkable role of DNA methylation in controlling queen-worker polymorphism ([26], see also Chap. 5.4). Structural plasticity in CA MG will provide an ideal substrate to investigate how epigenetic modifications may affect synaptic microcircuits.

Structural reorganization of MG is suggestive to play a role in long-term changes of behavior. Its enormous size in Hymenoptera brings the CA MG in a position to function as a neuronal substrate for "life-time memory" (or "life-history memory"). Whether structural rearrangements of MG are causal for the storage of long-term memory, however, remains to be shown.

The olfactory pathway in the honey bee is characterized by connections of a dual olfactory pathway to the MBs and lateral horn (LH) [10]. Long-term changes in the MB neuronal networks may contribute plastic components onto more rigid or "hard wired" parallel pathways to the LH, although it is not clear at this point (and should be investigated in the future) whether the LH neuronal circuits are more hardwired than those in the MBs. To induce changes in behavior, changes in the CA would have to be mediated to MB extrinsic (or output) neurons, like the PE1 neuron investigated by Mauelshagen [27] and Menzel and Manz [29]. In fact, these studies have demonstrated plasticity in the activity of MB output neurons associated with learning. An important future challenge will be to determine how the activity of such MB output neurons in turn will be affected by long-term structural plasticity at the MB input, in the CA MG, and whether plasticity in recurrent GABAergic neurons [18] mediates changes from the MB lobes back to the CA. These questions also represent interesting aspects for future neuroinformatics modeling approaches.

Structural plasticity in the CA MG may be related to stable long-term changes in behavior as indicated for olfactory LTM [21], whereas changes in KC output synapses in the MB lobes are thought to mediate short-term and intermediate-term memory phases (e.g. [20, 28]). Long-term plasticity does particularly make sense in insects with a long life span (months or years) and which express a high degree of behavioral plasticity (e.g. castes, age and task related polyethism, long-term memory) – both is the case in the honey bee and many other social Hymenoptera. Behavioral plasticity represents a most important aspect of social life. Whether the large size of the CA is related to such a function in long-term information storage needs to be further tested. Furthermore, central-place foragers like the honey bee (bees always return to the same place – the hive – after returning from foraging trips (see Chap. 2.5)) and other social insect colonies have to memorize profitable (or bad) food sources over extended periods of time, ideally over the entire life span. A similar argument was made for generalist beetles. Interestingly, these beetles were shown to possess large and doubled MB calyces compared to food-specialist beetles which posses only a single calyx [7]. Whether CA duplication is always correlated with a high number of KCs and associated MG needs to be shown in more comparative studies across insects.

A very interesting question is whether both genetically determined and environmentally induced differences in neuronal networks in the MBs contribute to division of labor, potentially by mediating different sensory thresholds [33] or differences in learning and memory abilities. Although the CA is likely to be not the exclusive place for mediating long-term neuronal-network changes, the substantial structural plasticity in MG and the enormous number of KC neuronal circuits are suggestive for being a suitable neuronal resource with large enough storage capacities for long-term adaptive adjustments. Future comparative studies are needed to find out where high numbers of KCs and associated MG together with a doubled CA have emerged during hymenopteran evolution.

Using concerted efforts and new tools like high resolution life- and molecular-imaging, 3D ultrastructural analyses, multi- and single unit electrophysiology, genetic and molecular interference in combination with well designed behavioral assays are promising approaches to address some of the above questions. The CA in the honey bee brain offers unique opportunities for the future to investigate neuronal plasticity and how it translates into changes in behavior. Compared to work on the mammalian hippocampus, for example, physiological imaging studies in the honey bee can be performed in live animals with intact brains (the MB calyces are close to the surface of the brain) plus using controlled sensory stimulation. New electrophysiological tools like simultaneous long-term multi-unit recordings from different brain sites will be extremely helpful to investigate temporal aspects of coding and how they may affect synaptic plasticity. The diversity of species within the Hymenoptera (social and solitary species) opens up unique opportunities to correlate CA attributes with sensory and cognitive capabilities and the evolution of sociality in comparative approaches. Finally, the already well investigated neuronal circuitry, the rich diversity in behavior, the availability of a sequenced genome, and findings on epigenetic mechanisms open up completely novel approaches for targeted interventions at the genetic, epigenetic, molecular, physiological and behavior levels. The large range of phenotypic plasticity in association with social organization make the honey bee a most promising model for future studies on the neuronal mechanisms underlying behavioral plasticity in a social context.

References

1. Abel R, Rybak J, Menzel R (2001) Structure and response patterns of olfactory interneurons in the honeybee, *Apis mellifera*. J Comp Neurol 437(3):363–383
2. Armstrong JD, Kaiser K, Müller A, Fischbach KF, Merchant N et al (1995) Flybrain, an on-line atlas and database of the *Drosophila* nervous system. Neuron 15(1):17–20
3. Aso Y, Grübel K, Busch S, Friedrich AB, Siwanowicz I et al (2009) The mushroom body of adult *Drosophila* characterized by GAL4 drivers. J Neurogenet 23(1–2):156–172
4. Becher MA, Scharpenberg H, Moritz RF (2009) Pupal developmental temperature and behavioral specialization of honeybee workers (*Apis mellifera* L.). J Comp Physiol A 195(7):673–679
5. Durst C, Eichmüller S, Menzel R (1994) Development and experience lead to increased volume of subcompartments of the honeybee mushroom body. Behav Neural Biol 62(3):259–263
6. Faber T, Menzel R (2001) Visualizing mushroom body response to a conditioned odor in honeybees. Naturwissenschaften 88(11):472–476
7. Farris SM, Roberts NS (2005) Coevolution of generalist feeding ecologies and gyrencephalic mushroom bodies in insects. P Natl Acad Sci USA 102(48):17394–17399
8. Farris SM, Robinson GE, Fahrbach SE (2001) Experience- and age-related outgrowth of intrinsic neurons in the mushroom bodies of the adult worker honeybee. J Neurosci 21(16):6395–6404
9. Frambach I, Rössler W, Winkler M, Schürmann FW (2004) F-actin at identified synapses in the mushroom body neuropil of the insect brain. J Comp Neurol 475(3):303–314
10. Galizia CG, Rössler W (2010) Parallel olfactory systems in insects: anatomy and function. Annu Rev Entomol 55:399–420
11. Ganeshina O, Menzel R (2001) GABA-immunoreactive neurons in the mushroom bodies of the honeybee: an electron microscopic study. J Comp Neurol 437(3):335–349
12. Ganeshina O, Vorobyev M, Menzel R (2006) Synaptogenesis in the mushroom body calyx during metamorphosis in the honeybee *Apis mellifera*: an electron microscopic study. J Comp Neurol 497(6):876–897
13. Giurfa M (2007) Behavioral and neural analysis of associative learning in the honeybee: a taste from the magic well. J Comp Physiol A 193(8):801–824
14. Groh C, Ahrens D, Rössler W (2006) Environment- and age-dependent plasticity of synaptic complexes in the mushroom bodies of honeybee queens. Brain Behav Evol 68(1):1–14
15. Groh C, Rössler W (2008) Caste-specific postembryonic development of primary and secondary olfactory centers in the female honeybee brain. Arthropod Struct Dev 37(6):459–468
16. Groh C, Tautz J, Rössler W (2004) Synaptic organization in the adult honey bee brain is influenced by brood-temperature control during pupal development. Proc Natl Acad Sci USA 101(12):4268–4273

17. Gronenberg W (2001) Subdivisions of hymenopteran mushroom body calyces by their afferent supply. J Comp Neurol 435(4):474–489
18. Grünewald B (1999) Physiological properties and response modulations of mushroom body feedback neurons during olfactory learning in the honeybee, *Apis mellifera*. J Comp Physiol 185:565–576
19. Hammer M (1993) An identified neuron mediates the unconditioned stimulus in associative olfactory learning in honeybees. Nature 366(6450):59–63
20. Heisenberg M (2003) Mushroom body memoir: from maps to models. Nat Rev Neurosci 4(4):266–275
21. Hourcade B, Muenz TS, Sandoz JC, Rössler W, Devaud JM (2010) Long-term memory leads to synaptic reorganization in the mushroom bodies: a memory trace in the insect brain? J Neurosci 30(18):6461–6465
22. Ismail N, Robinson GE, Fahrbach SE (2006) Stimulation of muscarinic receptors mimics experience-dependent plasticity in the honey bee brain. Proc Natl Acad Sci USA 103(1): 207–211
23. Kirschner S, Kleineidam CJ, Zube C, Rybak J, Grünewald B, Rössler W (2006) Dual olfactory pathway in the honeybee, *Apis mellifera*. J Comp Neurol 499(6):933–952
24. Kleineidam CJ, Rössler W (2009) Adaptations in the olfactory system of social hymenoptera. In: Gadau J, Fewell J (eds) Organization of insect societies. From genome to sociocomplexity. Harvard University Press, Cambridge/London, pp 195–219
25. Krofczik S, Khojasteh U, de Ibarra NH, Menzel R (2008) Adaptation of microglomerular complexes in the honeybee mushroom body lip to manipulations of behavioral maturation and sensory experience. Dev Neurobiol 68(8):1007–1017
26. Kucharski R, Maleszka J, Foret S, Maleszka R (2008) Nutritional control of reproductive status in honeybees via DNA methylation. Science 319(5871):1827–1830
27. Mauelshagen J (1993) Neural correlates of olfactory learning paradigms in an identified neuron in the honeybee brain. J Neurophysiol 69(2):609–625
28. Menzel R, Giurfa M (2001) Cognitive architecture of a mini-brain: the honeybee. Trends Cogn Sci 5(2):62–71
29. Menzel R, Manz G (2005) Neural plasticity of mushroom body-extrinsic neurons in the honeybee brain. J Exp Biol 208(Pt 22):4317–4332
30. Mobbs PG (1982) The brain of the honeybee *Apis mellifera*. 1. The connections and spatial-organization of the mushroom bodies. Philos Trans R Soc B 298(1091):309–354
31. Müller D, Abel R, Brandt R, Zockler M, Menzel R (2002) Differential parallel processing of olfactory information in the honeybee, *Apis mellifera* L. J Comp Physiol A 188(5):359–370
32. O'Donnell S, Donlan NA, Jones TA (2004) Mushroom body structural change is associated with division of labor in eusocial wasp workers (*Polybia aequatorialis*, Hymenoptera: Vespidae). Neurosci Lett 356(3):159–162
33. Pankiw T, Page RE (1999) The effect of genotype, age, sex, and caste on response thresholds to sucrose and foraging behavior of honey bees (*Apis mellifera* L.). J Comp Physiol A 185(2):207–213
34. Paulk AC, Phillips-Portillo J, Dacks AM, Fellous JM, Gronenberg W (2008) The processing of color, motion, and stimulus timing are anatomically segregated in the bumblebee brain. J Neurosci 28(25):6319–6332
35. Perisse E, Raymond-Delpech V, Neant I, Matsumoto Y, Leclerc C et al (2009) Early calcium increase triggers the formation of olfactory long-term memory in honeybees. BMC Biol 7:30
36. Rybak J, Menzel R (1993) Anatomy of the mushroom bodies in the honey-bee brain – the neuronal connections of the alpha-lobe. J Comp Neurol 334(3):444–465
37. Schröter U, Menzel R (2003) A new ascending sensory tract to the calyces of the honeybee mushroom, body, the subesophageal-calycal tract. J Comp Neurol 465(2):168–178
38. Schürman FW (1974) Functional anatomy of corpora pedunculata in insects. Exp Brain Res 19(4):406–432
39. Seid MA, Wehner R (2009) Delayed axonal pruning in the ant brain: a study of developmental trajectories. Dev Neurobiol 69(6):350–364

40. Smith D, Wessnitzer J, Webb B (2008) A model of associative learning in the mushroom body. Biol Cybern 99(2):89–103
41. Steiger U (1967) Über den Feinbau des Neuropils im Corpus Pedunculatum der Waldameise – elektronenoptische Untersuchungen. Z Zellforsch Mik Anal 81(4):511–536
42. Stieb SM, Muenz TS, Wehner R, Rössler W (2010) Visual experience and age affect synaptic organization in the mushroom bodies of the desert ant *Cataglyphis fortis*. Dev Neurobiol 70(6):408–423
43. Strausfeld NJ (2002) Organization of the honey bee mushroom body: representation of the calyx within the vertical and gamma lobes. J Comp Neurol 450(1):4–33
44. Tautz J, Maier S, Groh C, Rössler W, Brockmann A (2003) Behavioral performance in adult honey bees is influenced by the temperature experienced during their pupal development. Proc Natl Acad Sci USA 100(12):7343–7347
45. Tomer R, Denes AS, Tessmar-Raible K, Arendt D (2010) Profiling by image registration reveals common origin of annelid mushroom bodies and vertebrate pallium. Cell 142(5):800–809
46. Weaver N (1957) Effects of larval age on dimorphic differentiation of the female honeybee. An Entomol Soc Am 50:283–294
47. Weidenmüller A, Mayr C, Kleineidam CJ, Roces F (2009) Preimaginal and adult experience modulates the thermal response behavior of ants. Curr Biol 19(22):1897–1902
48. Witthöft W (1967) Absolute Anzahl und Verteilung der Zellen im Gehirn der Honigbiene. Zeitschrift für Morphologie der Tiere 61:160–184
49. Yasuyama K, Meinertzhagen IA, Schürmann FW (2002) Synaptic organization of the mushroom body calyx in *Drosophila melanogaster*. J Comp Neurol 445(3):211–226
50. Yuste R, Bonhoeffer T (2004) Genesis of dendritic spines: insights from ultrastructural and imaging studies. Nat Rev Neurosci 5(1):24–34

Chapter 3.3
Neurotransmitter Systems in the Honey Bee Brain: Functions in Learning and Memory

Monique Gauthier and Bernd Grünewald

Abstract Synaptic correlates of olfactory learning within the honey bee brain utilize several transmitters and receptors. Experiments unraveled distinct roles of these transmitter systems in cognitive processes. Cholinergic synaptic transmission is involved in acquisition and retrieval processes. At least two subtypes of nicotinic acetylcholine receptors exist in the honey bee brain, one involved in retrieval processes and another one linked to the formation of long-term memory. The electrophysiological and pharmacological properties of the underlying nicotinic acetylcholine receptors (nAChR) are well described whereas muscarinic acetylcholine receptors (mAChR) are physiologically unknown. The reward processing pathway largely depends on octopaminergic neuromodulation. Serotonin (5-HT) impairs the conditioned response during acquisition. Whether dopamine (DA) mediates aversive learning while octopamine (OA) mediates appetitive learning remains to be analyzed. Several studies indicated that GABA receptors play a role during odor learning, but the specific function of inhibition is not yet clear. Both inhibitory and excitatory glutamate receptors are required for certain forms of learning and for memory retrieval.

M. Gauthier (✉)
Centre de Recherches sur la Cognition Animale – UMR CNRS 5169, Université de Toulouse III,
118 route de Narbonne, F-31062 Toulouse Cedex 9, France
e-mail: gauthiem@cict.fr

B. Grünewald
Institut für Bienenkunde, Fachbereich Biowissenschaften, Institut für Zellbiologie und
Neurowissenschaften, Polytechnische Gesellschaft Frankfurt am Main, Goethe-Universität
Frankfurt am Main, Karl-von-Weg 2, D-61440 Oberursel, Germany
e-mail: b.gruenewald@bio.uni-frankfurt.de

C.G. Galizia et al. (eds.), *Honeybee Neurobiology and Behavior: A Tribute
to Randolf Menzel*, DOI 10.1007/978-94-007-2099-2_13,
© Springer Science+Business Media B.V. 2012

Abbreviations (Excluding Brain Areas)

α-BGT	α-Bungarotoxin
HA	Histamine
5-HT	Serotonin (5-hydroxy-tryptamine)
ACh	Acetylcholine
AChE	Acetylcholine esterase
AmTYR1	Honey bee TA receptor
CR	Conditioned response
CS	Conditioned stimulus
DA	Dopamine
GluCl	Glutamate-gated chloride channel
mAChR	Muscarinic ACh receptor (metabotropic G-protein-coupled receptors)
nAChR	Nicotinic ACh receptor
OA	Octopamine
PER	Proboscis extension reflex
siRNA	Small interfering RNA
TA	Tyramine
US	Unconditioned stimulus

3.3.1 Introduction

Neurons within the honey bee brain largely use the same neurotransmitters as those of the mammalian brain. The classical neurotransmitters, acetylcholine (ACh), glutamate and GABA as well as the neuromodulators, serotonin (5-HT) and DA, are present in both mammals and insects. The major excitatory neurotransmitters are ACh in the insect central nervous system (CNS) and glutamate at the neuromuscular junction. Inhibition is mediated via GABA and glutamate-gated chloride channels. DA, serotonin (5-HT), OA, tyramine (TA) and histamine (HA) are the known biogenic amines present in the insect CNS. For several transmitters, distinct roles during learning and memory formation have been identified in flies and bees. This review provides an overview on the honey bee receptor systems and their contribution to behavioral plasticity.

3.3.2 Functional Transmitter Neuroanatomy of the Honey Bee Brain

The olfactory conditioning of the proboscis extension reflex (PER) is largely used during behavioral pharmacological experiments. Bees can be conditioned to an odor as signaling stimulus to predict the occurrence of a sucrose reward. Pairing the odorant with the sucrose reward creates an association between the two stimuli (classical appetitive conditioning), leading bees to respond to the odor presented alone (conditioned stimulus, CS) by an extension of the proboscis (conditioned

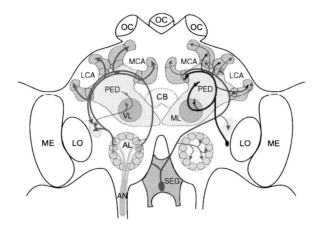

Fig. 3.3.1 Schematic drawing of the honey bee brain with some of the pathways involved in olfactory conditioning. Olfactory information enters the brain via the antennal nerve. From there projection neurons (*blue*) innervate the lateral horn, the calyces of the mushroom bodies and various other regions of the protocerebrum (cf. Kirschner et al. 2006). Within the mushroom bodies (*green*) feedback neurons (*black*) interconnect the output and input regions. Output neurons (*violet*) transmit Kenyon cell information to the lobes and to other regions of the protocerebrum. Reward information enters the antennal lobes, the calyces and the lateral horn via the VUM$_{mx1}$ neuron (*red*). Abbreviations: *AL* antennal lobe, *AN* antennal nerve, *CB* central body, *LCA* lateral calyx, *LO* lobula, *MCA* median calyx, *ML* medial lobe, *ME* medulla, *OC* ocelli, *PED* pedunculus, *SEG* subesophageal ganglion, *VL* vertical lobe

response, CR). The memory for the conditioned odorant lasts for a few hours after single-trial learning and for several days after multiple-trial learning (see Chaps. 6.2 and 6.3). This behavior is easy to obtain in harnessed honey bees and recordings from neurons mediating the olfactory and the reward information are possible while the honey bee is performing the behavioral task [24, 34]. Several studies demonstrated that biogenic amines are crucially involved during learning (review: [42]). In addition, the role of other transmitters such as glutamate or ACh was identified by pharmacological manipulations [7, 22, 33].

The main neuropils involved in olfactory learning are the antennal lobes (ALs), the mushroom bodies (MBs) and the subesophageal ganglion (SEG) (Fig. 3.3.1). Several of the putative transmitter receptors were immunocytochemically identified (see below, reviews: [2, 3]). The synapses from primary afferents of the olfactory receptor neurons onto local interneurons and projection neurons within the antennal lobe (AL) (see Chap. 4.1) are probably cholinergic, because the antennal nerve shows strong activity for the ACh degrading enzyme acetylcholine esterase (AChE) and functional ACh receptors are expressed within the AL glomeruli (see also Chap. 3.5). Most of the local AL interneurons are inhibitory. This is supported by immunocytochemistry and optophysiological studies [38]. However, in *Drosophila* some of the local neurons have been described to be excitatory cholinergic neurons. Glutamate-immunoreactive profiles are innervating the ALs where glutamate serves as an inhibitory transmitter (see Chap. 3.5). Finally, histamine-immunoreactive neurons innervate the whole ALs. From the different glomeruli the axons of the projection neurons form five different fascicles [31]. The lateral antennal lobe-protocerebral tract (l-APT) links the ALs to the lateral horn (LH) and the calyces (CA) of MBs. It is stained for taurine-like immunoreactivity [41]. The medio-lateral antennal lobe-protocerebral tract (ml-APT) transmits olfactory information to the lateral horn and various other protocerebral regions and is stained for GABA. The median antennal lobe-protocerebral tract (m-APT) sends information to the calyces of MBs and then to the lateral horn and has been stained for AChE. This indicates that the projection neurons running within the m-APT probably release ACh at their terminals. Accordingly, the postsynaptic Kenyon cells (KCs) are immunoreactive to an antiserum against nicotinic ACh receptors (nAChRs) [3] and express functional nAChRs (see Chap. 3.5). The transmitter of the MB intrinsic Kenyon cells is as yet unknown in honey bees as in *Drosophila*. Candidates include FRMFamide peptide [45, 47] (see also Chap. 3.8) and taurine [41]. Kenyon cell axons synapse onto MB output neurons and onto MB feedback neurons (see Chaps. 3.1 and 3.2). The feedback neurons are probably GABAergic and form a massive inhibitory loop, the proto-cerebro-calycal tract from the lobes to the calyces of the MBs [20].

Biogenic amines are synthesized by a relatively small number of neurons in the honey bee brain. Immunostaining showed that neurons releasing DA, serotonin (5-hydroxy-tryptamine, 5-HT), histamine (HA) and OA often possess widespread projections (for a review see [2]). *Dopamine*: Approximately 330 DA-immunoreactive somata have been identified in each brain hemisphere and the SEG. A dense network of DA-immunoreactive fibres surrounds the MBs and the central body. The fibres project into the MB neuropils and into the somata rind, where they synapse onto Kenyon cell bodies. The ALs are innervated by fine projections of DA-immunoreactive interneurons (see Chap. 3.7). *Serotonine*: All brain regions, except the protocerebral bridge, contain 5-HT-immunoreactive fibres. A total of 75 somata were identified with a dense, stratified staining in the optic lobes (OLs). A net of 5-HT-immunoreactive fibres innervates the MBs outside the calyces, the ALs and almost all parts of the central body. *Histamine*: About 150 HA-immunoreactive neurons innervate most parts of the protocerebrum except the MBs. The ALs are densely innervated by about 35 HA-immunoreactive fibres. HA is probably the transmitter of the visual system, because photoreceptor fibers terminating either in the lamina (LA) or in the medulla (ME) as well as axons from ocellar photoreceptors contain HA. *Octopamine*: OA is a key neuromodulatory transmitter mediating the reward information during classical appetitive conditioning in bees [24, 25]. The OA-immunoreactivity indicated five cell clusters containing just over 100 OA-immunoreactive somata. They comprise a number of neurosecretory cells of the pars intercerebralis, a cell cluster located mediodorsal to the ALs, a group of cells distributed on both sides of the protocerebral midline, another group between the lateral horn and the antennal mechanosensory and motor center (AMMC), and a single soma on either side of the central body. Within the SEG the cluster of ventral unpaired median neurons stains against OA-antisera. In this cluster lies the soma of identified neurons such as VUM_{md1} [44] or VUM_{mx1} [24]. VUM_{mx1} forms extensive axonal projections within both ALs, the lip and basal ring regions of the CAs, and the lateral horn. Among the 10 VUM neurons of the SEG six neurons innervate the central brain and four innervate the periphery [44]. *Tyramine*: Brain levels of tyramine (TA) were measured in normal and queenless honey bees [40] but the distribution of TA has not been studied at the cellular level and immunohistochemical studies are as yet missing. It is expected, however, that TA is present in all OA-containing cells, because it is a precursor of OA during biosynthesis [42]. A honey bee TA receptor was cloned (AmTYR1) [4], and is expressed in somata of most neuropils in the honey bee brain [36] and a second putative TA receptor was annotated from the honey bee genomic sequence [26]. Particularly the MB intrinsic Kenyon cells, the somata surrounding the ALs, the AMMCs, and the first and second optic chiasmata express *Amtyr1* mRNA [4].

3.3.3 Neurotransmitters Involved in Honey Bee Learning and Memory

3.3.3.1 *Acetylcholine*

ACh is widely distributed in the honey bee brain. Binding experiments have demonstrated the existence of cholinergic receptors with nicotinic and muscarinic pharmacology, and both support differential roles during honey bee associative and non-associative learning [21], memory formation and retrieval.

3.3.3.1.1 Nicotinic Receptors

Nicotinic binding sites are found in the OLs and the ALs, in the MBs, specifically in the lip region of the calyces, in the vertical and medial lobes and in the SEG. The physiology and pharmacology of the nAChRs are reviewed in Chap. 3.5. Briefly, the honey bee nAChRs are ionotropic, cation-selective receptors with a high Ca^{2+} permeability and capable of mediating fast excitatory synaptic transmission. The honey bee genome sequencing [Honeybee Genome Sequencing 10] indicated the presence of 2 β (β1-2) and nine α (α1-9) nicotinic acetylcholine subunits [29]. Five of them (α2, α4, α8, α7, β1) have been localized in olfactory neuropils using *in situ* hybridization and single-cell RT-PCR [15, 50, 51]. The presence of 11 genes for nicotinic subunits in the honey bee genome suggests that different subtypes of nAChRs with distinct pharmacological properties can result from the multiple combinations of subunit assembly. However, the subunit compositions of insect nAChRs remain as yet unknown. Interestingly, the response to nicotine recorded from cultured projection neurons of the l-APT differed from those of the m-APT [37]. This observation suggests that the two tested neuronal populations express different nAChR subtypes.

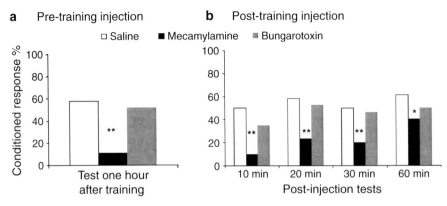

Fig. 3.3.2 Effects of mecamylamine and α-bungarotoxin on one-trial olfactory learning. (**a**) Effect on acquisition processes. The drugs were injected 20 min before training and the honey bees were tested 1 h after training. (**b**) Effects on retrieval processes. The drugs were injected 20 min after training. The conditioned response was tested at four different times after injection in independent groups. The saline and drug-injected groups comprised each 30 honey bees. Fisher χ^2 test (comparison to saline group): *$P<0.05$, **$P<0.01$ (Adapted from Gauthier 2010)

The roles of nAChRs in memory processes are rather complex. They were evaluated in harnessed honey bees using olfactory or tactile conditioning of the PER comprising one or multiple trials. The nicotinic antagonists, mecamylamine and α-bungarotoxin (α-BGT), were injected into the brain before or after the training session. In the one-trial learning situation, α-BGT had no effect on the different memory processes whatever the time of injection (Fig. 3.3.2a, b). Mecamylamine injections prior to one-trial learning impaired the acquisition processes (Fig. 3.3.2a). When the injection was performed 20 min after the single learning trial, it induced a decrease in retrieval performance for 1 h post training (Fig. 3.3.2b). The blocking of acquisition and retrieval of olfactory cue was probably not due to an impairment of olfactory perception by cholinergic blockade, because mecamylamine-injected bees could discriminate well between attractive and repellent odorants when tested in an olfactory Y-maze [7].

When mecamylamine was injected before multiple-trial tactile learning, a decrease of the CR rate was observed during the acquisition and at the short-term memory test but the long-term memory was intact (Fig. 3.3.3a). The same response profile was observed in olfactory experiments and indicated that mecamylamine-treated honey bees did indeed smell the odorants [23]. By contrast, injections of α-BGT before or after multiple-trial learning acquisition decreased memory retention 24 h after learning but not at shorter intervals of 1 and 3 h (Fig. 3.3.3a, b). A decrease of the 24 h memory retention was also observed after pre- or post-training injection of the potent nAChR antagonist methyllylcaconitine (MLA) (data not shown) [23].

From these results we hypothesized that at least two subtypes of nAChRs exist in the honey bee brain. Single stimulation as those prevailing during one-trial learning or retrieval tests will activate α-BGT-insensitive receptors whereas multiple-trial

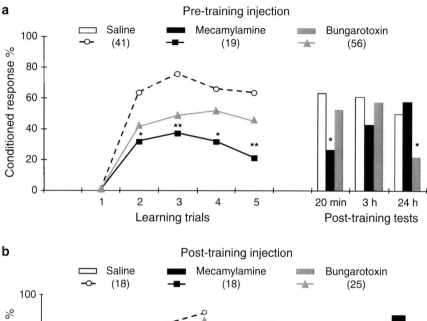

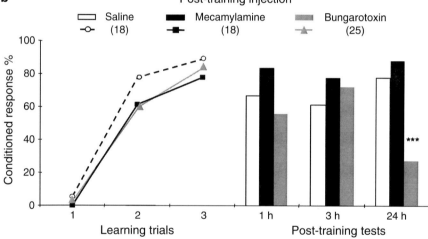

Fig. 3.3.3 Effects of mecamylamine and α-bungarotoxin on multiple-trial learning. (**a**) The drugs were injected 10 min before acquisition of five-trial tactile learning. Each trial consisted in presenting as a CS a metal plate to both antennae, followed by sucrose solution delivered to the proboscis. Each honey bee was tested at the three times after acquisition (Reprinted from Dacher et al. 2005). (**b**) The drugs were injected 20 min after the third acquisition trial of olfactory learning. Each honey bee was tested at the three times after acquisition (Reprinted from Gauthier et al. 2006). The numbers of honey bees in each group are presented in brackets. Fisher χ^2 test (comparison to saline group): *P<0.05, **P<0.01, ***P<0.001

learning will specifically activate α-BGT sensitive nAChRs. This in turn will trigger intracellular events leading to the formation of long-term memory [11, 12, 23].

Recently, we used small interfering RNA (siRNA) against α7 and α8 nicotinic subunits to decrease the expression of the target receptors during acquisition or retrieval of olfactory learning. Injections of siRNA against the α7 subunit totally disrupted the learning of the honey bees whereas siRNA against α8 induced retrieval impairment (Gauthier et al. *unpublished data*.). This suggests that the two α subunits do not co-assemble in the same nAChR. Because both mecamylamine and α8 siRNA injections impaired the ability of the honey bee to retrieve the learned odorant, we propose that α-BGT-insensitive receptors contain at least one α8 nicotinic subunit.

3.3.3.1.2 Muscarinic Receptors

Muscarinic AChRs are metabotropic, G-protein-coupled receptors. In the honey bee brain the localization of the muscarinic binding sites is mostly restricted to the central body (CB) and the pedunculi (PED) of the MBs [5]. A single gene coding for a muscarinic receptor (mAChR) has been identified in the honey bee genome [26] which shares homology with the gene coding for the *Drosophila* mAChR DM1. The muscarinic activation of the DM1 mAChR expressed in S2 cell line increases intracellular calcium through phospholipase C activation and cytosolic IP3 (inositol trisphosphate) increase. In the honey bee, applications of muscarine on cultured projection neurons induce an increase in intracellular calcium and the responses differ between neurons constituting the l-APT and the m-APT [37].

The behavioral effect of the muscarinic agonist pilocarpine in honey bees include a diminished aggressive behavior towards nestmates and improved kin recognition. These effects are antagonized by the muscarinic antagonist scopolamine [27]. Age-matched bees confined to the hive had enlarged MBs (comparable to those of foragers) after receiving pilocarpine treatments [28] (see Chap. 3.2). These two examples indicate that mAChRs are involved in structural and functional plasticity of the honey bee brain.

The effects of muscarinic antagonists were also studied on short-term memory using one-trial conditioning of the PER [8, 22]. Twenty-minute post-trial brain injection of the muscarinic antagonists scopolamine and atropine, but not pirenzepine, induced a decrease of the CR rates during the retrieval tests, an effect that lasted for 20 min, followed by a complete recovery of the conditioned response at 1 h (Fig. 3.3.4). No effects were observed on the memory tests when the drugs were injected 10 min before training. Thus, scopolamine and atropine, but not pirenzepine, induced a specific and temporary inhibition of retrieval processes.

We recently conducted similar experiments using multiple-trial learning session to test the effect of scopolamine on short- (1 and 3 h) and long- (24 h) term memory in independent groups. For the three tested intervals (1, 3 and 24 h) scopolamine injected 20 min prior to the test induced a transient decrease of the retrieval performance. Scopolamine injected before or immediately after the training session did not affect long-term memory at 24 h (unpublished observations). Together, these data indicate that muscarinic cholinergic signaling is important for memory retrieval but not for consolidation processes.

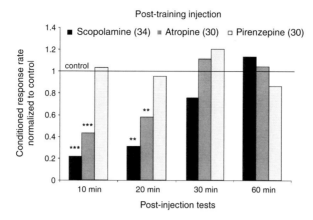

Fig. 3.3.4 Effect of scopolamine, atropine and pirenzepine on retrieval of one-trial olfactory learning. The drugs were injected 20 min after training. The conditioned response was tested at four different times after injection in independent groups. The conditioned response rate was normalized to saline control groups (y axis at the 1 value). The number of honey bees in each group is presented in brackets. Fisher χ^2 test (comparison to saline group): $**P < 0.01$, $***P < 0.001$ (Adapted from Gauthier et al. (1994) and Cano Lozano and Gauthier (1998))

Neuropharmacological approaches also identified the roles of the different parts of the MBs in olfactory learning and memory [6]. Mecamylamine injected before one-trial learning into the calyces impaired acquisition (Fig. 3.3.5a). Retrieval processes were disturbed by mecamylamine and scopolamine bilaterally injected 20 min post-training into the VLs but not into the calyces (Fig. 3.3.5b). These experiments indicate that the recall of memory requires both muscarinic and nicotinic receptors within or around the VLs. As yet, these cholinergic networks remain enigmatic because Kenyon cells are probably not cholinergic. We may hypothesize that putatively cholinergic extrinsic neurons projecting into the VLs are involved in retrieval processes.

Taken together, these experiments allowed us to establish a functional brain map of the formation and retrieval of olfactory memories. The MB calyces appear to be essential during acquisition and consolidation processes. During retrieval of simple CS-US associations, the cholinergic networks of MB calyces seem to be no longer necessary. This could also mean that the non-cholinergic l-APT and ml-APT neurons, are sufficient to activate the retrieval circuitry. As in *Drosophila*, MB output signaling is essential during retrieval but not during acquisition and consolidation of olfactory information. These data may suggest that a consolidated olfactory memory is – at least partially – located within the lobes of MBs in bees and flies. In addition, access to some consolidated memories requires the activation of VL networks. These results reinforce the hypothesis of a functional map of the brain with input regions (the ALs and the CAs) necessary for associating the CS and US and an output region (MB VLs) involved in retrieval processes.

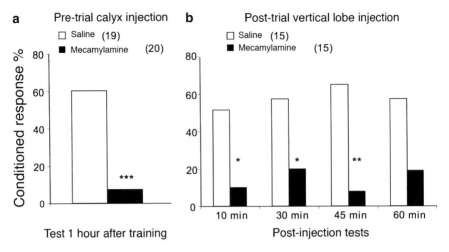

Fig. 3.3.5 Effects of localized brain injection of mecamylamine on one-trial olfactory learning. (**a**) Effect of mecamylamine injected 10 min before training between the median and lateral calyces of each MB. The honey bees were tested 1 h after training. (**b**) Effect of mecamylamine injected into the vertical lobe 20 min after training. The conditioned response was tested at different times after injection in independent groups. The number of honey bees in each group is presented in brackets. Fisher χ^2 test: *$P<0.05$, **$P<0.01$, ***$P<0.001$ (Reprinted from Cano Lozano et al. 2001)

3.3.3.2 Glutamate

In honey bees like in other insects, glutamate is an excitatory neurotransmitter at the neuromuscular junction but considerable evidence indicates that glutamate also acts as a neurotransmitter in the insect CNS. Glutamate-like immunoreactivity [3] and the expression pattern of the NMDA receptor subunit NR1 were described in the honey bee brain (see Chap. 3.4). Several evidences indicate a role of glutamate and NMDA receptors in learning and memory.

A metabotropic glutamate receptor AmGluRA identified in the honey bee genome is expressed in the brain of pupae and adult honey bees and is involved in olfactory memory formation since both agonists and antagonists of the receptor impair memory retention 24 h after training. However, a physiological characterization of AmGluRA has not yet been performed (see Chap. 3.4).

Glutamate also induces post-synaptic inhibition via activation of ionotropic glutamate-gated chloride channels (GluCl receptors) in cultured pupal honey bee AL neurons. The honey bee genome revealed the existence of a single GluCl subunit but alternative splicing leads to the formation of two variants [30]. The roles of these inhibitory glutamate receptors in olfactory learning were investigated by injecting the GluCl receptor agonist, ivermectin, into the brain. Low doses of ivermectin decreased LTM (long-term memory), higher doses rescued LTM performance. The precise mode of action needs to be analyzed, but injections under blockade of glutamate transporters using L-*trans*-PDC or the GABA analog, TACA, suggested that the rescue occurs through activation of GABA receptors [18].

3.3.3.3   GABA

Inhibitory neurotransmission in the insect as well as in the mammalian central nervous system is provided mainly by chloride channels gated by γ-amino-butyric acid (GABA$_A$ receptors). In the honey bee immunostaining showed the presence of GABA in all main neuropil areas with a high density in local interneurons of the ALs and in MB feedback neurons [3]. The functional properties of GABA$_A$ receptors were repeatedly analyzed in honey bee neurons (see Chap. 3.5). GABA receptors in the honey bee ALs appear to be required for fine odor discrimination [46]. Inhibition of the brain GABA receptors abolished the glomerular code for odor representation [38] and disrupted olfactory-induced synchronization of local field potentials in MBs [46].

The role of GABA in honey bee cognitive function was studied by behavioral pharmacological experiments. The insecticide fipronil acts as a chloride channel blocker and impaired tactile [1] and olfactory memory [16, 17], orientation abilities in a complex maze [14] and homing flight from a feeder to the hive [13]. The detrimental effects on olfactory memory of intra-thoracic fipronil injections could be rescued by co-injections of the glutamate transporter blocker L-*trans*-PDC indicating that both inhibitory GluCl and GABA$_A$ receptors may be involved in olfactory memory [17]. These experiments indicated that GABAergic inhibitory network contribution to associative learning might be mediated via olfactory information processing within the ALs [38, 46] or within the MBs.

GABAergic neurons provide feedforward inhibitory projections into the calyces [20]. In *Drosophila* these neurons are strongly activated by olfactory projection neurons, and maintain a tonic inhibition on Kenyon cells, explaining the sparse responses of Kenyon cells to odors. Such a gain-control and sparse odor coding by Kenyon cells have also been demonstrated in honey bees [48].

3.3.3.4   Biogenic Amines

The role of biogenic amines as neuromodulators in honey bees has been investigated in a number of studies (review: [42]) and has been demonstrated during habituation, sensitization and associative olfactory learning.

The first evidence for the pivotal role of OA came from electrophysiological experiments on the VUM$_{mx1}$ neuron of the SEG [24]. Depolarizations of VUM$_{mx1}$ may substitute for the reward property of the unconditioned stimulus (US) during classical conditioning [24] probably due to the depolarization-induced release of OA. Secondly, intraneuropilar OA injections into the ALs or the MB calyces can substitute for the US during classical conditioning [25]. Thirdly, silencing of OA receptor expression using siRNA injections impairs acquisition and retrieval of appetitive olfactory learning [19]. Early studies using OA injections into the brain indicated that OA increases the responsiveness to unconditioned odor stimuli. The specificity of OA during appetitive classical conditioning was proven by Menzel et al. [35]. Reserpine depletions of biogenic amines impair sensitization and appetitive conditioning [35]. Injections of OA into depleted animals rescue acquisition but not retrieval and sensitization. Although the intracellular pathway of the OA-mediated reward signal is as yet unclear, one mechanism may comprise intracellular Ca^{2+} signals via OA receptors (AmOA1) expressed by Kenyon cells (see also Chap. 3.5).

DA injections leave sensitization and conditioning unaffected but rescue motor performance in reserpine-treated animals [35]. DA inhibits the retrieval of learned odor information when injected after conditioning, but had no effects on acquisition when injected prior to conditioning. Interestingly, aversive olfactory conditioning of the sting extension reflex appears not to depend on OA but on DA. Recent evidence points to the involvement of dopaminergic neurons in aversive learning and octopaminergic neurons in appetitive learning (see Chap. 3.6).

Much less is known about the function of the other biogenic amines on learning-related processes. For example, the putative functional roles of HA during olfactory information processing or learning were not yet determined. However, bath applications of high concentrations of histamine reduced odor-induced neural activity in ALs [39]. Injections of 5-HT impair acquisition of appetitive olfactory conditioning [35]. Two 5-HT receptors were cloned, Am5-HT$_7$ [43] and Am5-HT$_{1A}$ [49]; the latter plays a role in phototactic behavior. Honey bee neurons express specific TA receptors, AmTYR1, which may also bind to OA [4] and another potentially more specific TA receptor [9]. However, the specific functional roles of TA still need to be analyzed in detail.

3.3.4   Conclusion and Outlook

Many transmitters and receptors are involved in honey bee learning. It is obvious that cholinergic synaptic transmission is involved in both memory formation and retrieval processes and the reinforcing processing pathways largely depend on octopaminergic and dopaminergic neuromodulations. The segregation of positive and negative reinforcing signals into two distinct aminergic systems could be a too simple scenario as abnormal DA receptor expression in the MBs of the fly impairs both appetitive and aversive learning. More information are needed as to the types of G-protein coupled receptors targeted by DA and OA, respectively, and the intra cellular signaling pathways they use [26].

Although anatomical [24] and pharmacological [25] data point to a convergence of aceylcholine and OA signals onto AL neurons and Kenyon cells, studies at the cellular level are missing to explain how the two concomitant signals can induce synaptic plasticity.

Several other aspects of the functional neurochemistry of the honey bee brain remain to be elucidated. For example, taurine is abundant in the insect central nervous system [41]. In honey bees, taurine is localized in Kenyon cells and neurons of the l-APT. Co-localization of taurine and FRMFamide was found in the vertical and medial lobes of the MBs in honey bee [47]. However, we still do not know how taurine or FRMFamide act in the honey bee brain. Recent findings indicate that FRMFamide, besides exerting its modulatory synaptic function through the activation of G-protein coupled receptors, can directly act on Na^+ channel, inducing a fast excitatory action on invertebrate neurons [32]. With the advent of new molecular techniques to selectively and temporarily silence gene expression of specific transmitter and receptor pathways we will refine our current knowledge and complete the picture of transmitter functions in learning and memory.

References

1. Bernadou A, Demares F, Couret-Fauvel T, Sandoz JC, Gauthier M (2009) Effect of fipronil on side-specific antennal tactile learning in the honeybee. J Insect Physiol 55(12): 1099–1106
2. Bicker G (1999) Biogenic amines in the brain of the honeybee: cellular distribution, development, and behavioral functions. Microsc Res Tech 44(2–3):166–178
3. Bicker G (1999) Histochemistry of classical neurotransmitters in antennal lobes and mushroom bodies of the honeybee. Microsc Res Tech 45:174–183
4. Blenau W, Balfanz S, Baumann A (2000) Amtyr1: characterization of a gene from honeybee (*Apis mellifera*) brain encoding a functional tyramine receptor. J Neurochem 74:900–908
5. Cano Lozano V, Armengaud C, Gauthier M (1995) Implication of the cholinergic system in memory processes in the honeybee: role and distribution of cholinergic receptors. In: 23rd Göttingen Neurobiology conference, Göttingen, p 79
6. Cano Lozano V, Armengaud C, Gauthier M (2001) Memory impairment induced by cholinergic antagonists injected into the mushroom bodies of the honeybee. J Comp Physiol A 187:249–254
7. Cano Lozano V, Bonnard E, Gauthier M, Richard D (1996) Mecamylamine-induced impairment of acquisition and retrieval of olfactory conditioning in the honeybee. Behav Brain Res 81:215–222
8. Cano Lozano V, Gauthier M (1998) Effects of the muscarinic antagonists atropine and pirenzepine on olfactory conditioning in the honeybee. Pharmacol Biochem Behav 59(4):903–907
9. Cazzamali G, Klaerke DA, Grimmelikhuijzen CJP (2005) A new family of insect tyramine receptors. Biochem Biophys Res Commun 338:1189–1196
10. Consortium (2006) Insights into social insects from the genome of the honeybee *Apis mellifera*. Nature 443:931–949
11. Dacher M, Gauthier M (2008) Involvement of NO-synthase and nicotinic receptors in learning in the honeybee. Physiol Behav 95:200–207
12. Dacher M, Lagarrigue A, Gauthier M (2005) Antennal tactile learning in the honeybee: effect of nicotinic antagonists on memory dynamics. Neuroscience 130:37–50
13. Decourtye A, Devillers J, Aupinel P, Brun F, Bagnis C et al (2011) Honeybee tracking with microchips: a new methodology to measure the effects of pesticides. Ecotoxicology 20(2):429–437
14. Decourtye A, Lefort S, Devillers J, Gauthier M, Aupinel P et al (2009) Sublethal effects of fipronil on the ability of honeybees (*Apis mellifera* L.) to orientate in a complex maze. Julius-Kühn-Archiv 423:75–83
15. Dupuis JP, Gauthier M, Raymond Delpech V (2011) Expression patterns of nicotinic subunits $\alpha2$, $\alpha7$, $\alpha8$ and $\beta1$ affect the kinetics and pharmacology of ACh-induced currents in the adult bee olfactory neuropiles. J Neurophysiol 106:1604–1613

16. El Hassani AK, Dacher M, Gauthier M, Armengaud C (2005) Effects of sublethal doses of fipronil on the behavior of the honeybee (*Apis mellifera*). Pharmacol Biochem Behav 82(1):30–39
17. El Hassani AK, Dupuis JP, Gauthier M, Armengaud C (2009) Glutamatergic and GABAergic effects of fipronil on olfactory learning and memory in the honeybee. Invert Neurosci 9(2):91–100
18. El Hassani AK, Giurfa M, Gauthier M, Armengaud C (2008) Inhibitory neurotransmission and olfactory memory in honeybees. Neurobiol Learn Mem 90(4):589–595
19. Farooqui T, Robinson K, Vaessin H, Smith BH (2003) Modulation of early olfactory processing by an octopaminergic reinforcement pathway in the honeybee. J Neurosci 23(12):5370–5380
20. Ganeshina O, Menzel R (2001) GABA-immunoreactive neurons in the mushroom bodies of the honeybee: an electron microscopic study. J Comp Neurol 437(3):335–349
21. Gauthier M (2010) State of the art on insect nicotinic acetylcholine receptor function in learning and memory. Adv Exp Med Biol 681:143–149
22. Gauthier M, Cano Lozano V, Zaoujal A, Richard D (1994) Effects of intracranial injections of scopolamine on olfactory conditioning retrieval in the honeybee. Behav Brain Res 63(2):145–149
23. Gauthier M, Dacher M, Thany SH, Niggebrügge C, Déglise P et al (2006) Involvement of α-bungarotoxin-sensitive nicotinic receptors in long-term memory formation in the honeybee (*Apis mellifera*). Neurobiol Learn Mem 86:164–174
24. Hammer M (1993) An identified neuron mediates the unconditioned stimulus in associative learning in honeybees. Nature 366:59–63
25. Hammer M, Menzel R (1998) Multiple sites of associative odor learning as revealed by local brain microinjections of octopamine in honeybees. Learn Mem 5:146–156
26. Hauser F, Cazzamali G, Williamson M, Blenau W, Grimmelikhuijzen CJP (2006) A review of neurohormone GPCRs present in the fruitfly *Drosophila melanogaster* and the honeybee *Apis mellifera*. Prog Neurobiol 80:1–19
27. Ismail N, Christine S, Robinson GE, Fahrbach SE (2008) Pilocarpine improves recognition of nestmates in young honey bees. Neurosci Lett 439(2):178–181
28. Ismail N, Robinson GE, Fahrbach SE (2006) Stimulation of muscarinic receptors mimics experience-dependent plasticity in the honey bee brain. Proc Natl Acad Sci USA 103(1):207–211
29. Jones AK, Raymond-Delpech V, Thany SH, Gauthier M, Sattelle DB (2006) The nicotinic acetylcholine receptor gene family of the honey bee, *Apis mellifera*. Genome Res 16:1422–1430
30. Jones AK, Sattelle DB (2006) The cys-loop ligand-gated ion channel superfamily of the honeybee, *Apis mellifera*. Invert Neurosci 6(3):123–132
31. Kirschner S, Kleineidam CJ, Zube C, Rybak J, Grünewald B et al (2006) Dual olfactory pathway in the honeybee, *Apis mellifera*. J Comp Neurol 499:933–952
32. Lingueglia E, Deval E, Lazdunski M (2006) FRMFamide-gated sodium channels and ASIC channels: a new class of ionotropic receptors for FRMFamide and related peptides. Peptides 5:1138–1152
33. Locatelli F, Bundrock G, Müller U (2005) Focal and temporal release of glutamate in the mushroom bodies improves olfactory memory in *Apis mellifera*. J Neurosci 25(50):11614–11618
34. Mauelshagen J (1993) Neural correlates of olfactory learning paradigms in an identified neuron in the honeybee brain. J Neurophysiol 69(2):609–625
35. Menzel R, Heyne A, Kinzel C, Gerber B, Fiala A (1999) Pharmacological dissociation between the reinforcing, sensitizing, and response-releasing functions of reward in honeybee classical conditioning. Behav Neurosci 113:744–754
36. Mustard JA, Kurshan PT, Hamilton IS, Blenau W, Mercer A (2005) Developmental expression of a tyramine receptor gene in the brain of the honey bee, *Apis mellifera*. J Comp Neurol 483(1):66–75
37. Raymond-Delpech V, Augier A, Paute S, Sandoz JC (2008) Functional study of two populations of projection neurons in the antennal lobe of adult honeybee. In: Club de Neurobiologie des Invertébrés, Toulouse

38. Sachse S, Galizia GC (2002) Role of inhibition for temporal and spatial odor representation in olfactory output neurons: a calcium imaging study. J Neurophysiol 87:1106–1117
39. Sachse S, Peele P, Silbering AF, Guhmann M, Galizia CG (2006) Role of histamine as a putative inhibitory transmitter in the honeybee antennal lobe. Front Zool 3:22
40. Sasaki K, Nagao T (2002) Brain tyramine and reproductive states of workers in honeybees. J Insect Physiol 48(12):1075–1085
41. Schäfer S, Bicker G, Ottersen OP, Storm-Mathisen J (1988) Taurine-like immunoreactivity in the brain of the honeybee. J Comp Neurol 268(1):60–70
42. Scheiner R, Baumann A, Blenau W (2006) Aminergic control and modulation of honeybee behaviour. Curr Neuropharmacol 4:259–276
43. Schlenstedt J, Balfanz S, Baumann A, Blenau W (2006) Am5-HT7: molecular and pharmacological characterization of the first serotonin receptor of the honeybee (*Apis mellifera*). J Neurochem 98(6):1985–1998
44. Schröter U, Malun D, Menzel R (2007) Innervation pattern of suboesophageal ventral unpaired median neurones in the honeybee brain. Cell Tissue Res 327:647–667
45. Schürmann FW, Erber J (1990) FMRFamide-like immunoreactivity in the brain of the honeybee (*Apis mellifera*). A light-and electron microscopical study. Neuroscience 38(3):797–807
46. Stopfer M, Bhagavan S, Smith BH, Laurent G (1997) Impaired odour discrimination on desynchronization of odour-encoding neural assemblies. Nature 390(6655):70–74
47. Strausfeld NJ, Homberg U, Kloppenburg P (2000) Parallel organization in honey bee mushroom bodies by peptidergic Kenyon cells. J Comp Neurol 424(1):179–195
48. Szyszka P, Ditzen M, Galkin A, Galizia CG, Menzel R (2005) Sparsening and temporal sharpening of olfactory representations in the honeybee mushroom bodies. J Neurophysiol 94(5):3303–3313
49. Thamm M, Balfanz S, Scheiner R, Baumann A, Blenau W (2010) Characterization of the 5-HT1A receptor of the honeybee (*Apis mellifera*) and involvement of serotonin in phototactic behavior. Cell Mol Life Sci 67:2467–2479
50. Thany SH, Crozatier M, Raymond-Delpech V, Gauthier M, Lenaers G (2005) Apisalpha2, Apisalpha7-1 and Apisalpha7-2: three new neuronal nicotinic acetylcholine receptor alpha-subunits in the honeybee brain. Gene 344:125–132
51. Thany SH, Lenaers G, Crozatier M, Armengaud C, Gauthier M (2003) Identification and localization of the nicotinic acetylcholine receptor alpha3 mRNA in the brain of the honeybee, *Apis mellifera*. Insect Mol Biol 12(3):255–262

Chapter 3.4
Glutamate Neurotransmission in the Honey Bee Central Nervous System

Gérard Leboulle

Abstract There is increasing evidence that a glutamatergic neurotransmission is present in the honey bee central nervous system. Besides the localization of glutamate in the brain, membrane and vesicular glutamate transporters as well as specific receptors have been characterized. Glutamate receptors homologous to their vertebrate counterparts (NMDA, non-NMDA and metabotropic) have been identified. In addition, there are inhibitory currents mediated by glutamate-gated chloride channels, specific to invertebrates. Glutamate neurotransmission is widespread in the brain, but it is probably less important in the mushroom body. Several studies show that the activation of different components of the neurotransmission is required during or shortly after conditioning for the formation of specific memory phases. In addition, different regions of the brain are differently implicated in memory processes.

Abbreviations (Excluding Brain Areas and Gene/Protein Names)

AMPA	Amino-3-hydroxy-5-methylisoxazole-4-propionic acid
CS	Conditioned stimulus
eLTM	Early long-term memory
GABA	γ-aminobutyric acid
Glu-ir	Glutamate-like immunoreactivity
GluCl	Glutamate-gated chloride
lLTM	Late long-term memory
LTM	Long-term memory
LTP	Long-term potentiation

G. Leboulle (✉)
Neurobiologie, Freie Universität Berlin, Berlin, Germany
e-mail: gerleb@zedat.fu-berlin.de

C.G. Galizia et al. (eds.), *Honeybee Neurobiology and Behavior: A Tribute to Randolf Menzel*, DOI 10.1007/978-94-007-2099-2_14,
© Springer Science+Business Media B.V. 2012

MTM   Mid-term memory
NMDA   N-Methyl-D-aspartic acid
PER   Proboscis extension reflex
RNAi   RNA interference
US   Unconditioned stimulus

3.4.1 Introduction

The honey bee is a valuable model for the study of learning and memory. It can be investigated under laboratory conditions using the proboscis extension reflex (PER) conditioning, a kind of classical conditioning [3]. In one of the versions of the PER conditioning, the animal learns to associate a neutral odor, the conditioned stimulus (CS), with a sucrose reward, the unconditioned stimulus (US). The olfactory pathway is well described in the honey bee (see Chaps. 3.3, 3.5, and 4.1). Olfactory sensory neurons project to the glomeruli of the antennal lobe (AL), the first relay station of the olfactory system. Most AL neurons, composed of local interneurons and projection neurons, express functional acetylcholine receptors. Projection neurons connect the AL with the lateral horn and the calyces of the mushroom body (MB) (see Chap. 3.3). The calyces are formed by the dendrites of the Kenyon cells (KCs), the intrinsic MB neurons that express functional acetylcholine receptors. The neurotransmitters of the KCs are not firmly identified. The calyces are divided in different regions receiving inputs from different sensory modalities: the lip, the collar and the basal ring. Olfactory projection neurons make synaptic connections in the lip and the basal ring (see Chap. 3.2). The US pathway in appetitive conditioning is well characterized (see Chaps. 3.3, 3.5, and 4.1). Notably, one octopaminergic neuron, VUM_{mx1}, is described. Its soma is located in the subesophageal ganglion and it projects bilaterally to the AL, the lateral horn and the calyces. It is expected that the simultaneous activation of neurons of the CS and of the US, during conditioning, induces a modification of the synaptic connections between these pathways. Indeed, it was shown that the AL and the MB, but not the lateral horn, are important sites for memory (see Chap. 6.2). It was also shown that the cholinergic and the octopaminergic neurotransmissions are implicated in this process (see Chap. 3.3).

However, recent advances have shown that learning and memory cannot be reduced to the study of the interface between the cholinergic and the octopaminergic pathways. Other regions within the brain and other neurotransmitters play an important role in these mechanisms. Studies in the honey bee and *Drosophila* showed that the central complex and the output sites of the MB, the lobes, are implicated in different aspects of the memory processes, like the formation of specific memory phases [33, 47, 49] and memory retrieval [21] (see Chap. 3.3). In addition, behavioral states, like motivation [7, 15, 20] or attention [44], represented by dopaminergic neurons and by specific signaling cascades, are also implicated in memory processes in flies and honey bees. Besides classical neurotransmitters, numerous neuropeptides are identified and play a role in olfactory memory as well (see Chap. 3.7). About 20 years ago, glutamate was proposed to be a neurotransmitter of the central nervous system (CNS) of the honey bee but it is only recently that its role in learning and memory formation was investigated.

3.4.2   Glutamate in the Honey Bee Nervous System

Glutamate-like immunoreactivity (Glu-ir) is found in motor neurons and in interneurons of the brain [2]. Glutamate is the neurotransmitter at the neuromuscular junction in insects [18]. Therefore, it was expected to detect Glu-ir in the axons and the somata of motor neurons and indeed glutamate-induced currents are characterized in honey bee muscles [6] (see Chap. 3.5). However, acetylcholine is the principal excitatory neurotransmitter in the CNS of insects [4], while glutamate fulfils this function in vertebrates [46]. For this reason, it was surprising to detect relatively high levels of Glu-ir in the brain. Since glutamate is an amino acid that is also implicated in universal metabolic processes, its function as neurotransmitter in the CNS is still debated. For example, it is known that glutamate is produced by photoreceptor neurons to be transported into glial cells to activate glycolysis [42]. However, there are good reasons to believe that glutamate is a neurotransmitter in the CNS of the honey bee: Specific glutamate transporters and receptors have been partially characterized at different levels in the brain.

3.4.3   Components of the Glutamatergic Neurotransmission

Neurotransmitters are released from synaptic vesicles at presynaptic sites. Glutamate is transported into synaptic vesicles by vesicular glutamate transporters called VGLUT. A *vglut* gene is identified in the honey bee genome [41]. Studies in *Drosophila* showed that a single *vglut* gene is abundantly expressed in the whole brain except in the MB [8, 39]. Therefore, it is reasonable to think that *vglut* is also expressed in the honey bee CNS.

A putative excitatory amino acid transporter, Am-EAAT, has been identified [22]. These membrane proteins are involved in the recycling of the neurotransmitter by removing it from the synaptic cleft. Am-EAAT shares the highest level of identity to the human EAAT2 glutamate transporter. It is predominantly localized in the brain in comparison with other body parts and expression levels are higher in late pupae than in adults. Several transcripts are probably produced from this gene and sequence analysis indicates that the ten trans-membrane domains, characteristic for these proteins, are conserved.

In vertebrates, glutamate induces exclusively excitatory currents mediated by amino-3-hydroxy-5-methylisoxazole-4-propionic acid (AMPA), kainate and N-methyl-D-aspartate (NMDA) glutamate receptors. Glutamate induced currents depend principally on the activation of non-NMDA receptors (AMPA and kainate), that are mainly permeable to Na^+ and K^+ ions. The NMDA receptor, which is principally permeable to Ca^{2+}, contributes only to a small extent to the glutamate-induced current. Cultured KCs of honey bee pupae express excitatory currents induced by a non-NMDA glutamate receptor (see Chap. 3.5). Beside this study, nothing is known about these receptors in the honey bee. Several subunits are identified in *Drosophila*, some of them are expressed at the neuro-muscular junction [28, 34, 35, 37], in the CNS [43, 45] or at both locations [14]. It is likely that a similar situation is found in the honey bee. Three genes (*nmdar1*, *nmdar2*, *nmdar3*) encode NMDA receptor subunits in the honey bee [41]. The AmNR1 subunit is encoded by *nmdar1* that comprises 17 exons covering about 8,000 bases pairs on the chromosome 3. Several variants of the mRNA are identified [38, 48]. One of them, AmNR1-1 encodes the complete subunit whilst the others encode truncated isoforms. It is not known whether these truncated isoforms have a biological function. Sequence analysis reveals that key regions of the NR1 subunit are conserved in AmNR1 [48]. The highest homology level is found in the membrane domains. In particular, an asparagine that influences the divalent cation affinity of the receptor is conserved. Specific PKA, PKC and PKG phosphorylation sites are also found.

In contrast to ionotropic glutamate receptors that mediate fast synaptic neurotransmission, metabotropic glutamate receptors are implicated in modulatory synaptic actions. Their activation induces a variety of effects like the indirect gating of ion channels and the activation of several signaling cascades that influence the properties of the cell. There are three types of identified glutamate metabotropic receptors in mammals. Type I is positively coupled to phospholipase C and increases inositol triphosphates and diacylglycerol levels. Types II and III are negatively coupled to adenylyl cyclase and diminish cAMP levels. In the honey bee, two metabotropic glutamate receptors, AmGluRA and AmGluRB, have been described [16, 23]. AmGluRA mRNA is detected in the brain, the abdomen and the thorax whilst AmGluRB mRNA is found exclusively in the brain. The expression of *amglura* is developmentally regulated, the mRNA levels are more important in adults than in pupae. Sequence analysis shows that the characteristic domains of metabotropic glutamate receptors, including a large N-terminal glutamate binding domain and a cysteine rich motif preceding the seven trans-membrane domains, are present. The *amglura* gene is relatively long, it spans at least 30,000 base pairs but its 6 introns cover more than 80% of the length. Functional AmGluRA receptors were expressed in HEK cells (Human embryonic kidney cells) with different G proteins α-subunits [23]. The receptor can be modulated by agonists and an antagonist of type II glutamate metabotropic receptor but not by type I or type III agonists. Thus, AmGluRA is probably a type II glutamate metabotropic glutamate receptor. Phylogenetic analysis also suggests that AmGluRA belongs to type II, while AmGluRB is an orphan receptor. The expression of both receptors in insect cells shows that AmGluRA has a higher affinity for glutamate than AmGluRB [16]. However, based on a study in *Drosophila*, it was proposed that the natural ligand of AmGluRB might not be glutamate [23].

In invertebrates, glutamate-gated chloride (GluCl) channels constitute a particular class of glutamate receptors that mediate inhibitory currents [36]. Thus, alongside γ-aminobutyric acid (GABA), glutamate is an inhibitory neurotransmitter identified in invertebrates (see Chaps. 3.3 and 3.5). Glutamate-induced chloride currents are expressed on the vast majority of cultured AL neurons of pupae and adult honey bees [1, 11]. Interestingly, beside glutamate-induced currents, most of the recorded cells also express currents induced by GABA and acetylcholine [1]. For this reason, the authors proposed that these cells might be connected to cholinergic sensory neurons and to local GABAergic neurons [1]. GluCl channels belong to the *cys*-loop ligand-gated ion channel superfamily, including among others nicotinic acetylcholine receptors and GABA receptors [19]. The annotation of the honey bee genome reveals only one gene, *am_glucl*, encoding two alternatively spliced variants. Nothing is known about the conformation of the functional receptor in the honey bee. Its physiological properties suggest that different subtypes are co-expressed but the GluCl subunit can probably not co-assemble with GABA receptor subunits, as it has been documented in *Drosophila* [1].

3.4.4   Structure of the Glutamate Neurotransmission

The mRNAs of AmGluRA and AmGluRB [16], of AmNR1 [48] and of Am-EAAT [22] were localized by *in situ* hybridization. In addition, the AmNR1 subunit was localized by immunohistochemistry. The similarity between the Glu-ir [2] and the AmNR1 subunit [48] signals in brain neuropiles speaks for the specificity of these detections.

AmGluRA and AmGluRB mRNAs are homogenously localized in all somata [16]. Glu-ir [2] and AmNR1 subunit [48] signals are most intensive in the protocerebral lobe, in the central complex, in the subesophageal ganglion and in the optic lobe (Figs. 3.4.1 and 3.4.2). High AmNR1 mRNA levels are found in the somata regions of these neuropiles [48] (Fig. 3.4.3). In the optic lobe, the most prominent Glu-ir signals are detected in the monopolar cells, in the lamina and the medulla, as the retinula cells are devoid of staining. Signals are also detected in the lobula and in the optic tubercle [2]. The AmNR1 subunit shows a similar expression pattern in specific layers of the optic lobe [48]. High Am-EAAT mRNA levels are detected in the somata of optic lobe neurons [22].

In the AL, Glu-ir [2] and the AmNR1 subunit [48] are localized in the glomeruli, where the signals are in general more intense at the periphery that is enriched with the terminals of olfactory sensory neurons (see Chap. 4.1) (Figs. 3.4.1 and 3.4.2). The AmNR1 subunit is probably expressed by a large number of projection neurons and local interneurons because the mRNA is detected in most AL somata (Fig. 3.4.3). Only a few somata of AL neurons are positive for Glu-ir. It is not known if these neurons provide input to AL neurons or if they project to other brain regions.

The calyces, the input site of the MB, show very low levels of Glu-ir [2] and AmNR1 subunit [48] compared with the surrounding neuropiles. The signals are restricted to the lip and the basal ring regions. In these regions, Glu-ir probably originates from unidentified projection neurons and AmNR1 from the dendrites of KCs because the NMDA receptor is mainly expressed postsynaptically [9].

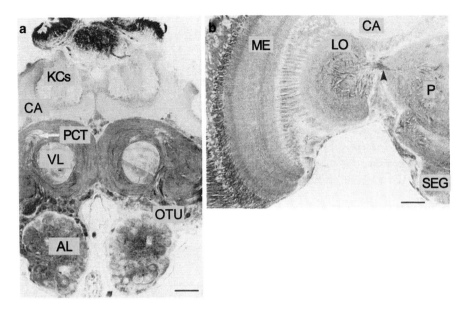

Fig. 3.4.1 Survey of Glu-ir in neuropiles of the central brain and optic lobes of the honey bee.
(**a**) The optic tubercles (OTU) contain Glu-ir fibers. The vertical lobe (VL), comprising the α and
the vertical γ lobes, contains rather low levels of staining. The calycal neuropil (CA) shows only
faint labeling except for the lip region, which is similar in staining intensity to that of the innermost
type I Kenyon cells (KCs). No labeling is found in the protocerebrocalycal tract (PCT). (**b**) Glu-ir
in the medulla (ME), lobula (LO), protocerebral lobe (P), and subesophageal ganglion (SEG).
Glu-ir fibers (*arrow*) extend from the lobula (LO) into the posterior protocerebrum. All scales,
100 μm (From Bicker et al. [2], Fig. 6)

The somata of type I and type II KCs are located in the inner and the outer part of
the calyces, respectively. In addition, KCs somata are located nearby the sub-region
of the calyces that they innervate. Highly variable AmNR1mRNA levels are found
in KCs [48]. In some sections of the calyces, the mRNA is restricted to the somata
of KCs innervating the lip and the basal ring, as in other sections it is detected in
almost all somata. It is surprising that the AmNR1 mRNA is detected in KCs that
do not innervate the lip and the basal ring regions (Fig. 3.4.3). In addition, the intensity of the mRNA signal appears very intensive in comparison with the AmNR1
subunit levels of the calyces. This indicates that the expression of the subunit is
either strongly post-transcriptionally regulated or that it is exported to other sites in
KCs. Only weak Glu-ir signals are detected in the somata of the innermost type I
KCs, innervating the basal ring, and in type II KCs [2] (Fig. 3.4.1). Interestingly,
the Am-EAAT mRNA levels are predominant in the innermost type I KCs [22].
Therefore, glutamate might be the neurotransmitter of only a small subset of KCs.
Recent studies in *Drosophila* support this assumption [8, 39].

    The axons of KCs project in the pedunculus and terminate in the lobes, the output
sites of the MB where they make up layers. The axons of type I KCs terminate in

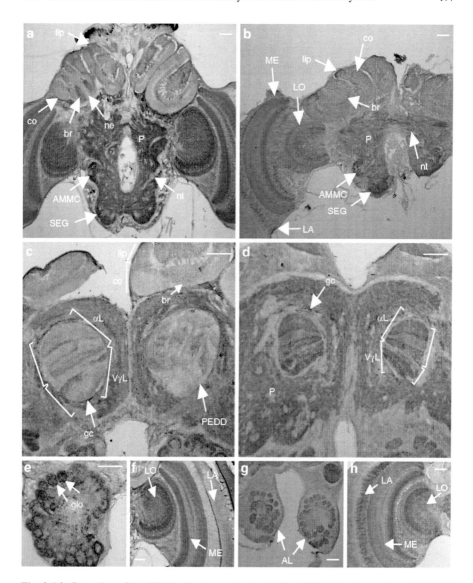

Fig. 3.4.2 Detection of AmNR1 in the honey bee brain. The NR1 subunit was detected with 2 antibodies directed against the NR1 subunit (NR1 pan (**a, c, e, f**) and mab363 (**b, d, g, h**)). Signals were found in the protocerebral lobe (P), the antennal mechanosensory and motor centre (AMMC) the subesophageal ganglion (SEG), and in the optic lobe, in the lobula (LO), medulla (ME) and lamina (LA) (**a, b, f, h**). In the MB signals were detected: in the lip (lip), in the basal ring (br) and in the neck (ne) and only limited signals were found in the collar (co) (**a–c**). The output region of the mushroom bodies revealed specific patterns in the α (αL) and the vertical γ (VγL) lobes, at the pedunculus devide (PEDD) between the vertical and the medial lobes and in glial cells (gc) (**c, d**). In the antennal lobes (AL) signals were found in the glomeruli (glo) (**e, g**). Scale bar = 100 μm (From Zannat et al. [48], Fig. 3)

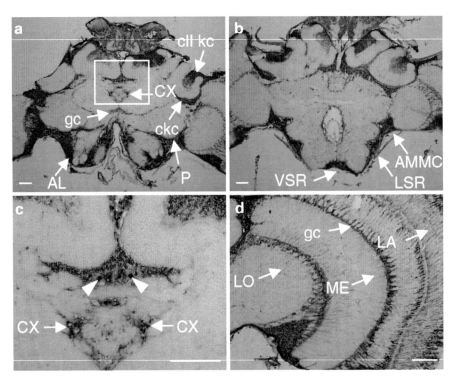

Fig. 3.4.3 Detection of *nmdar1* expression sites by *in situ* hybridization. Signals are detected in the somata region of the antennal lobes (AL), of the protocerebral lobe (P), of the central complex (CX), of the antennal mechanosensory and motor centre (AMMC), in the ventral (VSR) and lateral (LSR) soma rind of the subesophageal ganglion and in glial cells (gc). In the MB, type I (cII kc) and type II (ckc) KCs show heterogeneous signals (**a–c**). Enlargement of the window in (**a**) shows some isolated cells with intense staining (*white arrow* heads) (**c**). In the optic lobes, strong signals are detected in cell bodies of the lobula (LO), the medulla (ME) and the lamina (LA) (**d**). Scale bar = 100 μm (From Zannat et al. [48], Fig. 2)

the vertical and medial lobes and those of type II terminate principally in the vertical γ lobe [13, 40]. In addition, the calycal organization is represented in the lobes [40]. Glu-ir [2] and AmNR1 [48] are detected in several layers of the vertical α lobe, probably representing the lip, the collar and the basal ring, and in the vertical γ lobe (Figs. 3.4.1 and 3.4.2). It is expected that the Glu-ir detected in the lobes originates from KCs. However, only weak Glu-ir signals are detected in a subset of KCs somata. Thus, the origin of the Glu-ir and AmNR1 subunit signals in so many layers of the lobes remains to be determined. One possibility is that KCs receive input from extrinsic glutamatergic neurons in the lobes and that they express NMDA receptors at postsynaptic sites in this region. This would explain why high levels of AmNR1 mRNA and low Glu-ir levels are detected in KCs somata. It was already proposed that MB afferent neurons make connections with KCs in the lobes [40]. Alternatively, a study in *Drosophila* suggests that glutamate might be released as an autocrine or paracrine agent by KCs [39].

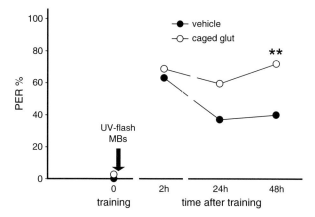

Fig. 3.4.4 Photorelease of glutamate in the MB immediately after training improves olfactory eLTM formation. Twenty minutes before training, the bees received systemic injections of vehicle or caged glutamate. The MB was photostimulated 5 s after training (UV flash, MBs). Vehicle, n = 35; caged glutamate, n = 32 (ANOVA repeated measurements, F = 4.86; p = 0.03 between groups and post hoc test; χ^2 test, **p < 0.01) (From Locatelli et al. [24], Fig. 2b)

3.4.5 The Glutamatergic Neurotransmission Is Important for Learning and Memory

The role of the glutamatergic neurotransmission in learning and memory was evaluated with the PER conditioning [3] (see Chap. 3.3). This procedure is extremely robust, one CS-US pairing leads to a conditioned response lasting several hours in more than 50% of the animals. Multiple conditioning trials induce the formation of a mid-term memory (MTM) immediately after learning and a consolidated long-term memory (LTM) lasting days. LTM can be subdivided into early LTM (eLTM, 1 and 2 days after conditioning) and late LTM (lLTM, from day 3 after conditioning).

Glutamate is important for the formation of olfactory memory [24]. Releasing a caged-glutamate complex on the MB of honey bees just after a single CS-US pairing consolidates a stable eLTM (Fig. 3.4.4). In addition, the specific agonists and the antagonist of type II metabotropic receptors were administered before multiple-trials conditioning to evaluate the role of AmGluRA in learning and memory [23]. Surprisingly, the agonists as well as the antagonist impair eLTM but have no effect during the acquisition and during the MTM test. In the same manner, L-trans-2, 4-pyrrolidine dicarboxylate (L-trans-PDC), an inhibitor of glutamate re-uptake acting on EAAT in vertebrates, induces a strong impairment of eLTM without affecting the acquisition and MTM when injected shortly before a multiple training procedure [26].

Several studies targeting GluCl channels suggest that they control other aspects of memory processes. Members of the *cys*-loop ligand-gated ion channel superfamily

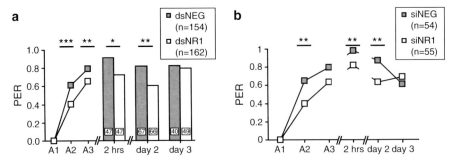

Fig. 3.4.5 The inhibition of the NR1 subunit selectively affects memory formation. One day after the injection of dsRNA (**a**), dsNEG, *gray*; dsNR1, *white*) or siRNA (**b**), siNEG, *gray*; siNR1, *white*), animals were subjected to three CS-US pairings (A1–A3). Memory was retrieved 2 hrs, on day 2 and on day 3 after conditioning. In the dsRNA experiment, the animals were tested only once. Data from the acquisition phase were pooled for all subgroups (n in the asset). The numbers on the bars represent the number of animals tested for each time point. In the siRNA experiment, n animals received multiple post-training tests. Asterisks indicate significant differences between groups (*p<0.05, **p<0.01, ***p<0.001, χ^2 test). PER represents the percentage of animals that showed a PER during the CS presentation (From Müßig et al. [31], Fig. 4)

are the target of pesticides, like fipronil and ivermectin. In the honey bee, fipronil inhibits principally chloride currents induced by glutamate and GABA [1]. Ivermectin is an agonist of GluCl channels well characterized in worms and flies [36]. These molecules were applied alone or in combination with drugs supposed to act either on the glutamatergic or on the GABAergic neurotransmissions to disentangle the possible role of these two kinds of chloride channels on the animal's behavior [10–12]. The application of the treatments before multiple training trials generally affects eLTM and sometimes the acquisition and MTM (see Chap. 3.3).

Pharmacological studies using NMDA receptor antagonists suggested a role in memory processes [25, 38]. Studies in *Drosophila* and mammals showed that the NR1 subunit is the obligatory subunit of NMDA receptors. For this reason, we used RNA interference (RNAi) against the AmNR1 subunit [31]. This technique induces the degradation of specific mRNA by double-stranded RNA having the sequence of the target mRNA [17]. The treatment led to a 30% reduction of the protein levels in the MB region. Interestingly, 2 hours after conditioning, the protein levels returned back to normal. This acute reduction of the NR1 subunit affected the acquisition phase, MTM and eLTM formation, but left lLTM intact (Fig. 3.4.5). The fact that lLTM was not affected by the treatment supports the previous idea that lLTM develops independently from earlier memory phases [15].

Some spatiotemporal aspects of glutamatergic neurotransmission have been deciphered. The studies on AmGluRA [23] and on the glutamate-release [24] show that memory formation is affected only when the treatments are applied during or shortly after conditioning. Releasing glutamate just before CS-US pairing or applying AmGluRA modulators 1 h after conditioning do not affect memory performances. In addition, application of AmGluRA modulators [23] or inhibition of

AmNR1 [31] during retrieval do not affect the performances of the animal. Thus, metabotropic and NMDA receptors are not required for the retrieval of memory. The studies on glutamate [24] and on AmNR1 [31] show the preponderant role of the glutamatergic neurotransmission in the MB. Interestingly, the release of glutamate [24] or the inhibition of the AmNR1 subunit (Gérard Leboulle, unpublished observation) in the AL at the moment of conditioning do not modulate memory formation. This shows that the glutamatergic neurotransmission in the AL and in the MB have different functions.

3.4.6   Outlook

Receptors homologous to the vertebrate NMDA, metabotropic and non-NMDA receptors, as well as a GluCl channel specific of invertebrates, are identified and partially characterized at the molecular, biochemical and physiological levels. This knowledge combined with the results of physiological studies help to better understand the structure of the glutamatergic networks. Almost all AL neurons express fast synaptic glutamate-induced currents principally mediated by GluCl channels. NMDA and metabotropic receptors are also localized in a majority of AL neurons and might thus be co-expressed with GluCl channels. Only a few AL neurons are glutamatergic and innervate unidentified cells. NMDA receptors, metabotropic receptors and excitatory currents induced by non-NMDA receptors were identified in KCs and might also be co-expressed in certain KCs. It would be interesting to determine if the glutamatergic neurotransmission found in different brain regions is dominated by inhibitory or excitatory currents. In addition, histological studies show that we have an incomplete understanding of the architecture of the glutamatergic neurotransmission. One important question is whether glutamatergic circuits are arranged in local networks in the different brain regions or if they are interconnected in a general circuitry.

Several studies show that glutamatergic neurotransmission plays a role in appetitive olfactory memory. Manipulating glutamate levels or glutamate receptor activity always modulate eLTM formation and in some cases (i.e. study of GluCl channel and NMDA receptors) acquisition and MTM. This suggests that the different components of neurotransmission play different roles in memory. It is worth noting that in some pharmacological studies, the specificity of the drugs used and the interpretation of their effects is questionable. In this regard, molecular tools, like RNAi, allow a more precise manipulation of the brain physiology, although they also have drawbacks (e.g. limited amplitude of the inhibition) and need to be improved. In this manner, we will be able to evaluate precisely the role of the different glutamate receptors in learning and memory.

The activation of the glutamatergic neurotransmission is required during or shortly after conditioning in the MB for memory formation. Although there is good evidence of glutamatergic neurotransmission in the AL, its modulation during or just after conditioning does not affect memory. It would be interesting to determine if the glutamate neurotransmission plays a role at a later moment of the procedure. The study on the AmNR1 subunit indicates that lLTM differs in its dependence on NMDA receptor activity from earlier memory phases. This is surprising because studies in other model systems show that the receptor is required during learning for the formation of all memory phases. It is known that eLTM is dependent on translation, and lLTM is dependent on transcription and translation in the honey bee [15]. Thus, it might be that NMDA receptors are differentially implicated in these spatially regulated processes. NMDA receptors are also studied for their implication in the induction of long-term potentiation (LTP), a cellular correlate of memory formation [5]. In vertebrates, LTP facilitates synaptic transmission by modifying the conformation and the composition of non-NMDA receptors expressed at the membrane [27]. Interestingly, LTP was successfully induced in the PE1 neuron, an extrinsic output neuron of the MB [30] (see also Chap. 3.1). However, this study and others showed that PE1 reduces its response to the CS after conditioning [29, 32]. It would be interesting to test if the glutamatergic neurotransmission is implicated in this particular kind of LTP.

References

1. Barbara GS, Zube C, Rybak J, Gauthier M, Grünewald B (2005) Acetylcholine, GABA and glutamate induce ionic currents in cultured antennal lobe neurons of the honeybee *Apis mellifera*. J Comp Physiol A 191:823–836
2. Bicker G, Schäfer S, Ottersen OP, Storm-Mathisen J (1988) Glutamate-like immunoreactivity in identified neuronal populations of insect nervous systems. J Neurosci 8:2108–2122
3. Bitterman ME, Menzel R, Fietz A, Schäfer S (1983) Classical conditioning of proboscis extension in honeybees (*Apis mellifera*). J Comp Psychol 97:107–119
4. Breer H, Sattelle DB (1987) Molecular-properties and functions of insect acetylcholine-receptors. J Insect Physiol 33:771–790
5. Cain DP (1997) LTP, NMDA, genes and learning. Curr Opin Neurobiol 7:235–242
6. Collet C, Belzunces L (2007) Excitable properties of adult skeletal muscle fibres from the honeybee *Apis mellifera*. J Exp Biol 210:454–464
7. Colomb J, Kaiser L, Chabaud MA, Preat T (2009) Parametric and genetic analysis of *Drosophila* appetitive long-term memory and sugar motivation. Genes Brain Behav 8:407–415
8. Daniels RW, Gelfand MV, Collins CA, DiAntonio A (2008) Visualizing glutamatergic cell bodies and synapses in *Drosophila* larval and adult CNS. J Comp Neurol 508:131–152
9. Dingledine R, Borges K, Bowie D, Traynelis SF (1999) The glutamate receptor ion channels. Pharmacol Rev 51:7–61
10. El Hassani AK, Dacher M, Gauthier M, Armengaud C (2005) Effects of sublethal doses of fipronil on the behavior of the honeybee (*Apis mellifera*). Pharmacol Biochem Behav 82:30–39
11. El Hassani AK, Dupuis JP, Gauthier M, Armengaud C (2009) Glutamatergic and GABAergic effects of fipronil on olfactory learning and memory in the honeybee. Invert Neurosci 9:91–100
12. El Hassani AK, Giurfa M, Gauthier M, Armengaud C (2008) Inhibitory neurotransmission and olfactory memory in honeybees. Neurobiol Learn Mem 90:589–595
13. Farris SM, Abrams AI, Strausfeld NJ (2004) Development and morphology of class II Kenyon cells in the mushroom bodies of the honey bee, *Apis mellifera*. J Comp Neurol 474:325–339
14. Featherstone DE et al (2005) An essential *Drosophila* glutamate receptor subunit that functions in both central neuropil and neuromuscular junction. J Neurosci 25:3199–3208

15. Friedrich A, Thomas U, Müller U (2004) Learning at different satiation levels reveals parallel functions for the cAMP-protein kinase A cascade in formation of long-term memory. J Neurosci 24:4460–4468
16. Funada M et al (2004) Characterization of the two distinct subtypes of metabotropic glutamate receptors from honeybee, *Apis mellifera*. Neurosci Lett 359:190–194
17. Hannon GJ (2002) RNA interference. Nature 418:244–251
18. Jan LY, Jan YN (1976) L-glutamate as an excitatory transmitter at the *Drosophila* larval neuromuscular junction. J Physiol 262:215–236
19. Jones AK, Sattelle DB (2006) The cys-loop ligand-gated ion channel superfamily of the honeybee, *Apis mellifera*. Invert Neurosci 6:123–132
20. Krashes MJ et al (2009) A neural circuit mechanism integrating motivational state with memory expression in *Drosophila*. Cell 139:416–427
21. Krashes MJ, Keene AC, Leung B, Armstrong JD, Waddell S (2007) Sequential use of mushroom body neuron subsets during *Drosophila* odor memory processing. Neuron 53:103–115
22. Kucharski R, Ball EE, Hayward DC, Maleszka R (2000) Molecular cloning and expression analysis of a cDNA encoding a glutamate transporter in the honeybee brain. Gene 242:399–405
23. Kucharski R, Mitri C, Grau Y, Maleszka R (2007) Characterization of a metabotropic glutamate receptor in the honeybee (*Apis mellifera*): implications for memory formation. Invert Neurosci 7:99–108
24. Locatelli F, Bundrock G, Müller U (2005) Focal and temporal release of glutamate in the mushroom bodies improves olfactory memory in *Apis mellifera*. J Neurosci 25:11614–11618
25. Lopatina NG, Ryzhova IV, Chesnokova EG, Dmitrieva LA (2000) N-Methyl-D-aspartate receptors in the short-term memory development in the honey bee *Apis mellifera*. Zh Evol Biokhim Fiziol 36:223–228
26. Maleszka R, Helliwell P, Kucharski R (2000) Pharmacological interference with glutamate re-uptake impairs long-term memory in the honeybee, *Apis mellifera*. Behav Brain Res 115:49–53
27. Malinow R, Malenka RC (2002) AMPA receptor trafficking and synaptic plasticity. Annu Rev Neurosci 25:103–126
28. Marrus SB, Portman SL, Allen MJ, Moffat KG, DiAntonio A (2004) Differential localization of glutamate receptor subunits at the *Drosophila* neuromuscular junction. J Neurosci 24:1406–1415
29. Mauelshagen J (1993) Neural correlates of olfactory learning paradigms in an identified neuron in the honeybee brain. J Neurophysiol 69:609–625
30. Menzel R, Manz G (2005) Neural plasticity of mushroom body-extrinsic neurons in the honeybee brain. J Exp Biol 208:4317–4332
31. Müßig L et al (2010) Acute disruption of the NMDA receptor subunit NR1 in the honeybee brain selectively impairs memory formation. J Neurosci 30:7817–7825
32. Okada R, Rybak J, Manz G, Menzel R (2007) Learning-related plasticity in PE1 and other mushroom body-extrinsic neurons in the honeybee brain. J Neurosci 27:11736–11747
33. Pascual A, Preat T (2001) Localization of long-term memory within the *Drosophila* mushroom body. Science 294:1115–1117
34. Petersen SA, Fetter RD, Noordermeer JN, Goodman CS, DiAntonio A (1997) Genetic analysis of glutamate receptors in *Drosophila* reveals a retrograde signal regulating presynaptic transmitter release. Neuron 19:1237–1248
35. Qin G et al (2005) Four different subunits are essential for expressing the synaptic glutamate receptor at neuromuscular junctions of *Drosophila*. J Neurosci 25:3209–3218
36. Raymond V, Sattelle DB (2002) Novel animal-health drug targets from ligand-gated chloride channels. Nat Rev Drug Discov 1:427–436
37. Schuster CM et al (1991) Molecular cloning of an invertebrate glutamate receptor subunit expressed in *Drosophila* muscle. Science 254:112–114
38. Si A, Helliwell P, Maleszka R (2004) Effects of NMDA receptor antagonists on olfactory learning and memory in the honeybee (*Apis mellifera*). Pharmacol Biochem Behav 77:191–197

39. Sinakevitch I, Grau Y, Strausfeld NJ, Birman S (2010) Dynamics of glutamatergic signaling in the mushroom body of young adult *Drosophila*. Neural Dev 5:10
40. Strausfeld NJ (2002) Organization of the honey bee mushroom body: representation of the calyx within the vertical and gamma lobes. J Comp Neurol 450:4–33
41. The Honeybee Genome Sequencing Consortium (2006) Insights into social insects from the genome of the honeybee *Apis mellifera*. Nature 443:931–949
42. Tsacopoulos M, Poitry-Yamate CL, Poitry S, Perrottet P, Veuthey AL (1997) The nutritive function of glia is regulated by signals released by neurons. Glia 21:84–91
43. Ultsch A et al (1992) Glutamate receptors of *Drosophila* melanogaster: cloning of a kainate-selective subunit expressed in the central nervous system. Proc Natl Acad Sci USA 89:10484–10488
44. van Swinderen B, Brembs B (2010) Attention-like deficit and hyperactivity in a *Drosophila* memory mutant. J Neurosci 30:1003–1014
45. Volkner M, Lenz-Bohme B, Betz H, Schmitt B (2000) Novel CNS glutamate receptor subunit genes of *Drosophila melanogaster*. J Neurochem 75:1791–1799
46. Weinberg RJ (1999) Glutamate: an excitatory neurotransmitter in the mammalian CNS. Brain Res Bull 50:353–354
47. Wu CL et al (2007) Specific requirement of NMDA receptors for long-term memory consolidation in *Drosophila* ellipsoid body. Nat Neurosci 10:1578–1586
48. Zannat T, Locatelli F, Rybak J, Menzel R, Leboulle G (2006) Identification and localisation of the NR1 sub-unit homologue of the NMDA glutamate receptor in the honeybee brain. Neurosci Lett 398:274–279
49. Zars T, Fischer M, Schulz R, Heisenberg M (2000) Localization of a short-term memory in *Drosophila*. Science 288:672–675

Chapter 3.5
Cellular Physiology of the Honey Bee Brain

Bernd Grünewald

Abstract Membrane-bound ion channels determine the electrical activity of excitable cells. In this respect honey bee neurons within the olfactory pathways are among the physiologically best studied insect cells. Several ionic currents were characterized from identified central neurons *in vitro*, in particular mushroom body Kenyon cells (KCs) and antennal lobe (AL) neurons. They express voltage-sensitive Na^+ and Ca^{2+} currents that depolarize the neurons upon activation. Outward K^+ currents are rather diverse. At least four types exist: a delayed rectifier, a rapidly inactivating A-type, a slowly inactivating and a Ca^{2+}-dependent K^+ current. This diversity of K^+ channels determines the threshold and shapes of single spikes and spike trains. Based on sequence analyses the honey bee genome contains genes coding for nine nicotinic acetylcholine receptor α-subunits, three GABA receptor subunits, one glutamate-chloride channel, three NMDA receptor subtypes, and two histamine-chloride channels. Acetylcholine-, GABA-, and glutamate-induced currents have been physiologically characterized. The ionotropic nicotinic cholinergic receptor is one of the major excitatory receptors of the olfactory pathway. It is involved during olfactory learning and therefore a good candidate for inducing learning-dependent synaptic plasticity (see Chap. 3.3). GABA-induced Cl^- currents provide the major inhibitory system. In addition, glutamate-sensitive Cl^- channels provide a parallel inhibitory network within the honey bee ALs. KCs express functional cation-selective AMPA-like receptors, whereas no physiological data exist on functioning NMDA-like receptors. Integrating the cell physiological data into a working model to explain experience-dependent neuronal plasticity is challenging, because the interactions of the various currents and signaling cascades and their contribution to experience-dependent plasticity remain to be analysed.

B. Grünewald (✉)
Fachbereich Biowissenschaften, Institut für Bienenkunde, Polytechnische Gesellschaft Frankfurt am Main, Goethe-Universität Frankfurt am Main, Karl-von-Frisch-Weg 2, D-61440 Oberursel, Germany
e-mail: b.gruenewald@bio.uni-frankfurt.de

C.G. Galizia et al. (eds.), *Honeybee Neurobiology and Behavior: A Tribute to Randolf Menzel*, DOI 10.1007/978-94-007-2099-2_15,
© Springer Science+Business Media B.V. 2012

Abbreviations (Excluding Anatomical Structures and Genes/Proteins)

4-AP	4-aminopyridine
AMPA	2-amino-3-(5-methyl-3-oxo-1,2-oxazol-4-yl) propanoic acid
cAMP	Cyclic adenosine monophosphate
MLA	Methyllycaconitine
NMDA	N-Methyl-D-aspartate
OA	Octopamine
TTX	Tetrodotoxin

3.5.1 Introduction

The electrical activity of excitable cells is determined by the ion channels that are expressed in the cell membrane. The generation and propagation of action potentials, synaptic transmission, and information processing of individual neurons or synaptic networks within the brain is mediated by ionic channel gating. Honey bee neurons have been analysed *in vivo* using intracellular recording techniques and optophysiological approaches, and *in vitro* using patch clamp recordings (see below, Table 3.5.1). These works make the bee brain one of the best studied insect systems for cellular physiology. This chapter reviews the physiology and pharmacology of ion channels and ionotropic transmitter receptors of central honey bee neurons, focussing on the olfactory pathways.

3.5.2 Voltage-Sensitive Currents

3.5.2.1 *Ionic Currents Recorded In Vitro*

Action potentials have been recorded *in vitro* from KCs [47], AL projection neurons and motoneurons [23], and olfactory receptor neurons [25]. Upon injection of depolarizing current, these cells generate overshooting action potentials that are sensitive to tetrodotoxine (TTX). To compare the ionic currents of honey bee neurons, they were taken into primary cell culture where they can be maintained for up to 2 weeks. In some cases the neurons were identified prior to recording by retrograde labelling using dextran-rhodamine injections [14]. Honey bee neurons express voltage-sensitive Na^+ (I_{Na}), K^+ (I_K) and Ca^{2+} (I_{Ca}) currents (Fig. 3.5.2).

   Activation of *voltage-gated Na^+* currents causes rapid membrane depolarization that forms the initial phase of action potentials. KCs and projection neurons express very similar I_{Na} currents [16]. I_{Na} activates at voltages more positive than −40 mV

Table 3.5.1 Physiologically characterized ionic currents of honeybee neurons

Current		Cell type	Method	References
Fast transient K$^+$	$I_{K,A}$	Kenyon cells	Patch clamp	[14, 31, 36, 47]
		Antennal motoneurons	*In situ* patch clamp	[23]
		ORN	Patch clamp	[25]
		Type 1 AL neurons	Patch clamp	[32]
Slow transient K$^+$	$I_{K,ST}$	Kenyon cells	Patch clamp	[47]
Sustained K$^+$	$I_{K,V}$	Kenyon cells	Patch clamp	[14, 31, 36]
		Antennal motoneurons	Patch clamp	[23]
		Antennal lobe neurons	Patch clamp	[14, 32]
		Skeletal muscle fibres	Patch clamp	[7]
		ORN	Patch clamp	[25]
Ca^{2+}-dependent K$^+$		Projection neurons	Patch clamp	[14]
		Type 1 AL neurons		[32]
Fast transient Na$^+$	I_{Na}	Kenyon cells	Patch clamp	[31, 36, 47]
		Olfactory receptor neurons	Patch clamp	[25]
Transient Ca^{2+}	I_{Ca}	Kenyon cells	Patch clamp	[31, 36]
		Projection neurons	Patch clamp	[14]
		ORN	Patch clamp	[25]
		Skeletal muscle fibres	Patch clamp	[7]
ACh receptors	I_{ACh}	Kenyon cells	Ca^{2+} imaging	[3]
		Kenyon cells	Patch clamp	[8, 11, 48]
		Antennal lobe neurons	Patch clamp	[1, 2, 30]
		Brain sections	In situ hybridisations, sequence	[42, 43]
GABA receptors	I_{GABA}	Kenyon cells	Patch clamp	[15]
		Antennal lobe neurons	Patch clamp	[2, 10]
Glutamate receptors	GluR$_{AMPA}$	Kenyon cells	Patch clamp	Grünewald, unpublished
	GluR	Skeletal muscle fibres	Patch clamp	[7]
	GluR$_{Cl}$	Antennal lobe neurons	Patch clamp	[2]

and the peak current amplitude of this fast transient, TTX-sensitive current is usually less than −1 nA. In addition, KCs express a small *sustained* I_{Na} (less than 1% of the total I_{Na}), which is voltage-sensitive and TTX-sensitive, but shows little or no inactivation during prolonged voltage command pulses [36, 48]. A comprehensive review comparing the voltage-gated currents in various insect species is presented by Wicher et al. [46].

Honey bee neurons express various I_k. Upon activation these currents hyperpolarize the membrane potential or keep it around the resting potential, because the flow of K$^+$ ions through the channels is usually outwardly directed at depolarized membrane potentials. All neurons tested possess a *delayed rectifier type K$^+$ current $I_{K,V}$*

[14, 23, 32, 36], which does not inactivate during prolonged depolarizing voltage commands. It is largely responsible for the repolarization of the membrane potential during a single spike [47]. Similar delayed rectifier K^+ currents were described in various other insects.

Transient K^+ currents have different cellular functions including spike repolarization, repetitive spiking or determining spike thresholds. The *shaker-like K^+ current* (also called A-current, $I_{K,A}$) is a fast activating, transient current, which is sensitive to the blocker 4-aminopyridine (4-AP) [31, 32]. *Shaker*-like K^+ currents are found in many different species across phyla and many different neuron types, where they influence major aspects of electrical activity such as spike duration during repetitive firing, firing frequency, synaptic transmission, or spike backpropagation (cf. [31, 46] for references). In Kenyon cells, $I_{K,A}$ plays only a minor role during spike repolarisation. Rather, $I_{K,A}$ modulates the spike duration and spike threshold [47]. Whereas KCs express pronounced *shaker*-like K^+ currents, such inactivating outward currents are much smaller in projection neurons [14].

A *Ca^{2+}-dependent K^+ current ($I_{K,Ca}$)* is expressed by projection neurons, but not by KCs [14, 32, 47]. Gating of these currents depends on the intracellular Ca^{2+} concentration. The Ca^{2+}-inflow at negative clamp potentials (below the Nernst potential of Ca^{2+}) activates an outward K^+ current. If the voltage-sensitive Ca^{2+} currents are blocked the K^+ current amplitude is decreased and the non-linear IV relationship is transferred into a linear one. The $I_{K,Ca}$ of AL neurons is modulated by dopamine (DA) [32]. Ca^{2+}-dependent K^+ currents play a major role in the control of neuronal excitability and, for example, mediate afterhyperpolarisation or spike repolarisation.

A fourth K^+ current was firstly indicated by mathematical simulations of the whole cell K^+ currents of honey bee KCs and subsequently experimentally identified [47]. It is a *slow transient K^+ current ($I_{K,ST}$)*. Unlike the *shaker*-like K^+ current, $I_{K,ST}$ is not sensitive to 4-AP and activates more slowly than A-type currents. Interestingly, two transient outward currents have been described in cultured *Drosophila* Kenyon cells, one of which was insensitive to 4-AP and might therefore correspond to the newly identified component in honey bee KCs. In *Drosophila,* currents with similar properties are based on genes of the *shab*-family (review: [46]). Computer modelling indicates that $I_{K,ST}$ is the primary determinant of the delayed spiking responses during constant current stimuli, and $I_{K,ST}$ prevented the model from responding to oscillatory stimuli. These findings suggest that the spiking characteristic of KCs *in vivo* could be profoundly altered by the modulation of $I_{K,ST}$.

Both the voltage-sensitive I_{Na} and I_K can be reversibly blocked by local anesthetics. Procaine as well as lidocaine reduces the current amplitude in a dose-dependent manner at concentrations between 0.1 and 10 mM [9]. They induce a very rapid block, which is readily reversible within a few minutes of wash. Consequently, local brain injections of procaine were used to study the dynamics and localisation of learning and memory formation within the honey bee brain [9].

Voltage-sensitive Ca^{2+} channels contribute to action potential generation, synaptic transmission or neuromodulation (review [46]). The voltage-sensitive Ca^{2+} currents of honey bee neurons are similar (with respect to steady-state activation, Cd^{2+}-sensitivity and inactivation) to those described in other insect preparations. They activate rapidly and show a slow inactivation [14, 36, 47].

3.5.2.2   *In Situ Patch Clamp Recordings*

Ionic currents recorded in the culture dish may differ from those of neurons within their normal brain environment. However, the rare examples of whole-cell currents recorded from bee brains *in situ* are similar to those *in vitro* (AL motoneurons: [23], projection neurons and Kenyon cells: [16]). Although the somata of insect neurons do not participate in action potential generation or propagation the overall shapes of the somata currents are conserved during *in situ* recordings. Outward currents of KCs show the typical transient K+ and delayed rectifier K+ currents. The same is true for AL neurons, which show significant inward currents and transients as well as sustained outward currents. Several technical problems, however, prevent the routine application of *in situ* recordings. First, the success rate of patch clamp recordings of neurons within the living brain is very low and the gigaseal is seldomly stable. Second, the space-clamping conditions of *in situ* recordings are largely compromised because the neurite is still attached to the soma, which hinders a detailed biophysical comparison between *in situ* and *in vitro* currents. Third, labelling of the recorded neurons for subsequent morphological identification and histological analysis is difficult, because the somata often tear off the remaining neurite during filling or during pipette retraction, which leads to rather weak staining intensities.

3.5.3   Mathematical Model of a Kenyon Cell

Using computer simulations of voltage-sensitive ionic currents one can analyse whether the experimentally derived kinetic and steady-state parameters are complete enough to be described in mathematical equations. Computer simulations can also be used to analyse the contributions of different currents to the generation of single action potentials and their behavior during repetitive spiking. Previous attempts to construct a honey bee Kenyon cell model based on voltage-clamp data [20, 31] were insufficient to correctly simulate the experimental data. The model by Pelz et al. [31] could not reproduce repetitive spiking. The Kenyon cell model published by Ikeno and Usui [20] spiked repetitively upon depolarization. The spike shape, however, was clearly different from the spike shape observed in Kenyon cells, both *in vivo* and *in vitro*. The partial weaknesses of these early models were due to the fact that no current clamp data were available at that time. This difficulty was overcome by measuring spike activity in cultured KCs [47].

Data from previous studies [14, 31, 36] together with new experiments [47] provided the bases for a Hodgkin-Huxley type model. The model consisted of a fast, transient Na+ current (I_{Na}), a fast, transient A-type K+ current ($I_{K,A}$), a delayed, non-inactivating K+ current ($I_{K,V}$), and a slow transient outward current ($I_{K,ST}$). The model was able to qualitatively reproduce the spiking behavior of the Kenyon cells. Simulations indicated that the primary currents that underlie spiking are I_{Na} and $I_{K,V}$, whereas $I_{K,A}$ and $I_{K,ST}$ modulate the spike shape and the characteristics of cellular responses to electrical stimulations. $I_{K,A}$ and $I_{K,ST}$ could be omitted from the cell model without affecting the general ability to spike repetitively. The model also mimicked the spike broadening that occurs in KCs when $I_{K,A}$ is blocked. This strongly expressed $I_{K,A}$ may prevent the KCs from firing action potentials and delay the onset of spiking upon experimental depolarisations. These data indicate that KCs have input resistances in excess of 1 GΩ and show little or no spontaneous activity, and no intrinsic bursting behavior *in vivo* as well as *in vitro*. These findings suggest that Kenyon cells *in vivo* are either constantly inhibited or inactive at resting potential, as they are in culture. Thus Kenyon cells may act as coincidence detectors, detecting simultaneous activity in projection neurons converging on the same KCs.

3.5.4 Synaptic Currents Within the Olfactory Pathway

The range of ionotropic transmitter receptors in insects is highly diverse and the physiological and pharmacological properties of the insect receptors are often very different from their vertebrate counterparts (reviews: [6, 17, 19, 45]). Based on sequence analyses, the honey bee genome contains genes coding for nine different α-subunits of the nicotinic AChR (Amelα1 – Amelα9) and 2 β-subunits (Amelβ1, Amelβ2), three different GABA$_A$ receptor subunits (Amel_GRD, Amel_RDL, Amel_LCCH3), one glutamate-gated chloride channel (Amel_GluCl), and two histamine-gated chloride channels (Amel_HisCl1, Amel_HisCl2). Currents through acetylcholine, GABA, and glutamate receptors have been physiologically characterized in honey bees (Figs. 3.5.1 and 3.5.2), physiological evidence for histamine-gated currents is still missing.

3.5.4.1 Acetylcholine

Immunocytochemistry indicates that acetylcholine (ACh) is the principal excitatory transmitter of honey bee olfactory receptor neurons (ORNs) and most AL projection neurons [24, 37]. Blockade of cholinergic synaptic transmission resulted in specific behavioral deficits in honey bees (see Chap. 3.3.3).

Functional nicotinic acetylcholine receptors (nAChR) were characterized in many insects (reviews: [17, 41]). Pressure applications of acetylcholine induce fast activating, desensitizing currents in cultured KCs from honey bee brains [11] or AL neurons [2]. This acetylcholine-induced current is a cation-selective current through a nAChR. The honey bee receptor is equally permeable for K$^+$ and Na$^+$ ions and has a high Ca^{2+} permeability [11]. This Ca^{2+}-permeability appears to be a general feature of insect nAChR. The receptor pharmacology shows a clear neuronal nicotinic profile [1, 2, 11, 48]. Acetylcholine and carbamylcholine are full agonists. Nicotine, epibatidine or cytisine are partial agonists. Based on the EC$_{50}$ values, the agonist with the highest affinity was nicotine, followed by acetylcholine, epibatidine, carbamylcholine and cytisine. The receptor is blocked by the classical nicotinergic blockers curare or

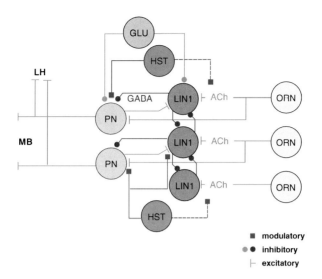

Fig. 3.5.1 Schematic diagram of a hypothetical synaptic wiring of the honey bee antennal lobe based upon the physiological identification of ligand-gated ionic currents and ultrastructural findings from other insect species. Olfactory receptor neurons (*ORN*) provide excitatory cholinergic input onto local inhibitory neurons (*LIN*) and projection neurons (*PN*). The LIN are GABAergic and presynaptic to PN and other LIN. They receive cholinergic input from PN, which are connecting the antennal lobe with the mushroom body (*MB*) and the lateral horn (*LH*). Glutamatergic neurons (*GLU*) provide a second inhibitory network and probably contact both PN and LIN. Finally, a group of histaminergic neurons (*HST*) may modulate the neural activity within the honey bee AL acting on the ionotropic ACh and glutamate receptors. Histamine may as well regulate the input into the AL by presynaptically inhibiting sensory afferences from the antenna (*dashed lines*); symbols, •=inhibitory, ⊥=excitatory, ■=modulatory synaptic connection (Modified after [2])

mecamylamine. However, dihydro-β-erythroidine (DHE) was the most potent antagonist of the honey bee nicotinic current, followed by methyllycaconitine MLA, and mecamylamine. The antagonist DHE has rarely been tested in insects. MLA, a plant alkaloid of larkspur (*Delphinium*) is a high-potency nicotinic blocker in KCs and several other insect preparations. In vertebrates MLA specifically blocks α7 containing neuronal receptors. This finding points to an interesting pharmacological similarity between the insect receptor and the vertebrate α7 receptor.

The honey bee nAChR is a target of neonicotinoid insecticides. These substances (like imidacloprid, thiacloprid or clothianidin) utilize the differences between the insect and vertebrate receptor to selectively interact with the insect receptor (review: [44]). Imidacloprid, for example, acts as a partial receptor agonist and elicits currents similar to those induced by nicotine or epibatidine [1, 2, 8, 30]. Although the affinity of the receptor to imidacloprid is comparably low, topical applications of imidacloprid affect learning capabilities of the honey bee and bees fed with imidacloprid showed reduced mobility (see Chap. 3.3).

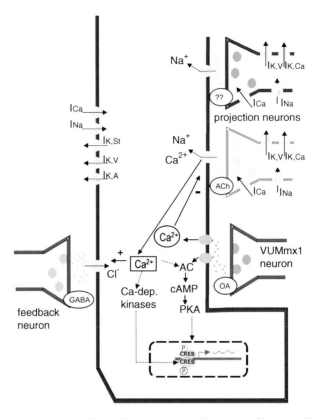

Fig. 3.5.2 Schematic diagram of the cellular physiology of honey bee Kenyon cells. The diagram includes all characterized ionic currents (see text for abbreviations). Glutamate-induced currents were omitted, because the releasing neurons are as yet unidentified. Odor-induced neural activity comprises release of acetylcholine from projection neurons onto Kenyon cells. This activates ionotropic nicotinic receptors, and in turn the membrane is depolarized by inflowing Na^+. In addition, the Ca^{2+} influx through the receptor may activate intracellular signalling cascades. Since only a subpopulation of projection neurons are cholinergic, a parallel, as yet unidentified transmitter is assumed. When an unpredicted unconditioned stimulus is perceived octopaminergic VUM_{mx1} neurons are activated. The released octopamine (*OA*) binds to metabotropic octopamine receptors on the Kenyon cell membrane. Specific OA receptor are coupled to an adenylyl cyclase (*AC*) that give rise to an elevated cAMP level, which in turn activates a PKA. In addition a distinct OA receptor subtype induces an intracellular Ca^{2+} signal probably through Ca^{2+} release from internal stores. The elevated Ca^{2+} level by the various sources (indicated by different symbols) may in turn act on transcription factors like CREB via Ca^{2+}-dependent kinases. Intracellular Ca^{2+} potentiates the current through the GABA receptor (+), while it reduces the peak amplitude of the ACh-induced current (−)

The stoichiometry of the functional receptor in honey bee tissue is still unknown. Thus, although pharmacological and physiological data suggest that the honey bee nAChR resembles the vertebrate neuronal receptors formed by α7, α9 or α10 subunits, the subunit composition of the physiologically characterized honey bee receptor awaits further analyses. To date, the expression of four honey bee nAChR subunits (α2, α4, α7, α8) have been localized in AL neurons and KCs [42, 43].

The honey bee nACh receptor probably mediates fast synaptic transmission, but its high Ca^{2+} permeability may induce modulatory effects as well. Nicotinic receptors in the vertebrate brain serve mainly modulatory functions and often act presynaptically. Each activation of the honey bee ACh receptor induces an inflow of Ca^{2+} ions which may activate Ca^{2+}-dependent intracellular pathways (Fig. 3.5.2). Thus, the *A. mellifera* nAChR may modulate the postsynaptic physiology like its vertebrate counterparts, a property that may contribute to its role during odor memory formation (see Chap. 3.3).

3.5.4.2   GABA

The inhibitory transmitter, γ-amino butyric acid (GABA), is very abundant within all neuropils of the honey bee brain. Within the AL probably most local interneurons are GABA-immunoreactive and may be inhibitory [35]. The mushroom body contains massive GABA-immunoreactive feedback neurons which connect the output lobes of this neuropil with its input regions, the calyces [4]. They may contribute to learning-induced synaptic plasticity ([13], see Chap. 3.2). Pressure applications of GABA induce rapidly activating Cl^- currents through ionotropic GABA receptors in cultured KCs and AL neurons [2, 10, 15]. Thus, the honey bee GABA receptor is functionally similar to the vertebrate $GABA_A$ receptor. The pharmacology of the honey bee GABA receptor identifies it as a typical insect receptor. (1) Muscimol and CACA act as agonists. (2) The receptor is sensitive to picrotoxine, but insensitive to bicuculline. (3) The insecticide, fipronil, a known GABA receptor blocker also blocks GABA-induced currents in honey bee neurons. The GABA-gated Cl^- channels of honey bee central neurons are probably composed of RDL and LCCH3 receptor subunits with an as yet unknown stoichiometry [10, 15]. Honey bee GABA-induced currents are modulated by intracellular Ca^{2+} [15]. This modulation may be mediated via Ca^{2+}-dependent phosphorylation at one of its multiple phosphorylation sites [22]. One functional role of GABAergic synaptic inhibition may be generating or shaping odor-induced spatio-temporal activity patterns within the ALs [33, 40].

3.5.4.3   Glutamate

Honey bee neurons express two different glutamate receptors: a cation-selective glutamate receptor with an AMPA-like pharmacology in cultured KCs ($GluR_{AMPA}$) and a chloride channel that is activated by glutamate in AL neurons ($GluR_{Cl}$) (see Chap. 3.4).

The excitatory glutamate current was identified by applications of glutamate onto cultured KCs that induced inward currents at a pulse potential of -110 mV (Wersing and Grünewald unpublished). The current reverses at a clamp potential of about 0 mV and is insensitive to the chloride channel blocker, picrotoxin. This glutamate-induced current can also be elicited by AMPA or kainate, but not by NMDA; the current is blocked by the AMPA receptor antagonist 6-cyano-7-nitroquinoxaline-2,3-dione (CNQX), but not by the NMDA blocker 2-amino-5-phosphonopentanoic acid (APV). These properties indicate a novel excitatory ligand-gated receptor, which has yet to be described physiologically in insect neurons. Published findings on the localization of NMDA receptor subunits [49, 50], and the effects of NMDA receptor antagonists [39] and NMDA receptor subunit down-regulation [28] on memory formation in honey bees (see Chap. 3.4) should be discussed in light of the fact that no insect glutamate receptor with an NMDA-like pharmacology have been identified in electrophysiological experiments so far.

Cultured honey bee AL neurons express glutamate-gated chloride currents. These glutamate currents comprise rapidly-activating and desensitizing (transient) and slowly desensitizing (sustained) components. The reversal potential of the glutamate-induced currents was close to the Nernst potential of Cl^- indicating that glutamate activates a chloride channel. This honey bee $GluR_{Cl}$ is partially sensitive to picrotoxin and bicuculline and is blocked by fipronil. It therefore shares several properties with the $GluR_{Cl}$ of other insects, crustaceans and nematodes (review: [6]). Thus, two independent inhibitory systems within the honey bee ALs may exist: a glutamatergic inhibitory network in parallel to the GABAergic network (see Chap. 3.4) (Fig. 3.5.1).

3.5.4.4 Histamine

Histamine is found in many neurons throughout the insect brain and is probably the transmitter released by insect photoreceptors (review: [29]). While no histamine-immunoreactive neurons have been detected within the mushroom bodies (MB), Bornhauser and Meyer [5] reported the presence of histamine-immunoreactive neurons within the honey bee ALs. Bath applied histamine blocks odor-induced calcium signals in the ALs Sachse et al. [34] and Jones et al. [21] found two putative histamine-gated Cl^- channel gene sequences within the honey bee genome. However, histamine fails to elicit any current in cultured AL neurons or in KCs [2].

3.5.5 Modulations of Ionic Currents

Biogenic amines are important modulators of neural functions within the insect nervous system (see Chap. 3.6; review: [38]). The role of octopamine (OA) is particularly interesting, because it regulates rather diverse behaviors such as flight, aggression, escape, sucrose responsiveness or division of labor. OA is essential for classical olfactory learning, memory recall and consolidation in honey bees (see Chap. 3.3). The neuronal representation of reward is mediated by OA release neurons in honey bees [18]. In contrast to the detailed knowledge on the localisation and biochemistry of the honey bee OA receptor, however, the cellular physiology of octopaminergic transmission is less well understood. KCs express OA receptors that are coupled to an adenylyl cyclase [27]. Activation of a distinct OA receptor expressed in human embryonic kidney cells (HEK cells) specifically induces an intracellular Ca^{2+} signal [12]. OA applications reversibly reduce currents through the nAChR of cultured KCs or AL neurons (Grünewald, unpublished). This OA modulation may be mediated by Ca^{2+}-dependent signalling cascades rather than by a cAMP/PKA-dependent pathway, because blockage of PKA does not block the OA modulations.

The GABA receptor of honey bee KCs is potentiated by intracellular Ca^{2+} [15]. When the intracellular Ca^{2+} rises the current amplitude reversibly increases. This Ca^{2+}-induced modulation of the GABA receptor may involve Ca^{2+}-dependent protein kinases. While it is known that many ionotropic receptors can be modulated by receptor-associated Ca^{2+}/calmodulin, the biochemical pathways modulating the honey bee GABA-induced currents are as yet unclear.

3.5.6 Interactions of Ionic Conductances Within Kenyon Cells – A Cellular Correlate of Odor Learning?

How do the various ionic currents interact within honey bee central neurons? Fig. 3.5.2 summarizes the identified conductances of KCs and their presynaptic projection neurons and feedback neurons. Activation of the CS pathway (represented by AL projection neurons) activates nicotinic receptors resulting in depolarization and Ca^{2+} influx into the Kenyon cell (I_{ACh}). Odor stimuli also lead to excitation of mushroom body feedback neurons [13], which project back onto KCs. These feedback neurons provide an inhibitory signal via Cl^- currents resulting from activation of ionotropic GABA receptors (I_{GABA}). This inhibitory input is regulated by the intracellular Ca^{2+} level. Calcium levels may rise as a result of I_{ACh} activation and Ca^{2+} entry through the nicotinic receptor itself or via activation of the metabotropic OA receptor (AmOA1) [12]. When an unpredicted reward is presented to the honey bee, OA is released from modulatory neurons such as VUM_{mx1}. Some insect OA receptors are coupled to adenylyl cyclase (AC) via stimulatory G-proteins, which ultimately leads to the activation of PKA [38]. The coincident activation of reward pathways together with the CS preceding the reward (forward pairing), probably leads to a pairing-specific effect in KCs. This may involve interactions between the signalling cascades induced by the ACh and the OA receptor. The dual activation of two independent Ca^{2+} signals (I_{ACh} and AmOA1-activation) is likely to result in a large rise in intracellular Ca^{2+} which may activate Ca^{2+}-dependent kinases and may be a signal for CREB-phosphorylation (Eisenhardt, pers. comm.) and the induction of immediate early genes. Alternatively, the adenylyl cyclase(s) may be dually regulated (by Ca^{2+} and by G-proteins activated by certain OA receptors) and may act as coincidence detector molecules.

3.5.7   Outlook

The characterizations of various voltage- and ligand-gated ionic currents of neurons within the honey bee brain contribute to our understanding of how these neurons generate action potentials and how these neurons form functional synaptic contacts. Together with mathematical models, we are building a physiologically realistic model of olfactory information processing within the honey bee brain. The long-term goal is to understand the biophysical and neuropharmacological mechanisms that underlie odor learning, memory formation and recall. This work is still in progress. For example, we still lack information on metabotropic ACh, GABA and glutamate receptors in bees. We still require more cell physiological data on synaptic transmission *in vivo*. We also need to identify the transmitter(s) of the insect KCs in order to better understand the output synapses of the MBs and their plasticity (e. g., [26]). With the description of the ionic currents of neurons that are essential components within this circuitry we have reached one milestone on our way. Consequently, honey bee KCs and AL neurons are to date among the best characterized **native** insect neurons as far as their cellular physiology is concerned. We ultimately wish to find out how the activation of these ionic currents and their modulation by biogenic amines translate into long-term changes in the cell physiology, biochemistry and morphology that are building blocks of the memory engram within the insect brain. For this link we should build on the vision to measure and manipulate ion channel activity in a behaving honey bee.

References

1. Barbara GS, Grünewald B, Paute S, Gauthier M, Raymond-Delpech V (2008) Study of nicotinic acetylcholine receptors on cultured antennal lobe neurones from adult honeybee brains. Invert Neurosci 8(1):19–29
2. Barbara GS, Zube C, Rybak J, Gauthier M, Grünewald B (2005) Acetylcholine, GABA and glutamate induce ionic currents in cultured antennal lobe neurons of the honeybee, *Apis mellifera*. J Comp Physiol A 191(9):823–836
3. Bicker G, Kreissl S (1994) Calcium imaging reveals nicotinic acetylcholine receptors on cultured mushroom body neurons. J Neurophysiol 71(2):808–810
4. Bicker G, Schäfer S, Kingan TG (1985) Mushroom body feedback interneurones in the honeybee show GABA-like immunoreactivity. Brain Res 360(1–2):394–397
5. Bornhauser BC, Meyer EP (1997) Histamine-like immunoreactivity in the visual system and brain of an orthopteran and a hymenopteran insect. Cell Tissue Res 287(1):211–221
6. Cleland TA (1996) Inhibitory glutamate receptor channels. Mol Neurobiol 13(2):97–136
7. Collet C, Belzunces L (2007) Excitable properties of adult skeletal muscle fibres from the honeybee *Apis mellifera*. J Exp Biol 210(Pt 3):454–464
8. Deglise P, Grünewald B, Gauthier M (2002) The insecticide imidacloprid is a partial agonist of the nicotinic receptor of honeybee Kenyon cells. Neurosci Lett 321(1–2):13–16
9. Devaud JM, Blunk A, Podufall J, Giurfa M, Grünewald B (2007) Using local anaesthetics to block neuronal activity and map specific learning tasks to the mushroom bodies of an insect brain. Eur J Neurosci 26(11):3193–3206

10. Dupuis JP, Bazelot M, Barbara GS, Paute S, Gauthier M et al (2010) Homomeric RDL and heteromeric RDL/LCCH3 GABA receptors in the honeybee antennal lobes: two candidates for inhibitory transmission in olfactory processing. J Neurophysiol 103(1):458–468
11. Goldberg F, Grünewald B, Rosenboom H, Menzel R (1999) Nicotinic acetylcholine currents of cultured Kkenyon cells from the mushroom bodies of the honey bee *Apis mellifera*. J Physiol 514(Pt 3):759–768
12. Grohmann L, Blenau W, Erber J, Ebert PR, Strunker T et al (2003) Molecular and functional characterization of an octopamine receptor from honeybee (*Apis mellifera*) brain. J Neurochem 86(3):725–735
13. Grünewald B (1999) Physiological properties and response modulations of mushroom body feedback neurons during olfactory learning in the honeybee, *Apis mellifera*. J Comp Phys A 185:565–576
14. Grünewald B (2003) Differential expression of voltage-sensitive K + and Ca2+ currents in neurons of the honeybee olfactory pathway. J Exp Biol 206(Pt 1):117–129
15. Grünewald B, Wersing A (2008) An ionotropic GABA receptor in cultured mushroom body Kenyon cells of the honeybee and its modulation by intracellular calcium. J Comp Physiol A 194(4):329–340
16. Grünewald B, Wersing A, Wustenberg DG (2004) Learning channels. Cellular physiology of odor processing neurons within the honeybee brain. Acta Biol Hung 55(1–4):53–63
17. Gundelfinger ED, Schulz R (2000) Insect nicotinic acetylcholine receptors: Genes, structure, physiological and pharmacological properties. In: Clementi F, Fornasari D, Gotti C (eds) Handbook of experimental pharmacology. Springer, Heidelberg, pp 497–521
18. Hammer M, Menzel R (1998) Multiple sites of associative odor learning as revealed by local brain microinjections of octopamine in honeybees. Learn Mem 5(1–2):146–156
19. Hosie AM, Aronstein K, Sattelle DB, Ffrench-Constant RH (1997) Molecular biology of insect neuronal GABA receptors. Trends Neurosci 20(12):578–583
20. Ikeno H, Usui S (1999) Mathematical description of ionic currents of the Kenyon cell in the mushroom body of honeybee. Neurocomputing 26–27:177–184
21. Jones AK, Raymond-Delpech V, Thany SH, Gauthier M, Sattelle DB (2006) The nicotinic acetylcholine receptor gene family of the honey bee, *Apis mellifera*. Genome Res 16(11):1422–1430
22. Jones AK, Sattelle DB (2006) The cys-loop ligand-gated ion channel superfamily of the honeybee, *Apis mellifera*. Invert Neurosci 6(3):123–132
23. Kloppenburg P, Kirchhof BS, Mercer AR (1999) Voltage-activated currents from adult honeybee (*Apis mellifera*) antennal motor neurons recorded in vitro and in situ. J Neurophysiol 81(1):39–48
24. Kreissl S, Bicker G (1989) Histochemistry of acetylcholinesterase and immunocytochemistry of an acetylcholine receptor-like antigen in the brain of the honeybee. J Comp Neurol 286(1):71–84
25. Laurent S, Masson C, Jakob I (2002) Whole-cell recording from honeybee olfactory receptor neurons: ionic currents, membrane excitability and odourant response in developing worker-bee and drone. Eur J Neurosci 15(7):1139–1152
26. Mauelshagen J (1993) Neural correlates of olfactory learning paradigms in an identified neuron in the honeybee brain. J Neurophysiol 69(2):609–625
27. Müller U (1997) Neuronal cAMP-dependent protein kinase type II is concentrated in mushroom bodies of *Drosophila melanogaster* and the honeybee *Apis mellifera*. J Neurobiol 33(1):33–44
28. Müssig L, Richlitzki A, Rössler R, Eisenhardt D, Menzel R et al (2010) Acute disruption of the NMDA receptor subunit NR1 in the honeybee brain selectively impairs memory formation. J Neurosci 30(23):7817–7825
29. Nässel DR (1999) Histamine in the brain of insects: a review. Microsc Res Tech 44(2–3):121–136

30. Nauen R, Ebbinghaus-Kintscher U, Schmuck R (2001) Toxicity and nicotinic acetylcholine receptor interaction of imidacloprid and its metabolites in *Apis mellifera* (Hymenoptera: Apidae). Pest Manag Sci 57(7):577–586

31. Pelz C, Jander J, Rosenboom H, Hammer M, Menzel R (1999) IA in Kenyon cells of the mushroom body of honeybees resembles shaker currents: kinetics, modulation by K+, and simulation. J Neurophysiol 81(4):1749–1759

32. Perk CG, Mercer AR (2006) Dopamine modulation of honey bee (*Apis mellifera*) antennal-lobe neurons. J Neurophysiol 95(2):1147–1157

33. Sachse S, Galizia CG (2002) Role of inhibition for temporal and spatial odor representation in olfactory output neurons: a calcium imaging study. J Neurophysiol 87(2):1106–1117

34. Sachse S, Peele P, Silbering AF, Guhmann M, Galizia CG (2006) Role of histamine as a putative inhibitory transmitter in the honeybee antennal lobe. Front Zool 3:22

35. Schäfer S, Bicker G (1986) Distribution of GABA-like immunoreactivity in the brain of the honeybee. J Comp Neurol 246(3):287–300

36. Schäfer S, Rosenboom H, Menzel R (1994) Ionic currents of Kenyon cells from the mushroom body of the honeybee. J Neurosci 14(8):4600–4612

37. Scheidler A, Kaulen P, Bruning G, Erber J (1990) Quantitative autoradiographic localization of [125I] alpha-bungarotoxin binding sites in the honeybee brain. Brain Res 534(1–2):332–335

38. Scheiner R, Baumann A, Blenau W (2006) Aminergic control and modulation of honeybee behaviour. Curr Neuropharmacol 4(4):259–276

39. Si A, Helliwell P, Maleszka R (2004) Effects of NMDA receptor antagonists on olfactory learning and memory in the honeybee (*Apis mellifera*). Pharmacol Biochem Behav 77(2):191–197

40. Stopfer M, Bhagavan S, Smith BH, Laurent G (1997) Impaired odour discrimination on desynchronization of odour-encoding neural assemblies. Nature 390(6655):70–74

41. Thany SH (2010) Insect nicotinic acetylcholine receptors, vol 683, Advances in experimental medicine and biology. Springer, New York

42. Thany SH, Crozatier M, Raymond-Delpech V, Gauthier M, Lenaers G (2005) Apisalpha2, Apisalpha7-1 and Apisalpha7-2: three new neuronal nicotinic acetylcholine receptor alpha-subunits in the honeybee brain. Gene 344:125–132

43. Thany SH, Lenaers G, Crozatier M, Armengaud C, Gauthier M (2003) Identification and localization of the nicotinic acetylcholine receptor alpha3 mRNA in the brain of the honeybee, *Apis mellifera*. Insect Mol Biol 12(3):255–262

44. Tomizawa M, Casida JE (2003) Selective toxicity of neonicotinoids attributable to specificity of insect and mammalian nicotinic receptors. Annu Rev Entomol 48:339–364

45. Usherwood PNR (1994) Insect glutamate receptors. Adv Insect Physiol 24:309–341

46. Wicher D, Walther C, Wicher C (2001) Non-synaptic ion channels in insects–basic properties of currents and their modulation in neurons and skeletal muscles. Prog Neurobiol 64(5):431–525

47. Wüstenberg DG, Boytcheva M, Grünewald B, Byrne JH, Menzel R et al (2004) Current- and voltage-clamp recordings and computer simulations of Kenyon cells in the honeybee. J Neurophysiol 92(4):2589–2603

48. Wüstenberg DG, Grünewald B (2004) Pharmacology of the neuronal nicotinic acetylcholine receptor of cultured Kenyon cells of the honeybee, *Apis mellifera*. J Comp Physiol A 190(10):807–821

49. Zachepilo TG, Il'inykh YF, Lopatina NG, Molotkov DA, Popov AV et al (2008) Comparative analysis of the locations of the NR1 and NR2 NMDA receptor subunits in honeybee (*Apis mellifera*) and fruit fly (*Drosophila melanogaster*, Canton-S wild-type) cerebral ganglia. Neurosci Behav Physiol 38(4):369–372

50. Zannat MT, Locatelli F, Rybak J, Menzel R, Leboulle G (2006) Identification and localisation of the NR1 sub-unit homologue of the NMDA glutamate receptor in the honeybee brain. Neurosci Lett 398(3):274–279

Chapter 3.6
Dopamine Signaling in the Bee

Julie A. Mustard, Vanina Vergoz, Karen A. Mesce, Kathleen A. Klukas, Kyle T. Beggs, Lisa H. Geddes, H. James McQuillan, and Alison R. Mercer

Abstract Dopamine (DA) is a signaling molecule derived from the amino acid tyrosine. It is an important neuromodulator, neurotransmitter and neurohormone in invertebrates as well as in vertebrates and numerous studies suggest roles for this amine in motor function, learning and memory, aggression, arousal and sleep, and in a number of other behaviors. A growing body of evidence suggests that DA plays a diversity of roles also in *Apis mellifera*. Three honey bee DA receptor genes have been cloned and characterized. In this chapter we focus on their likely involvement in the regulation of locomotor activity, ovary development, and olfactory learning and memory.

Abbreviations (Excluding Brain Areas)

cAMP   Cyclic AMP
DA   Dopamine
dNPF   *Drosophila* neuropeptide F
GPCR   G protein coupled receptor
HVA   Homovanillyl alcohol
OA   Octopamine
PKA   Protein kinase A
QMP   Queen mandibular pheromone

J.A. Mustard
School of Life Sciences, Arizona State University, Tempe, USA

V. Vergoz
School of Biological Sciences, University of Sydney, Sydney, Australia

K.A. Mesce • K.A. Klukas
Department of Entomology, University of Minnesota, Minneapolis, USA

K.T. Beggs • L.H. Geddes • H.J. McQuillan • A.R. Mercer (✉)
Department of Zoology, University of Otago, Dunedin, New Zealand
e-mail: alison.mercer@otago.ac.nz

C.G. Galizia et al. (eds.), *Honeybee Neurobiology and Behavior: A Tribute to Randolf Menzel*, DOI 10.1007/978-94-007-2099-2_16,
© Springer Science+Business Media B.V. 2012

3.6.1 DA Neurons Send Projections Throughout the Honey Bee Brain

Widefield arborizations of catecholamine-containing neurons in the bee brain were first detected using formaldehyde fluorescence histochemistry [21]. A decade later this pioneering work was verified and extended with the electrochemical detection of DA and mapping of DA-containing neurons in the brain [29, 41]. Using an antibody raised against DA, Schäfer and Rehder identified approximately 330 DA-immunoreactive somata in each brain hemisphere plus suboesophageal hemi-ganglion. Cell bodies of the majority of DA neurons identified were located in three clusters, one below the lateral calyx (LCA) of the mushroom bodies (MBs) and two in the anterior-vest (ventral protocerebum). However, DA-containing cells were also identified around the protocerebral bridge, below the anterior optic tubercles (AOTU), proximal to the central rim of the lobula and in the lateral and ventral somatal rind of the subesophageal ganglion (SEG) [41]. Fibres from the DA-labelled cluster beneath the LCA project into the mushroom body calyx (CA), pedunculus (PED), and the vertical (VL) and medial lobes (ML). A fine network of fibres runs through the neuropil of the lip, collar and basal ring. In addition, a few thin projections enter the region of the calyx containing the cell bodies of the intrinsic mushroom body neurons (the Kenyon cells) where they appear to make pre-synaptic connections [6]. The many discrete locations of DA neurons in the brain, their characteristic projection patterns and distinctive morphologies suggest that DA plays a wide variety of roles. Interestingly, recent evidence in *Drosophila* indicates that DA neurons located within the same cell cluster, and projecting to the same region of the brain, can be functionally heterogeneous [8, 23].

3.6.2 Receptors Mediating the Actions of DA Are Functionally Diverse

While relatively little is known, as yet, about the cellular mechanisms through which DA operates in the bee, the identification of receptors that mediate DA's actions is an important first step towards this goal. DA receptors are members of the G protein coupled receptor (GPCR) family of proteins. As described in detail elsewhere [33], honey bee DA receptor sequences share many features in common with other rhodopsin-like GPCRs, including seven transmembrane domains that form the ligand binding pocket, an extracellular amino-terminus, and an intracellular car-boxyl-tail. G proteins bind to the third intracellular loop between transmembrane domains five and six, and the interaction between the receptor and a G protein determines the signaling properties of the receptor. DA receptors are often classified into two groups based on their coupling to cAMP: D1-like receptors increase cAMP levels when activated, whereas D2-like receptors generally have either no effect on this signaling molecule or downregulate the production of cAMP. This classification system is somewhat limited, however, as some DA receptors have been found to

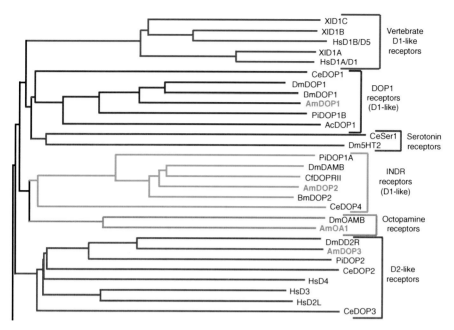

Fig. 3.6.1 Invertebrate dopamine (*DA*) receptor phylogeny. ClustalW2 (www.ebi.ac.uk/Tools/clustalw2/index.html) was used to construct a phylogram illustrating the homology between DA receptor subtypes. Selected octopamine and serotonin (5HT) receptors are also shown to demonstrate the relationships amongst the different receptors. Shown are: Human, Hs (NP_000789, NP_000785, NP_000788, NP_000787, NP_000786); *Xenopus*, Xl (P42289, P42290, P42291); Fruit Fly, Dm (NM_057659, NM_079824, NM_001014758, NM_169913, NM_169011); Honey Bee, Am (CAA73841, AAM19330, AY921573, CAD67999); *C. elegans*, Ce (AAO91736, BAD01495, AAR37416, NP_508238, NP_510684); *Bombyx*, Bm (NM_001114866, NM_001114987); Lobster, Pi (DQ295790, DQ295791, DQ900655); Flea, Cf (DQ459405); and *Aplysia*, Ac (AY918891) receptors

couple to multiple G protein subtypes and to regulate second messenger signaling molecules other than cAMP. In addition, DA receptors have been shown to change their coupling to different signaling pathways in response to ongoing activation, and to interact with other receptor proteins (review: [36]). The potential for plasticity at this level is enormous and represents an exciting area for future investigation.

As in other invertebrates, honey bee DA receptors cluster into three structurally and functionally distinct classes ([33]; Fig. 3.6.1). *AmDOP1*, which was cloned and characterized by Wolfgang Blenau and colleagues [5], is a member of the invertebrate DA receptor class (DOP1 receptor) that is most closely related to the vertebrate D1-like receptors [33]. Interestingly, this receptor is constitutively active and increases cAMP levels even in the absence of agonist [32], a property shared with some other members of this class of invertebrate DA receptors, as well as with some vertebrate D1-like DA receptors (review: [33]).

AmDOP2 is also positively coupled to cAMP [18, 32], but like its *Drosophila* orthologue, DAMB [12] the *Am*DOP2 receptor also couples to calcium (Beggs et al. in prep). Interestingly, the *Am*DOP2 receptor shows a closer phylogenetic relationship to the honey bee 'α-adrenergic-like' octopamine (OA) receptor, *Am*OA1, than to *Am*DOP1 or *Am*DOP3, placing it in a distinct group of "invertebrate type" DA receptors (*INDRs*; invertebrate DA receptor, [18, 33]; see Fig. 3.6.1).

AmDOP3 is a D2-like DA receptor [3]. However, while DA activation of *Am*DOP3 generally leads to a down-regulation of cAMP, constitutive activity of this receptor increases basal levels of cAMP [3].

3.6.3 Plasticity in DA Signaling Pathways of the Brain

Brain DA levels, levels of DA receptor gene transcript, and patterns of DA receptor gene expression in the brain change markedly during the lifetime of the bee (e.g. [18, 24, 43, 47]). In *Drosophila*, DA plays a developmental role [35] and this may be true also in the bee [20]. DA-immunoreactive processes originating from cell bodies located in the lateral deutocerebral soma rind, for example, invade the developing antennal lobes (AL) around pupal stage 3, prior to the formation of the glomerular (synaptic) neuropil [20]. The same cells extend processes into the AMMC (dorsal lobe) of the deutocerebrum, as well as to the protocerebrum and SEG [41]. Rapid growth of the synaptic neuropil around pupal stage 4 coincides with a surge in DA levels in the ALs and around this time antennal-lobe neurons *in vitro* respond to DA with enhanced cell body fiber outgrowth [20]. Calcium-activated potassium currents have been identified as targets of DA modulation in developing honey bee antennal-lobe neurons [37] and are likely to contribute to changes in cell excitability in this region of the brain.

Meredith Humphries and colleagues identified a strong correlation between the behavioral development of worker bees and changes in DA receptor gene expression in the MBs of the brain [18]. Levels of *Amdop2* expression in noncompact cells of the MBs are up-regulated with age suggesting that this receptor may contribute to the behavioral maturation of the bee. DA's influence on response thresholds for gustatory and olfactory stimuli may also affect the behavioral development of the bee [42, 50]. Interestingly, levels of DA (and other amines) in ALs of the brain have been found to be higher in foragers than in bees performing nursing duties regardless of age, suggesting that DA titres in this region of the brain may be linked to behavioral state [43]. In the MBs, however, Schulz and Robinson found changes in amine levels to be more strongly linked to age, than to behavior [43]. DA levels in the bee are affected also by environmental factors, such as stress [7], but what do we know about DA's functions in the brain?

3.6.4 DA Plays a Role in Aversive Learning

There is compelling evidence that DA plays a key role in aversive learning. Associative olfactory learning has been traced to the MBs [26, 31], regions of the brain that receive dense innervation from DA-containing cells (Fig. 3.6.2).

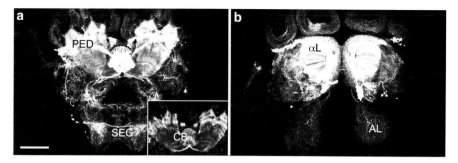

Fig. 3.6.2 Laser scanning confocal micrographs of a wholemounted 13 day-old honey bee brain immunolabeled for tyrosine hydroxylase (*TH*), the rate-limiting enzyme for the synthesis of dopamine (*DA*). The brain was fixed in 4% paraformaldehyde for 15 min followed by a series of buffer-based washes. The specimen was incubated for 98 h in a 1:400 dilution of a mouse monoclonal antiserum raised against TH (ImmunoStar Inc., Hudson, WI, USA). After various rinses, the brain was incubated for 64 h in a 1:200 dilution of a Cy-3-conjugated goat anti-mouse antiserum (Jackson ImmunoResearch, West Grove, PA, USA) and prepared for histological mounting between two coverslips. Images in (**a**) and (**b**) (frontal views) are from the same brain imaged at different depths; (**a**) and (**b**) are composites of ten and seven confocal sections respectively and acquired at 5 μm intervals. Inset of (**a**) (single section) shows the substructure of the *upper* and *lower* divisions of the CB. *αL* alpha lobe of the mushroom bodies, *AL* antennal lobe, *CB* central body, *PED* pedunculus of the mushroom bodies, *SEG* subesophageal ganglion. Scale bar = 160 μm for A and B

In *Drosophila*, DA neurons projecting to the MBs respond strongly to electric shock [39]. If output from these neurons is blocked, flies are unable to learn to associate an odor stimulus with punishment, but their appetitive memory remains intact [44]. These results have provided the first evidence that DA conveys the negative reinforcing properties of an aversive stimulus in the fly, a conclusion supported strongly by subsequent studies showing that activation of these same DA neurons can substitute for electric shock in an aversive conditioning paradigm [8].

DA signaling also appears to be critical for aversive learning in bees. A simple but robust paradigm for demonstrating aversive learning was developed by Martin Giurfa and his team in Toulouse [48]. Bees respond defensively to aversive stimuli by extending their sting. This reflexive response is highly predictable and can be used effectively to demonstrate aversive learning in bees. For example, if a bee is presented several times with an odor paired with a mild electric shock it learns to associate the odor with punishment; the bee will subsequently extend its sting in response to the odor alone in expectation of the punishment to follow [48]. Pharmacological studies provided the first hint of DA's involvement in the formation of aversive memories in the bee [48]. Vergoz and colleagues found that aversive learning could be blocked using the vertebrate DA receptor antagonists, spiperone and flupenthixol, compounds that they found did not impair a bee's ability to learn to associate an odor stimulus with a food reward.

3.6.5 Aversive Learning Can Be Blocked by Queen Mandibular Pheromone

A growing body of evidence suggests that queen mandibular pheromone (QMP) can block the negative reinforcing properties of an aversive stimulus by regulating DA signaling. Young worker bees exposed to QMP respond to electric shock with sting extension, but they are unable to learn an odor signal that predicts this negative outcome [49]. DA levels in the brain of young bees performing tasks within the colony are generally lower than the levels found in foragers [43, 47]. If a colony is made queenless, however, DA levels in the brain of young workers increase to a level similar to that seen in foragers [16, 38, 40]. The queen's influence on brain DA levels has been traced to QMP. Exposing young worker bees to this blend of pheromones lowers brain DA levels, transiently alters levels of DA receptor gene expression, and modifies cellular responses to this amine [4]. One of the receptors affected by QMP is *Am*DOP1. Levels of expression of this receptor are lower in young (2-day old) bees exposed to QMP than in bees reared without exposure to this pheromone [4]. However, the degree to which QMP's effects on *Am*DOP1 receptor expression contribute to QMP-induced impairment of aversive learning in young bees remains unclear as QMP also affects the function of the D2-like DA receptor, *Am*DOP3. *Am*DOP3 receptors are activated by homovanillyl alcohol (HVA) [2], which is one of the key components of QMP [46], and young bees exposed to HVA alone also show impaired aversive learning [49].

The *Drosophila* orthologue of *Am*DOP1 (dDA1) has been strongly implicated in aversive learning. In mutant flies that show abnormal expression of the dDA1 receptor in the MBs aversive learning is severely impaired [19, 45]. Interestingly, appetitive learning is also compromised in these flies, suggesting that this D1-like DA receptor may be involved in establishing or retrieving appetitive olfactory memories, as well as aversive memories. Evidence suggests that DAMB receptors (the *Drosophila* orthologue of *Am*DOP2) also play a role in aversive learning, but at least in *Drosophila* larvae this D1-like DA receptor type appears not to be involved in appetitive learning [45]. DA's involvement in modulating the expression of learned responses to both rewarding and punishing stimuli is discussed in detail in an excellent review by Barron [1].

3.6.6 Motivation and Memory

DA injected into the brain of the bee has no apparent effect on the acquisition of appetitive memories, but inhibits appetitive memory recall [25, 27]. Recent studies may explain these effects. Hungry bees learn to associate odor signals with a food reward much better than satiated bees [10]; it turns out that the same is true in the fly. Recently, Scott Waddell and colleagues identified a subpopulation of DA neurons projecting to the MBs of the fly brain that inhibit the expression of appetitive memory in satiated flies [23]. Their results suggest that in hungry flies activity in these neurons is inhibited by neuropeptide F (dNPF), the insect orthologue of neuropeptide Y (NPY), which regulates food seeking behavior in mammals (see Chap. 3.7). Thus the presence of dNPF releases the flies from the inhibitory effects of DA, promoting appetitive memory performance. DA may play a similar role in the bee. Friedrich and colleagues in Uli Müller's group have shown that cAMP-dependent protein kinase A (PKA) activity in the MBs is strongly influenced by satiation [10]. Furthermore, the effects of satiation can be used to tease apart different and parallel functions of the cAMP-PKA cascade during conditioning [10]. In this context, it will be interesting to see whether DA signaling contributes to these effects. Consistent with this possibility, DA has been found to increase PKA activity in Kenyon cells *in vitro* [30]. DA applied iontophoretically into the mushroom body neuropil reduces and frequently reverses the sign of olfactory-evoked potentials recorded in this region of the brain [28], but the functional significance of these effects has yet to be revealed.

The studies described above highlight several important challenges in evaluating DA's roles in the brain. Firstly, DA neurons are functionally heterogeneous [8, 23, 45] and thus should ideally be analysed as individuals, or perhaps small groups. Secondly, most if not all mushroom body intrinsic neurons express more than one DA receptor type [3, 12, 19, 24, 45]. DA's actions on individual cells, as well as on neural networks, will depend not only on the complement of DA receptor types expressed, but potentially also on the functional state of the receptors and their interaction with other receptor proteins. Finally, DA's actions centrally have yet to be clearly differentiated from the peripheral effects of this amine. Evidence suggests that DA affects sensory perception in the fly [45] and in bees also, recent data suggest that DA acts at the level of the antennal sensory neurons [50]. A better understanding of DA's actions peripherally and how they influence learning, memory and motivation in the bee awaits further investigation.

3.6.7   A Role in Locomotion and Motor Control

It is likely that DA acts at multiple sites to influence behavior in the bee; affecting sensory information in the periphery, the regulation of central pattern generators and probably also higher order processing of information in the brain. Motor behavior may be modulated via two different processes: changes in the amount of DA present and changes in DA receptor expression levels. In the honey bee, both of these situations have been shown to be correlated with changes in behavior. The switch from in-colony tasks to foraging is accompanied by significant increases in DA levels in the worker bee brain [43, 47]. Harano and colleagues have conducted several studies investigating the relationship between DA levels and locomotion in honey bee queens and drones [13, 15]. Virgin queens have higher levels of DA in their brains than age matched mated queens [13]. The higher DA levels correlate well with elevated levels of locomotion observed for young queens in a line crossing assay [15] and both in queens and drones, injection of the DA receptor agonist 6,7-ADTN increased activity levels while injection with the antagonist flupenthixol resulted in decreased locomotion [15]. The role of DA in motor behavior in worker bees was examined using a detailed analysis of the behavioral profile of pollen foragers [34]. Bees injected with the general DA receptor antagonist flupenthixol spent less time walking than buffer-treated bees, and also spent less time engaged in other active behaviors such as flying. Instead, bees in the flupenthixol groups spent more time standing still (a behavior rarely seen in control bees) or upside down, due to the loss of coordination necessary for bees to right themselves.

Changes in DA receptor expression that correspond with changes in behavior have also been observed. Expression levels of *Amdop2* in the Kenyon cells of the MBs are lower in young bees than in older forager bees [18]. Using RNA interference, levels of this receptor in the MBs of worker bees were reduced to determine whether motor function was altered [34]. The behavioral profiles of foragers with reduced *Amdop2* transcript levels were similar to those injected with the antagonist flupen-thixol. Specifically, bees spent less time walking and more time standing still in comparison to control bees. However, unlike the case with flupenthixol, the decrease in *Amdop2* levels in the brain did not affect the amount of time bees spent upside down. This result suggests that either the *Am*DOP2 receptor does not play a role in this behavior or that a central pattern generator in the thoracic ganglion may contribute to coordinating the leg movements necessary for righting behavior. Either way, the shifts in behavior observed for bees with reduced *Amdop2* levels confirm that changes in the DA receptor expression levels in the brain can significantly impact honey bee behavior. Given QMP's effects on DA signaling it is perhaps no surprise that this pheromone also reduces locomotor activity in young worker bees [4], as well as their mean maximum speed [50].

3.6.8   What Role Does DA Play in Ovary Development?

Interesting correlations have been identified between shifts in DA levels and changes in the reproductive status of worker bees [9, 16, 40]. Dombroski and colleagues [9], for example, found that feeding DA to queenless worker bees increases the percentage of bees with developed ovaries and in *Drosophila*, inhibition of DA synthesis results in a significant decrease in fertility in newly emerged female flies [35]. Analysis of DA levels suggests a potential gonadotropic effect of DA also in honey bee queens. Queens not only have larger ovaries than workers but also higher levels of DA in the brain [40]. Interestingly, Harano and colleagues found that brain DA levels are lower in mated queens than in virgin queens of the same age [13]. They suggest that in virgin queens high levels of DA in the brain might be related to the activation or maturation of ovarian follicles in previtellogenic stages. Correlations between DA levels and changes in the reproductive organs of drones have also been identified [14]. The significance of these correlations remains to be determined but they suggest a possible role for DA in altering the reproductive status of the bee. Intriguingly,

Kocher and colleagues [22] have recently shown that worker attraction to QMP is negatively correlated with ovariole number in worker bees. QMP is known to inhibit ovary development [17] and to have significant effects on gene expression in the honey bee brain [11]. If the colony loses its queen, workers are released from the inhibitory effects of this pheromone and begin to develop ovaries (review: [46]). As QMP affects DA signaling, a better understanding of the mechanisms through which QMP operates may help to reveal the role that DA plays in ovary development in the bee.

3.6.9 Conclusions and Future Directions

As in other animals, DA signaling affects many aspects of honey bee behavior. A number of studies in the bee suggest a role for DA in learning and memory, particularly in aversive learning. In addition, several studies show that DA signaling modulates motor function and activity levels. Although a direct connection has not yet been established, several lines of evidence suggest that DA may play an important role in the reproductive development of the bee. The ability of DA to influence a number of processes is undoubtedly due to its activation of distinct receptors, which allows for modulation of different signaling cascades.

Significant headway has been made in our understanding of the mechanisms through which DA acts; however, many questions remain. For example, how does DA signaling in the MBs lead to changes in motor output that reflect association of olfactory or visual stimuli with punishment or reward? What is DA's role in the central bodies of the brain, and what are the consequences of DA's actions in the periphery? Answers to these questions will provide a deeper understanding of the behavior of the bee and undoubtedly other organisms as well.

References

1. Barron AB, Søvik E, Cornish JL (2010) The roles of dopamine and related compounds in reward-seeking behavior across animal phyla. Front Behav Neurosci 4. doi:10.3389/fnbeh.2010.00163
2. Beggs KT, Mercer AR (2009) Dopamine receptor activation by honey bee queen pheromone. Curr Biol 19:1206–1209
3. Beggs KT, Hamilton IS, Kurshan PT, Mustard JA, Mercer AR (2005) Characterization of a D2-like dopamine receptor (*AmDOP3*) in honey bee, *Apis mellifera*. Insect Biochem Mol Biol 35:873–882
4. Beggs KT, Glendining KA, Marechal NM, Vergoz V, Nakamura I, Slessor KN, Mercer AR (2007) Queen pheromone modulates brain dopamine function in worker honey bees. Proc Natl Acad Sci USA 104:2460–2464
5. Blenau W, Erber J, Baumann A (1998) Characterization of a dopamine D1 receptor from *Apis mellifera*: cloning, functional expression, pharmacology, and mRNA localization in the brain. J Neurochem 70:15–23

6. Blenau W, Schmidt M, Faensen D, Schurmann F-W (1999) Neurons with dopamine-like immunoreactivity target mushroom body Kenyon cell somata in the brain of some hymenopteran insects. Int J Insect Morph Embryol 28:203–210
7. Chen YL, Hung YS, Yang EC (2008) Biogenic amine levels change in the brains of stressed bees. Arch Insect Biochem Physiol 68:241–250
8. Claridge-Chang A, Roorda RD, Vrontou E, Sjulson L, Li H, Hirsh J, Miesenböck G (2009) Writing memories with light-addressable reinforcement circuitry. Cell 139:405–15
9. Dombroski TCD, Simões ZLP, Bitondi MMG (2003) Dietary dopamine causes ovary activation in queenless *Apis mellifera* workers. Apidologie 34:281–289
10. Friedrich A, Thomas U, Müller U (2004) Learning at different satiation levels reveals parallel functions for the cAMP-protein kinase A cascade in formation of long-term memory. J Neurosci 24:4460–4468
11. Grozinger CM, Sharabash NM, Whitfield CW, Robinson GE (2003) Pheromone-mediated gene expression in the honey bee brain. Proc Natl Acad Sci USA 100(Suppl 2):14519–14522
12. Han KA, Millar NS, Grotewiel MS, Davis RL (1996) DAMB, a novel dopamine receptor expressed specifically in *Drosophila* mushroom bodies. Neuron 16:1127–1135
13. Harano K, Sasaki K, Nagao T (2005) Depression of brain dopamine and its métabolite after mating in the European honeybee (*Apis mellifera*) queens. Naturwissenschaften 92:310–313
14. Harano K, Sasaki M, Nagao T, Sasaki K (2008) Dopamine influences locomotor activity in honeybee queen: implications for a behavioural change after mating. Physiol Entomol 33:395–399
15. Harano K-I, Sasaki K, Nagao T, Sasaki M (2008) Influence of age and juvenile hormone on brain dopamine level in male honeybee (*Apis mellifera*): association with reproductive maturation. J Insect Physiol 54:848–853
16. Harris JW, Woodring J (1995) Elevated brain dopamine levels associated with ovary development in queenless worker honey bees (*Apis mellifera* L.). Comp Biochem Physiol Part C Comp Pharmacol Toxicol 111:271–279
17. Hoover SER, Winston ML, Oldroyd BP (2005) Retinue attraction and ovary activation: responses of wild type and anarchistic honey bees (*Apis mellifera*) to queen and brood pheromones. Behav Ecol Sociobiol 59:278–284
18. Humphries MA, Mustard JA, Hunter SJ, Mercer A, Ward V, Ebert PR (2003) Invertebrate D2 type dopamine receptor exhibits age-based plasticity of expression in the mushroom bodies of the honeybee brain. J Neurobiol 55:315–330
19. Kim Y-C, Lee G-H, Han K-A (2007) D_1 dopamine receptor dDA1 is required in the mushroom body neurons for aversive and appetitive learning in *Drosophila*. J Neurosci 27:7640–7647
20. Kirchhof BS, Homberg U, Mercer AR (1999) Development of dopamine-immunoreactive neurons associated with the antennal lobes of the honey bee, *Apis mellifera*. J Comp Neurol 411:643–563
21. Klemm N (1976) Histochemistry of putative transmitter substances in the insect brain. Prog Neurobiol 7:99–169
22. Kocher SD, Ayroles JF, Stone EA, Grozinger CM (2010) Individual variation in pheromone response correlates with reproductive traits and brain gene expression in worker honey bees. PLoS One 5(2):e9116. doi:10.1371/journal.pone.0009116
23. Krashes MJ, DasGupta S, Vreede A, White B, Armstrong JD, Waddell S (2009) A neural circuit mechanism integrating motivational state with memory expression in *Drosophila*. Cell 139:416–427
24. Kurshan PT, Hamilton IS, Mustard JA, Mercer AR (2003) Developmental changes in expression patterns of two dopamine receptor genes in mushroom bodies of the honeybee, *Apis mellifera*. J Comp Neurol 466:91–103
25. Macmillan CS, Mercer AR (1987) An investigation of the role of dopamine in the antennal lobes of the honeybee, *Apis mellifera*. J Comp Physiol A 160:359–366
26. Menzel R, Müller U (2001) Neurobiology. Learning from a fly's memory. Nature 411:433–434
27. Mercer AR, Menzel R (1982) The effects of biogenic amines on conditioned and unconditioned responses to olfactory stimuli in the honeybee, *Apis mellifera*. J Comp Physiol A 145:363–368

28. Mercer AR, Erber J (1983) The effects of amines on evoked potentials recorded in the mushroom bodies of the bee brain. J Comp Physiol 151:469–476
29. Mercer AR, Mobbs PG, Davenport AP, Evans PD (1983) Biogenic amines in the brain of the honeybee, *Apis mellifera*. Cell Tiss Res 234:655–677
30. Müller U (1997) Neuronal cAMP-dependent protein kinase type II is concentrated in mushroom bodies of *Drosophila melanogaster* and the honeybee, *Apis mellifera*. J Neurobiol 33:33–44
31. Müller U (2006) Memory: cellular and molecular networks. Cell Mol Life Sci 63:961–962
32. Mustard JA, Blenau W, Hamilton IS, Ward VK, Ebert PR, Mercer AR (2003) Analysis of two D1-like dopamine receptors from the honey bee, *Apis mellifera*, reveals agonist-independent activity. Mol Brain Res 113:67–77
33. Mustard JA, Beggs KT, Mercer AR (2005) Molecular biology of the invertebrate dopamine receptors. Arch Insect Biochem Physiol 59:103–117
34. Mustard JA, Pham PM, Smith BH (2010) Modulation of motor behavior by dopamine and the D1-like dopamine receptor AmDOP2 in the honey bee. J Insect Physiol 56:422–430
35. Neckameyer WS (1996) Multiple roles for dopamine in *Drosophila* development. Dev Biol 176:209–219
36. Neve KA, Seamans JK, Trantham-Davidson H (2004) Dopamine receptor signaling. J Rec Signal Trans 24:165–205
37. Perk CG, Mercer AR (2006) Dopamine modulation of honey bee (*Apis mellifera*) antennal-lobe neurons. J Neurophysiol 95:1147–1157
38. Purnell MT, Mitchell CJ, Taylor DJ, Kokay IC, Mercer AR (2000) The influence of endogenous dopamine levels on the density of [^{3}H]SCH23390-binding sites in the brain of the honey bee, *Apis mellifera* L. Brain Res 855:206–216
39. Riemensperger T, Voller T, Stock P, Buchner E, Fiala A (2005) Punishment prediction by dopaminergic neurons in *Drosophila*. Curr Biol 15:1953–1960
40. Sasaki K, Nagao T (2001) Distribution and levels of dopamine and its metabolites in brains of reproductive workers in honeybees. J Insect Physiol 47:1205–1216
41. Schäfer S, Rehder V (1989) Dopamine-like immunoreactivity in the brain and suboesophageal ganglion of the honeybee. J Comp Neurol 280:43–58
42. Scheiner R, Plückhahn S, Öney B, Blenau W, Erber J (2002) Behavioural pharmacology of octopamine, tyramine and dopamine in honey bees. Behav Brain Res 136:545–553
43. Schulz DJ, Robinson GE (1999) Biogenic amines and division of labor in honey bee colonies: behaviourally related changes in the antennal lobes and age-related changes in the mushroom bodies. J Comp Physiol A 184:481–488
44. Schwärzel M, Monastirioti M, Scholz H, Friggi-Grelin F, Birman S, Heisenberg M (2003) Dopamine and octopamine differentiate between aversive and appetitive olfactory memories in *Drosophila*. J Neurosci 23:10495–10502
45. Selcho M, Pauls D, Han K-A, Stocker FR, Thum AS (2009) The role of dopamine in *Drosophila* larval classical olfactory conditioning. PLoS One 4(6):e5897
46. Slessor KN, Winston ML, Le Conte Y (2005) Pheromone communication in the honeybee (*Apis mellifera* L.). J Chem Ecol 31:2731–2745
47. Taylor DJ, Robinson GE, Logan BJ, Laverty R, Mercer AR (1992) Changes in the brain amine levels associated with the morphological and behavioral development of the worker honeybee. J Comp Physiol A 170:715–721
48. Vergoz V, Roussel E, Sandoz J-C, Giurfa M (2007) Aversive learning in honeybees revealed by the olfactory conditioning of the sting extension reflex. PLoS One 2(3):e288
49. Vergoz V, Schreurs HA, Mercer AR (2007) Queen pheromone blocks aversive learning in young worker bees. Science 317:384–386
50. Vergoz V, McQuillan HJ, Geddes LH, Pullar K, Nicholson BJ, Paulin MG, Mercer AR (2009) Peripheral modulation of responses to honey bee queen pheromone. Proc Natl Acad Sci USA 106:20930–20935

Chapter 3.7
Neuropeptides in Honey Bees

C. Giovanni Galizia and Sabine Kreissl

Abstract Neuropeptides may be the most ancient chemical messengers between neurons. As all insects, bees have a large number of different putative neuropeptides and peptide receptors, most of which have been characterized only poorly, if at all. Therefore, we briefly review the role that neuropeptides play in insect nervous systems, and then review the specific occurrence of peptides in honey bee. A few exemplary peptide families are treated with greater detail, including FMRFamide related peptides (FaRPs), SIFamide, allatostatin A (AST A). While the role of several peptides may or may not correspond to that reported for other insects, but has not yet been investigated in bees specifically (e.g. bursicon and corazonin involved in molting), a few peptides have been analyzed in honey bees (e.g. tachykinin, PBAN, sNPF, which are involved in nectar and pollen foraging). Immunostainings against neuropeptides are also a powerful tool for anatomical studies, because they can be used to characterize small populations of neurons based on their neuropeptide expression patterns.

In addition to a short overview about neuropeptides we will include a few unpublished observations about neuropeptide organization in the honey bee brain, especially in the honey bee antennal lobe (AL).

Abbreviations

AL	Antennal lobe
CoAl	Corpora allata
CoCa	Corpora cardiaca
GABA	Gamma-amino-butyric-acid

C.G. Galizia (✉) • S. Kreissl
Department of Biology, University of Konstanz, D-78457 Konstanz, Germany
e-mail: giovanni.galizia@uni-konstanz.de

C.G. Galizia et al. (eds.), *Honeybee Neurobiology and Behavior: A Tribute to Randolf Menzel*, DOI 10.1007/978-94-007-2099-2_17,
© Springer Science+Business Media B.V. 2012

KC   Kenyon cells
Maldi-TOF   Matrix assisted laser desorption/ionisation-time of flight
MB   Mushroom bodies
OL   Optic lobes

3.7.1   Neuropeptides in the Insect Nervous System

3.7.1.1   *What Are Neuropeptides*

Information transfer across neurons occurs via several channels. The most ubiquitous is the point-to-point release of signaling molecules via clear, small (30–50 nm) synaptic vesicles at synaptic terminals, with small classical transmitters such as acetylcholine, glutamate or GABA. Similarly, biogenic amines (octopamine (OA), serotonin) are released at synaptic sites or extrasynaptically, but close to it, from large dense core vesicles (>50–200 nm). Peptides are also released from large dense core vesicles, either at the synaptic cleft or extrasynaptically but still in the neural tissue, or at neurohemal organs into the hemolymph, where they then act as hormones on distal targets. Due to the combinatorial nature of short peptide chains, the number of possible peptides is virtually infinite, potentially allowing for specific communication across cells even in situations where the messenger substance (the peptide) is released into a larger volume (reviews: [18, 27, 28]).

Peptides are often localized in neurons with either local projections confined to a small part of the brain or with global projections in large neuropil areas. An example for the first is the presence of Tachykinin in Kenyon cells and local interneurons of the antennal lobe (AL) of the honey bee [42] (Fig. 3.7.1). An example for the latter is provided by four SIFamide-immunoreactive neurons in the pars intercerebralis of honey bees, which innervate almost the entire brain (Fig. 3.7.1) as in other insects [45].

In many cases, peptidergic neurons also release a classical fast acting transmitter, so that neuropeptides act as co-transmitters. Generally, neuropeptide release has a higher calcium threshold at the synaptic terminal, meaning that a weakly active neuron will release its "main" transmitter already when a few spikes reach the synaptic terminal, and will co-release its neuropeptide only when there is a large train of spikes inducing massive calcium influx to the presynaptic terminal [1, 47]. After release neuropeptides are degraded by ubiquitous, extracellular peptidases, with different time constants than the reuptake mechanisms of classical neurotransmitters. Thus, the information that a postsynaptic neuron receives from a classical transmitter and from a neuropeptide is quite different, both in terms of necessary presynaptic activity, and in terms of temporal structure. In addition, many neurons release neuropeptides extrasynaptically, adding the difference that there is no point-to-point transmission, but rather a local-area-transmission [18, 28, 29], not to be mistaken with the global release of neuropeptides at neurohemal organs.

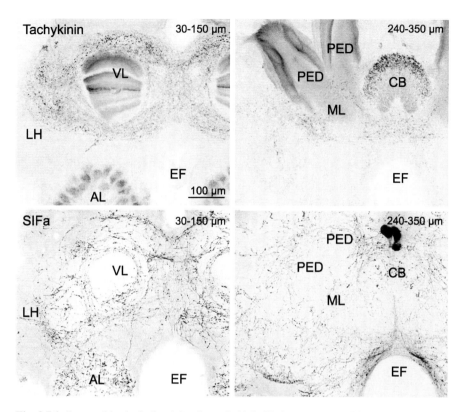

Fig. 3.7.1 Immunohistological staining for tachykinin-IR (*upper row*), and SIFamid-IR (*bottom row*) in the central bee brain area. Confocal stacks of fluorescent images at depths as indicated in the *upper right* corners have been projected to allow for a frontal view in an anterior (*left column*) and a posterior (*right column*) position. In the anterior position, the vertical lobes (*VL*), the lateral horn (*LH*), a portion of the antennal lobes (*AL*), and the esophageal foramen (*EF*) are indicated. The posterior position covers the medial lobes (*ML*), pedunculi (*PED*), the central body (*CB*), and the posterior surface of the brain. The figure shows how distinct types of neurons can be visualized with antiserum staining against two different neuropeptides. Staining for tachykinin labels strata in the MB lobes and pedunculi. Note the clear staining in the AL glomeruli. In both neuropils immunoreactivity was found in local neurons. Staining is also present in the CB. Staining for SIFamide labels four very large somata in the pars intercerebralis. Their arborization covers almost all areas of the brain, including the ring neuropil and the AL glomeruli. Most of the vertical lobes are devoid of staining, but the ventral part is well innervated with stained profiles

The most important neurohemal organ in the insect brain is located in the retrocerebral complex, with the corpora cardiaca (CoCa) and corpora allata (CoAl) which connect to the dorsal heart. Similar to the situation in the vertebrate hypothalamus-pituitary system, various peptides belonging to different peptide families are released from centrifugal neurons in the retrocerebral complex, where they then either reach the hemolymph directly, or induce peptide and/or hormone release into the circulatory system from local secretory cells in these structures. Important mechanisms regulated by CoCa and CoAl peptides are the control of molting in larvae and other developmental processes. Peptides involved in many insects include prothoracotropic hormone PTTH, a peptide dimer, and corazonin [13].

3.7.1.2 Identification and Nomenclature of Neuropeptides

Peptide genes encode for prepropeptides, long chains often containing the code for several peptides. The prepropeptide RNA is translated by ribosomes, just like for proteins, and the prepropeptide is then cleaved to propeptides. Propeptides are generally modified into the functional peptide by further cleavage and by amidation of the c-terminus [18]. The amidating enzyme PHM has been studied in bees [49]. Peptides vary in amino acid length between five and a few dozens amino acids. Some neuropeptides consist of peptide dimers, i.e. they consist of two peptide chains.

Since the sequencing of the honey bee genome [43] peptides in the honey bee can be studied by searching genome information (which peptides are potentially possible?), by using proteomics tools, such as peptide mass spectroscopy, which is increasingly sensitive also for small tissue samples, allowing for brain-area specific resolution (which peptides are really expressed?), and by immunohistochemical analysis (what is the morphology of peptidergic neurons?). The latter approach gives the best resolution, but needs sufficiently specific antibodies. Because many neuropeptides are considerably conserved across species, an antiserum created for, say, a particular locust peptide may also work in bees. However, because even conserved neuropeptides differ in their detailed sequence, and individual peptides of a given family often have similar subsequences, antibodies may cross-react or fail in binding and thus produce false positive or false negative results. Thus appropriate controls are mandatory and results should carefully be interpreted. It should be noted that peptide research is strongly gaining momentum due to new technological developments, in particular in genomics and proteomics [45].

Peptides are classified into families. The names may be related to the sequence. An example is provided by the FMRFamide like neuropeptides, i.e. peptides that have the c-terminal sequence FMRF (phenylalanine-methionine-arginine-phenylalanine-NH_2). This family is included in the even larger family of FMRFamide related peptides (FaRPs, members of which share just the last two amino acids -RF), that contains, among others, the myosuppressins, sulfakinins, NVPIYQEPRF, sNPF, FMRFamide-like and FLRFamide-like peptides [30]. Other names of neuropeptides are related to the history of the discovery of their functions. Many peptides have been originally found in insects by screening fractions of the hemolymph for their physiological action on various organs *in vitro* or *in vivo*. Tachykinins (Greek: tachy=rapid, kinin=motion) were first identified as substances that activate gut movement. In insects they form a large family terminating with the amino acids phenylalanine-X-glycine-X-methionine-NH_2, where X is either an aliphatic or an aromatic amino acid [44]. The nomenclature of insect neuropeptides is further complicated because some peptides have been characterized separately in different species, giving them different names based on their function, only to be later found to be homologous (MIP, AST-B). Some peptides serve the same function in different arthropods, and therefore share a similar name, even though they do not serve that function across all insects which have the peptide. Allatostatins (AST) were discovered as peptides that inhibit the release of juvenile hormones from the CoAl (hence the name), but different species use different peptides for this function, and the same peptides are used for other functions in different areas [39]. Thus, we now have three different AST neuropeptide families, all acting on juvenile hormone synthesis in some but not all insects. However, each of these three families display their own characteristic sequence: AST-A was discovered in cockroaches; its members share the C-terminal sequence -YXFGL-NH$_2$ with X being a variable amino acid. AST-B acts on CoAl in some cricket, and its sequence corresponds to the *Manduca* myoinhibitory peptide, MIP. AST-C, which acts on CoAl in other crickets and in *Manduca* as well as in other leptidopterans, again has a characteristic sequence – PISCF – and is completely different from the other two AST families [27, 29].

3.7.1.3   Neuropeptide Receptors and Functions in Insects

Every peptide has (at least) one cognate receptor, which is responsive to this peptide (but may, in some cases, also respond to other peptides at much higher concentrations). Most of these receptors are G-Protein-Coupled Receptors (GPCR), that activate a second messenger cascade upon binding the peptide [27]. Some peptides activate membrane-bound tyrosine kinases, and one ionotropic peptide receptor for FMRFamide is known [26]. Within the target cells, peptides can elicit all possible effects that are mediated by second messenger cascades: from opening ion channels to their modification (e.g. phosphorylation), to activation of gene transcription. While genome analysis gives the full complement of a species' peptides, it also tells us that there is a large family of orphan GPCRs, i.e. GPCRs for which the ligand is not yet known. It is likely that at least some of these respond to neuropeptides [16]. Here, substantial research is still needed, because understanding peptidergic transmission will only be possible when we know not only the presynaptic, peptidergic neurons, but also the postsynaptic, peptide-sensitive neurons.

To study these questions in the case of the honey bee, researchers have to create hypotheses by extrapolating from functions in other species, because little research has been done about the specific effects of neuropeptides in bees. The very nature of neuropeptides – their large diversity and specific cognate GPCRs – should teach us modesty. The role of neuropeptides is so diverse across species that we need great care in extrapolating from one species to another.

Thus, as an example, we know from several studies that AST-A inhibits visceral and skeletal muscles [29]. AST-A decreased synaptic transmission by pre- and postsynaptic mechanisms at neuromuscular junctions in crustaceans [25] and inhibited the pyloric rhythm of the stomatogastric ganglion in locusts [50] while increasing spike-time precision, again in crustaceans [6]. The physiological action of AST-A in bees has not been studied yet. In another example, we know that tachykinin is excitatory in most systems, and inhibitory in others [21, 44]. Its action in bees (or in different locations of the bee brain) remains to be elucidated.

Despite this functional diversity of some neuropeptides, other neuropeptides also reveal stunning examples of conservation, sometimes only visible via indirect comparisons. For example, the insect diuretic hormone-I (DH-I) and its receptor share similarities with the corticotropin releasing hormone (CRH) in vertebrates, but these similarities are not sufficient to prove their common ancestry. However, CRH signaling is regulated by a binding protein (CRH-BP) in vertebrates, and this binding protein has an unambiguous ortholog in honey bees, suggesting that the entire system (peptide, binding protein, and receptor) share a common ancestry at least to the basis of Bilateria [19].

3.7.2 Neuropeptides and Their Receptors in Honey Bees

Based on the complete sequencing of the bee's genome [43], it was possible to search for all potential neuropeptides by doing sequence analyses [8, 20, 34]. This resulted in a total of 36 genes coding for putative neuropeptide precursors containing 100 peptide sequences.

Mass spectroscopy of brain tissue (e.g. Maldi-TOF, Maldi-TOF-TOF) revealed the presence of many peptides, including in several studies the localization of these peptides to particular brain areas [5, 8, 20, 34]. The total count of peptides from these studies amounts to 67 peptides from 20 families [8]. Thus, proteomics gives smaller numbers than genomics. The difference has several causes: some peptides might only be expressed during very short periods of development, or in very few cells, making their detection difficult. Several peptides are only expressed in peripheral organs that are often not included in proteomics studies. Other sequences might indeed be evolutionary remnants that are not expressed at all.

The main areas to be studied are the brain itself (i.e. the supraesophageal ganglion, consisting of protocerebrum, deutocerebrum and tritocerebrum), the subesophageal ganglion, the retrocerebral complex (CoAl/CoCa), and the ventral nerve chord. Within the brain, subdivisions often isolated in proteomic analyses include the mushroom bodies (MB) area, the optic lobes (OL), and the ALs.

Table 3.7.1 collects the current knowledge about neuropeptide distribution in the honey bee brain. Here, we have formed groups for clarity. For example, allatostatin A comprises a group of six members, tachykinin a group of nine members. Again, it is clear from the table that some families are named based on their initial functional description, others based on their sequence, if that was known before a functional role has been described. Our grouping is based on their sequence, which does not imply that they perform the same function. Each one of the nine tachykinins might have very different tasks, either because it may be expressed in different neurons, or because it activates different target receptors and/or target cells. As a consequence, the table is intended as a coarse overview: please refer to the original literature cited in the table.

Importantly, the table shows that neuropeptides are clearly selective in their expression patterns. The CoAl/CoCa only exhibit myosuppressin, corazonin, PBAN, AST-C, MVPVPVHHMADELLRNGPDTV, NVPIYQEPRF and sNPF within the list. Similarly, any of the other subregions listed in the table has its own combinatorial expression pattern. Furthermore, the lacking entries in the table are also informative: AST-B, for example, is missing in honey bees (and hence in the table), even though it is a widespread neuropeptide in insects. Allatotropin may be present in larvae only [15], highlighting that neuropeptides can play an important role in development. Although orcokinin has been detected by MS in the brain [8], no localization could be detected by immunolabeling [17]. Gastrin/CCK (cholecystokinin) is also missing in the bee genome.

3.7.2.1 Localization of Neuropeptides in Specific Neuropils

Genomics and proteomics can characterize the presence of neuropeptides, but to study their fine-scale distribution, it is necessary to label the peptidergic neurons using immunohistochemistry, at least as long as no transgenes can be created that would express molecular tags in the respective neurons. The number of neuropeptides that has been thoroughly investigated is limited as compared to the total of peptides present. We review a few of the best studied examples.

FMRFamide has been studied in the bee brain by immunostaining [12, 38, 40]. These studies showed about 120 somata in the brain, and additionaly 30 in the subesophageal ganglion, clustered in 13 paired cell groups [38]. Many are associated with the CoAl/CoCa system, with likely neurosecretory function [12]. Kenyon Cell (KC) somata are hardly stained, but a subgroup of KCs forms distinct strata in MB stalks and the output lobes [38, 40]. Although we know that a gene coding for the precise sequence for which the antiserum was raised (FMRFamide) does not exist in the honey bee genome, these stainings are conceptually important as what we see is most likely an inclusive (combinatorial) pattern due to an expected cross-reaction with FaRPs of the honey bee.

In Fig. 3.7.2, the three rows show immunohistochemical stainings for FMRFamide, myosuppressin, and sulfakinin, with one frontal view of the anterior brain, and one frontal view of the posterior brain including the medial lobes (ML), the central body (CB) area, and the posterior slope region. Note the stained somata and distinct bands within the vertical lobe, indicating selective subpopulations of KCs. The precise location of the labeled somta varies considerably across individuals. The FMRFamide-antiserum recognizes several peptides that end in –RFamide. Therefore, the neurons labeled in the FMRFamide panel comprise neurons containing any of the five FaRPs identified in honey bees (see Table 3.7.1). *Apis mellifera* myosuppressin ends with –FLRFamide, and a subgroup of FMRFir neurons is selectively stained with the polyclonal antiserum against *Leucophea madeirae* myosuppressin (a gift of Hans Agricola). The third row, sulfakinin, shows staining using an

Table 3.7.1 Honey bee neuropeptides

Peptide (family)	Synonyms	No.	Brain			MB			AL		Receptors (5)
			MS	IHC	isH	MS (calyx)	IHC	isH	IHC	isH	
Occurence in the brain (after 1–6)											
Allatostatin A		6	+	+ (7)		–	+		+		Am 30
Tachykinin		9	+ (18)	+ (x)	+(18, 19)	+	+	+ (18)	+	+ (18)	Am 42
SIFa		3	+	+ (x)		+	+		+		Am 35
PDF	Pigment dispersing factor	1	+	+ (8)							Am 56
Myosuppressin (FLRFa containing)		2	+	+ (x)		–					Am 21
Sulfakinin (LRFa containing)	FLRFa like	1	+	+ (9, x)		–					Am 37
sNPF (one LRFa containing)	Short neuropeptide F	2			+ (21)	–					Am 32
TWKSPDIVIRFa (RFa containing)	"FMRF-like", only one RFa	4	+			–					
NVPIYQEPRF (RF containing)		3	+			+					
FMRF		5	+	+ (12, 13, 14)		–	+		+		Am 22
PBAN	Pheromone biosynthesis activating neuropeptide										
Allatostatin C		3	+			+					Am 31
Orcokinin		5	+			+					
Calcitonin-like DH	Calcitonin-like diuretic hormone	1	+								
IDLSRFYGHFNT		4	+			+					
ITGQGNRIF		1	+			+					
MVPVPVHHMADELLRNGPDTV		1	+			+					
NP-like precursor	Neuropeptide-like precursor	8	+			–					
CAPA		2	+			–					Am 27, Am 28
Corazonin	Periviszerokinin-like	2	(+), 11	+ (10)							Am 45
Apidaecin-1		3	(+)								

Peptide	Full name					Am
Pyrokinin						Am 25, Am 26
CCAP	Crustacean cardioactive peptide	+ (6)	+ (x)			Am 46
Allatotropin						
AKH	Adipokinetic hormone, Neurohomone D					Am 44
Allatostatin B						Am 41
Genes for peptides present by homology search (after 2)						
Bursicon						Am 47
CCAP						Am 46
ITP	Ion tranport peptide					Am 53, Am 55
DH	Diuretic hormone					Am 29
ETH	Ecdysis triggering hormone					
EH	Eclosion hormone					
Insulin			+ (ELISA, 17, 20)			
NPF	Neuropeptide F			+ (21)		
NPFF	Neuropeptide FF					
Neuroparsin					+	
IHC Vertebrate-like peptides						
Gastrin/CCK			+ (15)			
Prolactin-like			+ (16)			

(1) [8]; (2) [20]; (3) [34]; (4) [5]; (5) [16]; (6) [43]; (7) [24]; (8) [7]; (9) [2]; (10) [7]; (11) [46]; (12) [38]; (13) [12]; (14) [40]; (15) [31]; (16) [37]; (17) [33]; (18) [42]; (19) [41]; (20) [4]; (21) [3]; (x) unpublished Konstanz results

MS mass spectroscopy; *IHC* immunohistochemistry; *isH* in situ hybridization

An regulary updated version of this table can be downloaded at http://neuro.uni-konstanz.de/peptides

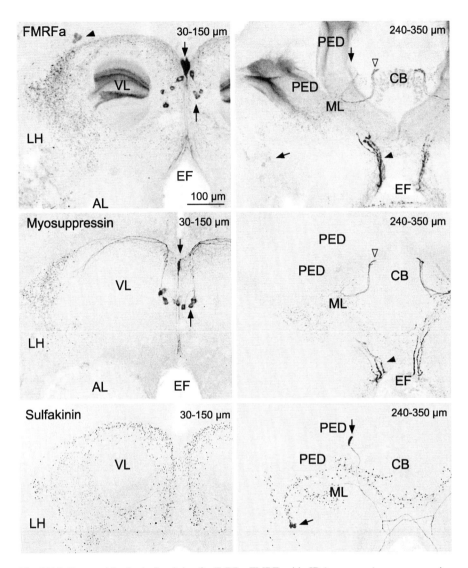

Fig. 3.7.2 Immunohistological staining for FaRPs: FMRFamide-IR (*upper row*), myosuppressin-IR (*middle row*), and sulfakinin-IR (*bottom row*), showing the vertical lobes (*VL*), the lateral horn (*LH*), a portion of the antennal lobes (*AL*), medial lobes (*ML*), pedunculi (*PED*), the central body (*CB*) and the nervi corporis cardiaci (*arrowhead*). Confocal stacks of fluorescent images at depths as indicated in the *upper right corners* have been projected to allow for a frontal view in an anterior (*left column*) and a posterior (*right column*) position. The figure shows that cross-talk of the FMRF antiserum leads to inclusive staining of different FaRPs. Staining for FMRFamide-IR is thought to reveal neurons containing several different –RFamides. This staining includes the neurons stained by the two more specific antisera against myosuppressin (*arrows* in the *left columns* of the *upper* and the *middle rows*) and sulfakinin (*arrows* in the *right columns* of the *upper* and the *bottom rows*). In addition, other somata (*arrowhead* in the *left column*) and very distinct bands of KCs in the VLs and the MLs are stained. A columnar structure in the CB, a prominent projection from the anterior somata (*right column*, open *arrowhead*) and the nervi corporis cardiaci (*arrowhead*) are

antiserum against *Periplaneta americana* sulfakinin, a peptide that ends (in the honey bee) with –LRFamide (a gift of Hans Agricola). Note the very distinct somata and neurites in the posterior slope region and the crisp staining in the anterior protocerebrum around the vertical lobe. The six neurons visible in these stainings have previously been described [2] and are also included in the FMRFamide stainings. NPF and sNPF gene expression has recently been localized to neurons in the neurosecretory cell clusters [3]. It should be noted, that the FaRP(s) contained in Kenyon cells remain to be determined. However, with the very limited knowledge about neuropeptide receptors of bees, our understanding of the functional relevance of neuropeptides is restricted to date. Therefore, the greater insight gained so far, is about the increasing number of histochemically diverse populations of neurons previously being treated as homogeneous groups.

Gastrin/CCK: Staining against CCK showed distinct subpopulations of KCs [31, 40]. In this case the precise target of the antiserum is unknown due to sequence differences, CCK not being among the peptides found in bees.

Tachykinin has been investigated using in situ hybridization, showing its localization to some neurons in the OL and local neurons of the AL, and to many KCs in the MBs [42]. This expression was complemented by immunocytochemistry (Fig. 3.7.1) and was reported as unpublished data [22]. Within the MB, stainings indicate a clear subdivision of KCs into biochemically distinct subpopulations, suggesting that their contribution to brain activity, learning and memory and plasticity is diverse. Moreover, in situ hybridization shows that even small and large KCs are not homogeneous populations but at least the large KCs comprise preprotachykinin gene expressing and nonexpressing somata [42].Thus, stainings for tachykinin, gastrin/CCK and FMRFamid tell us that there are these distinct populations of neurons that express a particular peptide, and therefore have some distinct physiological properties.

AST-A: Allatostatin has been thoroughly analyzed recently [24]. There are about 500 AST-immunoreactive (ASTir) neurons in the brain, scattered in 18 groups. Interestingly, the precise location of the labeled soma clusters varies considerably across individuals. ASTir fibers innervate almost all areas of the brain. At the same time, ASTir neurites generally form networks within functionally distinct areas, e.g. the ALs, the MBs or the OLs, rather than spanning across areas, indicating local functions of the peptide. Only a few very large neurons have widespread arborizations,

---

Fig. 3.7.2 (continued) stained in the more posterior part of the brain. Staining for myosuppressin labels 12 neurons. Four large somata are stained in the pars intercerebralis (*upper arrow* in the *left column*) that send axons to the retrocerebral complex, and four other somata are stained on each side in the anterior protocerebrum (*lower arrows* in the *left column*). The latter neurons innervate the lateral protocerebrum (left column) and a part of the posterior protocerebral neuropil (*right column, open arrowhead*). They do not enter into the MB lobes. Innervation of the retrocerebral complex is supplied by the nervus corporis cardiaci (*arrowhead*). Staining for sulfakinin-IR labels three somata on each side in the posterior protocerebrum. These neurons innervate the neuropil around the vertical lobes and a part of the lateral protocerebrum. Based on their projection pattern, these neurons are a subpopulation of the neurons visualized in the staining for FMRFamide-IR but distinct from the neurons labeled in the myosuppressin staining

including those projecting into the CoAl/CoCa, suggesting a neurohemal release. Some, but not all AST neurons express GABA as their classical transmitter. These data show that peptides from the AST-A family have several distinct roles in different neurons: either with GABA or without, either in small local neurons, most likely with a point-to-point transmission, or in large, global neurons with volume transmission, or even into the circulation via a neurohemal organ (the latter has not yet been shown directly).

PDF: This peptide (pigment-dispersing factor, probably homologous to the crustacean pigment-dispersing hormone PDH) has been found in the bee brain [7] but not in the rest of the body (see also Table 3.7.1). A single group of approx. 20 neurons (in each brain hemisphere) is labeled at the medial margin of the medulla (ME) in the OL. These neurons target specific neuropils in the OLs and the protocerebrum, but not in the AL and the antennal-mechanosensory motor centers (AMMCs). In *Drosophila*, PDF is related to the circadian clock and coexpressed with the gene *per*, while its role in the honey bee is as yet unclear. While PDF-labeling does not vary with the time of day or the age of the bee in some instances [7], other observations of cycling PDF still need to be confirmed. In this respect, it is an interesting observation that there are more PDF-immunoreactive cells in large bumblebees which also have stronger rhythms than in small bumblebees [48].

3.7.2.2 Functions of Neuropeptides in Honey Bees

The function of a few peptides have been studied in honey bees. A better studied example concerns insulin-like peptides, ILP. Generally, these peptides have a strong role in metabolism, growth, reproduction and aging [29]. In many respects, feeding-related behavior control in insects is functionally similar to the situation in mammals: upon food intake or high levels of nutrients, insulin is released in mammals, or ILP is released in insects. These peptides repress the synthesis of glucagon (in mammals) or of adipokinetic hormone, AKH (in insects) [29]. In bees, ILP influences juvenile hormone release, adding, among others, to the longevity of queens [11]. Royal jelly induces ILP release in queen larvae [43]. In queen workers, ILP signaling appears to be reversed: nurses have higher nutrient levels than foragers, but foragers express more ILP [4]. Indeed, inhibiting the ILP- related rapamycin pathway in nurse bees delayed their caste development to becoming foragers, suggesting that this peptide also plays a role in the regulation of social behavior. This is a nice example of the intricate relationship of complex behaviors (such as social division of nursing/foraging behavior) and basic metabolic regulation (satiety and blood-sugar levels). One example comes from opioid system: bees injected with morphine, an opiate receptor agonist, show a weaker sting response to electric shock. The morphine effect can be prevented by injecting the competitive opiate receptor antagonist naloxone. While morphine itself is not a peptide, the natural ligands for opioid receptors in vertebrates are endorphins. These endorphins comprise enkephalin, kyotorphin, and similar peptides.

These specific substances were not active in a physiological test in bees [32], and in fact their exact sequence does not appear in the list of expressed honey bee peptides. Which – if any – natural neuropeptides might play a role in a putative honey bee opiate system remains to be elucidated.

The levels of expressed tachykinin, PBAN and sNPF are modified in nectar and pollen foraging bees, suggesting a role in food intake [10], similar to what has been shown in other species. Many other peptides decrease or increase their levels depending on the animal's behavioral experience.

Apart from these few examples, honey bee researchers still have to look at other insect species for possible peptide action (reviews: [13, 28, 29]). Several peptides that act metabolically may fulfill the same function as in other insects, e.g. a complex sequence of activity via eclosion hormone (EH) and eclosion triggering hormone (ETH), crustacean cardioactive peptide (CCAP), myosuppressin and other peptides regulates the complex behavioral sequences during pupation and molt, with bursicon being likely involved in the tanning of the cuticle after molting of the adult animal [23, 27]. However, these studies were mostly done on moths or *Drosophila*, and have not been replicated in bees. The information about functions within the brain is even sparser.

3.7.2.3   Neuropeptides in the Honey Bee Olfactory System

Even when their function is not understood, peptides can help understand brain circuitry. Thus, by labeling populations of neurons using peptide-selective antibodies, different populations of neurons can be characterized. Function follows form in many cases, and a thorough morphological analysis can guide in developing hypotheses about brain circuits. Here we review the situation in the honey bee olfactory system, including some as yet unpublished data from our own lab.

Approximately 60,000 olfactory receptor axons innervate the AL glomeruli in the olfactory system of the worker honey bee (see also Chap. 4.1, Sandoz). No peptide has been reported in olfactory receptor neurons in honey bees so far. Within the AL, there are approx. 160 glomeruli, forming four distinct groups based on the antennal nerve branch that innervate them: T1-T4 [14]. With respect to peptide immunohistochemistry, glomeruli in T1-T3 appear uniform, while T4 glomeruli have distinct localization patterns. A large number of local neurons (LN) interconnect glomeruli (in the range of approx. 3,000 following our own counts). Many LNs express GABA [36]; in our as yet unpublished counts GABAergic neurons make up about a half of all LNs. A small population of 35 neurons expresses histamine [9], and the remainder may have one or several as yet unidentified neurotransmitters maybe also including glutamate. Within the GABAergic population, there is a small subpopulation of 20 neurons that are also immunoreactive against the neuropeptide AST-A [24], and another, more sizable population of approx. 420 GABA immunoreactive neurons that are also immunoreactive against the neuropeptide tachykinin (unpublished observations).

A group of four centrifugal neurons located in the pars intercerebralis innervates the AL glomeruli with a fine network of neurites, exhibiting immunoreactivity against the neuropeptide SIFamide (Fig. 3.7.1, bottom row). The role of SIFamide is unknown in bees; in flies SIFamide is involved in sexual behavior [45].

Projection neurons (PNs, approx. 800 in number) project from the ALs to the lateral protocerebrum and to the MBs. No neuropeptide has been found in PNs in honey bees so far. PNs innervate the KCs (approx. 170,000 in number) in the MBs. As reported above, KCs can be subdivided into several groups based on peptide immunostaining and gene expression (Figs. 3.7.1 and 3.7.2) [40, 42], suggesting that these subgroups may play different functional roles in olfactory and/or multimodal processing.

3.7.3 Outlook

Neuropeptide research is an important area of honey bee neuroscience that is currently growing rapidly, due to a series of new technological developments. Already, it is thanks to immunohistochemical stainings for peptides that we know many more neuron types than we would know from intracellular stainings alone. These data are important for our understanding of brain circuitry. It is also important to realize that at the current stage, information appears sometimes contradictory or inconclusive, because we do not always understand how much similarity and how much diversity there is across insect species. New developments will hopefully quickly add to our understanding of peptide distribution and their function. The big open questions are both functional (what is the role of each peptide in a particular developmental stage and/or for a particular behavior, and how do they act physiologically) and evolutionary (how did the peptide diversity, peptide receptors, and additional proteins involved in the system, such as binding proteins, evolve). Honey bees may play a particularly interesting role in elucidating brain functions controlled by peptides and related to their social behavior.

References

1. Adams ME, O'Shea M (1983) Peptide cotransmitter at a neuromuscular junction. Science 221:286–289
2. Agricola HJ, Bräunig P (1995) Comparative aspects of peptidergic signaling pathways in the nervous systems of arthropods. In: Breidbach OKW (ed) The nervous system of invertebrates: an evolutionary and comparative approach. Birkhäuser, Basel, pp 303–327
3. Ament SA, Velarde RA, Kolodkin MH, Moyse D, Robinson GE (2011) Neuropeptide Y-like signalling and nutritionally mediated gene expression and behaviour in the honey bee. Insect Molecular Biology 20(3):335–345
4. Ament SA et al (2008) Insulin signaling is involved in the regulation of worker division of labor in honey bee colonies. Proc Natl Acad Sci USA 105(11):4226–4231

5. Audsley N, Weaver RJ (2006) Analysis of peptides in the brain and corpora cardiaca-corpora allata of the honey bee, *Apis mellifera* using MALDI-TOF mass spectrometry. Peptides 27(3):512–520
6. Billimoria CP et al (2006) Neuromodulation of spike-timing precision in sensory neurons. J Neurosci 26(22):5910–5919
7. Bloch G et al (2003) Patterns of PERIOD and pigment-dispersing hormone immunoreactivity in the brain of the European honeybee (*Apis mellifera*): age- and time-related plasticity. J Comp Neurol 464(3):269–284
8. Boerjan B et al (2010) Mass spectrometric profiling of (neuro)-peptides in the worker honeybee, *Apis mellifera*. Neuropharmacology 58(1):248–258
9. Bornhauser BC, Meyer EP (1997) Histamine-like immunoreactivity in the visual system and brain of an orthopteran and a hymenopteran insect. Cell Tissue Res 287:211–221
10. Brockmann A et al (2009) Quantitative peptidomics reveal brain peptide signatures of behavior. Proc Natl Acad Sci USA 106(7):2383–2388
11. Corona M et al (2007) Vitellogenin, juvenile hormone, insulin signaling, and queen honey bee longevity. Proc Natl Acad Sci USA 104(17):7128–7133
12. Eichmüller S, Hammer M, Schäfer S (1991) Neurosecretory cells in the honeybee brain and suboesophageal ganglion show FMRFamide-like immunoreactivity. J Comp Neurol 312:164–174
13. Gäde G, Hoffmann KH (2005) Neuropeptides regulating development and reproduction in insects. Physiol Entomol 30(2):103–121
14. Galizia CG, McIlwrath SL, Menzel R (1999) A digital three-dimensional atlas of the honeybee antennal lobe based on optical sections acquired by confocal microscopy. Cell Tissue Res 295(3):383–394
15. Glasscock J, Mizoguchi A, Rachinsky A (2005) Immunolocalization of an allatotropin in developmental stages of *Heliothis virescens* and *Apis mellifera*. J Insect Sci 5:7–7
16. Hauser F et al (2006) A review of neurohormone GPCRs present in the fruitfly *Drosophila melanogaster* and the honey bee *Apis mellifera*. Prog Neurobiol 80(1):1–19
17. Hofer S et al (2005) Novel insect orcokinins: characterization and neuronal distribution in the brains of selected dicondylian insects. J Comp Neurol 490(1):57–71
18. Hökfeld T et al (2000) Neuropeptides – an overview. Neuropharmacology 39(8):1337–1356
19. Huising MO, Flik G (2005) The remarkable conservation of corticotropin-releasing hormone (CRH)-binding protein in the honeybee (*Apis mellifera*) dates the CRH system to a common ancestor of insects and vertebrates. Endocrinology 146(5):2165–2170
20. Hummon AB et al (2006) From the genome to the proteome: uncovering peptides in the *Apis* brain. Science 314(5799):647–649
21. Ignell R et al (2009) Presynaptic peptidergic modulation of olfactory receptor neurons in *Drosophila*. Proc Natl Acad Sci USA 106(31):13070–13075
22. Johard HAD et al (2008) Intrinsic neurons of *Drosophila* mushroom bodies express short neuropeptide F: relations to extrinsic neurons expressing different neurotransmitters. J Comp Neurol 507(4):1479–1496
23. Kim YJ et al (2006) A command chemical triggers an innate behavior by sequential activation of multiple peptidergic ensembles. Curr Biol 16(14):1395–1407
24. Kreissl S, Strasser C, Galizia CG (2010) Allatostatin immunoreactivity in the honeybee brain. J Comp Neurol 518(9):1391–1417
25. Kreissl S et al (1999) Allatostatin modulates skeletal muscle performance in crustaceans through pre- and postsynaptic effects. Eur J Neurosci 11(7):2519–2530
26. Lingueglia E et al (1995) Cloning of the amiloride-sensitive FMRFamide peptide-gated sodium channel. Nature 378:730–732
27. Mercier J, Doucet D, Retnakaran A (2007) Molecular physiology of crustacean and insect neuropeptides. J Pestic Sci 32(4):345–359
28. Nässel DR (2009) Neuropeptide signaling near and far: how localized and timed is the action of neuropeptides in brain circuits? Invertebr Neurosci 9(2):57–75

29. Nässel DR, Winther AME (2010) *Drosophila* neuropeptides in regulation of physiology and behavior. Prog Neurobiol 92(1):42–104
30. Nichols R (2003) Signaling pathways and physiological functions of *Drosophila melanogaster* FMRFamide-related peptides. Ann Rev Entomol 48:485–503
31. Noble MGL (1987) Immunohistochemical localization of agastrin/CCK like peptide in the brain of the honeybee. In: Eder JRH (ed) Chemistry and biology of social insects. Pepperny, München
32. Núñez J et al (1997) Alarm pheromone induces stress analgesia via an opioid system in the honeybee. Physiol Behav 63(1):75–80
33. Oconnor KJ, Baxter D (1985) The demonstration of insulin-like material in the honey bee, *Apis mellifera*. Comp Biochem Physiol B-Biochem Mol Biol 81(3):755–760
34. Predel R, Neupert S (2007) Social behavior and the evolution of neuropeptide genes: lessons from the honeybee genome. Bioessays 29(5):416–421
35. Roller L et al (2006) Molecular cloning of [Thr(4), His(7)]-corazonin (Apime-corazonin) and its distribution in the central nervous system of the honey bee *Apis mellifera* (Hymenoptera: Apidae). Appl Entomol Zool 41(2):331–338
36. Schäfer S, Bicker G (1986) Distribution of GABA-like immunoreactivity in the brain of the honeybee. J Comp Neurol 246(3):287–300
37. Schmid KP et al (1989) Immunocytochemical localization of prolactin-like antigenic determinants in the neuro-endocrine system of the honeybee (*Apis mellifica*). Histochemistry 91(6):469–472
38. Schürmann FW, Erber J, Schürmann FW, Erber J (1990) FMRFamide-like immunoreactivity in the brain of the honeybee (*Apis mellifera*). A light- and electron microscopic study. Neuroscience 38:797–807
39. Stay B, Tobe SS (2007) The role of allatostatins in juvenile hormone synthesis in insects and crustaceans. Ann Rev Entomol 52:277–299
40. Strausfeld NJ, Homberg U, Kloppenburg P (2000) Parallel organization in honey bee mushroom bodies by peptidergic Kenyon cells. J Comp Neurol 428(4):760–760
41. Takeuchi H et al (2003) Identification of a tachykinin-related neuropeptide from the honeybee brain using direct MALDI-TOF MS and its gene expression in worker, queen and drone heads. Insect Mol Biol 12(3):291–298
42. Takeuchi H et al (2004) Prepro-tachykinin gene expression in the brain of the honeybee *Apis mellifera*. Cell Tissue Res 316(2):281–293
43. The Honeybee Genome Consortium (2006) Insights into social insects from the genome of the honeybee *Apis mellifera*. Nature 443(7114):931–949
44. Van Loy T et al (2010) Tachykinin-related peptides and their receptors in invertebrates: a current view. Peptides 31(3):520–524
45. Verleyen P, Huybrechts J, Schoofs L (2009) SIFamide illustrates the rapid evolution in Arthropod neuropeptide research. Gen Comp Endocrinol 162(1):27–35
46. Verleyen P et al (2006) Cloning and characterization of a third isoform of corazonin in the honey bee *Apis mellifera*. Peptides 27(3):493–499
47. Vilim FS et al (1996) Release of peptide cotransmitters in *Aplysia*: regulation and functional implications. J Neurosci 16:8105–8114
48. Weiss R et al (2009) Body size-related variation in Pigment Dispersing Factor-immunoreactivity in the brain of the bumblebee *Bombus terrestris* (Hymenoptera, Apidae). J Insect Physiol 55(5):479–487
49. Zabriskie TM et al (1994) Peptide amidation in an invertebrate – purification, characterization and inhibition of peptidylglycine alpha-hydroxylating monooxygenase from the heads of honeybees (*Apis mellifera*). Arch Insect Biochem Physiol 26(1):27–48
50. Zilberstein Y et al (2004) Neuromodulation for behavior in the locust frontal ganglion. J Comp Physiol A 190(4):301–309

Chapter 3.8
Brain Anatomy and Physiology: Commentary

Randolf Menzel

3.8.1 Anatomy

Revolutions in science are often connected to unique researchers. In neuroscience such a unique researcher was Ramon y Cajal. Perfecting Golgi's silver stain and interpreting what he saw in his microscope in an innovative way, he catapulted the understanding of the nervous system to new horizons. Today's neuroanatomy benefits from the revolutions introduced by powerful microscopes, computers and software, and these revolutions have many fathers and mothers. It is hard to believe that 3D images of neurons and neural nets are available to us only one to two decades. Camera-lucida pictures of neurons still impress us by their structural complexity and beauty, but considering the enormous loss of information by such artistic drawings one wonders what more than just documentation and classification of structural features could be reached by this method. Scientific reports have now to be based on 3D images of segmented structures incorporated into appropriate reference systems. Neurons can be quantified with respect to their real branching patterns, the length components of all branches and their spatial relations. Neural nets can be reconstructed in 3D from multiple single neurons. Neuropils and tracts as well as somata clusters can be composed in 3D and provide the intrinsic landmarks for embedding neural elements into a 3D atlas. Digital neuroanatomy provides us already with the tools for such an endeavor although the necessary software is still not satisfactory and requires a large amount of tedious hand work. Digital neuroanatomy requires segmentation of reference structures, tracing neurons, bridging between large ranges of spatial resolution, registering the structures in a virtual reference system (the atlas) and using this information for ontologies of neuron related information. The first steps in this direction have been done, and insect brains are at the forefront of this endeavor taking advantage of their suitable size for digital microscopy.

R. Menzel (✉)
Institut für Biologie, Neurobiologie, Freie Universität Berlin, Berlin, Germany
e-mail: menzel@neurobiologie.fu-berlin.de

C.G. Galizia et al. (eds.), *Honeybee Neurobiology and Behavior: A Tribute to Randolf Menzel*, DOI 10.1007/978-94-007-2099-2_18,
© Springer Science+Business Media B.V. 2012

Let's look into the future by imagining that we had the tools available already 50 years ago when the first intracellular markings were accomplished, and the new fluorescence tracing techniques and immuno stainings were developed. Thousands of single neurons would gather in the bee brain atlas. Sometimes I dream about an intelligent data bank, an ontology of the bee brain that stores and makes accessible the anatomical, physiological and molecular characteristics of all the neurons that had been recorded and marked in the last 50 years. Such a rich body of information would revolutionize the way we think about the structure, function and cognition of this little brain. Such an ontology would not only capture and store the data in a relevant way (in 3D, time-resolved recordings, documents of the steps during data analysis) but even more importantly would help to link these vast amount of data in reference to the structure of the respective neurons and neuropils. We cannot blame the researchers in the past for publishing their data in traditional way but we would reduce the power of our future research if we ignore the potentials provided by the standard brain atlas and the development of a related ontology. These goals will only be reached if we join forces and work together on improvements of digital anatomy and physiology. Ideally, there should be no neuroanatomy study without digital neuron tracing and registering into the brain atlas! Functional data related to neural structures should be compiled in an archive available to all of us. Too much has been lost over the last 50 years. We should do it better, now.

Given the neuroanatomical organization of the olfactory input to the mushroom body, the convergence with the reinforcing pathway (VUMmx1) in the lip region and the dependence of coincident excitation as a requirement of Kenyon cell firing it is obvious that the lip should house at least part of the olfactory memory trace. Why don't we see it? Ca^{2+}-imaging of PN boutons did not indicate learning-related functional changes (unpublished observations from our lab). The most reliable effect of associative olfactory learning in Kenyon cells is expressed in the compensation of the non-associative depression caused by stimulus repetition [7], a surprisingly small and rather unspecific "memory trace" indeed. It is still a mystery why Ca^{2+}-imaging of both the pre- and postsynaptic sites in the lip region does not register specific associative plasticity. One possibility may relate to the spatial sparseness of the effects and the problem that the lip does not provide intrinsic structures for localizing recording sites between preparations. Analyzing single preparations is bound to the problem that the necessary number of repetitions of stimulations changes the responses due to habituation and extinction effects.

Another mystery relates to the large volume (both in absolute and relative terms) of the mushroom body calyces. Comparing solitary and social flying hymenoptera [1] found already 100 years ago a correlation with social life style. Although more data on the effect of age and experience were collected, we still have not identified the social factors that drove the evolution of the mushroom bodies. Life in a large social community may require more elaborate olfactory communication and related adaptive neural processing. Potentially longer individual life time and highly flexible dealing with a much larger range of environmental conditions may constitute another set of factors affecting mushroom body volume. Whether the presence of rather elaborate mushroom bodies in parasitizing pre-social hymenoptera constitutes a pre-adaptation for those functions of the mushroom body in current social hymenoptera is an interesting speculation [3] but the correlations found let us uninformed why the mushroom body should be involved in finding hosts.

A striking property of the calyx organization is its ordered representation of processed sensory inputs. An integration across all sensory inputs, that is possibly related to context dependent forms of learning and to the extraction of rules defined by the sequence of events may be of particular importance. In my view we have not applied yet the appropriate paradigms in searching for functional and structural plasticities in the calyx. The two extremely different test conditions used so far, simple olfactory PER conditioning, and undefined experiences of foraging, appear not to allow any conclusions and are very limited in their predictive power. We need to know more precisely what kind of experience shapes the internal organization of the calyx, and I expect closer analysis of exploratory flights and navigation experience at the individual level will provide cues. Also controlled laboratory protocols in which bees are not restrained so that they can move yet in a tethered way may provide important clues to study calyx plasticity upon different forms of experience.

Finally I want to raise a third point. So far we work with correlations and tend to interpret them as causal relations. We record from neurons and correlate their responses and responses changes to the experimental procedure, we manipulate behavior and relate structural changes in the brain, we apply pharmacological and to a limited amount molecular tools to interfere with ongoing processes. How do we reach beyond this level which could be seen as unsatisfactorily vague? Reaching beyond correlations require manipulation of neural mechanisms at several levels, knock-down of particular neurons or neural nets, recovery from knock-down, targeted activation and inactivation of the suspected neurons or neural nets. In the future it will hopefully be possible to follow the *Drosophila* way, maybe not by creating transgenic lines but rather by somatic transfection of single animals.

3.8.2   Physiology

NMDA glutamate receptors and LTP induction are closely related in hippocampal and cortical neurons of the mammalian brain. Is there LTP in bee brain neurons, and if so, is it connected to NMDA receptors? The answer to the first question is yes, to the second, we do not know. Associative LTP could be induced by pairing depolarization of the PE1 neuron with tetanic stimulation of the presynaptic mushroom-body neurons, the Kenyon cells [4]. These experiments were carried out in an in-vivo preparation. Surprisingly, we did not find LTP in all preparations. Probably different modulatory states of the whole animal may have set different conditions, which either permitted LTP to develop or block it. It will be exciting to search for these modulatory conditions.

Although the insect NMDA receptor may well be a double regulated channel as in the mammalian brain, associative plasticity during the acquisition phase appears to be independent of NMDA receptors. This surprising result is described in Gérard Leboulle's chapter. It will be necessary to record over days from identified neurons, e.g. the PE1 in order to correlate LTP induction and its possible effect on memory consolidation. Extracellular recordings from the PE1 are possible [5], and an extension to long lasting recordings may well be possible.

A major limitation in all our attempts to unravel the mechanisms of associative plasticity, LTP and the potential role of NMDA receptors in memory consolidation is our ignorance about the transmitter(s) of Kenyon cells. The very small population of glutamate-ir Kenyon cells makes it unlikely that glutamate is a major transmitter of Kenyon cells. The situation may well be different for mushroom body extrinsic neurons, and in such a case NMDA receptors on the PE1 neuron or on other neurons of the lobes may be controlled by other extrinsic neurons. The picture emerging from these speculations relate associative plasticity in PE1 (and possibly in other mushroom body extrinsic neurons) to a transmitter system(s) of Kenyon cells that is yet unknown, and to a modulatory effect that is mediated by NMDA receptors and controls memory consolidation. In such a situation associative LTP in PE1 should be independent of NMDA receptors but dependent on receptors for the transmitters of Kenyon cells.

The spatial separation of somata from the excitation flow in central insect neurons potentially provides us with a unique access to the anatomical and functional characterization of the related axons and dendrites. As Bernd Grünewald points out, filling central neurons by patch electrodes via their somata with neuroanatomical tracer dyes has not been successful so far in the bee brain. This is a great pity because the often peripheral localization of somata and their spatial arrangements in well defined clusters should allow us to target specific neurons even on the level of single identified neurons. We should not give up searching for ways of overcoming these problems. In a longer run one may envisage a map of somata of identified neurons combined with the standard atlas of the living bee brain as a tool for direct access to sets of neurons for opto- and electrophysiological recordings. A range of exciting questions come to mind that could be approached this way. For example, structural changes of dendritic arbors may be followed during development, ageing and experience; the variance of structural features of single identified neurons could be quantified; neurons combined in networks could be filled with different functional dyes for simultaneous recording.

Monique Gauthier and Bernd Grünewald also addressed the question of how pesticides directed towards insects act at the cellular level. This is an extremely important topic, and we should realize that the sub-lethal effects of these molecules on honeybees in their natural environment are usually not examined in any of the test procedures used by pharmaceutical companies or official control institutions. The neonicotinoid imidacloprid, and particularly clothianidine, have both caused serious problems to bee colonies on a large scale, and sub-lethal effects are practically unknown. In our research community, we have the necessary tools to address the questions of the sub-lethal effects, their underlying mechanisms and their real impact on colony survival. We are also independent from any commercial or industrial lobbying group so that we can provide the necessary information to the science community and the public. In my view we also have the moral and ethic responsibility to become involved in the difficult process of outbalancing the pros and cons of pesticide application in the environment. This responsibility is not just motivated by the care and preservation of honeybees but rather for other insects particularly solitary hymenoptera that will be affected by such pesticides but whose damage or even extinction will stay mostly unrecognized. From my personal experience I can only advise everybody against establishing research contracts with pharmaceutical companies. I made mistakes in this field myself. It should not be accepted that such research cannot be published and made public by other means.

Alison Mercer's discovery that the queen controls dopamine levels within the brain of young bees via the action of queen mandibular pheromone challenges in my view the current picture of social organization in the honey bee colony and possibly in other Hymenopteran societies. It has been suggested mainly through modeling approaches that the queen does not rule by direct actions on the ongoing behavior of colony members but rather sets the stage for self-organizing and emerging properties of social organization [2]. The rather vague concept of emerging social regulatory processes can now be replaced at least partly by direct actions of the queen on the value system (i.e. the neural system underlying aversive reinforcement, [8]) of young workers. Social coherence results from an agreement of what needs to be done and why this is of benefit for the individual member. Thus, revised modeling of insect societies will have to include knowledge about the neural mechanisms of decision-making processes at the level of the individual.

It appears to me that the current models of insect societies assume stimulus–response connections as the basic principle of behavioral control at the individual level. Colony members are viewed as response elements that switch on and off their behavioral programs according to the external stimuli surrounding them. The conditions under which actions are selected from a repertoire of potential actions are usually not considered or are related to genetically-controlled developmental processes. This is certainly only a part of the picture. Actions are selected in addition and most importantly in reference to an internal evaluation process of the individual. However, what is good for the society may not be good for the individual, or vice versa. Indeed the queen does not control or manipulate directly the assumed stimulus-control connections but rather the evaluation system of the individual during a particular time of its function within the society, and that is a much more effective way of keeping control over colony members, because it reaches a whole range of potential behaviors and not just isolated stimulus–response connections. I have asked above (commentary to section 1): Are bees switching off their central brain when entering the hive? Are they just rather stupid members in a network of community functions, e.g. a reproductive network? The evaluation system is a major component of the central brain modulating higher order functions in an orchestrated way. The results from Alison Mercer's work demonstrated nicely that indeed the central brain stays in action also inside the hive.

One of the many mysteries in the neural organization of behavior is context dependence and in particular neural functions under the control of body conditions. Animals sense stimuli differently, create body-state related search behavior and form different stimulus–response connections. A large number of neurons all over the brain and the ventral cord from sensory integration to motor control have to be orchestrated accordingly. Neuroscientists usually relate such phenomena to modulatory functions of aminergic and peptidergic networks. Some correlations between aminergic modulation and behavioral switching have been documented in insects [6], but even in the case of aminergic modulation our knowledge is rather limited. Peptidergic modulation is an unexplored topic. The impressive number of up to 100 peptides acting as transmitters, modulators and centrally regulating hormones suggest a major impact of peptidergic modulation of neural functions in the bee brain. How may we proceed in analyzing such functions? To reach beyond correlations it will be necessary to manipulate aminergic and/or peptidergic neuro-transmission by molecular techniques in defined groups of animals or single animals within a colony, a task far beyond current bee research methodology, but as stated above somatic transfection of individuals may well be possible rather soon.

References

1. Alten HV (1910) Zur Phylogenie des Hymenopterengehirns. Naturwissenschaften 46:511–590
2. Camazine S, Deneubourg JL, Franks N, Sneyd J, Theraulaz G et al (2003) Self-organization in biological systems. Princeton University Press, Princeton, USA
3. Farris SM, Schulmeister S (2011) Parasitoidism, not sociality, is associated with the evolution of elaborate mushroom bodies in the brains of hymenopteran insects. Proc R Soc B 278(1707): 940–951
4. Menzel R, Manz G (2005) Neural plasticity of mushroom body-extrinsic neurons in the honeybee brain. J Exp Biol 208(Pt 22):4317–4332
5. Okada R, Rybak J, Manz G, Menzel R (2007) Learning-related plasticity in PE1 and other mushroom body-extrinsic neurons in the honeybee brain. J Neurosci 27(43):11736–11747
6. Pflüger HJ, Stevenson PA (2005) Evolutionary aspects of octopaminergic systems with emphasis on arthropods. Arthropod Struct Dev 34(3):379–396
7. Szyszka P, Galkin A, Menzel R (2008) Associative and non-associative plasticity in kenyon cells of the honeybee mushroom body. Front Syst Neurosci 2:3
8. Vergoz V, Schreurs HA, Mercer AR (2007) Queen pheromone blocks aversive learning in young worker bees. Science 317(5836):384–386

Part IV
Sensory Systems

Chapter 4.1
Olfaction in Honey Bees: From Molecules to Behavior

Jean-Christophe Sandoz

Abstract For more than a century, honey bees have constituted a major model for the study of olfactory detection, processing, learning and memory. This chapter reviews major advances based on three main approaches. Firstly, we address the experimental study of bees' olfactory behavior, from early experiments on free-flying workers until laboratory-based training protocols on restrained individuals. We describe bees' impressive discrimination and generalization abilities depending on odor quality and quantity, their capacity to grant special properties to olfactory mixtures as well as to recognize individual components. Secondly, we provide a detailed description of the olfactory pathways of the bee brain that subtend these behaviors, based on anatomical and immunochemical studies. We show how odors are detected by olfactory receptors carried by receptor neurons in the antenna, which convey information to a first processing relay, the antennal lobe (AL). We describe processing circuits within this structure and show how olfactory information is then conducted to higher-order centres, the mushroom bodies (MBs) and the lateral horn (LH), following different pathways through the brain. We finish by discussing the structure of the MBs, their local circuits and output connections and how they may be linked to motor output. Thirdly, we show how functional approaches based on the recording of odor-evoked activity in the bee brain allow following the series of transformations of the olfactory representation through its different centers. Data from electrophysiological and optical imaging approaches are reviewed. Doing so, we explain how coupling behavior with functional approaches allows understanding the perceptual representation of odors.

J.-C. Sandoz (✉)
LEGS-CNRS, 1 avenue de la Terrasse, 91198 Gif-sur-Yvette cedex, France
e-mail: sandoz@legs.cnrs-gif.fr

C.G. Galizia et al. (eds.), *Honeybee Neurobiology and Behavior: A Tribute to Randolf Menzel*, DOI 10.1007/978-94-007-2099-2_19,
© Springer Science+Business Media B.V. 2012

Abbreviations (excluding anatomical structures)

CS   Conditioned stimulus
GABA   Gamma-aminobutyric acid
PER   Proboscis extension reflex
US   Unconditioned stimulus

4.1.1   The Olfactory Task: From Molecules to Percept

In the environment, chemical molecules are the vessel of crucial information for animals, determining their eventual survival and reproductive success. Perhaps for this
reason, the sense of chemoreception is ubiquitously represented in the animal king-
dom. The role of the olfactory system is to decode the complex eddies of volatile molecules in the environment and shape them into pieces of relevant information that will
allow the animal to make decisions and engage in adapted behaviors. Major tasks of
the olfactory system include identification of food sources, detection of dangers,
recognition of potential mates and social interactions. How the nervous system
operates this transformation, from the detection of chemical molecules, through the
processing of neural representations, until the formation of percepts has been the focus
of intense research especially in vertebrates and in insects. A general finding of these
studies is that the basic rules underlying olfactory processing are highly similar in
both groups probably through convergence due to similar constraints.

Odor molecules exist in a myriad of chemical compositions, three-dimensional
shapes, vibration properties, etc. In olfaction, the first transformation is thus the
mapping of the complex physicochemical space of odor molecules onto an olfactory
receptor space. This step involves the detection of particular features of the molecules
by dedicated receptor (and associated) proteins, leading through transduction of the
signal to the activation of a subset of receptor cells. This combinatorial code will then
be conveyed to a series of brain structures, undergoing intense processing and a
reformatting of odor representation. Lastly, these representations will map onto the
perceptual space, involved in behavioral decision, linking odor quality with hedonic
value and learned relationships between odors and probable outcomes. Among
insects, the honey bee *Apis mellifera* L. has been for more than a century, a key insect
model for studying olfaction, as it presents a wide range of behaviors relying on
olfaction both within the colony (pheromone communication) and outside (foraging),
and readily allows behavioral, neuroanatomical and neurophysiological approaches.

4.1.2   Behavioral Study of Olfaction in Honey Bees

Honey bees are generalist pollinators and are not bound to a limited number of
plants for gathering food. However, individually, bees are 'flower constant', memo-
rising the features of a given floral species and exploiting it as long as profitable.

Learned floral cues include color, odor, shape and texture, but odors play the most prominent role, being most readily associated with nectar or pollen reward [25]. Floral aromas are mixtures of many volatiles that vary with respect to genotype, developmental stage and environmental conditions [31]. To maximize profit from foraging, honey bees have to show good *olfactory discrimination* capacity. Indeed, they can differentiate between very subtle differences in floral odor blends, as between genotypes or flowering stages [31]. At the same time, many variations in floral scent are not indicative of any difference in reward quality, and therefore, another key ability is *olfactory generalization.* This ability allows extending a behavior learned for a given stimulus to other stimuli, which are perceived as different, but sufficiently similar, to the learned one. Both abilities were recognized experimentally by Karl von Frisch [47] in a pioneering investigation, in which free-flying bees visited an artificial feeder presenting several essential oils (odor mixtures). Von Frisch observed that after learning one odor, bees tended to prefer this odor over others, clearly *discriminating* among odors, although they also sometimes visited other odors that were, to the human nose, similar to the rewarded one, thus displaying clear *generalization* behavior. Following von Frisch's seminal work, many experiments were performed with free-flying bees visiting scented feeders (e.g. [23, 32]). To provide more controlled conditions, an appetitive olfactory learning protocol on restrained individuals was developed. In the Pavlovian conditioning of the proboscis extension reflex (PER) (see Chaps. 3.3 and 6.2), an odor (conditioned stimulus, CS) is associated with sucrose solution (unconditioned stimulus, US) and gradually gains control over the bees' PER [3]. More recently, an aversive olfactory learning protocol was developed, in which an odor CS is associated with a mild electric shock (US) and gradually controls the bees' sting extension reflex [46]; (see Chap. 3.6). Hence, olfactory processing, detection and learning can be studied in laboratory conditions and compared with respect to different reinforcement modalities.

4.1.2.1 Olfactory Learning, Discrimination and Generalization

The olfactory abilities and behavior of honey bees are the fruit of millions of years of co-evolution between hymenoptera and angiosperms. However, bees' learning abilities are not limited to floral odors, and they are able to learn to associate with sucrose reward even repulsive odors (propanol, 3-methyl indole aka 'skatol') [45]. Bees can even learn to associate pheromonal odors with sucrose reinforcement, as shown for aggregation pheromones (citral, geraniol, [23, 45]) and more surprising for alarm pheromones (IPA and 2-heptanone, [23, 39, 41]). However, when learned, pheromones are not treated like general odors and alarm pheromones for instance produce very high generalization to other odors [39].

To study discrimination, differential conditioning is used: bees are repeatedly presented with two odors, one being associated with reinforcement (CS+), the other being presented without reinforcement (CS−). Successful discrimination is shown if bees respond significantly more to the CS + than to the CS−. To study generalization, bees are first conditioned to one odorant CS, and are then presented with novel odorants without reinforcement. The perceived similarity between the CS and each novel odorant is measured as the level of response to this odorant relative to the CS. Vareschi [45] was the first to use PER conditioning to study the honey bees' discrimination capacities with a wide range of odors. He used a kind of differential conditioning, in which one odor (CS) is rewarded and 27 other odors are presented without reinforcement in-between CS trials. A total of 1,816 odor pairs were tested, and bees were found to differentiate >95% of these odor pairs. Free-flying bees show the same discrimination efficiency (97% of 1,848 tested odor pairs [23]).

Olfactory perception has been intensively studied using aliphatic odor molecules, as they can be easily described by two main characteristics: their chemical group and their carbon chain length [15, 41]. Guerrieri et al. [15] systematically studied the generalization behavior of bees with 16 odorants presenting all combinations of 4 possible functional groups (primary and secondary alcohols, aldehydes, ketones) and 4 chain lengths (6–9 carbons). They found that the first factor determining generalization was a molecule's chain length, followed by the chemical group. Thus on a simple set of odor molecules, chemical dimensions appeared clearly encoded in the bee brain [15], in accordance with neurophysiological recordings of odor-evoked activity (see below). Bees' natural environment provides a wealth of odor molecules and we are still far from knowing all the encoding dimensions in the bee brain.

4.1.2.2 Concentration Coding and Concentration Invariance

Honey bees can learn absolute odor concentrations. Kramer [20] trained workers to follow odor gradients on a locomotion compensator. When rewarded at a particular odor concentration, bees showed a typical upwind walk in a range of concentrations relatively close to the learned one (20–180%), but walked downwind outside of these boundaries. Free-flying bees visiting a vertical odor array usually choose the trained odor at the right concentration and reject higher or lower concentrations [7]. However, differential conditioning between two concentrations of the same odor is very difficult in harnessed bees [29]. These contrasting accounts may relate to different sensory and motivation states in free and restrained honey bees.

Concentration influences the salience of olfactory stimuli. Generally, odors are learned more quickly at a higher concentration and support better memory consolidation [14, 29, 50]. Moreover, the discrimination power between odorants increases with concentration [14, 50]. Lastly, bees generalize more from low to high concentrations, than from high to low concentrations [14, 29]. Overall, odor identity is thus not invariant as a function of concentration, and bees can both differentiate between concentrations of the same odorant, and generalize between them. Such capacities may be crucial for identifying and locating floral source.

4.1.2.3   Mixture Perception

In nature, floral odors are not single molecules but complex mixtures. Honey bees must therefore both discriminate among complex blends and recognize the same floral source independently of variations in its composition. Many experiments have studied the learning of olfactory mixtures with whole floral extracts (e.g. [31]) or synthetic mixtures of 6–14 components (e.g. [32, 33, 49]). Generally, when bees learn a mixture they usually respond to some of its components (termed key-compounds) much more than to others. What determines that a component is a key-compound? Neither relative quantity nor volatility alone are predictive but it depends on other components in the mixture [33, 49]. Thus, processing of different odorants simultaneously produces unpredictable outcomes, a phenomenon termed *mixture interaction*. Due to the apparent complexity of mixture processing, research on mixture interactions has focussed on binary mixtures [4, 14, 40]. Generally an odor A is better learned when presented alone, than when together with a second odorant B [40]. When learning a mixture AB, bees can recognize the components, but one component is better learned than the other, a phenomenon called *overshadowing* [40].

Concepts and methods from experimental psychology have been used to test two main theories of mixture representation and learning. First, the *elemental approach* assumes that a compound AB will be represented in the brain as two elements, A and B, each of which can be associated with the US. In other terms, "the whole equals the sum of its parts". Radically different, the *configural approach* assumes that the representation of AB is a different entity from the representations of A and B ("the whole is different from the sum of its parts"). None of these accounts alone explained bees' behavior in patterning experiments, in which bees are trained to differentiate between two single odorants A and B and the mixture AB. Only a configural-like expansion of elemental models called the *unique cue hypothesis* predicted bees' responses [4]. In addition to the representations of the elements, the compound would give rise to a supplementary (internal) representation, the *unique cue*, which corresponds to the *synthetic properties* of the mixture.

4.1.3   The Honey Bee Olfactory System

Olfactory processing follows different steps, from the detection of molecules, via primary processing by AL networks, until the establishment of olfactory representations in higher-order brain centers. A simplified model of olfactory pathways is provided in Fig. 4.1.1.

4.1.3.1   Peripheral Odor Detection: The Antenna

Peripheral odor detection starts at the level of olfactory receptor neurons (ORNs), which are located below cuticular structures on the antennae, called sensillae.

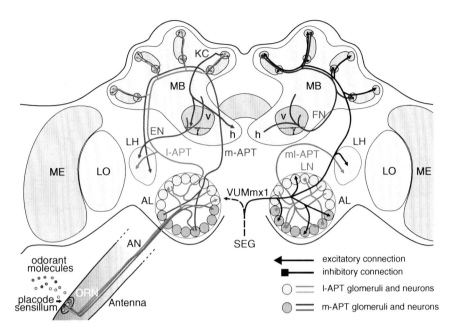

Fig. 4.1.1 The honey bee brain and the olfactory pathway. For clarity different neuron types are presented separately in the two brain hemispheres. On the *left*, major excitatory pathways involved in the transmission of olfactory information in the brain are shown. On the *right*, mostly inhibitory connections and modulatory neurons are shown. The AL, first-order olfactory neuropil, receives input from ~60,000 olfactory receptor neurons (ORNs) which detect odorants within placode sensilla in the antenna. Within the AL's anatomical and functional units, the 165 glomeruli, ORNs contact ~4,000 inhibitory local neurons (LNs) which carry out local computations and ~800 projection neurons which further convey processed information via different tracts. The lateral antenno-cerebralis tract (l-APT) projects first to the lateral horn (LH) and then to the mushroom body (MB) calyces (lips and basal ring), while the medial tract (m-APT) projects to the same structures, but in the reverse order. Both tracts are uniglomerular, each neuron taking information within a single glomerulus, and form two parallel, mostly independent olfactory subsystems (in green and in magenta), from the periphery until higher-order centers. Multiglomerular projection neurons form a medio-lateral tract (ml-APT) which conveys information directly to the medial protocerebrum and to the LH. The dendrites of the Kenyon cells (KCs), the MBs' 170,000 intrinsic neurons, form the calyces, while their axons form the pedunculus. The output regions of the MB are the vertical and medial lobes, formed by two collaterals of each KC axon. Within the MBs, feedback neurons (FN) project from the *pedunculus* and lobes back to the calyces, providing inhibitory feedback to the MB input regions. Extrinsic neurons (ENs) take information from the pedunculus and the lobes and project to different parts of the protocerebrum and most conspicuously to the LH. It is thought that descending neurons from these areas are then involved in the control of olfactory behavior. The neuron VUM_{mx1}, which provides appetitive reinforcement to the brain, projects from the subesophageal ganglion (SEG), where it gets gustatory input from sucrose receptors, to the brain and converges with the olfactory pathway in three areas, the AL, the MB calyces and the LH

In honey bees, poreplate sensilla (*sensilla placodea*) are the main olfactory sensilla [8]. Each sensillum placodeum (see Chap. 4.2, Fig. 4.1.1) is an oval-shaped 9×6 μm thin cuticular plate with numerous minute pores innervated by 5–35 ORNs [8]. Odorant molecules reach the dendrites of ORNs by diffusing through a receptor hemolymph located in the sensillum cavity. Odorant binding proteins (OBPs) may help transporting the odorants through the hemolymph, however their role in bees has not been confirmed *in vivo* yet. When reaching the ORN membrane, odorant molecules interact with olfactory receptor proteins (OR). Insect ORs belong to a family of highly divergent proteins with seven-transmembrane domains, which are very different from vertebrate OR families. The functional receptor is a heteromeric complex of an OR and a broadly expressed co-receptor AmOr2, the honey bee ortholog to the co-receptor Or83b of *Drosophila* [34]. Honey bees present a remarkable expansion of the insect odorant receptor family relative to other insects with 163 potentially functional OR genes [34].

4.1.3.2   Primary Olfactory Centre: The Antennal Lobe

ORN axons form the antennal nerve to reach the AL, primary olfactory centre of the insect brain (Fig. 4.1.1). The antennal nerve splits into six sensory tracts upon entrance into the AL. Four of these tracts (T1–T4) innervate distinct portions of the AL while the two remaining tracts (T5–T6) bypass the AL. The bee AL is compartmentalized in 165 anatomical and functional units, the glomeruli. Glomeruli can be recognized based on their position, size and shape [13]. In *Drosophila*, axons of ORNs expressing the same OR converge onto the same glomerulus [48]. Thus, the array of AL glomeruli would correspond to the array of ~163 OR types found in the genome [34]. Two main neuron types compose the AL:

Local neurons (LNs) have branching patterns restricted to the AL. They are especially numerous in the honey bee with ~4,000 LNs, compared to only between ~100 and ~360 in other insects [11]. In bees, LNs can be classified in two main types [9, 10]. Homogeneous LNs (homo-LNs) innervate most if not all glomeruli in a uniform manner; heterogeneous LNs (hetero-LNs) innervate one dominant glomerulus with very dense innervation and a few other glomeruli with sparse processes. ~750 LNs are GABAergic, a rather low proportion when comparing to other insects in which almost all LNs are GABAergic [11]. Glutamate, histamine and several peptides were also identified in the AL (see Chap. 3.7). To this day no evidence of excitatory LNs exists in honey bees.

Projection neurons (PNs) connect the AL with higher-order brain areas (Fig. 4.1.2), following five different pathways, called *antenno-protocerebral tracts* (APTs - [1, 26]). PNs can be classified in two types. Uniglomerular projection neurons (uPNs) branch in a single glomerulus and project to the MBs and to the LH using the two major APT tracts. Multiglomerular projection neurons (mPNs) branch in most glomeruli. Their axons form three lesser tracts, the medio-lateral (ml) APTs, leading not to the MBs but to different regions of the lateral protocerebrum [1, 19].

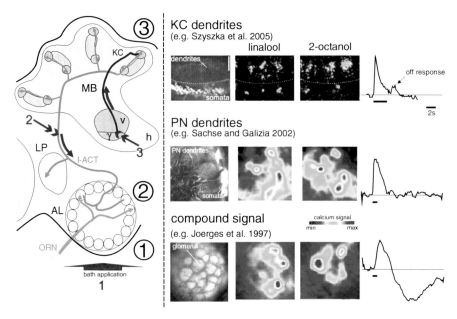

Fig. 4.1.2 Optical imaging of odor representations in the bee brain. Odor representation was studied at different levels of olfactory processing thanks to *in vivo* calcium imaging. On the left staining strategy and imaged neuron types are shown, while on the right, activity maps evoked by two sample odorants as well as an exemplary time course are presented. (1) Using bath-application of a calcium-sensitive dye (Calcium-Green 2 AM), a compound signal can be recorded in the AL in response to odors. This signal is thought to represent mostly olfactory input from the ORN population (see text). Different odors induce different, but overlapping, multiglomerular activity patterns. Bath application signals are temporally slow and biphasic. (2) Using retrograde staining with a migrating dye (Fura-2 dextran), projection neurons can be selectively stained. A dye-coated electrode is inserted into the PNs axon tract (*arrow 2*). The dye is taken up by the neurons and migrates back to their dendrites in AL glomeruli. Such staining allows the selective recording of AL ouput information sent to higher-order centers. Odors also induce multiglomerular activity patterns, but these are scarcer (less glomeruli are activated) and more contrasted than the compound signals. The time course is mostly phasic-tonic, but also presents some complex temporal patterns and inhibitions. (3) Using a similar strategy, inserting the electrode into the vertical lobe allowed recording activity from Kenyon cell dendrites and somata. Olfactory representation becomes even sparser in the MBs as few KCs respond to each odorant. Responses are phasic and often present off-responses at stimulus offset (Recordings 1 and 2, from [5, 6] – Recordings 3 from [44])

The more numerous uPNs (~800) form two roughly equal tracts towards higher brain centers, the lateral (l-APT) and medial (m-APT) tract. The l-APT runs on the lateral side of the protocerebrum, forming collaterals in the LH and continues to the MB calyces (CA). The m-APT runs along the brain midline first towards the MBs where collaterals enter into the calyces, and then travels laterally to the LH [1, 19]. L-APT neurons take information from glomeruli receiving input from the T1 tract of ORNs, while m-APT neurons receive input from T2, T3 and T4 glomeruli [1]. This corresponds to non-overlapping groups of 84 and 77 glomeruli, respectively

[19] so that each tract conveys information about two independent portions of the honey bee odor detection repertoire.

Summarizing the complexity of connections within the AL, one can estimate that a single honey bee glomerulus is innervated on average by ~400 ORNs (~60,000 divided by 165 glomeruli), ~1,000 LNs (~4,000 innervating each an average of 40 glomeruli), and 5 PNs (~800 divided by 165 glomeruli) [11].

4.1.3.3   Higher-Order Olfactory Centres: The Mushroom Bodies and the Lateral Horn

MB-intrinsic neurons are the Kenyon cells (KCs), which form two cup-shaped regions called calyces in each hemisphere. MB calyces are anatomically and functionally subdivided into the lip, the collar and the basal ring ([26, 43], Chap. 3.2). The lip region and the inner half of the basal ring receive olfactory input, whereas the collar and outer half of the basal ring receive visual input [26], in addition to mechanosensory and gustatory pathways. The projections of individual PNs extend in most parts of each calyx, but l- and m-APT project to different subregions ([19], see Chap. 3.1). PN boutons form multisynaptic microcircuits in the MB lips, with GABAergic input and KC output connections arranged to form particular structures termed microglomeruli (see Chap. 3.2). KC axons project in bundles into the central brain, forming the *pedunculus* and the vertical and medial lobes (also called α– and β-lobes). The calyx is topologically represented in the lobes [26, 43]. About 55 GABAergic feedback neurons from the MB output lobes project back to the calyces (Fig. 4.1.2). KCs mostly provide bifurcating axons to both vertical and medial lobes. In bees, ~800 PNs diverge onto a major proportion of the 170,000 KCs of each MB (olfactory KCs). Each PN contacts many KCs and each KC receives input from many PNs. If figures calculated in other insects (locust *Schistocerca Americana*) were to apply to the honey bee, each KC would contact ~400 PNs (50% of PN number). This organization appears ideal for a combinatorial readout across PNs.

The second major target area of all PNs is the LH. The LH shows relative PN tract-specific compartmentalization [19]. Processing within this structure as well as the connectivity between PNs and other neurons are still mostly unknown.

4.1.3.4   From Olfaction to Multimodal Representation and Behavior: MB Output Neurons

Several neuron populations and single neurons project from the MBs towards other brain areas (Fig. 4.1.1), with major output regions in the α and β lobes [26]. About 400 extrinsic neurons (ENs) from the α-lobe have been studied in details ([35],

Chap. 3.1). Some are unilateral neurons with projection fields restricted to the ipsilateral protocerebrum, while others are bilateral neurons connecting both α–lobes, or projecting from one lobe to the contralateral protocerebrum around the α–lobe [35]. A single neuron in each MB, called PE1, forms a major output pathway from the MB *pedunculus* [24], and projects to the LH where it synapses directly or via interneurons onto descending neurons. Some centrifugal neurons project back from the MBs towards the AL [19, 35].

4.1.3.5 Modulation and Reinforcement

The olfactory pathway also receives input from several modulatory systems. In particular, the formation of a neural association between odor and sucrose reinforcement relies on the co-activation of the olfactory pathway and a pathway representing appetitive reinforcement. A single, putatively octopaminergic, neuron in the bee brain, VUM-mx1 (Fig. 4.1.1) was shown to represent the sucrose US, because the forward (but not backward) pairing of an odor CS with an artificial depolarization of this neuron produces an associative memory trace [16]. VUM-mx1, has its cell body in the SEG, and converges with the olfactory pathway in both brain hemispheres at three sites, the AL, the CA and the LH. As for aversive reinforcement, it depends on dopamine but its neural substrate is still currently unknown [46].

4.1.4 Functional Analysis of Olfactory Processing

4.1.4.1 Peripheral Odor Detection: The Antenna

The search for the neural correlates of olfaction started at the periphery, using extracellular recordings of single placodes [2, 22, 45]. In a pioneer work, Lacher and Schneider [22] recorded responses to benzylacetate in workers, and caproic acid in drones, but found no answers to light, sound, water vapour or CO_2. Vareschi [45] proposed the existence of 10 different sensory cell types, based on cross-adaptation. One type responded mostly to the queen component 9-ODA and was found in 70% of the drones. Workers showed a higher probability of cells responding to the aggregation pheromone than to 9-ODA. Moreover, this work showed that odors that were classified in different groups in the electrophysiology were actually better differentiated by bees in their behavior. However, even odors belonging to the same group could be discriminated behaviorally. Akers and Getz [2] found that units with similar odor-response spectra were more likely to be found in different placodes than within the same placode.

4.1.4.2   The Antennal Lobe

Thanks to the technique of *in vivo* optical imaging, neural activity could be recorded at the glomerular level ([18], Fig. 4.1.2). Calcium imaging in particular uses fluorescent dyes to measure the increase of intracellular calcium (from the extracellular medium and/or released from internal stores) following neuronal excitation. This allows recording activity maps from a whole glomerular array simultaneously. The first developed technique used bath-application of Calcium Green 2-AM, recording a composite calcium signal potentially originating from all cell populations within the AL: ORNs, LNs, PNs and glial cells [18]. Due to the numerical preponderance of ORNs and the very stereotypical time course of calcium signals, without any spontaneous activity or inhibitory responses (the hallmarks of LNs and PNs), these recordings are thought to emphasize presynaptic calcium variations from ORNs, with possibly a significant contribution from glial cells surrounding each glomerulus. This compound signal would thus be representative of sensory input [5, 37].

Odors elicit combinatorial activity patterns across glomeruli ([18]; Fig. 4.1.2) according to a specific distributed code conserved between individuals. Odor coding in the bee AL thus corresponds to an across-fibre pattern, each glomerulus (representing an ORN type expressing a given OR) showing a rather broad molecular receptive range. What is the significance of this code? Guerrieri et al. [15] studied the generalization behavior of honey bees among a panel of 16 odorants (Fig. 4.1.3a) for which the activity patterns were known [38]. This study demonstrated a significant correlation between behavioral similarity among odors and neurophysiological similarity (Fig. 4.1.3b). Thus, calcium signals, even obtained from ~30 glomeruli accessible to imaging, predicted the generalization behavior of honey bees. However, the correlation showed some scatter and extending neurophysiological recordings to other glomeruli (comparing l- and m-APT regions, for instance) or to other parts of the brain may ameliorate this prediction.

LNs belonging to both anatomical types, homo-LNs and hetero-LNs, produce spikes in honey bees [9]. LNs can be odor-specific, responding differentially to odors, with excitatory responses to some odors and inhibitory responses to others. Generally, the response profile of a recorded hetero-LN corresponds to the known response profile of the innervated glomerulus, suggesting that hetero-LNs take their input in this glomerulus [12]. LNs tend to show a shorter latency than PNs, which allows them to rapidly and efficiently inhibit the PN firing [21].

PN responses are the product of direct excitation from ORNs, direct inhibition from LNs and possibly disinhibition from LN-LN connections and can therefore be temporally complex [27]. PNs are usually spontaneously active and may change their responses upon odor presentations in an either excitatory or inhibitory manner [1, 27]. PNs respond rather non-specifically to many odors, exhibiting phasic–tonic response patterns usually outlasting the stimulus [21]. Based on retrograde staining with a dextran-coupled dye, imaging of PN population activity allowed

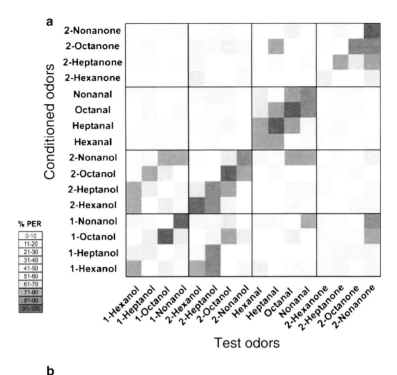

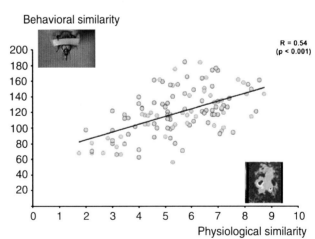

Fig. 4.1.3 Neural representations and olfactory behavior. (**a**) Using a generalization experiment, the perceptual similarity among all possible pairs of 16 aliphatic odorants was measured. Odors used for conditioning are presented vertically, and odors used in the generalization test are presented horizontally. Bees respond preferentially to the learned odor (*main diagonal*), but also to other – perceptually similar – odors. For instance, they generalize between odors sharing the same carbon chain length (*smaller diagonals*) or the same functional group (Boxes along diagonals). (**b**) Similarity among odors at the neural level (measured in the AL using bath-application of the calcium dye, method 1 in Fig. 4.1.2) significantly correlates with similarity at the behavioral level, as measured in a (Data from [15])

comparing glomerular activity patterns between the AL input and output representations, even within the same animal [37]. Such comparison showed that most glomeruli which are intermediately or weakly active in the compound signal, do not present any calcium increase in PNs [36]. Thus, PN patterns are sparser than input patterns. Moreover, AL networks improve the separability of odor representations, as shown both with single odors over a wide concentration range [37] and with mixtures [6].

AL processing performs mainly two operations: *gain control*, which quantitatively controls the overall amount of PN activity and *contrast enhancement* which qualitatively modifies the activity patterns [5, 6, 36, 37]. The first operation would be the result of a global inhibitory network, corresponding to homo-LNs and possibly GABAergic [36]. The second network would be an asymmetrical inhibitory network, corresponding to hetero-LNs, whose neurotransmitter is still unknown.

4.1.4.3   The Mushroom Bodies

In locusts, KCs often respond with a single or very few spikes, do not show any spontaneous activity, and respond to very few odorants [30]. Calcium imaging in honey bees confirmed these properties ([44], Fig. 4.1.2). Thus, olfactory representation is subject to progressive sparsening. In the CA it involves several mechanisms. First, the low synaptic strength between PNs and KCs would imply that coherent input from many PNs simultaneously is needed to excite a KC [30]. Second, KCs would detect coincidence among many PNs thanks to odor-driven inhibition produced by LH inhibitory neurons locked in anti-phase to PN oscillations [30]. Third, local microcircuits involving GABA processes in the MB microglomeruli would also shape KC responses ([44] see Chap. 3.2). Such sparseness of odor representation would allow specific coding of odorant concentrations or mixtures.

4.1.4.4   Mushroom Body Output and the Lateral Horn

The most studied MB output neuron, PE1, is easily recognizable thanks to a characteristic firing pattern in doublets or triplets of action potentials [24, 28]. This wide-field neuron is multimodal and in conditioning experiments reduces its response to the rewarded odor [28]. The function of neurons like PE1 might be to integrate information over several sensory modalities and indicate that a particular stimulus combination has been learned. Other (non-PE1) output neurons respond to odors, but only the responses of a small proportion are modified by learning.

Practically nothing is known about odor processing and representation in the LH. In *Drosophila*, recent neuroanatomical work could reconstruct putative maps of olfactory input to the LH and predict a clear segregation between candidate pheromone responsive PNs and fruit odor responsive PNs [17]. Thus, particular subregions of the LH may code the biological nature of olfactory stimuli. If a similar organization exists in the honey bee, one could expect the LH to exhibit pheromone processing regionalization. Because the LH receives input from associative neurons like PE1, it may also represent a pre-motor centre for both innate behavior (pheromones) and acquired behavior (associative learning).

4.1.4.5   Concentration Coding and Concentration Invariance

Odor concentration strongly affects odor maps in the AL, increasing the number of glomeruli in the pattern [37]. Neurons, such as multiglomerular PNs, integrating overall excitation over many glomeruli may be adequate for monitoring absolute stimulus concentration. However, how can both odor-specific concentration coding and concentration invariance be achieved given the changing nature of the odor representation with concentration? The identity of an odorant is combinatorial and resides more in the *relative* activation of different glomeruli (or PNs) than in the absolute activation of individual glomeruli. Therefore, neurons recognizing a particular pattern of inputs, such as KCs, could perform both operations: while some KCs appear tuned to a narrow concentration range of one particular odorant, other KCs recognize the same odorant on a wide concentration scale. Some concentration invariance can also be achieved earlier in the olfactory pathway, mainly through gain control mechanisms. Indeed, imaging experiments show that processing within the AL makes odor representation more reliable over a broader concentration range [37]. Moreover, the two PN subsystems may provide differential information to higher-order centres: while l-APT neurons display low concentration dependency, m-APT neurons change their response strongly with concentration [51].

4.1.4.6   Mixture Processing

A glomerulus is generally activated by a mixture when at least one of its components activates this glomerulus [5, 18]. The more components a mixture contains, the more *suppression* phenomena, are observed, i.e. cases in which the glomerular response to a mixture is lower than its response to the components. AL processing via LN networks increases suppression cases, allowing the emergence of synthetic properties, i.e. a neural representation that cannot be predicted based only on component information [6]. Doing so, reformatting by LNs increases separability among odor mixture representations. However, PN boutons in the MB lips show strong suppression in l-APT PNs but not in m-APT PNs, suggesting that one PN population could be involved in synthetic processing, while the other would conserve component information [51].

4.1.4.7 Temporal Aspects of Olfactory Coding

From the periphery [2] until PNs [21, 27] electrophysiological recordings show complex temporal response patterns in response to odors. Moreover odor stimulation gives rise to field potential oscillations in response to odors [42]. Considerable work, especially in locusts suggested that such temporal patterning and synchrony may play a major role in coding odor identity [27]. This possibility was explicitly tested in honey bees by removing oscillations using pharmacological application of picrotoxin [42], a blocker of GABA but also of other chloride channels. After learning an odor, injected bees generalized more than control bees [42]. This change was interpreted as the indication that synchrony is necessary for fine odor discrimination, but we now know that picrotoxin also changes the spatial response patterns, so that this interpretation remains open.

4.1.5 Conclusion and Outlook

A century of experiments have been performed on the behavior, neuroanatomical organization and functional responses of the honey bee brain to odors. The honey bee olfactory system is tuned for performing a number of operations that are crucial for meeting the demands of social life, food search and mating. These are to (1) detect and identify odor stimuli, allowing graded responses to increasingly similar odors; (2) measure stimulus concentration allowing both concentration invariant and concentration-specific odor recognition; (3) detect components as well as extract mixture-unique properties within a mixture; (4) learn relationships between almost any odor and appetitive or aversive outcomes. Although our understanding of odor representation at the different levels of the bee brain has greatly improved in the last years, entire regions have yet to be explored. The most prominent are the m-APT dependent part of AL and MBs, and the whole LH. Thanks to optical imaging, the spatial representation of odors has been intensively studied, but temporal aspects are yet poorly understood. Even in such simple systems as compared to vertebrates, olfactory coding involves complex interactions between different neuron types, so that only computational approaches feeding on experimental data may help understanding the dynamics and processing rules of the system. Lastly, plasticity appears in multiple regions of the olfactory pathway, but their implication for tuning the olfactory system or for storing outcome-related memories is still far from understood. It shall be the goal of future research to progress in these questions, so that a comprehensive model of olfactory detection, processing and learning in the honey bee can be constructed, the ultimate goal of sensory neuroscience.

References

1. Abel R, Rybak J, Menzel R (2001) Structure and response patterns of olfactory interneurons in the honeybee, *Apis mellifera*. J Comp Neurol 437:363–383
2. Akers RP, Getz WM (1992) A test of identified response classes among olfactory receptor neurons in the honey-bee worker. Chem Senses 17:191–209
3. Bitterman ME, Menzel R, Fietz A, Schäfer S (1983) Classical conditioning of proboscis extension in honeybees. J Comp Psychol 97:107–119
4. Deisig N, Lachnit H, Sandoz JC, Lober K, Giurfa M (2003) A modified version of the unique cue theory accounts for olfactory compound processing in honeybees. Learn Mem 10:199–208
5. Deisig N, Giurfa M, Lachnit H, Sandoz JC (2006) Neural representation of olfactory mixtures in the honeybee antennal lobe. Eur J Neurosci 24:1161–1174
6. Deisig N, Giurfa M, Sandoz JC (2010) Antennal lobe processing increases separability of odor mixture representations in the honeybee. J Neurophysiol 103:2185–2194
7. Ditzen M, Evers JF, Galizia CG (2003) Odor similarity does not influence the time needed for odor processing. Chem Senses 28:781–789
8. Esslen J, Kaissling KE (1976) Zahl und Verteilung antennaler Sensillen bei der Honigbiene (*Apis mellifera* L.). Zoomorphology 83:227–251
9. Flanagan D, Mercer AR (1989) Morphology and response characteristics of neurones in the deutocerebrum of the brain in the honeybee *Apis mellifera*. J Comp Physiol A 164:483–494
10. Fonta C, Sun XJ, Masson C (1993) Morphology and spatial distribution of bee antennal lobe interneurones responsive to odours. Chem Senses 18(2):101–119
11. Galizia CG (2008) Insect olfaction. In: Smith DV, Firestein S, Beauchamp GK (eds) The senses, a comprehensive reference. Elsevier, London, pp 725–769
12. Galizia CG, Kimmerle B (2004) Physiological and morphological characterization of honeybee olfactory neurons combining electrophysiology, calcium imaging and confocal microscopy. J Comp Physiol A 190:21–38
13. Galizia CG, McIlwrath SL, Menzel R (1999) A digital three-dimensional atlas of the honeybee antennal lobe based on optical sections acquired using confocal microscopy. Cell Tissue Res 295:383–394
14. Getz WM, Smith KB (1991) Olfactory perception in honeybees: concatenated and mixed odorant stimuli, concentration, and exposure effects. J Comp Physiol A 169:215–230
15. Guerrieri F, Schubert M, Sandoz JC, Giurfa M (2005) Perceptual and neural olfactory similarity in honeybees. PLoS Biol 3:e60
16. Hammer M (1993) An identified neuron mediates the unconditioned stimulus in associative olfactory learning in honeybees. Nature 366:59–63
17. Jefferis GS, Potter CJ, Chan AM, Marin EC, Rohlfing T, Maurer CR Jr, Luo L (2007) Comprehensive maps of *Drosophila* higher olfactory centers: spatially segregated fruit and pheromone representation. Cell 128:1187–1203
18. Joerges J, Küttner A, Galizia CG, Menzel R (1997) Representations of odours and odour mixtures visualized in the honeybee brain. Nature 387:285–288
19. Kirschner S, Kleineidam CJ, Zube C, Rybak J, Grünewald B, Rössler W (2006) Dual olfactory pathway in the honeybee, *Apis mellifera*. J Comp Neurol 499:933–952
20. Kramer E (1976) The orientation of walking honeybees in odour fields with small concentration gradients. Physiol Entomol 1:27–37
21. Krofczik S, Menzel R, Nawrot MP (2009) Rapid odor processing in the honeybee antennal lobe network. Front Comput Neurosci 2:9
22. Lacher V, Schneider D (1963) Elektrophysiologischer Nachweis der Riechfunktion von Porenplatten (Sensilla placodea) auf den Antennen der Drohne und der Arbeitsbiene (*Apis mellifera* L.). Z vergl Physiol 47:274–278
23. Laska M, Galizia CG, Giurfa M, Menzel R (1999) Olfactory discrimination ability and odor structure-activity relationships in honeybees. Chem Senses 24:429–438

24. Mauelshagen J (1993) Neural correlates of olfactory learning paradigms in an identified neuron in the honeybee brain. J Neurophysiol 69:609–625
25. Menzel R, Greggers U, Hammer M (1993) Functional organization of appetitive learning and memory in a generalist pollinator, the honey bee. In: Lewis AC (ed) Insect learning. Chapman & Hall, New York/London, pp 79–125
26. Mobbs PG (1982) The brain of the honeybee *Apis mellifera* I.The connections and spatial organization of the mushroom bodies. Philos Trans R Soc Lond B 298:309–354
27. Müller D, Abel R, Brandt R, Zockler M, Menzel R (2002) Differential parallel processing of olfactory information in the honeybee, *Apis mellifera* L. J Comp Physiol A 188:359–370
28. Okada R, Rybak J, Manz G, Menzel R (2007) Learning-related plasticity in PE1 and other mushroom body-extrinsic neurons in the honeybee brain. J Neurosci 27:11736–11747
29. Pelz C, Gerber B, Menzel R (1997) Odorant intensity as a determinant for olfactory conditioning in honeybees: roles in discrimination, overshadowing and memory consolidation. J Exp Biol 200:837–847
30. Perez-Orive J, Mazor O, Turner GC, Cassenaer S, Wilson RI, Laurent G (2002) Oscillations and sparsening of odor representations in the mushroom body. Science 297:359–365
31. Pham-Delègue MH, Etiévant P, Guichard E, Masson C (1989) Sunflower volatiles involved in honeybee discrimination among genotypes and flowering stages. J Chem Ecol 15:329–343
32. Pham-Delègue MH, Bailez O, Blight MM, Masson C, Picard-Nizou AL, Wadhams LJ (1993) Behavioral discrimination of oilseed rape volatiles by the honeybee *Apis mellifera* L. Chem Senses 18:483–494
33. Reinhard J, Sinclair M, Srinivasan MV, Claudianos C (2010) Honeybees learn odour mixtures via a selection of key odorants. PLoS One 5:e9110
34. Robertson HM, Wanner KW (2006) The chemoreceptor superfamily in the honey bee, *Apis mellifera*: expansion of the odorant, but not gustatory, receptor family. Genome Res 16:1395–1403
35. Rybak J, Menzel R (1993) Anatomy of the mushroom bodies in the honey bee brain: the neuronal connections of the alpha-lobe. J Comp Neurol 334:444–465
36. Sachse S, Galizia CG (2002) The Role of inhibition for temporal and spatial odor representation in olfactory output neurons: a calcium imaging study. J Neurophysiol 87:1106–1117
37. Sachse S, Galizia CG (2003) The coding of odour-intensity in the honeybee antennal lobe: local computation optimizes odour representation. Eur J Neurosci 18:2119–2132
38. Sachse S, Rappert A, Galizia CG (1999) The Spatial representation of chemical structures in the antennal lobe of honeybees: steps towards the olfactory code. Eur J Neurosci 11:3970–3982
39. Sandoz JC, Pham-Delegue MH, Renou M, Wadhams LJ (2001) Asymmetrical generalisation between pheromonal and floral odours in appetitive olfactory conditioning of the honey bee (*Apis mellifera* L.). J Comp Physiol A 187:559–568
40. Smith BH (1998) Analysis of interaction in binary odorant mixtures. Physiol Behav 65:397–407
41. Smith BH, Menzel R (1989) The use of electromyogram recordings to quantify odourant discrimination in the honey bee, *Apis mellifera*. J Insect Physiol 35:369–375
42. Stopfer M, Bhagavan S, Smith BH, Laurent G (1997) Impaired odour discrimination on desynchronization of odour-encoding neural assemblies. Nature 390:70–74
43. Strausfeld NJ (2002) Organization of the honey bee mushroom body: representation of the calyx within the vertical and gamma lobes. J Comp Neurol 450:4–33
44. Szyszka P, Ditzen M, Galkin A, Galizia CG, Menzel R (2005) Sparsening and temporal sharpening of olfactory representations in the honeybee mushroom bodies. J Neurophysiol 94:3303–3313
45. Vareschi E (1971) Duftunterscheidung bei der Honigbiene - Einzelzell-Ableitungen und Verhaltensreaktionen. Z vergl Physiol 75:143–173
46. Vergoz V, Roussel E, Sandoz JC, Giurfa M (2007) Aversive learning in honeybees revealed by the olfactory conditioning of the sting extension reflex. PLoS One 2:e288

47. von Frisch K (1919) Über den Geruchsinn der Biene und seine blütenbiologische Bedeutung. Zoologisches Jahrbuch Teil Physiologie 37:1–238
48. Vosshall LB, Wong AM, Axel R (2000) An olfactory sensory map in the fly brain. Cell 102(2):147–159
49. Wadhams LJ, Blight MM, Kerguelen V, Métayer ML, Marion-Poll F, Masson C, Pham-Delègue MH, Woodcock CM (1994) Discrimination of oilseed rape volatiles by honey bee: novel combined gas chromatographic-electrophysiological behavioral assay. J Chem Ecol 20:3221–3231
50. Wright GA, Smith BH (2004) Different thresholds for detection and discrimination of odors in the honey bee (*Apis mellifera*). Chem Senses 29:127–135
51. Yamagata N, Schmuker M, Szyszka P, Mizunami M, Menzel R (2009) Differential odor processing in two olfactory pathways in the honeybee. Front Syst Neurosci 3:16

Chapter 4.2
Taste Perception in Honey Bees

Maria Gabriela de Brito Sanchez

Abstract The sense of taste is of fundamental importance for honey bees both in a foraging and in a social context. Tastes are crucial for choosing profitable food sources, resins, water sources and for nestmate recognition. Peripheral taste detection occurs within cuticular hairs, the chaetic and basiconic sensilla, which host gustatory receptor cells and, usually a mechanoreceptor cell. Gustatory sensilla are mostly located on the distal segment of the antennae, on mouthparts and on the tarsi of forelegs. These cells respond with varying sensitivity to sugars, salts, and possibly amino acids, proteins and water. So far, no cellular responses to bitter substances were found although inhibitory effects of these substances on sucrose receptor cells could be recorded. When bees are free to express avoidance behaviors, they reject highly concentrated bitter and saline solutions. However, such avoidance disappears when bees are immobilized in the laboratory. In this case, they ingest these solutions, even if they suffer afterwards a malaise-like state or even die from such ingestion. Central processing of taste occurs mainly in the subesophageal ganglion (SEG) but the nature of this processing remains unknown. We suggest that coding tastes in terms of their hedonic value, thus classifying them in terms of their palatability, is a basic strategy that a central process of taste should achieve for survival. Furthermore, we highlight important areas for future research in the biology of honey bee taste.

Abbreviations

Grs Gustatory receptor genes
MG Methathoracic ganglion

M.G. de Brito Sanchez (✉)
Research Center on Animal Cognition, CNRS – University Paul Sabatier,
118 route de Narbonne, 31062 Toulouse Cedex 9, France
e-mail: debrito@cict.fr

C.G. Galizia et al. (eds.), *Honeybee Neurobiology and Behavior: A Tribute to Randolf Menzel*, DOI 10.1007/978-94-007-2099-2_20,
© Springer Science+Business Media B.V. 2012

PE1 neuron   Pedunculus extrinsic neuron 1
PER   Proboscis Extension Reflex
RT-qPCR   Real-time quantitative PCR
SCT   Subesophageal calycal tract
SEG   Subesophageal ganglion
VUM$_{mx1}$ neuron Ventral unpaired median neuron of the maxillary neuromere 1

4.2.1 Honey Bee Taste in an Ecological and Social Context

Although the processing of sensory information of flowers (e.g. colors, odors) by honey bees has been intensively studied in the last decades (vision: see Chaps. 4.4, 4.5 and 6.6; olfaction: see Chap. 4.1), less is known about the processing of gustatory stimuli by honey bees. Yet, gustatory stimuli play a fundamental role in a honey bee's life. Taste is a crucial modality in a foraging context as honey bee foragers collect nectar and pollen, which provide carbohydrates and proteins, respectively, necessary for individual and collective survival. Nectar presents not only different types of sugars such as sucrose, glucose and/or fructose but also organic acids, lipids, minerals, vitamins and aromatic compounds, even if these substances constitute a low percentage of nectar contents [18]. Pollen contains proteins but also lipids, mineral salts, albumin, vitamins, amino acids, growth regulator factors, folic acid and enzymes among others [18]. Furthermore, besides foraging for nectar and pollen, bees collect water and in this context they respond to salts. Additionally, they collect resin for elaborating propolis and are then in contact with compounds such as prenylated and nonprenylated phenylpropanoids, terpenoids and anthracene derivatives, which have been identified in the resin loads transported in the corbiculae [45]. Finally, bees chew and process wax with their mouth parts and may therefore taste the chemicals in it.

   Although the examples provided above refer essentially to adult bees that engage in different foraging activities outside the hive, younger bees within the hive may also use their gustatory senses for different purposes. Besides olfaction, taste may allow intracolonial recognition within the dark world of a hive. It has been repeatedly shown that cuticular hydrocarbons confer a chemical signature allowing nestmate recognition (e.g. [5, 6]). A tight interaction between wax comb and cuticular hydrocarbons has been shown [4] so that both may constitute a continuous medium for any hydrocarbon-soluble substances used by honey bees in nest-mate recognition. Cuticular hydrocarbons are usually high-molecular weight compounds so that airborne detection may not be the primary detection channel; contact chemoreceptors may be involved and gustatory detection may be the privileged channel for nestmate recognition.

4.2.2 Peripheral Processing of Taste: The Gustatory Sensilla

In the honey bee, the antennae, mouth parts and distal segments of the forelegs constitute the main chemosensory organs. Gustatory receptor cells are located within specialized cuticular structures called sensilla (Fig. 4.2.1a), which often take the form of hairs (chaetic sensilla) or pegs (basiconic sensilla) [12]. Chaetic sensilla of different sizes can be found on the glossa, labial palps, galeae, antennae and tarsi of honey bee workers. Basiconic sensilla can also found on labial palps, galeae and tarsi but not on the antennae and glossa [50].

Gustatory sensilla have a characteristic aperture at the apex (a pore or a papilla) through which gustatory substances can penetrate after contacting the hair or peg. Gustatory receptor cells, usually from three to five [11], are located within each sensillum and bath in a receptor hemolymph (Fig. 4.2.1b). Sensilla on the mandibles have only one sensory neuron but the gustatory role of these sensilla is unclear. Each neuron projects a dendrite up the shaft of the hair or peg to the apex. Such a

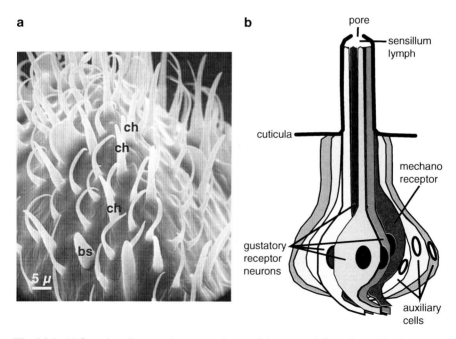

Fig. 4.2.1 (**a**) Scanning-electron-microscope picture of the antennal-tip surface of the honey bee showing chaetic (ch) and basiconic (bs) sensilla. (**b**) Schematic drawing of a chaetic sensillum. Four gustatory receptor cells (*grey, violet, blue, orange*) bathing in a cavity defined by auxiliary sensillar cells (*green, olive, white*) and filled with sensillum lymph extend their dendrites towards the apex of the cuticular hair. A mechanoreceptor cell (*red*) is attached to the basal wall of the hair. Tastants penetrate into the sensillum through a pore at the apex

branch bears the molecular receptors to which the appropriate molecules will bind. Gustatory molecular receptors are thought to be G-protein coupled proteins. In most cases, a mechanoreceptor cell terminating at the base of the shaft can also be found (Fig. 4.2.1b). This neuron is stimulated not by gustatory stimuli but by the movement experienced by the sensilla upon contact with a gustatory stimulus. As gustatory organs have to explore and manipulate food, evaluating the position and density of the food is facilitated by the presence of these mechanoreceptor cells associated with gustatory receptor cells within the same sensilla.

4.2.2.1   Peripheral Processing of Taste: Gustatory Sensilla on the Antennae

Gustatory antennal perception plays a role in appetitive food sensing as shown by the fact that stimulation of the antennae with sucrose solution elicits the reflex of extension of the proboscis (henceforth PER; [3, 43]). Approximately 300 chaetic sensilla were found distributed over the antennal flagellum [12]. An important concentration of these sensilla was found on the ventral surface of the distal segment of the antennae, which constitutes the primary antennal contact region with tastants. About half of the sensilla observed on the antennae present five gustatory receptor neurons and one mechanoreceptor neuron.

Electrophysiological, extracellular recordings of single sensilla were used to characterize the gustatory sensitivity of receptor neurons hosted in antennal sensilla located on the tip of the antennae. Haupt [19] showed that antennal chaetic sensilla (which he termed 'trichoid') are very sensitive to sucrose stimulation. All intact antennal taste hairs showed responses at the lowest sucrose concentration tested (0.1% w/w or 2.9 mM) and therefore their response threshold was below 0.1%. Sensitivity of these sensilla is higher than that of taste hairs of the proboscis where thresholds of about 0.34% (10 mM) were estimated in earlier studies [48, 49]. This high sensitivity highlights the fundamental role of antennal gustatory receptors in locating a potential food source.

Sucrose responses of antennal sensilla are dose-dependent [7, 19]. It seems that in most cases, only a single cell type is activated by sucrose stimulation although relying on spike amplitude is not always a consistent criterion in the case of taste cells. Indeed, it is a common observation that electrophysiological responses of gustatory receptor cells are not always regular and may even vary in spike amplitude or interspike intervals [21]. Sucrose responses between different hairs on the same antenna showed a high degree of variability in spike frequency. Such variability allows extending the dynamic range of sucrose perception in an individual bee [19].

Antennal chaetic sensilla recorded in two different studies [7, 19] did not respond to a solution of 10 mM KCl, thus suggesting that these sensilla do not have a cell responding to water, as found in other insects. Although very sensitive bees respond with PER to water vapor [22] this response may be elicited by antennal

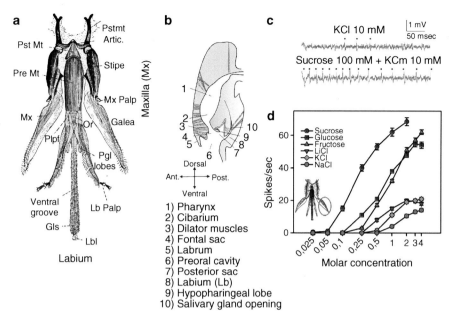

Fig. 4.2.2 (**a**) Mouth pieces of a honeybee worker. Ventral view of parts forming the proboscis, labium (Lb) in the *middle* and maxillae (Mx) at sides, flattened out. Lbl: labellum; Gls: glossa; Pgl lobes: paraglossal lobes; Pre Mt: prementum; Pst Mt: postmentum; Pstmt Artic: postmental articulation; Plpf: palpiger; Or: salivarium opening. (**b**) Side view of the oral cavities of a honeybee worker. The food first enters the preoral cavity formed from the labrum and the bases of the mouthparts (e.g. labium); the cavity is divided into a frontal and a posterior sac by the hypopharingeal lobe. Salivary glands open into the posterior sac or salivarium. The preoral cavity continues into the cibarium and then into the pharynx. (**c**) Examples of chaetic-sensilla recordings at the level of the galeae. The *upper trace* shows a cell response to KCl 10 mM, the *lower trace* another response to sucrose 100 mM and KCl 10 mM. Asterisks indicate action potentials. (**d**) Chaetic sensilla on the galeae respond linearly to the log of solute concentrations of sucrose, glucose, fructose, NaCl, KCl, and LiCl (From [49])

hygroreceptors. Responses to a solution of NaCl 50 mM were recorded at the level of antennal chaetic sensilla thus indicating that receptor cells tuned to salts exist on the antennae [7].

4.2.2.2   Peripheral Processing of Taste: Gustatory Sensilla on the Mouth Parts

The mouth parts are constituted by the mandibles, maxillae and the labium. The maxillae and the labium form the proboscis (Fig. 4.2.2a). Each maxilla is constituted by a broad, flat plate, the stipe, and by an elongated lobe, the galea. The labium is made from a small plate, the postmentum, a broad plate, the prementum (together they form the mentum), and a glossa. Labial palps, together with the galeae, surround the tongue to form a food canal groove through which liquids can be sucked up into the mouth. The whole structure is folded against the head when not in use.

At the base of the mouthparts the preoral cavity forms a sac where the food is first ingested (Fig. 4.2.2b). This cavity is divided into frontal and posterior sacs by the central hypopharingeal lobe. Salivary glands open into the posterior sac or salivarium. The preoral cavity is prolonged into the cibarium, a cavity whose muscles in its walls form a suction pump, which facilitates food ingestion through the proboscis. The cibarium continues into the pharynx. The true mouth lies at the intersection of both; from there the food passes into the pharyngeal tube and then into an esophagus, which leads to a crop, whose capacity can reach 60 µl [31].

As mentioned above, sensilla on the mandibles have a unique receptor cell besides a mechanosensory cell. Their role in gustation is not clear and there are no studies implicating them in taste detection. The proboscis, on the other hand, presents many sensilla that have been related to gustatory processes. Single-sensilla recordings (Fig. 4.2.2c) showed that chaetic sensilla on the galeae respond linearly to the log of solute concentrations of sucrose, glucose, fructose, NaCl, KCl, and LiCl but not to $CaCl_2$ or $MgCl_2$, which fail to give consistent responses [49] (Fig. 4.2.2d). These sensilla exhibit much higher firing rates for sugar than salt solutions. Four different spike types can be seen. The first type has the highest amplitude and is seen upon sugar stimulation. The second type has lower amplitude and occurs in the first 30 s of salt stimulation. A third type with the lowest amplitude appears with spikes of the second type after prolonged stimulation with KCl. A fourth type with a high amplitude results from mechanical stimulation. It was concluded that from the five neurons present in each galeal chaetic sensilla, one is mechanosensory, and the other four respond to tastants, one definitely to sugars and two to electrolytes. The gustatory tuning of the fourth cell remains unknown. Whitehead and Larsen [49] suggested that it may be responsive to proteins [9], aminoacids [14, 39, 40], "natural foods" [10], or glandular secretions. Responses of the sensilla to mechanical stimulation show phasic-tonic characteristics. None of the sensilla tested by Whitehead and Larsen [49] responded to water.

At the level of the labium, chaetic sensilla are concentrated on the glossa. Each of these sensilla also presents four gustatory receptor cells and a mechanosensory cell. Other taste sensilla are located on the distal segments of the labial palps. Chaetic sensilla on these segments were also investigated [48]. Their spike responses correlate with the log of the concentrations of sucrose, glucose, fructose, NaCl, KCl and LiCl, but not with $CaCl_2$ or $MgCl_2$, which give inconsistent responses. The firing rates are higher and thresholds lower to the sugars than to the electrolytes. None of the sensilla tested responded to water. As for the galeae, basiconic sensilla are also present in the distal segments of the labial palps but no study has been performed to characterize their gustatory responses.

Sensilla are also present in the oral cavity. Food entering this cavity contacts approximately 50–60 hypopharingeal sensilla, which are located on the basis of the cibarium. Light-microscope observations suggest that these sensilla contain four neurons [11]. Although functional studies on these sensilla have not been performed, they resemble cibarial contact chemoreceptors known from other insects. Thus, they would process food before it passes on into the esophagus. These receptors could also sample brood food and solutions regurgitated by worker bees.

4.2.2.3   Peripheral Processing of Taste: Gustatory Sensilla on the Forelegs

Taste sensilla are located on the tarsus and pretarsus of the forelegs. Sensilla are mostly chaetic and are distributed evenly between the five subsegments of the tarsus, with a high concentration on the terminal claw-bearing pretarsus. Chaetic sensilla share similarities with those found in the mouth parts, with a mechanosensory cell ending at their base and four cells running to the tip of the shaft [50]. PER can be elicited upon sucrose stimulation of the tarsi, thus indicating that sugar receptors may be present within tarsal gustatory sensilla. Marshall [27] found that bees exhibited PER at a concentration of 2.85% when stimulated at the antennae but that a concentration of 34% was required to elicit PER when the tarsi were stimulated. Over a wide range of sucrose concentrations sucrose responsiveness is always significantly higher for antennal than for tarsal stimulation [8]. Whitehead and Larsen [48] reported 318 chaetic sensilla but no basiconic sensilla on the antennae and 10–20 chaetic sensilla and 0–6 basiconic sensilla per tarsomere of the forelegs. Thus, a simple numeric comparison shows that, at least for chaetic sensilla, the antennae are equipped with 15–30 times more receptors than the tarsi. Such a comparison is, however, senseless without a functional characterization of the specificity and sensitivity of tarsal taste receptor cells by means of electrophysiological recordings.

Dose-dependent responses for sucrose on tarsal sensilla were found by Lorenzo [25] which correspond with the known sucrose sensitivity recorded in behavioral experiments (see above). Moreover, contrarily to antennal chaetic sensilla (see above), responses were found for very low concentrations of KCl (0.1 mM) thus suggesting that a water cell may exist within chaetic sensilla of the tarsi. Besides, a dose-response curve was obtained for KCl, thus demonstrating the presence of a cell responding to electrolytes. As for the mouth parts, these studies were exclusively performed on chaetic sensilla. Basiconic sensilla, present on the distal segments of the legs, have not been recorded so far, due to the technical difficulty associated with their reduced size. This means that as for the other gustatory appendages, only a partial view of the gustatory sensitivity of the forelegs is available.

4.2.3   Molecular Studies on Honey Bee Gustation

Since the decoding and publication of the genome of the honey bee [46], researchers interested in different aspects of the biology of the honey bee have access to bioinformatics tools that allow performing comparative genomic research using as a model the other insect for which most is known in terms of genetic architecture, the fruit fly *Drosophila melanogaster*. Such identification of gustatory receptor genes yielded a surprising result: only 10 gustatory receptor genes (*Grs*) were found [35], which was taken as a proof of a rather limited taste repertoire, at least compared to that of fruit flies, which possess 68 gustatory receptors encoded by 60 genes through alternative splicing (review: [30]), and mosquitoes, which possess 76 gustatory receptors encoded by 52 genes [20]. From the 10 gustatory receptor genes of honey bees, two (*AmGr1* and *AmGr2*) seem to correspond to the eight candidate sugar receptors identified in the fly [30]. The specificity of the other eight remains to be determined.

Robertson and Wanner [35] hypothesized that such a limited number of Grs in bees is due to the fact that these insects have little need for gustatory receptors to locate and recognize food because flowering plants have evolved mechanisms to attract and reward bees for pollination services. They argued that bees do not require the ability to detect and discriminate between the numerous plant secondary chemicals and toxins usually deployed in the chemical ecological arms races between most plants and many insect herbivores so that there is no need for the bees to develop additional taste receptors. Although no functional study is so far available to determine the tastant specificity of any of the 10 *Grs* of the honey bee, *RT-qPCR* and *in situ* hybridization studies, combined with electrophysiological analyses of receptor sensitivity in heterologous systems could soon provide some answers about their functional value. In this way, a fundamental step towards understanding the gustatory world of honey bees would be achieved.

4.2.4 The Case of Bitter Taste Perception in Honey Bees

An argument used to justify the scarceness of gustatory receptor genes in the honey bee is that this insect would not have the ability to detect and discriminate between numerous plant secondary chemicals and toxins usually employed as defense by some plants. This statement contrasts with behavioral responses of foraging bees to natural nectars and pollens, which may contain phenolic compounds and other secondary compounds such as nicotine and caffeine [23, 41]. Concentrations of deterrent compounds in nectar and pollen are usually low. For instance, naturally occurring concentrations of amygdalin are between 4 and 10 ppm [24], which correspond to 8.75×10^{-6} M and 2.19×10^{-5} M, respectively. Honey bees seem to cope efficiently with this natural range of concentrations. Whereas high concentrations of phenolic substances deter them [16], low concentrations are attractive to them [41]. Honey bees even prefer solutions with low concentrations of nicotine and caffeine over a control (20% sucrose) solution. A similar but non-significant pattern was detected also for all concentrations of amygdalin [41]. It seems, therefore, that nectars containing substances that are considered deterrent due to their unpalatable taste (to humans) are in fact preferred by honey bees although if concentrations of such substances are too high, nectars may be rejected. Selectivity may also depend on the resources that are effectively available to bees. Toxic honey may become acceptable in the absence of other nectar sources [44]. This observation may be related to Karl von Frisch's statement on honey bee's reactions towards bitter substances [47]. He wrote that *"bees are much less sensitive to bitter substances than we"* and that *"it is possible to contaminate sugar with a bitter substance that does not interfere with its being taken up by bees but that renders it unacceptable to man"*. In fact, sensitivity or lack of it with respect to aversive compounds may depend not only on what is available to forager bees, as shown by field experiments, but also on the specific experimental context used to ask the bee about its taste capabilities.

For instance, although free-flying bees reject sucrose solutions with high concentrations of bitter substances in the field (see above), this rejection is not visible when bees are harnessed in the laboratory and presented with different kinds of bitter substances [2]. In this case they ingest without reluctance a considerable volume (20 µl, i.e. one third of their crop capacity) of various aversive substances, including concentrated saline solutions and substances that taste bitter to humans, even if some of them induce a high post-ingestional mortality and reduce, therefore, their probability of survival. These substances do not seem, therefore, to be unpalatable to harnessed bees. However, they induce a post-ingestional malaise-like state that in some cases results in death [2]. These results indicate that deterrent substances have an aversive effect on harnessed bees, which is based on the physiological consequences that their ingestion generates rather than on distasteful sensory experiences.

Neither quinine nor salicine inhibited the proboscis extension reflex elicited by previous antennal stimulation with sucrose solution when delivered at the level of the antennae at different concentrations [7]. Similar results were obtained when quinine, salicine and caffeine when delivered at the level of the tarsi [25]. Focusing on the mouth parts showed that harnessed bees that extended the proboscis when stimulated with sucrose, and that received different concentrations of quinine or salicine on the mouth parts upon PER, showed a significant proboscis retraction only for the highest concentration of bitter substance (100 mM). However, retraction only reached 20–30%, thus showing that the effect of bitter substances delivered at the level of the of mouth parts is moderate (de Brito Sanchez et al. unpublished). Bees that extended massively their proboscis to sucrose 1 M responded only partially when stimulated with a mixture of sucrose 1 M and quinine 100 mM. The mixture of sucrose 1 M and salicine 100 mM had no such suppressive effect. This implies that quinine, but not salicine, may interfere with sucrose reception.

These results are confirmed by electrophysiological investigations on different body appendages. On one hand, electrophysiological recordings of taste sensilla performed at the level of the antennal tip (chaetic sensilla; [7]), mouth parts (chaetic and basiconic sensilla on the galeae, labial palps and glossa; de Brito Sanchez et al. unpublished) and distal segments of the forelegs (chaetic sensilla; [25]) could not reveal sensilla that respond specifically to bitter substances such as quinine and salicine at different concentrations. Depending on the appendages considered, other deterrent substances were also assayed with the same result. However, when given together with sucrose the picture changed: electrophysiological responses of chaetic sensilla to sucrose solution 15 mM were inhibited upon stimulation with a mixture of sucrose 15 mM and quinine 0.1 mM, but not with a mixture of sucrose 15 mM and salicine 1 mM [7]. This means that interference of quinine with sucrose reception may occur within the same sensilla responding to sucrose solution. It was concluded that a honey bee could detect the presence of quinine solution due to its peripheral, within-sensillum inhibitory effect on sugar receptor cells [7].

A new twist into this story has been introduced by recent experiments that used freely-flying instead of harnessed honey bees [1] Freely-flying honey bees were trained to discriminate two similar colors in a Y-maze in which the rewarded color (the target) was paired with sucrose solution, as usual, while the alternative color (the distracter) was associated either with 60 mM quinine solution or with water (see Chap. 4.5). These experiments showed that the presence of quinine solution on a visual distracter promoted its rejection, thus improving discrimination of perceptually similar stimuli. In other words, a difficult visual discrimination was rendered possible by the penalizing, aversive effect of the concentrated quinine solution [1]. The results of these experiments with freely-flying bees show a surprising difference with the responses exhibited by harnessed bees in the laboratory for which the same quinine solution does not seem to have an unpalatable effect [2]. It therefore appears that the critical aspect for uncovering the aversive nature of a deterrent compound is the possibility of freely moving that was available in one case [1] but not in others [2, 7, 25]. A fundamental goal would be to determine the reason for such a change in gustatory thresholds for aversive substances once bees are immobilized.

4.2.5 Central Processing of Taste

In the honey bee, as in other insects [29], primary projections of taste neurons on head appendages reach the central nervous system mostly at the level of the subesophageal ganglion (SEG) (Fig. 4.2.3). Besides motor control of the mouth parts and mechanosensory information processing, gustatory processing is one of the major roles of the SEG. The SEG results from the fusion of the mandibular, maxillary, and labial neuromeres (Fig. 4.2.3a). These are arranged sequentially with the mandibular neuromere being anterior and the labial posterior. The more anterior mandibular and maxillary neuromeres successively decrease in volume compared with the posterior labial neuromere. Eight longitudinal tracts run through each half of the ganglion. Dorsal and ventral commissures have been described for the three different neuromeres [33].

Axons of gustatory neurons and mechanosensory neurons hosted in gustatory sensilla project to the mandibular, maxillary, and labial neuromeres via the mandibular nerve (probably mechanosensory neurons, see above), the maxillary nerve, and the labial nerve, respectively [34] (Fig. 4.2.3b, c). Projections of gustatory and mechanosensory neurons hosted in gustatory sensilla on the antennae also project

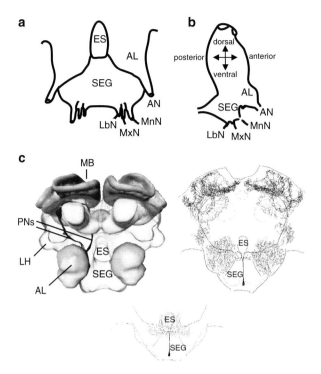

Fig. 4.2.3 (**a**) Schematic frontal view of the subesophageal ganglion (SEG) region showing the afferences of the labial nerves (LbN), the mandibular nerves (MnN) and the maxillary nerves (MxN). AL: antennal lobe; AN: antennal nerve; ES: esophagus. (**b**) Side view of the brain showing the SEG. (**c**) The VUM$_{mxl}$ neuron (ventral unpaired median cell of the maxillary neuromere) (courtesy of R. Menzel). *Left*: three-dimensional reconstruction of the honey bee brain in frontal view without the optic lobes, showing the main stages of the olfactory circuit: antennal lobes (AL), lateral horn (LH) and mushroom bodies (MB) (via projection neurons PNs). *Right*: Morphology of VUM$_{mxl}$ showing the connectivity with the key stages of the olfactory circuit: ALs, lateral horn and lips and basal rings of the mushroom body calyces. *Bottom*: In the SEG, the primary neurite projects dorsally from the ventral median soma and bifurcates beyond the esophagus (ES). Dendritic arborizations occur in the dorsal SEG and tritocerebrum

to the SEG [32, 42]. Mechanosensory and gustatory neurons project to different regions of the SEG. Sensory projections from the proboscis are confined to the ventral portions of the maxillary and labial neuromeres of the SEG, overlapping with the arborizations of neurons of the subesophageal calycal tract (SCT). The SCT links the ventral SEG to the calyces of the mushroom bodies (MBs) [38], suggesting that they also receive mechanosensory and/or gustatory input from the SEG (see Chap. 3.2).

The first-described ventral unpaired median neuron of the maxillary neuromere (VUM$_{mxl}$) has been characterized in great detail, both at the physiological and morphological levels [17] (see also Chap. 3.1). Its cell body lies in a median position within the ventral cell cluster of the SEG and its primary neurite innervates the antennal lobes (AL), the lateral horn (LH), and the lip and basal ring of the MBs, all key-structures of the bee olfactory circuit (Fig. 4.2.3d). Such a neural connectivity and the fact that VUM_{mx1} is activated upon sucrose stimulation of the antennae and proboscis led to the hypothesis that VUM_{mx1} mediates the appetitive properties of sucrose reinforcement. Indeed, VUM_{mx1} activity is sufficient to mediate the reward in olfactory conditioning [17]. In other words, pairing of an odorant with an artificial depolarization of VUM_{mx1} generated by injecting current into the neuron is the equivalent of having experienced an odorant followed by sucrose. As a consequence, a bee treated in this way learns to respond with a PER to the odorant even if it had never experienced real sucrose associated to it. How gustatory sucrose receptors convey information to VUM_{mx1} is still unknown but it is thought that they project to the SEG where they would synapse directly or indirectly onto VUM_{mx1}.

In the central ventral portion of the SEG, Schröter et al. [37] identified ten different VUM neurons including six VUM neurons innervating neuropile regions of the brain and the SEG exclusively (central VUM neurons) and four VUM neurons with axons in peripheral nerves (peripheral VUM neurons). The role of these neurons is still unclear. Some of them respond to sucrose but also to water and salt thus making the question of taste encoding in the SEG complex [37].

No systematic study has tried so far to uncover whether there are organizational functional principles in the architecture of the honey bee SEG. In *Drosophila*, afferences from neurons expressing molecular receptors for sweet substances are spatially separated from those expressing receptors for bitter substances at the SEG [26]. It was concluded that the hedonic value of the tastes (palatable vs. non-palatable) corresponds to a neural classification principle achieved in the SEG. However, these results refer to the receptor neuron level and not to 2nd order interneurons. In the locust *Schistocerca gregaria*, Rogers and Newland [34] recorded 2nd order interneurons in the metathoracic ganglion (MG). These interneurons receive afferences from gustatory receptor neurons on the hindlegs and are broadly tuned to different chemical stimuli. The duration of their response to different chemicals provides a direct measure of aversiveness and strongly correlates with behavioral withdrawal responses [36]. Thus, 2nd order interneurons in the MG of the locust may also encode the hedonic value of the tastants perceived.

Other neurons in the central nervous system of honey bees exhibit significant responses upon antennal and proboscis stimulation with sucrose. For instance, the PE1 neuron [28], a neuron arising from the pedunculus of the MBs and which has extensive arborizations in the median and lateral protocerebrum exhibits increased spiking activity upon sucrose stimulation; yet, this neuron also responds to odors and mechanical stimulations and no other tastants have been assayed to determine its gustatory tuning so that its role in gustatory coding is unclear. The same applies to the so-called feedback neurons [15], which connect the output regions of the MBs (α and ß lobes, and pedunculus) with their ipsilateral input region (ipsilateral calyx). These neurons also respond to odors and sucrose stimulation, but as for the PE1 neuron, these responses reflect the multimodal and integrative nature of MBs, from which they take the information, rather than providing a precise gustatory code.

4.2.6   Conclusion and Outlook

Research on honey bee gustation is still in its infancy compared, for instance, to honey bee vision and olfaction. Yet, important progresses have been made in the last decade. A fundamental advance has been the sequencing of the honey bee genome which allowed determining that honey bees possess, in principle, ten gustatory receptor genes [35]. However, the ligands of these receptors remain unknown. Research should therefore concentrate on determining the natural ligands of these receptors in order to understand the gustatory world of a honey bee.

Molecular gustatory receptors are hosted by gustatory sensilla and even if there has been some electrophysiological work to characterize taste processing at the level of these sensilla, studies on peripheral processing are extremely scarce. From the two typical gustatory sensilla, chaetic and basiconic, single-sensilla recordings have only analyzed neuronal responses of receptor cells hosted in chaetic sensilla. As already explained, this choice is based on the small size and difficulty of accessing the short pegs of basiconic sensilla. Yet, recording from these sensilla is possible and should be achieved in a systematic way. Otherwise, peripheral analyses on honey bee gustation represent only a partial view of what honey bees could detect in gustatory terms.

The most important endeavor, however, is to determine the kind of processing occurring at the central level as no perceptual phenomenon, in this case taste perception, can be directly derived from receptor responses. Whether or not the SEG of the honey bee encodes tastes following their hedonic value is still an open question. Multi-electrode recording techniques, allowing to measure populational codes upon gustatory stimulation, could represent an important endeavor to decipher the principles of central gustatory processing in the honey bee.

References

1. Avarguès-Weber A, de Brito Sanchez MG, Giurfa M, Dyer AG (2010) Aversive reinforcement improves visual discrimination learning in free-flying honeybees. PLoS One 5(10):e15370
2. Ayestaran A, Giurfa M, de Brito Sanchez MG (2010) Toxic but drank: gustatory aversive compounds induce post-ingestional malaise in harnessed honeybees. PLoS One 5(10)
3. Bitterman ME, Menzel R, Fietz A, Schäfer S (1983) Classical conditioning of proboscis extension in honeybees (*Apis mellifera*). J Comp Psychol 97:107–119
4. Breed MD, Williams KR, Fewell JH (1988) Comb wax mediates the acquisition of nest-mate recognition cues in honey bees. Proc Natl Acad Sci USA 85(22):8766–8769
5. Chaline N, Sandoz JC, Martin SJ, Ratnieks FL, Jones GR (2005) Learning and discrimination of individual cuticular hydrocarbons by honeybees (*Apis mellifera*). Chem Senses 30(4):327–335
6. Dani FR, Jones GR, Corsi S, Beard R, Pradella D et al (2005) Nestmate recognition cues in the honey bee: differential importance of cuticular alkanes and alkenes. Chem Senses 30(6):477–489
7. de Brito Sanchez MG, Giurfa M, Mota TRD, Gauthier M (2005) Electrophysiological and behavioural characterization of gustatory responses to antennal 'bitter' taste in honeybees. Eur J Neurosci 22(12):3161–3170

8. de Brito Sanchez MG, Chen C, Li JJ, Liu FL, Gauthier M et al (2008) Behavioral studies on tarsal gustation in honeybees: sucrose responsiveness and sucrose-mediated olfactory conditioning. J Comp Physiol A 194(10):861–869
9. Dethier VG (1961) Beharioral aspects of protein ingestion by blowfly *Phormia regina* Meigen. Biol Bull 121(3):456–470
10. Dethier VG (1974) Specificity of labellar chemoreceptors of blowfly and response to natural foods. J Insect Physiol 20(9):1859–1869
11. Dostal B (1958) Riechfähigkeit und Zahl der Riechsinneselemente bei der Honigbiene. Z vergl Physiol 41(2):179–203
12. Esslen J, Kaissling KE (1976) Number and distribution of sensilla on antennal flagellum of honeybee (*Apis mellifera* L). Zoomorphologie 83(3):227–251
13. Galic M (1971) Die Sinnesorgane an der Glossa, dem Epipharynx und dem Hypopharynx der Arbeiterin von *Apis mellifica* L. (Insecta, Hymenoptera). Z Morph Tiere 70:201–228
14. Goldrich NR (1973) Behavioral responses of *Phormia regina* (Meigen) to labellar stimulation with amino acids. J Gen Physiol 61(1):74–88
15. Grünewald B (1999) Physiological properties and response modulations of mushroom body feedback neurons during olfactory learning in the honeybee, *Apis mellifera*. J Comp Physiol A 185(6):565–576
16. Hagler JR, Buchmann SL (1993) Honey bee (Hymenoptera, Apidae) foraging responses to phenolic-rich nectars. J Kans Entomol Soc 66(2):223–230
17. Hammer M (1993) An identified neuron mediates the unconditioned stimulus in associative olfactory learning in honeybees. Nature 366:59–63
18. Harborne JB (1994) Introduction to ecological biochemistry, 4th edn. Academic, London
19. Haupt SS (2004) Antennal sucrose perception in the honey bee (*Apis mellifera* L.): behaviour and electrophysiology. J Comp Physiol A 190(9):735–745
20. Hill CA, Fox AN, Pitts RJ, Kent LB, Tan PL et al (2002) G protein coupled receptors in *Anopheles gambiae*. Science 298(5591):176–178
21. Hiroi M, Marion-Poll F, Tanimura T (2002) Differentiated response to sugars among labellar chemosensilla in *Drosophila*. Zool Sci 19(9):1009–1018
22. Kuwabara M (1957) Bildung des bedingten Reflexes von Pavlovs Typus bei der Honigbiene. *Apis mellifica* J Fac Sci Hokkaido Univ Ser VI Zool 13:458–464
23. Liu FFW, Yang D, Peng Y, Zhang X, He J (2004) Reinforcement of bee–plant interaction by phenolics in food. J Apicult Res 43:153–157
24. London-Shafir I, Shafir S, Eisikowitch D (2003) Amygdalin in almond nectar and pollen – facts and possible roles. Plant Syst Evol 238(1–4):87–95
25. Lorenzo E (2009) Electrophysiological characterization of bitter taste perception at the level of the tarsi in the honey bee *Apis mellifera*. University Paul Sabatier, Toulouse
26. Marella S, Fischler W, Kong P, Asgarian S, Rueckert E et al (2006) Imaging taste responses in the fly brain reveals a functional map of taste category and behavior. Neuron 49(2):285–295
27. Marshall J (1935) On the sensitivity of the chemoreceptors on the antenna and fore-tarsus of the honey-bee, *Apis mellifica* L. J Exp Biol 12(1):17–26
28. Mauelshagen J (1993) Neural correlates of olfactory learning paradigms in an identified neuron in the honeybee brain. J Neurophysiol 69(2):609–625
29. Mitchell BK, Itagaki H, Rivet MP (1999) Peripheral and central structures involved in insect gustation. Microsc Res Tech 47(6):401–415
30. Montell C (2009) A taste of the *Drosophila* gustatory receptors. Curr Opin Neurobiol 19(4):345–353
31. Núñez JA (1982) Honeybee foraging strategies at a food source in relation to its distance from the hive and the rate of sugar flow. J Apicult Res 21(3):139–150
32. Pareto A (1972) Die zentrale Verteilung der Fühlerafferenz bei Arbeiterinnen der Honigbiene, *Apis mellifera* L. Z Zellforsch 131:109–140
33. Rehder V (1988) A neuroanatomical map of the suboesophageal and prothoracic ganglia of the honeybee. Proc R Soc B 235:179–202

34. Rehder V (1989) Sensory pathways and motoneurons of the proboscis reflex in the suboe-sophageal ganglion of the honey bee. J Comp Neurol 279:499–513
35. Robertson HM, Wanner KW (2006) The chemoreceptor superfamily in the honey bee, *Apis mellifera*: expansion of the odorant, but not gustatory, receptor family. Genome Res 16(11):1395–1403
36. Rogers SM, Newland PL (2002) Gustatory processing in thoracic local circuits of locusts. J Neurosci 22(18):8324–8333
37. Schröter U, Malun D, Menzel R (2007) Innervation pattern of suboesophageal ventral unpaired median neurones in the honeybee brain. Cell Tissue Res 327(3):647–667
38. Schröter U, Menzel R (2003) A new ascending sensory tract to the calyces of the honeybee mushroom, body, the subesophageal-calycal tract. J Comp Neurol 465(2):168–178
39. Shimada I (1975) Chemical treatments of labellar sugar receptor of fleshfly. J Insect Physiol 21(9):1565–1574
40. Shiraish A, Kuwabara M (1970) Effects of amino acids on labellar hair chemosensory cells of fly. J Gen Physiol 56(6):768–782
41. Singaravelan N, Nee'man G, Inbar M, Izhaki I (2005) Feeding responses of free-flying honey-bees to secondary compounds mimicking floral nectars. J Chem Ecol 31(12):2791–2804
42. Suzuki H (1975) Antennal movements induced by odor and central projection of antennal neurons in honeybee. J Insect Physiol 21(4):831–847
43. Takeda K (1961) Classical conditioned response in the honey bee. J Insect Physiol 6:168–179
44. Tan K, Guo YH, Nicolson SW, Radloff SE, Song QS et al (2007) Honeybee (*Apis cerana*) foraging responses to the toxic honey of *Tripterygium hypoglaucum* (Celastraceae): changing threshold of nectar acceptability. J Chem Ecol 33(12):2209–2217
45. Teixeira EW, Negri G, Meira R, Message D, Salatino A (2005) Plant origin of green propolis: bee behavior, plant anatomy and chemistry. Evid-Based Compl Alt Med 2(1):85–92
46. The Honeybee Genome Sequencing Consortium (2006) Insights into social insects from the genome of the honeybee *Apis mellifera*. Nature 444(7118):512–512 (vol 443, pg 931, 2006)
47. von Frisch K (1967) The dance language and orientation of honey bees. Belknap, Cambridge
48. Whitehead AT (1978) Electrophysiological response of honey bee labial palp contact chemore-ceptors to sugars and electrolytes. Physiol Entomol 3:241–248
49. Whitehead AT, Larsen JR (1976) Electrophysiological responses of galeal contact chemore-ceptors of *Apis mellifera* to selected sugars and electrolytes. J Insect Physiol 22:1609–1616
50. Whitehead AT, Larsen JR (1976) Ultrastructure of the contact chemoreceptors of *Apis mellifera* L. (Hymenoptera: Apidae). Int J Insect Morphol Embryol 5(4/5):301–315

Chapter 4.3
The Auditory System of the Honey Bee

Hiroyuki Ai and Tsunao Itoh

Abstract The auditory organ of honey bee is the "Johnston's organ (JO)" on the antennae which detects airborne vibration during waggle dance communication and also detects air current during flight. The sensory afferents of the JO send their axons to two distinct areas of the bee brain, the Antennal mechanosensory centers (AMMC) and the Superior posterior slope (SPS). Within these termination fields sensory axons in the ventro-medial SPS are characterized by both thick processes with large varicosities and somatotopy, while those in the AMMC by both thin processes with small varicosities and no somatotopy, suggesting that vibratory signals detected by the JO are processed in dual parallel pathways in these primary sensory centers. In order to clarify the characteristics of auditory processing, the response properties of the interneurons to the vibration stimuli, arborizing in these primary sensory centers have been investigated. AMMC-Int-1 and AMMC-Int-2 densely arborize in AMMC and respond stimulus-phase-dependently to the vibratory stimulation on the ipsilateral antenna with high sensitivity in the range of 250–300 Hz, which is the main airborne vibration frequencies generated by the waggle dance. While SPS-D-1 has dense arborizations in the SPS and sends axons into the ventral nerve cord, with blebby terminals in the contralateral dSEG and SPS, and respond to the vibratory stimulation on the ipsilateral antenna with long-lasting excitation during olfactory stimulation on the contralateral antenna. The possible roles of the parallel systems in the primary auditory centers are discussed.

Keywords Honey bee Standard Brain (HSB) • Integration • Audition • Olfaction • Waggle dance • Parallel processing • Identified neuron

H. Ai(✉) • T. Itoh
Division of Biology, Department of Earth System Science, Fukuoka University,
8-19-1 Nanakuma, Jonan-ku, Fukuoka 814-0180, Japan
e-mail: ai@fukuoka-u.ac.jp

C.G. Galizia et al. (eds.), *Honeybee Neurobiology and Behavior: A Tribute to Randolf Menzel*, DOI 10.1007/978-94-007-2099-2_21,
© Springer Science+Business Media B.V. 2012

Abbreviations

AMMC   Antennal mechanosensory and motor center
dSEG   dorsal region of subesophageal ganglion
GABA   Gamma amino butyric acid
HBS   Honey bee standard brain
JO   Johnston's organ
SPS   Superior posterior slope

4.3.1 Introduction

Animals communicate with each other for sharing various informations. Sound is suitable for sending complicated information, and auditory communication has evolved in higher order species, insects, amphibian, birds and mammals. In insects the auditory communications in cricket and katydid are most familiar. The sound they use is a stridulation caused by scraping the bilateral wings against each other. These insects are one of the groups with most species in Polyneoptera and have evolved tympanal organs with high sensitivity. For example, the primitive katydid *Buccaris membra-cionides* can detect a faint sound (12.8 dB) by their tympanal organs and the spatial range for detecting the conspecific courtship song is about 2 km in radius. On the other hand, air particle movements are also caused by the sound sources. Some flying insects use the "airborne vibration" caused by wingbeats for their communication. For example, male mosquitoes *Aedes aegypti* are very sensitive to the wingbeats (the frequency is ca. 380 Hz) caused by the conspecific female and approach to the female for courtship [15]. Moreover male mosquitoes can discriminate between mature or immature females by detecting fine differences in frequencies.

Honey bees learn the place of nectar or pollen-bearing flowers not from "seeing" but from "hearing and smelling" the dance performed by foragers (see Chap. 2.2). The foragers returning to the hive display waggle dances in order to inform the dance-attendees (followers) of the floral odor, the direction and distance from their hive to a site of the flowers and the followers are recruited to a remote food source according to these informations [12]. This finding was recently proved directly by following the flight paths of recruits using harmonic radar recording [23, 29] (see also Chap. 2.3). Karl von-Frisch [12] also analyzed which elements of the waggle dance were correlated for the distance from the hive to the flower, and suggested the duration of waggling is one of the index of distance. Since then, ethological studies on waggle-dance communication have suggested that "airborne vibration" generated by the wing vibration during waggling are important cues in dance communication [7, 24]. These results suggest the duration of airborne vibration could be the index of distance. Airborne vibrational signals emitted by the waggle-dancer have been studied extensively; they consist of roughly 30 pulses per second; each pulse lasting for about 20 ms with a carrier frequency of about 265 Hz [19].

Behavioral experiments have demonstrated that honey bees are able to detect air-particle movements with the Johnston's organ (JO) located at the second segment (pedicel) of the antenna (Fig. 4.3.1) [7]. The mechanical sensitivities of the antennal flagellum are specifically high in response to low intensity stimuli of 265–350 Hz frequencies [37]. The primary sensory neurons in the JO transduce the mechanical vibration of the flagellum (the third antennal segment) into neural excitation. The sensory neurons in the JO of the aged forager are specialized for detecting vibrations of the antennae with frequencies in the range of 250–300 Hz, which is the normal range of the main airborne vibration frequencies generated by the waggle-dancer. The JO has also been suggested to play a role in detecting air current during flight [34].

Recently it was found that the followers orientate toward the dancer using the floral odor carried by the dancer [11], and a pheromone which is produced and released by the dancer [36]. These results suggest that the floral and dancer's odors accelerate the dance communication (see Chap. 2.4) and lead the nestmates to foraging, and therefore might trigger or accelerate vibration processing in the follower's brain. However, the neural correlates of the integration of airborne vibratory stimuli and olfactory stimuli have been examined only partially [2–4]. These studies could be an important step to understand how the follower decodes the dancer's message in her brain. The previous studies have revealed the morphology of the JO, the structure of the primary auditory centers projected by the afferents of JO [3] and the morphologies and response properties to vibration and olfactory stimuli of several types of interneurons with dense arborizations in the primary auditory centers, Antennal mechanosensory centers (AMMC) [4] and the superior posterior slope (SPS) [2]. These studies suggest that there are at least two parallel systems for auditory processing, the first one through the AMMC and the second one through the SPS and that both of them are modulated by olfactory stimulation.

4.3.2 Central Projection of JO Afferents in the Honey Bee Brain

JO exists in the second segment of the antenna (pedicel) and the sensory afferents send their axons into the brain through the antennal nerve (Fig. 4.3.1a). JO is a multicellular mechanosensory organ which is composed of approx. 240 scolopidia which are stimulated by movement of the flagellum (Fig. 4.3.1b). The airborne vibration deflects the flagellum from side to side. The deflections of the flagellum stretch the scolopales at the opposite side of the pedicel. About 720 somata of the JO are divided into three subgroups based on their location: a dorsal group (dJO), a ventral group (vJO), and an anterior group (aJO) (Fig. 4.3.1c, d). These soma groups send axons to different branches (N2 to N4) diverging from the antennal nerve. All sensory afferents send collateral axon branches with fine terminals to the dorsal lobe (DL), while the main axons trifurcate into the fascicles T6I, T6II and T6III [3, 27, 35]. The axons in T6I, T6II and T6III terminate ipsilaterally in the ventro-medial superior posterior slope (vmSPS), the Antennal

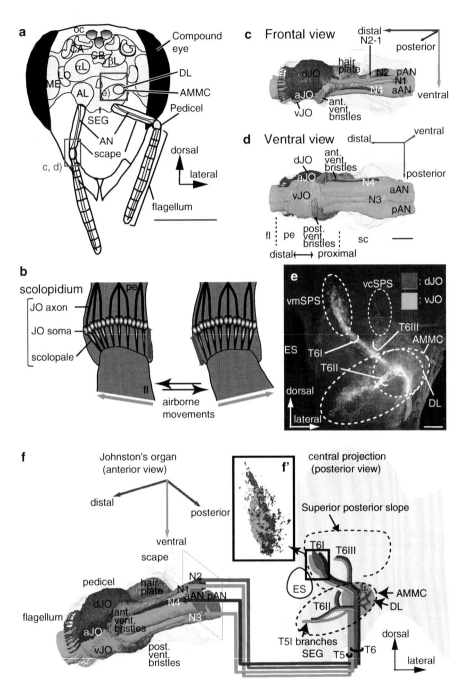

mechanosensory and motor center (AMMC, it includes DL and its extension to the dorsal region of the subesophageal ganglion) and ventro-central superior posterior slope (vcSPS), respectively (Fig. 4.3.1e, f). Axon terminals in vmSPS possess thick processes with large presynaptic varicosities, while those in AMMC and vcSPS have thin processes with small presynaptic boutons [3]. Moreover the axon terminals in vmSPS stemming from the three soma groups aJO, vJO, dJO in the pedicel are spatially segregated, showing some degree of somatotopy (Fig. 4.3.1f'). This spatial segregation is not observed in axon terminals running in AMMC and vcSPS [3]. Kamikouchi et al. [18] established a comprehensive projection map of the JO afferents from the antenna to the primary auditory center of the *Drosophila* brain by using GAL4 enhancer-trap strains and identified five groups of these JO afferents (JON A-E), which show somatotopic organization in the primary mechanosensory center, AMMC. The somatotopic organization of JO afferents are also observed in mosquito [16], suggesting it is a common characteristics of the central projection of JO afferents for detecting the direction of the mechanical stimuli. Moreover the AMMC and vcSPS which have no somatotopy might be responsible for vibration processing specific in honey bee, but not in *Drosophila* and mosquitoes.

Fig. 4.3.1 (a) Schematic drawing of antennae and brain in the head capsule viewed frontally. The antenna is composed of three parts, the scape (*sc*), pedicel (*pd*) and flagellum (*fl*). The Johnston's organ (*JO*) is in the pedicel. The left hemisphere shows the main neuropils on the anterior plane of the brain through the antennal lobe (*AL*) and the right one those on the middle plane of the brain through the dorsal lobe (*DL*). Two red boxes indicate the areas of the images shown in (**c**), (**e**). (**b**) Schematic drawing of structures in the JO. The JO is composed of about 240 scolopidia in which each scolopidium has a few JO sensory cells. Airborne movements cause the deflections of the flagellum (*blue arrows*) which stretch the scolopales in the opposite side against the deflections (*red arrows*). (**c**) and (**d**) 3D reconstructions of JO and the related sensory structures in the antenna viewed frontally (**c**), and ventrally (**d**). a (or p) AN, anterior (or posterior) antennal nerve; a (d or v) JO, anterior (dorsal or ventral) Johnston's organ; ant. (or post.) vent bristles, anterior (or posterior) ventral bristles; N2 (2-1, 3 or 4), nerve 2 (2-1, 3 or 4). (**e**), Central projection of the JO afferents. The photo shows differential staining of ventral JO (*green*) and dorsal JO (*magenta*). The sensory afferents of ventral JO and dorsal JO send axons to the dorsal lobe (DL) through T6 and trifurcate into ventro-medial Superior posterior slope (vmSPS), Antennal mechanosensory and motor center (AMMC) and ventro-central SPS (vcSPS) through T6I, II and III, respectively. Those of anterior JO are also the same as those of dorsal and ventral JO. (**f**), Central projection of three JO groups. The left image shows three-dimensional reconstruction of sensory organs in pedicel, including JO (blue, anterior JO; green, ventral JO; magenta, dorsal JO) and their innervating nerves (N1-4). The right image shows the schematic drawing of the brain showing the central projection of JO subgroups and exteroceptors. The sensory axons of JO subgroups through T6I terminate in the vmSPS, forming collectively tulip-petal like termination profiles (**f'**). The exteroceptors afferents send axons through T5 and terminate in the dorsal lobe and dorsal SEG through T5I branches. Orientation of antenna and brain in the chapter is given according to the body axis. AN, antennal nerve; αL, alpha lobe of mushroom body (MB); βL, beta lobe of MB; CA, calyx of MB; CB, central body; ES, esophagus; LO, lobula; ME, medulla; oc, ocelli; SEG, subesophageal ganglion. Scale bars; 1 mm in (**a**), 100 μm in (**d**) and 50 μm in (**e**) (Modified from Ai et al. (2007) [3])

4.3.3 Morphology and Physiology of Vibration-Sensitive Interneurons

Several interneurons have been identified that arborize densely in the AMMC [4] and the SPS [2] and respond to vibratory stimulation applied to the JO with specific spike patterns. Three groups of interneurons can be distinguished: AMMC interneuron type 1 (AMMC-Int-1, Fig. 4.3.2), AMMC interneuron type 2 (AMMC-Int-2, Fig. 4.3.3), and SPS descending neuron type 1 (SPS-D-1, Fig. 4.3.4). The morphological interactions between each identified interneuron and JO afferents are visualized within the Honey bee standard brain (HSB) [5] (see Chap. 3.1).

4.3.3.1 AMMC-Int-1

AMMC-Int-1 is a local interneuron in the primary auditory center. This neuron has a dense arborization in the AMMC and thin arborizations in the ventral protocerebrum (Fig. 4.3.2a, b). The arborizations in the ventral protocerebrum comprise a small number of fine spines, while those in the AMMC possess not only spines but also fine blebs (presumable presynaptic boutons). The JO afferents run close to the AMMC-Int-1 branches in the DL (Fig. 4.3.2b).

The patterns of responses of the AMMC-Int-1 to the vibratory and olfactory stimulation are very interesting. The AMMC-Int-1 are spontaneously active and show on-off phasic excitation to vibratory stimuli (arrowheads in Fig. 4.3.2c-upper record). Olfactory stimuli applied to the contralateral antenna cause a long-lasting excitation on this neuron (Fig. 4.3.2c-middle record). The spike frequency gradually increases and doubles the spontaneous frequency at its peak response. When the vibratory stimulation is applied again after the olfactory stimulation, the neuron shows a typical tonic inhibition (arrow in Fig. 4.3.2c-lower record). One individual AMMC-Int-1, which has responded with the typical on-off phasic excitation to vibratory stimulation, changed its responses to tonic inhibition when depolarizing current is injected in order to simulate synaptic input [4]. The changes of the response patterns, therefore, depend on the membrane potentials and reflect differences in the state of an AMMC-Int-1.

4.3.3.2 AMMC-Int-2

AMMC-Int-2 is a type of projection neurons in the primary auditory center. This neuron projects to the AMMC, the lateral SPS, and the lateral protocerebrum (LP) (Fig. 4.3.3, [4]). The neuron responds with phasic-tonic excitation to vibratory stimulation of relatively high amplitude (e.g. 30 μm). The stimulus-induced spike activity depends on both amplitude and frequency of the vibration stimulus, though the

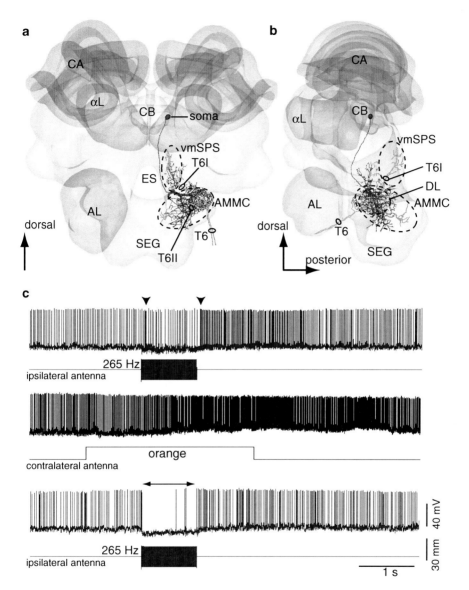

Fig. 4.3.2 (a) and (b) Spatial relationship between an AMMC-Int-1 (*magenta*) and the Johnston's organ (JO) afferents (*green*) (A, frontal view; B, lateral view). The intracellularly marked AMMC-Int-1 and dye-injected JO afferents from two different specimens are reconstructed and registered into the Honeybee Standard Brain (HSB). The soma of the AMMC-Int-1 neuron is located in the dorsal, most posterior region of the protocerebrum and posterior to the central body (CB). AMMC-Int-1 has dense arborizations in the AMMC and sends a small branch into ventral protocerebrum. (**c**), The vibration and olfactory response patterns of an AMMC-Int-1. (**c**) *Upper;* The vibration applied to the ipsilateral antenna causes an on-off phasic excitation (*arrowheads*). *Middle;* Olfactory stimulation, orange odor, applied to the contralateral antenna cause a long-lasting excitation. *Lower;* When the vibratory stimulation is applied again after the olfactory stimulation, the neuron shows a typical tonic inhibition (*arrow*) (This figure is modified from Ai et al. (2009) [4])

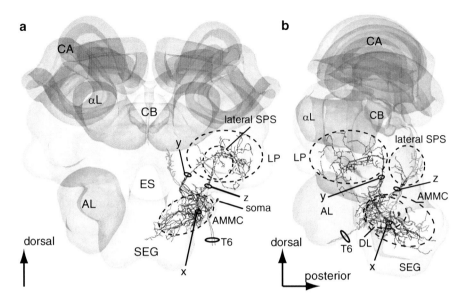

Fig. 4.3.3 Spatial relationship between a AMMC-Int-2 (magenta) and the Johnston's organ (JO) afferents (green), (**a**) frontal view; (**b**) lateral view. The soma is located in the posterolateral region of the dorsal lobe (DL). The neuron has three major ramifications (x, y, and z). The most strongly ramified arborizations (x) are arborizing in the AMMC with numerous fine spines. A long process (y) terminates in the lateral protocerebrum (LP) with fine presynaptic terminals. A small branch (z) emanates from the major DL branch and projects into the lateral portion of the superior posterior slope (lateral SPS) with fine presynaptic terminals (Modified from Ai et al. (2009) [4])

responses are saturated for vibrations above 30 μm in amplitude, both at 265 and 300 Hz. The sensitivity is maximal at 265 Hz [4] which is in the range of the waggle dance frequency.

4.3.3.3 SPS-D-1

SPS-D-1 is a descending neuron of the SPS. This neuron projects to the ipsi-lateral and contra-lateral SPS as well as to the contra-lateral SEG (Fig. 4.3.4a, b, [2]).

The neuron does not respond to vibratory stimulation alone at 265 Hz, but responds with long-lasting excitation to the vibration at 265 Hz when an olfactory stimulation is applied to the contralateral antenna (Fig. 4.3.4c).

4.3.4 Parallel Pathways of JO Afferents

The sensory axons of the JO run into the brain and trifurcate into three fascicles T6I, II and III. Especially the axon terminals running in sensory fascicle T6I terminate in the ventro-medial superior posterior slope (vmSPS) where they are closely appositioned

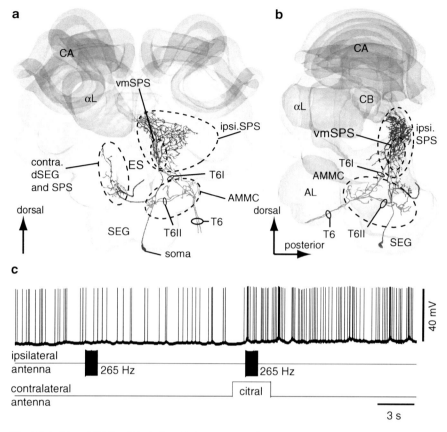

Fig. 4.3.4 (**a**) and (**b**) Spatial relationship between a SPS-D-1 (*magenta*) and the Johnston's organ (JO) afferents (*green*) (**a**) frontal view; (**b**) lateral view. The soma of the SPS-D-1 is located in the ventral median region of the SEG. SPS-D-1 overlaps with the terminal branches of the JO in the ventromedial superior posterior slope (vmSPS). The SPS-D-1 has dense and broad arborizations in ipsilateral SPS (ipsi. SPS) and sparse and fine arborizations with presynaptic terminals in the contralateral dorsal SEG (contra. dSEG) and SPS. (**c**) On the first vibratory stimulation, the SPS-D-1 does not respond to vibration applied to the antenna. Then second vibration stimulus during olfactory stimulation (citral) applied to the contralateral antenna causes a long-lasting excitation. The excitation lasts for several minutes

with those of secondary interneurons from the ocelli [3, 25, 27]. T6I is also known to be in close proximity to termination fields of visual projection neurons from the lobula (LO) (the neuropil specialized for movement detection) [21]. The vmSPS also has the dendrites of motion-sensitive descending interneurons [14]. Thus, the targeting of JO axons into the vmSPS may permit simultaneous sampling of external mechanosensory (vibrational) signals detected by the JO as well as visual signals. Considering the participation of the JO in flight control [22], the JO may modify the visuomotor coordination during flight by detecting air current [34] (Fig. 4.3.5d).

Axons of JO afferents terminating in the AMMC are characterized by thin processes with small presynaptic terminals. Axon terminals of the hair plate on the

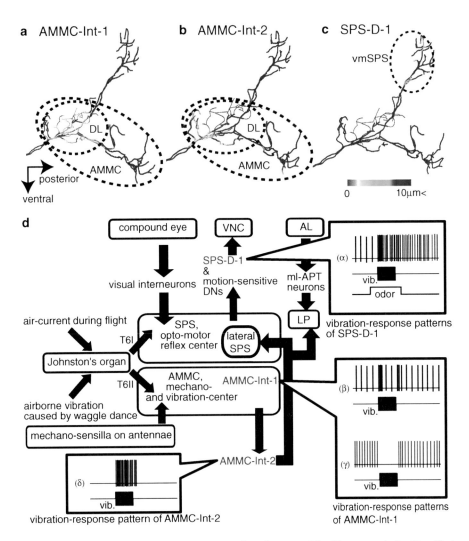

Fig. 4.3.5 (**a–c**) Possible synaptic contact regions between JO afferents and the identified vibration-sensitive interneurons (lateral view). (**a–c**) shows the areas close to AMMC-Int-1, AMMC-Int-2 and SPS-D-1 on the JO afferents in the HSB, respectively. Distances between these interneurons and JO afferents are calculated on the superimposed three-dimensional image. Red spots on the skeletonized fibers of JO afferents indicate close appositions to these interneurons. In (**a**) the areas of red spots on the JO afferents exist in the central region of the dorsal lobe (DL). In (**b**) the areas of red spots on the JO afferents exist in the anterior region of the DL. In (**c**) the areas of red spots on the JO afferents exist in the ventro-medial Superior posterior slope (vmSPS). (**d**) A scheme of processing pathways of mechanosensory signals detected by Johnston's organ (JO). The JO detects airborne vibration caused by waggle dance and also air-current during flight. The JO afferents send this information to the SPS and AMMC, through T6I and T6II, respectively. The SPS is thought to be an opto-motor reflex center, which has axon terminals of visual interneurons and also has a dendritic arborization of motion-sensitive descending neurons (DNs). SPS-D-1 also has a dendritic arborization in the SPS and responds with a long-lasting excitation to vibration stimuli to the ipsilateral antenna during olfactory stimuli to the contralateral antenna (α). On the other hand, the

pedicel (Fig. 4.3.1c), a kind of exteroceptors detecting the movement of antennal-joint between the pedicel and flagellum, are closely appositioned to those of JO afferents in the AMMC [3]. In the waggling phase of the bee's dance, not only the airborne vibration but also the jet air flow are caused by the dancer's wing vibration at the tip of abdomen of the dancer [24] (see also Chap. 2.2), suggesting the airborne vibration and the air flow include some information about the food source. Therefore the dance-followers tend to attend the dancing bee at the rear end of the dancer for receiving the mechanical signal [30] (see also Chap. 2.4). Since the following bees always extend their antennae toward the dancing bee during the waggle dance communication, the followers must use the antennal mechanosensors, including the hair-plates for tactile-sensing the waggling body or JO for detecting the air-particle movements caused by waggle dance. Moreover the orientation of the dancers during waggle phase relative to gravity codes the direction to the flower and the followers decode this direction also in relation to gravity. It has been revealed that the sensory hairs on the neck detecting the orientation of body axis against gravity project to the dorsal SEG (dSEG) [6], suggesting that the dSEG is one of the primary gravity centers and AMMC, close to the dSEG may serve to integrate the signals originating from the mechanosensory organs of the antenna (JO and exteroceptors on the antennae) and from the sensory hairs on the neck during waggle dance communication (Fig. 4.3.5d).

4.3.5   Possible Neural Circuits for Processing Vibration

The anatomical results suggest that the AMMC-Int-1 and -2 are closely appositioned to the JO afferents in the AMMC (Figs. 4.3.2 and 4.3.3). AMMC-Int-1 responds with an on-off-phasic excitation to vibration applied to the JO with the high sensitivity at 265 Hz, irrespective of the duration of vibration [4]. The AMMC-Int-2 responds with a tonic excitation most sensitive to 265-Hz vibration, which corresponds to the peak frequency of airborne vibrations caused by waggle dance [4]. These response patterns of the interneurons to the vibratory stimulation are closely related with those of the JO; the JO neurons are most sensitive at 265 Hz [37]. Moreover, by registering AMMC-Int-1 and AMMC-Int-2 into the HSB [2], it was postulated that AMMC-Int-1 arborizes close to JO afferents in the central region of the DL (Fig. 4.3.5a), while AMMC-Int-2 does so in the anterior region of

---

Fig. 4.3.5 (continued) AMMC is thought to be a mechanosensory center, which has axon terminals of mechano-sensilla on the antenna and also has those of Johnston's organ. AMMC-Int-1 is a local interneuron in the primary centers and responds to the vibration with on- and off-phasic excitation (β). After olfactory stimulation to the contralateral antenna, the AMMC-Int-1 responds as a tonic inhibition to vibratory stimulation (γ). AMMC-Int-2 sends its axon to the LP and lateral SPS and responds to the vibration with tonic excitation (δ). The LP also has axon terminals of ml-APT neurons, which is one of olfactory interneurons originating from the AL. VNC, ventral nerve cord

the DL (Fig. 4.3.5b). In the antennal lobe (AL), olfactory sensory neurons synapse with local interneurons and projection neurons. It is very interesting that there are similar morphological types, local interneurons (AMMC-Int-1) and projection neurons (AMMC-Int-2), also in the primary auditory center.

AMMC-Int-1 receives bimodal inputs [4]. One is an on-off excitatory input when the vibration stimulus is applied to the ipsilateral antenna, and the other is a tonic inhibitory input when the vibration stimulus is applied to the ipsilateral antenna after the olfactory stimulus is applied to the contralateral antenna (Fig. 4.3.2c). Because the dendrites of AMMC-Int-1 are very close to the JO sensory axons in the DL (Fig. 4.3.5a), the on-off excitatory input may be caused by direct excitatory synapses from the JO axons on the AMMC-Int-1. The tonic inhibitory response of the AMMC-Int-1 suggests that AMMC-Int-1 may receive inhibitory synaptic input from some theoretical interneurons that are excited by JO afferents. GABA-like immunoreactive profiles in the DL and the SEG [32] may support this assumption, though inhibitory synaptic inputs to AMMC-Int-1 neurons have not yet been confirmed. The response pattern of AMMC-Int-1 to vibration changes corresponds to the magnitude of spontaneous activity and can also be changed by depolarizing current injection into the neuron [4]. The effect of depolarizing current injection may simulate the effect of additional inputs through other sensory systems (e.g., olfactory input). Thus, the neural activities appear to be regulated by direct or indirect synaptic inputs from as yet unknown interneurons. In our study the vibration applied immediately after odor stimulation induces tonic inhibition on the AMMC-Int-1, while the same vibration induces on-off phasic excitation before the olfactory stimulation (Fig. 4.3.2c). These results suggest that the AMMC-Int-1's responses to the vibratory stimulation might be modulated by the odor stimulation. There is also evidence for modulatory effects in the DL from experiments showing visual conditioning, antennal motor learning, and operant conditioning of an identified motoneuron in the DL [8–10], and for the modulation of the DL motoneurons by biogenic amines [28].

AMMC-Int-2 receives an excitatory input when the vibration is applied to the ipsilateral antenna [4]. Because the dendrites of AMMC-Int-2 are very close to the JO sensory axons in the DL (Fig. 4.3.5b), the sensory neurons of the JO may have direct excitatory synapses on the AMMC-Int-2. Moreover, the AMMC-Int-2 neurons send axons to the lateral protocerebrum (LP) thus overlapping with neurons of the ml-APTs [1, 20, 31] and to the SPS exhibiting fine presynaptic boutons. The LP is one of the second-order centers of olfaction [20], and the SPS is the second-order center of vision [21]. The AMMC-Int-2 may have a role in sending the vibratory information to these other neuropils in which the visual and olfactory signals are processed (Fig. 4.3.5d).

The SPS-D-1 has a fan-shaped projection pattern with presynaptic terminals all over the SPS (Fig. 4.3.4a, b) and does not overlap with the AMMC-Int-1 and -2. From the results of the registration of the SPS-D-1 and JO afferents into the HSB, the JO afferents in T6I closely approach the SPS-D-1 in the vmSPS (Fig. 4.3.5c). The sensory neurons of the JO may have direct synapses on SPS-D-1. SPS-D-1 does not respond to the vibration to the ipsilateral JO, but responds to vibration with a long-lasting excitation, when olfactory stimulation is applied to the contralateral antenna (Fig. 4.3.4c). Some olfactory interneurons, which originate from the contralateral AL, and project to the SPS may modulate the synaptic transmission from JO afferents to the SPS-D-1.

4.3.6   Outlook on the Studies of "Auditory System of the Honey Bee"

As shown in Fig. 4.3.5d there are parallel pathways from the sensory afferents of the JO to the vibration-sensitive interneurons in the primary auditory center. One is a pathway running into the SPS, which is thought to be an opto-motor reflex center in the brain. Many motion-sensitive interneurons and descending neurons having arborization in the SPS have been identified. The axon afferents through T6I into vmSPS suggest that the vmSPS is a region for integrating visual signals detected by ocelli or compound eyes and mechano-sensory signals detected by JO. SPS-D-1 is one of the candidates for integrating such multimodal signals (α in Fig. 4.3.5d). The other is a pathway running into the AMMC, which receives the sensory axon terminals of antennal mechano-sensilla for tactile perception and of Johnston's organ. These mechano-perceptions have been suggested to contribute to the waggle dance communication (tactile perception to the dancer and airborne vibration caused by dancers). AMMC-Int-1 and AMMC-Int-2, which have dendritic arborizations in the AMMC, can monitor the duration of the vibratory stimulation by these response patterns (β-δ in Fig. 4.3.5d). Thus the bee's auditory system has dual parallel pathways: vmSPS pathway and AMMC pathway. Such dual parallel pathways have already been revealed in the olfactory system in the bee brain [13, 20, 26] and also in the auditory system in cricket [17, 33]. These studies suggest that dual parallel processing is a common characteristic in the olfactory and auditory systems in insects.

References

1. Abel R, Rybak J, Menzel R (2001) Structure and response patterns of olfactory interneurons in the honeybee, *Apis mellifera*. J Comp Neurol 437(3):363–383
2. Ai H (2010) Vibration-processing interneurons in the honeybee brain. Front Syst Neurosci 3:19
3. Ai H, Nishino H, Itoh T (2007) Topographic organization of sensory afferents of Johnston's organ in the honeybee brain. J Comp Neurol 502(6):1030–1046
4. Ai H, Rybak J, Menzel R, Itoh T (2009) Response characteristics of vibration-sensitive interneurons related to Johnston's organ in the honeybee, *Apis mellifera*. J Comp Neurol 515(2):145–160
5. Brandt R, Rohlfing T, Rybak J, Krofczik S, Maye A, Westerhoff M, Hege HC, Menzel R (2005) Three-dimensional average-shape atlas of the honeybee brain and its applications. J Comp Neurol 492(1):1–19

6. Brockmann A, Robinson GE (2007) Central projections of sensory systems involved in honey bee dance language communication. Brain Behav Evol 70(2):125–136

7. Dreller C, Kirchner WH (1993) How honeybees perceive the information of the dance language. Naturwissenschaften 80:319–321

8. Erber J, Schildberger K (1980) Conditioning of an antennal reflex to visual stimuli in bees (*Apis mellifera* L.). J Comp Physiol 135:217–225

9. Erber J, Pribbenow B, Grandy K, Kierzek S (1997) Tactile motor learning in the antennal system of the honeybee (*Apis mellifera* L.). J Comp Physiol A 181:355–365

10. Erber J, Pribbenow B, Kisch J, Faensen D (2000) Operant conditioning of antennal muscle activity in the honey bee (*Apis mellifera* L.). J Comp Physiol A 186(6):557–565

11. Farina WM, Grüter C, Diaz PC (2005) Social learning of floral odours inside the honeybee hive. Proc R Soc B 272(1575):1923–1928

12. Frisch Kv (ed) (1967) The tail-wagging dance as a means of communication when food sources are distant. *The Dance Language and Orientation of Bees*, vol 1, 1st edn. Belknap Press of Harvard University Press, Cambridge

13. Galizia CG, Rössler W (2010) Parallel olfactory systems in insects: anatomy and function. Annu Rev Entomol 55:399–420

14. Goodman LJ, Fletcher WA, Guy RG, Mobbs PG, Pomfrett CDJ (eds) (1987) Motion sensitive descending interneurons, ocellar L_D neurons and neck motoneurons in the bee: a neural substrate for visual course control in *Apis mellifera.*, Neurobiology and behavior of honeybees, vol 1. Springer-Verlag, Berlin

15. Gopfert MC, Briegel H, Robert D (1999) Mosquito hearing: sound-induced antennal vibrations in male and female *Aedes aegypti*. J Exp Biol 202:2727–2738

16. Ignell R, Dekker T, Ghaninia M, Hansson BS (2005) Neuronal architecture of the mosquito deutocerebrum. J Comp Neurol 493(2):207–240

17. Imaizumi K, Pollack GS (2005) Central projections of auditory receptor neurons of crickets. J Comp Neurol 493(3):439–447

18. Kamikouchi A, Shimada T, Ito K (2006) Comprehensive classification of the auditory sensory projections in the brain of the fruit fly *Drosophila melanogaster*. J Comp Neurol 499(3):317–356

19. Kirchner WH, Lindauer M, Michelsen A (1988) Honeybee dance communication: Acoustical indication of direction in round dances. Naturwissenschaften 75:629–630

20. Kirschner S, Kleineidam CJ, Zube C, Rybak J, Grünewald B, Rössler W (2006) Dual olfactory pathway in the honeybee, *Apis mellifera*. J Comp Neurol 499(6):933–952

21. Maronde U (1991) Common projection areas of antennal and visual pathways in the honeybee brain, *Apis mellifera*. J Comp Neurol 309(3):328–340

22. McIver SB (ed) (1985) Mechanoreception. *Comprehensive insect physiology, biochemistry and pharmacology*, vol 6. Pergamon Press, Oxford

23. Menzel R, De Marco RJ, Greggers U (2006) Spatial memory, navigation and dance behaviour in *Apis mellifera*. J Comp Physiol A 192(9):889–903

24. Michelsen A (2003) Karl von Frisch lecture. Signals and flexibility in the dance communication of honeybees. J Comp Physiol A 189(3):165–174

25. Mobbs PG (1982) The brain of the honeybee *Apis mellifera*. I, The connections and spatial organization of the mushroom bodies. Philos Trans E Soc Lond B 298:309–354

26. Müller D, Abel R, Brandt R, Zöckler M, Menzel R (2002) Differential parallel processing of olfactory information in the honeybee, *Apis mellifera* L. J Comp Physiol A 188(5):359–370

27. Pareto A (1972) Spatial distribution of sensory antennal fibres in the central nervous system of worker bees. Z Zellforsch Mikrosk Anat 131(1):109–140

28. Pribbenow B, Erber J (1996) Modulation of antennal scanning in the honeybee by sucrose stimuli, serotonin, and octopamine: behavior and electrophysiology. Neurobiol Learn Mem 66(2):109–120

29. Riley JR, Greggers U, Smith AD, Reynolds DR, Menzel R (2005) The flight paths of honeybees recruited by the waggle dance. Nature 435(7039):205–207

30. Rohrseitz K, Tautz J (1999) Honey bee dance communication: waggle run direction coded in antennal contacts? J Comp Physiol A 184:463–470
31. Rybak J, Kuss A, Lamecker H, Zachow S, Hege HC, Lienhard M, Singer J, Neubert K, Menzel R (2010) The digital bee brain: integrating and managing neurons in a common 3D reference system. Front Syst Neurosci 4:30
32. Schäfer S, Bicker G (1986) Distribution of GABA-like immunoreactivity in the brain of the honeybee. J Comp Neurol 246(3):287–300
33. Schildberger K (1984) Temporal selectivity of identified auditory neurons in the cricket brain. J Comp Physiol A 155:171–185
34. Srinivasan MV, Zhang S (2004) Visual motor computations in insects. Annu Rev Neurosci 27:679–696
35. Suzuki H (1975) Antennal movements induced by odour and central projection of the antennal neurones in the honey-bee. J Insect Physiol 21:831–847
36. Thom C, Gilley DC, Hooper J, Esch HE (2007) The scent of the waggle dance. PLoS Biol 5(9):e228
37. Tsujiuchi S, Sivan-Loukianova E, Eberl DF, Kitagawa Y, Kadowaki T (2007) Dynamic range compression in the honey bee auditory system toward waggle dance sounds. PLoS One 2(2):e234

Chapter 4.4
Honey Bee Vision in Relation to Flower Patterns

Misha Vorobyev and Natalie Hempel de Ibarra

Abstract Bees form an important and representative group of insect pollinators. Because vision of the honey bee, *Apis mellifera*, has been studied in detail and its compound eyes are similar to those of other bees, the honey bee is a useful model for studying the evolutionarily relationship between flower displays and vision of bees. Three streams of research allow us to understand the relationship between flower displays and vision of pollinators: (i) optical, anatomical and physiological studies of the eye, (ii) behavioral studies of vision and (iii) analysis of multispectral images of flowers. The combination of these approaches allows us to apprehend flower perception by bees. This is achieved in two steps: (1) reconstruction of views of flowers as they are seen through insect eyes; (2) reconstruction of processing of these images by bees. This process allowed us to demonstrate that flower patterns have been evolutionarily adapted for being efficiently detected by bees.

4.4.1   Optics and Physiology of the Honey Bee Eye

4.4.1.1   Optics of the Eye

Bees have apposition compound eyes. In this type of eye, each ommatidium acts as an individual optical unit. The optical resolution of the apposition compound eye is determined by the angle between the optical axes of adjacent ommatidia,

M. Vorobyev (✉)
School of Optometry and Vision Science, University of Auckland,
Auckland, New Zealand
e-mail: m.vorobyev@auckland.ac.nz

N. Hempel de Ibarra
Centre for Research in Animal Behaviour, School of Psychology,
University of Exeter, Perry Road, Devon EX4 4QG, UK

C.G. Galizia et al. (eds.), *Honeybee Neurobiology and Behavior: A Tribute to Randolf Menzel*, DOI 10.1007/978-94-007-2099-2_22,
© Springer Science+Business Media B.V. 2012

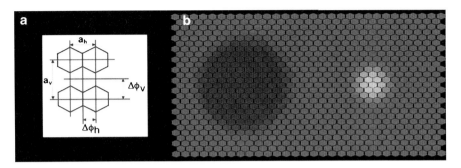

Fig. 4.4.1 Projection of two circular stimuli onto the frontal region of the honey bee compound eye. (**a**) Facet lens pattern, a_h and a_v are the primitive translation vectors in the horizontal and vertical direction respectively. $\Delta\Phi_H$ and $\Delta\Phi_V$ are the interommatidial angles in the horizontal and vertical directions respectively. (**b**) The stimuli subtend visual angles of 5° (*yellow*) and 15° (*violet*) at the achromatic and chromatic detection thresholds respectively. Relative excitations of ommatidia with respect to that of the ommatidium projecting onto the centre of the stimuli are shown: the stronger the coloration the higher the excitation of the ommatidia

$\Delta\Phi$, and by the acceptance angle of each ommatidium, $\Delta\rho$, which is defined as a half width of an ommatidium's directionality sensitivity (Fig. 4.4.1a). The angles between optical axes of adjacent ommatidia are inversely proportional to the radius of curvature of the eye, R, and can be calculated as: $\Delta\Phi = \Delta X/R$, where ΔX is the distance between centres of adjacent ommatidia. The lower limit of the acceptance angle can be estimated from the diffraction theory - in the diffraction limit the acceptance angle is proportional to the diameter of the opening aperture of an ommatidium, D, and can be estimated as: $\Delta\rho = \lambda/D$, where λ denotes wavelength of light. In the eye of the honey bee-worker, $\Delta\rho$ is only slightly greater than the value predicted from the diffraction limit [25, 26]. To adequately sample the optical image, the interommatidial angle, $\Delta\Phi$, must be similar to the ommatidium acceptance angle, $\Delta\rho$. Therefore a better optical resolution is achieved by increasing the size of ommatidia (decreasing of $\Delta\rho$) and the radius of curvature of an eye (decreasing $\Delta\Phi$), i.e. by increasing the size of the eye. A compromise between eye size and optical resolution is achieved by having good resolution in the frontal region of the eye and poor resolution elsewhere. The frontal eye of the honey bee worker can be approximated by an ellipsoid with the vertically oriented long axis and with ommatidia forming a nearly ideal hexagonal lattice. Values of interommatidial and acceptance angles are known: the interommatidial angles in vertical and horizontal directions are $\Delta\Phi_V = 0.9°$ and $\Delta\Phi_H = 1.6°$, respectively (Fig. 4.4.1) [24, 35]; also, according to electrophysiological measurements, $\Delta\rho = 2.6°$ [26]. The geometry of compound eyes is almost identical between individual honey bee workers [24, 35]. Therefore, based on the values presented above, an optical modeling of the honey bee eye is possible and gives useful predictions, which can be related with behavioral measures of visual resolution and vice versa.

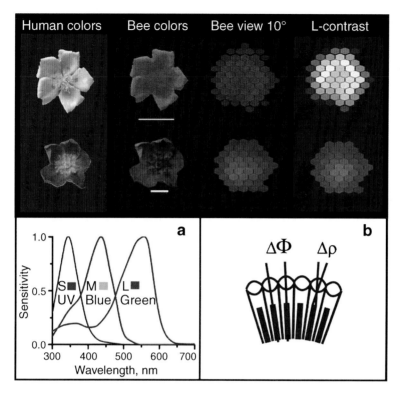

Fig. 4.4.2 Flowers as seen through the honey bee eyes. *Left panel* (scale 1 cm) shows displays (*human colors*) of a small flower (*Helianthemum nummularia, upper row*) and of a larger flower (*Rosa acicularis, lower row*) on *green* background. Spectral sensitivities of the S, M, and L receptors of honey bees (**a**) were used to calculate quantum catches from multispectral images. To show '*bee colors*', we used three primary colors of a computer monitor (*blue* for S, *green* for M and *red* for L). To show '*bee views*' for flowers subtending 10°, we projected the images onto ommatidial lattice of the honey bee (**b**). The right panel shows the distribution of L signals (brightness for bees) in ommatidia using the scale of grays. In a smaller flowers, *Helianthemum nummularia* (*upper row*), the center has lower L-receptor signal than the surround, i.e. it has a pattern that can be detected when it subtends an angle larger than 6.5°, which for this flower having a diameter of 1.7 cm corresponds to a distance of 15 cm. In a bigger flower, *Rosa pimpinellifolia* (*lower row*), the centre has a stronger L-receptor signal than the surround, i.e. it has a pattern that can be detected when it subtends an angle larger than 10°, which for this flower having a diameter of 4.2 cm corresponds to a distance of 24 cm (Reproduced from [14], Fig. 1)

4.4.1.2   Spectral Sensitivities of Honey Bee Photoreceptors

Honey bees have in their eyes three spectral types of photoreceptor cells, peaking in the UV, blue and green parts of the spectrum (*S,M,L* for short-, middle- and long-wavelength), with peak sensitivities at 344 nm (UV), 436 nm (blue) and 556 nm (green), respectively (Fig. 4.4.2) [1, 30, 31]. The study of the spectral sensitivities of a large number species of hymenopteran insects revealed a similarity between the location of spectral sensitivity peaks in a wavelength scale [34]. Therefore it is likely that color vision of hymenopteran pollinators is similar to color perception of the honey bee.

The absorption of visual pigments can be accurately modeled using standard templates [e.g. 11]. Therefore spectral sensitivity of vertebrates can be predicted from the results of molecular genetic studies of visual pigments. However, modeling of spectral sensitivity among invertebrates is a more challenging task. In the compound eyes of insects and crustaceans, rhabdomeric photoreceptors with different visual pigments are usually fused into a common light guide. This may lead to mutual filtering of light and narrowing of receptor spectral sensitivity [36]. In addition, the screening pigment may alter the spectral sensitivity.

4.4.1.3   Distribution of Photoreceptors in the Honey Bee Eye

In our eyes, each photoreceptor cell samples a different point in the space. Since the receptive fields for color should include different spectral types of photoreceptor cells, the random distribution of cones in the human retina leads to a poor spatial resolution of chromatic vision (review: [27]). However, because in compound eyes the rhabdomers are fused together into a common light guide, such eyes, theoretically allow for higher spatial resolution of color vision, by virtue of locating all spectral types of receptors within each ommatidium. Each ommatidium in the honey bee eye contains nine photoreceptor cells, R1-R9. Early studies of the honey bee eye suggested that each ommatidium contains all three spectral types of photoreceptor (UV, blue and green), indicating that insects have an eye design that optimizes the spatial resolution of color vision [12, 29]. This conclusion has been recently shown to be wrong through the discovery of a random arrangement of receptors with different spectral sensitivities in the honey bee [45]. Three spectral types of ommatidia have been found using *in situ* hybridization of the three mRNAs encoding the opsins of the UV-, blue- and green-absorbing visual pigments. Type I ommatidia contain one UV, one blue and six green receptors, type II ommatidia contain two UV and six green receptors, and type III ommatidia contain two blue and six green receptors; in all cases, the resolution of the labeling method was not enough to determine the opsin present in the ninth short photoreceptor cell [45].

4.4.1.4   Color and Polarization Vision

The rhabdomeric photoreceptors of arthropods and mollusks are inherently polarization-sensitive by virtue of their microvillar design [25]. Because a single polarization sensitive photoreceptor cannot distinguish between the orientation of polarization and the spectral composition of light, the neural signals caused by changes in polarization cannot be distinguished from those caused by the change of spectral composition of light. However, in insects, with a notable exception of a butterfly *Papilio augeus* [20], polarization vision is separated from color vision. This is achieved by twisting photoreceptors along their longitudinal axis in such a way that microvilli of individual rhabdomers are not aligned [46]. In this way, the twisting rhabdom cannot act as an analyzer of polarized light orientation. The honey bee eye is composed of twisted photoreceptors except for the uppermost dorsal rim of the eye, which is sensitive to polarization [47].

4.4.2 Modeling the Signals of Photoreceptors: Flowers as Seen Through the Eyes of the Honey Bee

The appearance of flowers as they are seen through the eyes of the honey bee can be reconstructed using a computer model of the honey bee eye (Figs. 4.4.2 and 4.4.3). The method is based on calculating the quantum catches of photoreceptors for each ommatidia using multispectral images of flowers [14, 41, 44].

Quantum catches, Q_i, of the honey bee receptors ($i = S, M, L$) were calculated for each point of the image as a linear combination of the camera signals, A_k:

$$Q_i = \sum_k C_{ik} A_k \tag{4.4.1}$$

where C_{ik} are the coefficients whose values depend on the spectral sensitivity of the honey bee photoreceptors (Fig. 4.4.2a), on statistics of flower spectra and on the illumination spectrum [41]. Illumination was measured using a S2000 spectrometer (Ocean Optics) or estimated from color temperatures of scales recorded through a broadband filter [41].

To show flowers as they are seen by bees from different distances, we projected flower images onto the compound eye of the honey bee (Figs. 4.4.1 and 4.4.2) [41]. The ommatidial lattice is described by the interommatidial angles $\Delta\Phi_H = 0.9°$ and $\Delta\Phi_V = 1.6°$ and by the acceptance angle of an ommatidium ($\Delta\rho = 2.6°$) [24, 26] (Fig. 4.4.1b). Φ_H and Φ_V are the angular coordinates relative to the centre of the visual field of the ommatidium in the vertical and horizontal directions.

To simulate the image as it is seen through the optics of the insect eye, we calculated the total quantum catch, Q^{om}, for each ommatidium of the lattice as:

$$Q^{om} = \iint A(\Phi_V, \Phi_H) Q(\Phi_V, \Phi_H) \cos(\beta) \, d\Phi_V \, d\Phi_H \tag{4.4.2}$$

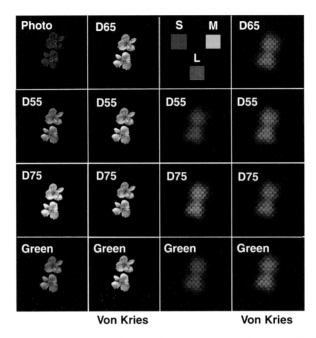

Fig. 4.4.3 Flowers of *Veronica chamaedrys* seen through the eye of a honey bee. *Left upper corner:* A colored photo-graph. Reconstructed images show the quantum catches of the S, M, and L receptors in false colors. *Left panels:* Flower as it can be seen from a very close distance. *Right panels:* A projection onto the ommatidium array 9 of flowers viewed from a distance of 8 cm. Each hexagon indicates an ommatidium. Illuminations are indicated in each panel. Compare the images obtained under the standard D65 illumination with those obtained at different illuminations before and after correction by von Kries transformation (Reproduced from [44], Fig. 4.4.1)

where Φ_H and Φ_V are the angular coordinates relative to the centre of the visual field of the ommatidium in vertical and horizontal directions. $Q(\Phi_V, \Phi_V)$ is the quantum catch as calculated for each point of the image, and β the angle with respect to the normal to the image plane. The angular sensitivity function, $A\left(\Phi_V, \Phi_H\right)$, of an ommatidium is given by Snyder et al. [36]:

$$A\left(\Phi_V, \Phi_H\right) = Exp\left[-2.77\left(\left(\Phi_V^2 + \Phi_H^2\right)/\Delta\rho^2\right)\right] \qquad (4.4.3)$$

4.4.3   Coding of Visual Information – Behavioral Studies

Color vision of the honey bee has been investigated in more details than that of any animal other than man. In color matching experiments Daumer [5] has shown that the honey bee has trichromatic vision. Analysis of behavioral spectral sensitivity and of wavelength discrimination has confirmed the trichromacy of the honey bee color vision [13]. Moreover, behavioral spectral sensitivity and wavelength discrimination can be quantitatively predicted from the spectral sensitivities of the three types of photoreceptors found in the honey bee eyes [2, 4, 30, 43] (for details of modeling see Appendix).

4.4.3.1   Color Constancy

Color constancy, the ability to perceive colors constantly in conditions of changing illumination, has been demonstrated in many animals including the honey bee [32, 48]. One of the first proposed models of color constancy, a von Kries transformation, assumes that signals of photoreceptors are scaled so that the color of illumination remains invariant. Such an algorithm can be implemented by receptor adaptation, and so invokes the simplest physiological mechanism as no special-purpose neural circuitry is required. Although it is not known which algorithm of color constancy animals use, the von Kries model yields predictions that agree with results of behavioral experiments [32, 48]. Von Kries transformation does not lead to perfect color constancy. However, this algorithm compensates well for the changes of color appearance (Fig. 4.4.3) [44].

4.4.3.2   Chromatic Vision

By definition, chromatic vision is sensitive to the changes in the spectral composition of light stimuli, but is not sensitive to the variations in intensity. In contrast, achromatic vision is sensitive to the variations in stimulus intensity. Chromatic vision is mediated by color opponent neurons computing differences between signals of photoreceptors of different spectral types. Achromatic vision is mediated by neurons that depend either on the signals of a single spectral type of photoreceptor, or sum signals of different photoreceptors types. Analysis of a large body of behavioural experiments suggested that the honey bee does not use the intensity of light stimuli for color discrimination [30]. This indicates that in honey bee color is coded by color opponent chromatic mechanisms. Electrophysiological recordings from the honey bee brain have revealed the presence of color opponent neurons [18, 22, 23]. These neurons have large receptive fields (ca 30°) and therefore their function may be related to chromatic adaptation and color constancy rather than to coding chromatic aspects of color *per se*.

To reveal the inputs of photoreceptors onto color opponent neurons, Backhaus [2] used multidimensional scaling of honey bee color choices recorded upon multiple-choice tests after single-color training [3]. He concluded that honey bees discriminate colors using two color opponent mechanisms, whose signals are combined to provide color distance using city-block metric. The directions of the mechanisms agree with those found in physiological recordings [2]. However, a subsequent analysis of color discrimination in the honey bee has shown that a variety of directions of the color-opponent mechanisms describe behavioral data equally well [4, 39], indicating that it is not possible to determine inputs of color opponent mechanisms from behavioral color discrimination experiments.

Color discrimination in the honey bee can be described by a receptor-noise limited color opponent model [40]. The only parameters of the model are the levels of noise in receptor mechanisms. The noise in the three types of photoreceptor cells of the honey bee has been measured electrophysiologically and the results of the model predictions have been compared with behavioral color thresholds [43]. The model describes the shape of behavioral spectral sensitivity of a honey bee [13] with a remarkable accuracy. The absolute thresholds are lower than those predicted from the measured values of noise in single photoreceptor cells, which indicates that signal-to-noise ratio is increased by summation of signals of single photoreceptor cells [43].

4.4.3.3   Behavioral Resolution of the Honey Bee Eye

The term resolution covers grating resolution and single object resolution. The limit for grating resolution is set by the sampling frequency of the mosaic of ommatidia, $v_s = 1/(2\Delta\Phi)$, where $\Delta\Phi$ is the interommatidial angle. Srinivasan and Lehrer [37] trained honey bees to distinguish horizontal and vertical grating at different distances. They found that the highest spatial frequency that the honey bees can reliably resolve was 0.26 cycles/deg with no evidence of a difference between horizontal and vertical gratings. This grating has half a period of 1.9°, which is very close to the limit set by interommatidial angles in the frontal eye of the honey bee ($\Delta\Phi_V = 0.9°$ and $\Delta\Phi_H = 1.6°$, [24]).

The ability of bees to detect and discriminate colored targets from a distance has been investigated by training honey bees to detect and discriminate vertically presented colored disks [6–10, 28] or disks surrounded by a ring of different color [15–17]. Bees detected and discriminated such stimuli using the high-resolution frontal part of their eye. The performance of bees has been assessed in terms of minimal angles that the stimuli must subtend in order to be detected or discriminated (limiting angle, α_{min}). The limiting angles fall into two categories – the targets that have contrast for the L (green) receptor can be detected until they subtend an angle of 5° while the targets that lack contrast to the L receptor can be detected when they subtend an angle of at least 15° (Fig. 4.4.1b) Bees cannot detect large stimuli (visual angle >15°) if these stimuli do not have chromatic contrast to the background [7]. Interestingly, the limiting visual angle does not depend on the magnitude of contrast to the background, as bees detect equally well targets that have different chromatic contrasts or L-receptor contrasts from at least as low as 1.3 (30% difference from background) to at least as high as 4.1 (310% difference from background) [8].

An implication of these findings is that bees detect stimuli using two largely separated visual pathways. The high resolution vision is mediated by the L-receptor alone and is analogous to the luminance pathway of humans. Therefore, in honey bees, the L receptor mediates achromatic vision and is often referred to as bright-ness pathway. The low resolution vision is mediated by chromatic visual pathways, which receive inputs from S, M and L photoreceptors. These pathways are used alternatively – bees discriminate stimuli subtending large angles using only chro-matic vision, while small stimuli are detected and discriminated using L receptor alone [6, 7, 9]. The smallest stimulus that can be detected by achromatic vision (5°) covers seven ommatidia on the frontal eye of the honey bee. The smallest stimulus that can be detected by chromatic vision alone (15°) covers 67 ommatidia [8]. Theoretically it is possible to detect and discriminate stimuli using the signal of only one ommatidium and honey bee drones can detect a queen (or a dummy) using a signal of only one ommatidium [38]. It is not clear why honey bee workers require signals from many ommatidia for detection and discrimination of stimuli. The involvement of many ommatidia may indicate complex visual processing of omma-tidial signals. This processing cannot be explained by linear summation of signals of different ommatidia because the limiting visual angle does not depend on the magnitude of contrast [7]. In the case of chromatic vision, the processing involves but is not restricted to summation of ommatidial signals because the signal-to-noise ratio of chromatic pathways is greater than that predicted from physiological record-ing [43]. The involvement of several ommatidia in the detection by achromatic vision may serve for identification of objects by resolving their boundaries and honey bees can probably assign a common border to groups of objects, as they detect groups of circles from further distance than individual circles [49]. Thus it is predicted that honey bees can detect patches of flowers from further distances than individual flowers.

The ability of bees to detect and discriminate circular patterns has been studied using stimuli presented vertically [16, 17]. A critical parameter determining the detect-ability of targets from long distance is the distribution of L-receptor contrasts within the target [16, 17]. The distribution of M- and S-receptor contrasts does not affect the detectability [8, 16]. A ring having strong L-receptor contrast (bright) around a disk with weak L-receptor contrast (dim) yields a detection limit of 6.5°, whereas a dim ring around a bright disk is detected only when it subtends more than 10° [16]. When viewed through the low-resolution eyes of bees, stimuli having a bright surround and dim centre have enhanced edges, whilst stimuli with dim surround and bright centre have blurred edges [16]. This impaired detectability of targets with blurred edges is a likely consequence of processing of visual information by neurons having centre-surround organization [7]. Such neurons are found in the visual pathways of many animals and are probably a general feature of neural coding in vision [19].

While the behavioral resolution of the frontal eye of the honey bee has been studied in detail, little is known about the resolution of other parts of the eye and it is unclear if the results obtained in the studies of frontal eye can be extrapolated to the ventral part of the eye [10]. Behavioral resolution of the ventral eye appears to be coarser than that of the frontal eye [10] which agrees with differences in optical resolution between these parts of the eye. It is possible that, in the ventral part of the eye, in addition to the L receptor, the M (blue) receptor is involved in high resolution vision [10]. This hypothesis converges with the finding that M receptors are more frequent in the anterior ventral region of the honey bee eye [45].

4.4.4   The Relation of Flower Patterns to Vision of the Honey Bee

Many bee-pollinated flowers are approximately radially symmetric with the color of the central part being different from the color of the surrounding part. Hempel de Ibarra and Vorobyev [14] reconstructed views of such flowers from multispectral images and calculated the signals from the S, M and L receptors of the honey bee eye when looking at the flowers' central (inner) and surrounding regions (outer) (Fig. 4.4.2). The majority of flowers had a dim center and a bright surround for L receptor, i.e. they had patterns that are easy to detect for bees (see above). The size of a flower is likely to be determined by a trade-off between the needs to increase flower detectability and the costs of increasing the display. Hence, flowers whose concentric patterns are difficult to detect (bright centre and dim surround) are likely to have larger diameters than those whose concentric patterns that are easy to detect (dim centre and bright surround). In accordance with this hypothesis, in the majority of plant orders, flowers whose centers are brighter than surrounds for the L-receptor type tend to have larger diameters than flowers with dim centers and brighter surrounds (Fig. 4.4.4).

Flower size is predicted to depend on the distribution of L contrast, but not on the distributions of S and M contrasts, because bees do not use S and M receptors to detect small targets from far away (see above). In addition, flower size is predicted to be independent of L, M and S signals both in the inner and outer parts of the flower, because detectability does not depend on the signals of receptors *per se* (see above). In accordance with this prediction, L, M and S signals both from the inner and outer parts of the flowers do not depend on flower diameter. The ratio of L-receptor signals for the inner to the outer part is strongly correlated with flower diameter; the correlation was weaker for the M receptor and inexistent in the case of the S receptor (Fig. 4.4.4). The correlation between M-receptor ratio and flower diameter is a likely consequence of the overlap between the spectral sensitivities of L and M receptors, which, in turn, leads to strong correlation of the M and L signals. The relationship between flower size and flower pattern can be explained as an adaptation to pollinator visual capacities because those aspects of flower patterns that do not affect the detectability are not correlated with flower diameter. This relationship cannot be explained by phylogenetic constraints because in the majority of plant orders widely separated in phylogeny, flowers with bright centers tend to have larger diameters.

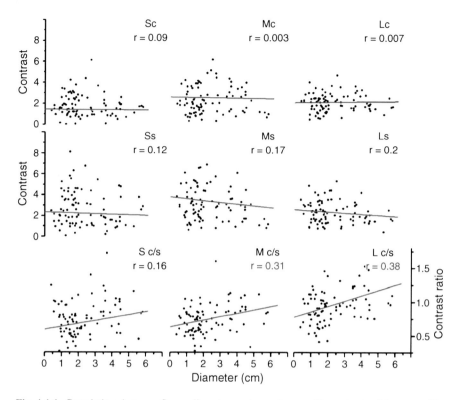

Fig. 4.4.4 Correlations between flower diameters and receptor-specific contrasts of the center (Sc, Mc Lc) and surround (Ss, Ms, Ls) regions of flower corollas, and the ratio of the contrasts of center to the surround (S c/s, M c/s, L c/s). Color symbols indicate plant orders. Patterns that are difficult to detect for bees (*bright center*) have Lc/s > 1, while patterns that are easy to detect (*dim center*) have Lc/s < 1. Note positive correlation for L c/s and absence of correlation for receptor specific contrasts (Reproduced from [14], Fig. 4.4.1)

4.4.5   Concluding Remarks and Outlook

Analysis of multispectral images of bee-pollinated flowers suggests that flower patterns have been evolutionarily adapted to pollinator's spatial vision. This conclusion has been reached on the basis of the assumption that processing of visual information in the honey bee is similar to that of other hymenopteran pollinators. Is this assumption justified? Can the honey bee be used as a model species that helps us to understand processing of visual information in many hymenopteran pollinators? Future studies of vision of many species of bees, including bumblebees and solitary bees, may help us to answer these questions.

While behavioral, physiological and optical studies of the honey bee vision allow us to relate spatial vision of the honey bee to achromatic aspects of flowers patterns, the relationship between color vision of the honey bee and chromatic aspects of flower patterns remains mysterious. We can accurately predict the ability of honey bees to discriminate colors [43], we have investigated the ability of the honey bees to detect and discriminate colorful patterns [16, 17] and we have collected and analyzed a large body of multispectral images of flowers [14], but we do not know what strategies do flowers use to combine colours in their patterns. Do specific combinations of flower colors attract pollinators better than a random combination of colors? If color combinations that attract pollinators best exist, can the existence of such color combinations be explained as an adaptation to color vision of bees? Note that bees can be attracted to certain color combinations simply because flowers with certain patterns offer a high-quality reward. These are the questions that we hope will be answered during the next 20 years.

Appendix: Quantitative Theory of Honey Bee Color Vision

Color can be described by photoreceptor quantum catches, Q_i, which are calculated as:

$$Q_i = C_i \int I^e(\lambda) R_i(\lambda) d\lambda, \tag{4.4.4}$$

where $i = S, M, L$ denotes the spectral type of a photoreceptor, λ is the wavelength, $I^e(\lambda)$ is the spectrum of the light entering the eye, $R_i(\lambda)$ is the relative spectral sensitivity of a receptor of a spectral type i, and C_i the scaling factors (constants) describing absolute sensitivity of photoreceptors. Quantum catches are measured as the number of absorbed photons per integration time. Because the illumination spectrum is given in number of photons/(nm m^2 radian^2 s^2), the units for scaling factors C_i are m^2 radian^2 s^2.

Of practical importance is the case when surface colors are considered. The spectrum of the light reflected from a surface with spectral reflectance, $S(\lambda)$, illuminated by a light source with spectrum $I(\lambda)$ equals to $I(\lambda)S(\lambda)$. Therefore, for a visual system viewing a surface with spectral reflectance $S(\lambda)$ illuminated by a light source with spectrum $I(\lambda)$, photoreceptor quantum catches are given by:

$$Q_i = C_i \int I^e(\lambda) S(\lambda) R_i(\lambda) d\lambda. \tag{4.4.5}$$

Instead of absolute quantum catches, often relative quantum catches, q_i, are used

$$q_i = \frac{Q_i}{Q^w_i}, \tag{4.4.6}$$

where, $Q^w_i = C_i \int I^e(\lambda) R_i(\lambda) d\lambda$ are quantum catches describing color of illumination, i.e. quantum catches corresponding to an ideal white surface. Note that relative quantum catches, q_i, do not depend on scaling factors, C_i. Relative quantum catches remain largely invariant in conditions of changing illumination and Eq. 4.4.6 describes von Kries color constancy.

Photoreceptors are biological light-measuring devices. For a light-measuring device with no internal noise, the error is determined solely by fluctuations of the number of absorbed photons per integration time. The number of absorbed photons has a Poisson distribution and therefore the fluctuations (standard deviation) of the number of absorbed quanta, δQ_i^R, is equal to the square root of the number of absorbed quanta (Rose – de Vries law):

$$\delta Q_i^R = \sqrt{Q_i} \qquad (4.4.7)$$

In addition to the fluctuations of the number of absorbed quanta, other sources of noise affect the accuracy of measuring the light [43]. At very low light levels, the dark noise, d_i, plays important role. This noise does not depend on the number of absorbed photons:

$$\delta Q_i^d = d_i. \qquad (4.4.8)$$

As the intensity of illumination increases the noise increases faster than Eq. 4.4.7 predicts. This noise can be described by Weber law, which states that the noise is proportional to the signal, i,e.

$$\delta Q_i^w = \omega_i Q_i, \qquad (4.4.9)$$

where, ω_i is equivalent to Weber fraction. Combination of the three sources of noise gives:

$$\delta Q_i = \sqrt{d^2 + Q_i + \omega_i^2 Q_i^2} \qquad (4.4.10)$$

The relative noise is defines as:

$$\frac{\delta Q_i}{Q_i} = \sqrt{\frac{d^2}{Q_i^2} + \frac{1}{Q_i} + \omega_i^2} \qquad (4.4.11)$$

The relative noise can be decreased by summation of signals of n photoreceptor cells by a factor $\sqrt{n}$.

The Weber law gives a fair approximation of the noise to signal ratio over a wide range of light intensities because photoreceptors adapt to changing illumination conditions (review: [33]). The Weber law implies that relative values of the receptor quantum catches, q_i, are sufficient to describe the noise in receptor channels.

When the Weber law is valid it is practical to describe photoreceptor signals, f_i, using logarithmic transformation of quantum catches

$$f_i = ln(q_i) = ln(Q_i) - ln(Q^w{}_i)$$ (4.4.12)

From Eqs. 4.4.9, 4.4.12 it follows that the variation of receptor signal so defined is equal to the Weber fraction because

$$\delta f_i = \frac{df_i}{dq_i} \delta q_i = \frac{d \ln(q_i)}{dq_i} \omega_i q_i = \frac{\omega_i q_i}{q_i} = \omega_i$$ (4.4.13)

Color can be represented as a point in color space where the separation of any two points in this space can be assigned a distance, ΔS. When the distance is less than a certain threshold distance, ΔS^t, the stimuli are not discernable. The receptor-noise limited color opponent model [40, 43] is based on the following assumptions.

1. In a color vision system with three spectral types of photoreceptors, color is coded by at least two unspecified color opponent mechanisms.
2. Color opponent mechanisms are not sensitive to changes in the stimulus intensity.
3. The sensitivity of these mechanisms is set by noise originating in photoreceptors.

Let f_i be receptor signals scaled so that the increase of the intensity of a light stimulus adds the same value to all receptor signals (e.g. $f_i = ln(q_i)$, see Eq. 4.4.12) and ω_i be the noise in the receptor mechanisms, Δf_i be the difference in receptor signals between two stimuli, then the distance between two color stimuli can be calculated as:

$$\Delta S = \sqrt{\frac{\omega_S^2(\Delta f_L - \Delta f_M)^2 + \omega_M^2(\Delta f_L - \Delta f_S)^2 + \omega_L^2(\Delta f_M - \Delta f_S)^2}{\omega_L^2\omega_M^2 + \omega_L^2\omega_S^2 + \omega_M^2\omega_S^2}}$$ (4.4.14)

Note that when Weber law applies all ω_i remain constant. Electrophysiological measurements of the signal-to-noise ratio in the honey bee photoreceptors give the following results: $\omega_S = 0.13$, $\omega_M = 0.06$, $\omega_L = 0.12$ [43]. Predictions based on Eq. 4.4.14 are in excellent agreement with behavioural data [43].

Equation 4.4.14 describes distance in chromatic plane. The following transformation of receptor signals allows us to construct an equal distance color space, where Euclidean distance corresponds to the distance given by Eq. 4.4.14 [21]:

$$X_1 = A(f_L - f_M)$$
$$X_2 = B(f_S - (af_L + bf_M)),$$ (4.4.15)

Where

$$A = \sqrt{\frac{1}{\omega_L^2 + \omega_M^2}}$$

$$B = \sqrt{\frac{\omega_L^2 + \omega_M^2}{\omega_L^2\omega_M^2 + \omega_L^2\omega_S^2 + \omega_M^2\omega_S^2}}$$

$$a = \frac{\omega_M^2}{\omega_L^2 + \omega_M^2}$$

$$b = \frac{\omega_L^2}{\omega_L^2 + \omega_M^2}$$

The first theory of color vision has been proposed by Backhaus [2] on the basis of multidimensional scaling of the honey bee color choices. He assumed that receptor signals (E_i) can be described as:

$$E_i = \frac{q_i}{1+q_i} \tag{4.4.16}$$

Coding is performed by two color opponent mechanisms termed A and B, whose outputs are calculated as:

$$A = -9.86E_s + 7.70E_b + 2.16E_L$$
$$B = -5.17E_s + 20.25E_b - 15.08E_L \tag{4.4.17}$$

The axes corresponding to the directions A and B are assumed to define a 'perceptual color' space of the honey bee ('color opponent coding space', [2]). The distance in this space is calculated using city-block metric as:

$$\Delta S = |A| + |B| \tag{4.4.18}$$

The model describes a large body of behavioural data [39]. It is important to note that the proposed 'perceptual color space' does not correspond to a chromatic diagram, i.e. the position of a light stimulus in this space depends on its intensity. Therefore the model incorrectly predicts that color discrimination deteriorates as the intensity o light stimuli increases [15, 42]. In particular, the model predicts that bees are not able to discriminate bright UV reflecting white flowers, from green leaves. This prediction has been shown to be incorrect, as bees can easily discriminate bright UV reflecting white from leaf-like green [15, 42].

References

1. Autrum H, Zwehl V (1964) Die spektrale Empfindlichkeit einzelner Sehzellen des Bienenauges. J Comp Physiol A 48(4):357–384
2. Backhaus W (1991) Color opponent coding in the visual system of the honeybee. Vision Res 31(7–8):1381–1397
3. Backhaus W, Menzel R, Kreißl S (1987) Multidimensional scaling of color similarity in bees. Biol Cybern 56(5):293–304
4. Brandt R, Vorobyev M (1997) Metric analysis of threshold spectral sensitivity in the honeybee. Vision Res 37(4):425–439
5. Daumer K (1956) Reizmetrische Untersuchung des Farbensehens der Bienen. J Comp Physiol A 38(5):413–478
6. Giurfa M, Vorobyev M (1997) The detection and recognition of colour stimuli by honeybees: performance and mechanisms. Israel J Plant Sci 45:129–140
7. Giurfa M, Vorobyev M (1998) The angular range of achromatic target detection by honey bees. J Comp Physiol A 183(1):101–110
8. Giurfa M, Vorobyev M, Kevan P, Menzel R (1996) Detection of coloured stimuli by honeybees: minimum visual angles and receptor specific contrasts. J Comp Physiol A 178(5):699–709
9. Giurfa M, Vorobyev M, Brandt R, Posner B, Menzel R (1997) Discrimination of coloured stimuli by honeybees: alternative use of achromatic and chromatic signals. J Comp Physiol A 180(3):235–243
10. Giurfa M, Zaccardi G, Vorobyev M (1999) How bees detect coloured targets using different regions of their compound eyes. J Comp Physiol A 185(6):591–600
11. Govardovskii VI, Fyhrquist N, Reuter T, Kuzmin DG, Donner K (2000) In search of the visual pigment template. Vis Neurosci 17(4):509–528
12. Gribakin FG (1969) Cellular basis of colour vision in the honey bee. Nature 223(5206):639–641
13. Helversen O (1972) Zur spektralen Unterschiedsempfindlichkeit der Honigbiene. J Comp Physiol A 80(4):439–472
14. Hempel de Ibarra N, Vorobyev M (2009) Flower patterns are adapted for detection by bees. J Comp Physiol A 195(3):319–323
15. Hempel de Ibarra N, Vorobyev M, Brandt R, Giurfa M (2000) Detection of bright and dim colours by honeybees. J Exp Biol 203(21):3289–3298
16. Hempel de Ibarra N, Giurfa M, Vorobyev M (2001) Detection of coloured patterns by honeybees through chromatic and achromatic cues. J Comp Physiol A 187(3):215–224
17. Hempel de Ibarra N, Giurfa M, Vorobyev M (2002) Discrimination of coloured patterns by honeybees through chromatic and achromatic cues. J Comp Physiol A 188(7):503–512
18. Hertel H (1980) Chromatic properties of identified interneurons in the optic lobes of the bee. J Comp Physiol A 137(3):215–231
19. Hubel D (1988) Eye, brain and vision. Scientific American Library. W.H. Freeman, New York
20. Kelber A (1999) Why 'false' colours are seen by butterflies. Nature 402(6759):251
21. Kelber A, Vorobyev M, Osorio D (2003) Animal colour vision–behavioural tests and physiological concepts. Biol Rev Camb Philos Soc 78(1):81–118
22. Kien J, Menzel R (1977) Chromatic properties of interneurons in the optic lobes of the bee I. Broad band neurons. J Comp Physiol A 113(1):17–34
23. Kien J, Menzel R (1977) Chromatic properties of interneurons in the optic lobes of the bee II. Narrow band and colour opponent neurons. J Comp Physiol A 113(1):35–53
24. Kirschfeld K (1973) Optomotorische Reaktionen der Biene auf bewegte "Polarisations-Muster". Z Naturforsch 28c:329–338
25. Land M, Nilsson D-E (2002) Animal eyes. Oxford University Press, Oxford
26. Laughlin SB, Horridge GA (1971) Angular sensitivity of the retinula cells of dark-adapted worker bee. J Comp Physiol A 74(3):329–335
27. Lee BB (2004) Paths to colour in the retina. Clin Exp Optom 87(4–5):239–248

28. Lehrer M, Bischof S (1995) Detection of model flowers by honeybees: The role of chromatic and achromatic contrast. Naturwissenschaften 82(3):145–147
29. Menzel R (1979) Spectral sensitivity and color vision in invertebrates. In: Autrum H (ed) Handbook of sensory physiology, vol VII, 6c. Springer, Berlin, pp 503–580
30. Menzel R, Backhaus W (1991) Colour vision in insects. In: Gouras P (ed) Vision and visual disfunction. Macmillan, London, pp 262–288
31. Menzel R, Blakers M (1976) Colour receptors in the bee eye—morphology and spectral sensitivity. J Comp Physiol A 108(1):11–13
32. Neumeyer C (1981) Chromatic adaptation in the honeybee: successive color contrast and color constancy. J Comp Physiol A 144(4):543–553
33. Osorio D, Vorobyev M (2005) Photoreceptor spectral sensitivities in terrestrial animals: adaptations for luminance and colour vision. Proc R Soc B 272(1574):1745–1752
34. Peitsch D, Fietz A, Hertel H, de Souza J, Ventura DF et al (1992) The spectral input systems of hymenopteran insects and their receptor-based colour vision. J Comp Physiol A 170(1):23–40
35. Seidl R (1980) Die Sehfelder und Ommatidien-Divergenzwinkel der drei Kasten der Honigbiene (*Apis mellifica*). Verh Dtsch Zool Ges 1980:367
36. Snyder AW, Menzel R, Laughlin SB (1973) Structure and function of the fused rhabdom. J Comp Physiol A 87(2):99–135
37. Srinivasan MV, Lehrer M (1988) Spatial acuity of honeybee vision and its spectral properties. J Comp Physiol A 162(2):159–172
38. Vallet AM, Coles JA (1991) A method for estimating the minimum visual stimulus that evokes a behavioural response in the drone, *Apis mellifera male*. Vision Res 31(7–8):1453–1455
39. Vorobyev M, Brandt R (1997) How do insect pollinators discriminate colours. Israel J Plant Sci 45:103–114
40. Vorobyev M, Osorio D (1998) Receptor noise as a determinant of colour thresholds. Proc R Soc B 265(1394):351–358
41. Vorobyev M, Gumbert A, Kunze J, Giurfa M, Menzel R (1997) Flowers through the insect eyes. Israel J Plant Sci 45:93–102
42. Vorobyev M, Hempel de Ibarra N, Brandt R, Giurfa M (1999) Do "white" and "green" look the same to a bee? Naturwissenschaften 86(12):592–594
43. Vorobyev M, Brandt R, Peitsch D, Laughlin SB, Menzel R (2001) Colour thresholds and receptor noise: behaviour and physiology compared. Vision Res 41(5):639–653
44. Vorobyev M, Marshall J, Osorio D, Hempel de Ibarra N, Menzel R (2001) Colourful objects through animal eyes. Color Res Appl 26:S214–216
45. Wakakuwa M, Kurasawa M, Giurfa M, Arikawa K (2005) Spectral heterogeneity of honeybee ommatidia. Naturwissenschaften 92(10):464–467
46. Wehner R, Bernard GD (1993) Photoreceptor twist: a solution to the false-color problem. Proc Natl Acad Sci USA 90(9):4132–4135
47. Wehner R, Bernard GD, Geiger E (1975) Twisted and non-twisted rhabdoms and their significance for polarization detection in bee. J Comp Physiol A 104(2):225–245
48. Werner A, Menzel R, Wehrhahn C (1988) Color constancy in the honeybee. J Neurosci 8(1):156–159
49. Wertlen AM, Niggebrügge C, Vorobyev M, Hempel de Ibarra N (2008) Detection of patches of coloured discs by bees. J Exp Biol 211(Pt 13):2101–2104

Chapter 4.5
Psychophysics of Honey Bee Color Processing in Complex Environments

Adrian G. Dyer

Abstract Psychophysics examines the relationship between test stimuli specified in physical terms and the behavioral responses of animals evoked by these stimuli. This chapter explores the psychophysics of how color information is processed by free flying honey bees, and how individual bees exhibit remarkable behavioral flexibility depending upon the type of conditioning procedure applied during training. Specifically, honey bees that learn color information in isolation, which is termed absolute conditioning, only demonstrate a coarse level of color discrimination. In contrast, bees that learn a target color in the context of perceptually similar distractor stimuli, termed differential conditioning, can learn to make relatively fine color discriminations. However, with decreasing color separation between target and distractor stimuli there is a soft sigmoidal function describing honey bee discrimination, which explains why flowers have evolved saliently different colors so as to minimise perceptual errors by bee pollinators. For perceptually difficult color tasks bees trade-off speed for accuracy, suggesting that studies that wish to link the behavior of individual bees to physiological mechanisms need to control for this factor. Indeed, if honey bees are provided with differential conditioning where distractor choices are penalized, then there is a significant improvement in the frequency of accurate choices. The chapter thus underlines how very careful conditioning techniques are required for future work to link psychophysics testing of free flying honey bees to the physiological mechanisms that facilitate color perception.

A.G. Dyer (✉)
Department of Physiology, Monash University, Clayton 3800, Vic, Australia

Media and Communication, RMIT University, 124 La Trobe Street,
Melbourne 3000, Vic, Australia
e-mail: adrian.dyer@rmit.edu.au

C.G. Galizia et al. (eds.), *Honeybee Neurobiology and Behavior: A Tribute to Randolf Menzel*, DOI 10.1007/978-94-007-2099-2_23,
© Springer Science+Business Media B.V. 2012

4.5.1 Introduction

Color perception is a construct of the brain [34]. Once we appreciate this fact it is possible to understand why different species may perceive the same object as being of a different color appearance [30]. Color vision is a process of sampling parts of the electromagnetic spectrum, and for animal vision the range of radiations that may be sampled extends from about 300 to 700 nm. For example, human trichromatic color vision is based on three cone photoreceptor types maximally sensitive to blue (420 nm), green (534 nm) and red (564 nm) radiation [4], and humans do not typically see ultraviolet radiation (300–400 nm) [13].

The spectral sensitivities of photoreceptors for a number of different hymenopteran insects have been measured [39], and the sensitivity of the different receptor classes show a high degree of coincidence in terms of the wavelengths at which they peak [8, 12]. The sensitivities shown in Fig. 4.5.1 describe the probability with which the honey bee photoreceptors absorb photons of particular wavelengths. The spectral positions of the respective photoreceptors along the wavelength scale is thus important because the brain creates the perception of color by comparing the signals from the different receptors [3, 34]. The physiological basis for such comparison is provided by color opponency, a neural mechanism by which the input of different receptor types is antagonized (i.e. generates different excitatory vs. inhibitory responses) at the level of color opponent neurons. Two main classes of color opponent neurons were initially reported for the bee brain [31], but recent work shows that there might be up to ten different types of color opponent neurons in the brain [48]. Currently it is not clear how all these different types of neurons act to enable color perception, or the possible effects that learning phenomena may have on the perceptual capabilities driven by color opponency [21]. This leads to the following important questions; is color perception an invariant capability resulting from the neural machinery involved in the processing of colors, or does it allow flexible modulations based on individual experience? This current study thus concentrates on the discussion of color learning of free flying honey bees from a psychophysics approach, and aims to answer the previous questions based on results from several recent studies.

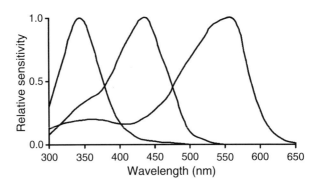

Fig. 4.5.1 Spectral sensitivity of honey bee photoreceptors (Data from [40]) normalised to 1.0. Similar sets of photoreceptors are found in most hymenopteran trichromats [8, 40]

4.5.2 Color Learning in Bees

The first demonstration of color discrimination in the honey bee used a training method of rewarding bees with sucrose solution for landing on colored surfaces [45] (see Chap. 6.6). The first quantifications on color acquisition curves for colors paired with sugar solution showed that some wavelengths may be learned faster than others [36]. These experiments were performed by rewarding the experimental bees on a monochromatic light of given wavelength and testing it afterwards with that light presented against an alternative one. The training was therefore a case of absolute conditioning (see Chap. 6.6), where the test presented a discrimination situation that was not available during the training [36]. Under these test conditions it was shown that variations in color acquisition depended exclusively on the wavelength that was rewarded and not on the alternative wavelength presented during the tests [36]. A notable point about this type of color learning is that the information is learnt very quickly, typically in one to eight learning trials [36, 37]. In particular, it was shown that wavelengths that appear violet to humans (413–428 nm) determine higher response levels in a subsequent color presentation compared to other wavelengths of the bee spectrum. A secondary peak of correct responses could also be found around 530 nm. These findings were interpreted as the reflect of innate color preferences, which were later demonstrated by testing color choices of naïve honey bee foragers in their first foraging flights [25]. The findings were also suggestive that honey bees may generalize stimuli that share a dimension of perceptual similarity, which has since been demonstrated for both honey bee spatial [47] and color vision [17], as well as for bumblebee color vision [26].

4.5.3 The Importance of Absolute and Differential Conditioning

In natural foraging environments individual bees may encounter a range of different scenarios for learning color information. For example, in scenario (i) most flowers of a particular color may present a nutritional reward like nectar, whilst in scenario (ii) only some flowers of a particular may color contain a reward, as non-rewarding flowers mimicking the model flower may also be present. A biologically relevant example of scenario (ii) is termed Batesian mimicry by some flowering plants [40]. Does color learning depend upon the context of a particular scenario? In honey bees this question was answered by Martin Giurfa by training individual honey bees to enter a Y-maze apparatus (see Chap. 6.6) and choose the arm containing a rewarding color, whilst the alternative arm of the Y-maze only presented a non-rewarded neutral grey background color. This models scenario (i) and is called absolute conditioning as the 'target' color is learnt in isolation (the neutral background color is saliently different). In this experiment phase bees rapidly learn the target color (Fig. 4.5.2a) and can very reliably discriminate it in tests from a perceptually dissimilar color, however, a perceptually similar color stimulus is not discriminated at a level significant from chance expectation (Fig. 4.5.2b). One possibility for the bees not

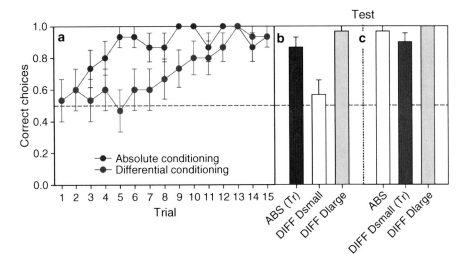

Fig. 4.5.2 Performance of honey bees trained with either absolute or differential conditioning. A perceptually small color difference (DIFF Dsmall) is only learnt with differential conditioning. (**a**) Acquisition (mean ± SE; n = 15 bees for each curve) with *red circles* for absolute conditioning and *blue circles* for differential conditioning. (**b**) Performance in the tests of the group trained in absolute conditioning. ABS(Tr) test which presented the target versus *grey background colors* used in the trained situation. DIFF Dsmall test presented the trained 'target' stimulus versus a perceptually similar color. DIFF Dlarge test presented the 'target' and a perceptually different color. (**c**) Performance in the tests of the group trained with differential conditioning. ABS test presenting the target color versus the *gray background color*. DIFF Dsmall(Tr) test presented the trained 'target' stimulus versus a perceptually similar color. DIFF Dlarge test presented the 'target' and a perceptually different color (Data from [2, 23])

discriminating the perceptually similar colors is that these stimuli may be below the discrimination threshold of the honey bee visual system. However, in an experiment phase where one arm of the Y-maze contains a rewarded target color and the alternative arm contains a non-rewarded perceptually similar color (Fig. 4.5.2a), honey bees slowly learn to discriminate between the colors. This is called differential conditioning (see Chap. 6.6), and essentially models scenario ii described above. An important difference between absolute conditioning and differential conditioning procedures is that the acquisition rate is significantly different in both cases; absolute conditioning is fast, whilst differential conditioning is slow [21, 23]. It was hypothesized that the relatively fine color discrimination observed with differential conditioning may be a result of the development of selective attention mechanisms in the honey bee brain [23], although empirical evidence to test this hypothesis is still outstanding and would be a high value topic for future research.

The finding that absolute- or differential-conditioning results in dramatically different levels of color learning in honey bees allows the separation of the possibilities that either (a) the perceived difference between two color stimuli is an immutable property that is constrained by the visual machinery of a honey bee, or

(b) visual discrimination is not an absolute phenomenon constrained by visual machinery and color learning can be modulated by individual experience. The data clearly show that individual experience is critical to color learning, and is not entirely constrained by visual machinery [21, 23].

The finding that absolute- or differential- conditioning is important for color learning has been confirmed in several other studies in honey bees [17], bumble-bees [15, 16], ants [7] and moths [29], suggesting that this is a widespread phenom-enon in color perception by insects. Thus it can be concluded that individual experience is very important for how visual systems in insects learn to use color information, which in turn has important implications for understanding angio-sperm pollination depending upon both the type and level of experience of a par-ticular insect [16, 21].

4.5.4   The Importance of Aversive Conditioning

Now that we can see that differential conditioning is important to how honey bees learn color information, another consideration has to be addressed to understand how color is learnt by bees. Recent work has shown that for honey bees aversive reinforcement of stimuli can promote learning in its own right [44], and previous work on bumblebees indicated that a combination of both an appetitive sucrose solution with a target conditioned stimulus (CS+) and an aversive or bitter tasting quinine solution with the distractor stimulus (CS−) can significantly improve color discrimination [9]. The question of whether aversive reinforcement improves color discrimination in free flying honey bees was recently addressed by training separate groups of honey bees in a Y-maze (see Chap. 6.6) to learn a fine or coarse color discrimination task [1]. The type of task was quantified by plotting the loci of stim-uli colors in two different models of bee color space and confirmed that one color task was coarse and described by a large color difference, whilst the other color task was fine and could be described by a small color difference [1]. For bees that were trained to a coarse color difference there was no significant difference in learning performance depending upon whether or not an aversive quinine solution was asso-ciated with the distractor stimuli (Fig. 4.5.3a, b). Importantly, this result shows why many previous studies on bee color discrimination were able to demonstrate a certain level of color learning with the classical appetitive conditioning technique [3, 36, 37]. However, for bees trained to discriminate a fine color difference there was a significantly different result; bees provided with aversive reinforcement on the distractor stimulus could learn the fine color discrimination task, whilst bees for which the distractor only contained plain water, there was no evidence of color learning (Fig. 4.5.3c, d). The mechanism for how a quinine solution could influence bee choices as an aversive agent could be due to either a post-ingestional malaise or a distasteful sensory experience. Control experiments showed that bees could not remotely sense quinine via olfaction as they sampled sucrose solution, water or quinine with equal frequency [1]. In addition, honey bees frequently drank 2μL

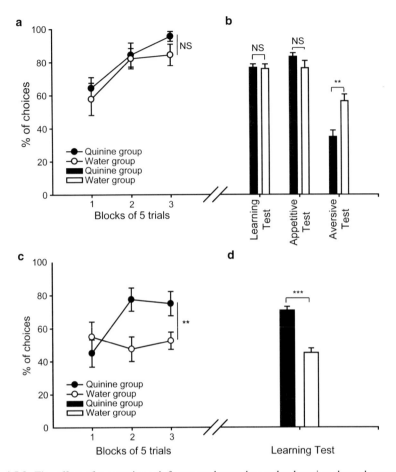

Fig. 4.5.3 The effect of a negative reinforcer on honey bee color learning depends upon the perceptual similarity of target and distractor stimuli. (**a**) Learning acquisition of dissimilar color stimuli [correct choices (%) in blocks of 5 trials; means ± s.e.m.; N = 9 honey bees for each curve. The curve with black dots represents acquisition by the quinine group (CS– reinforced with quinine); the curve with white dots represents acquisition by the water group (CS– reinforced with water)]. (**b**) Performance (means + s.e.m. of percentages of 'CS+' choices ('learning' and 'appetitive' test) or CS– choices ('aversive' test); N = 9 for each bar) in non-rewarded tests. Black bars represent the results of the quinine group; white bars represent the results of the water group. The learning performance in this easy color discrimination task was not significantly different between test groups. However, bees from the quinine groups avoided the stimulus (CS–) that was associated with quinine during training when presented versus a neutral stimulus; this avoidance was significantly different to the water group (**: p < 0.005) (**c**) Learning acquisition for similar color stimuli (correct choices (%) in blocks of 5 trials; means ± s.e.m.; N = 8 for each curve). The curve with black dots represents acquisition by the quinine group (CS– reinforced with quinine); the curve with white dots represents acquisition by the water group (CS– reinforced with water) and is significantly different. (**d**) Performance [means + s.e.m. of percentages of CS+ choices ('learning' and 'appetitive' test) or CS– choices ('aversive' test); N = 8 for each bar] in non-rewarded tests. Black bars represent the results of the quinine group; white bars represent the results of the water group, and only bees from the quinine group solved this difficult discrimination task (**: p < 0.005; ***: p < 0.001) (Modified from data in [1])

droplets of water, but rarely consumed 1µL droplets of quinine solution, and thus the improvement in visual learning was due to a distasteful sensory experience [1].

The data on how aversive reinforcement improves color discrimination [1] thus strongly support the finding that that individual experience is very important for how visual systems in honey bees [23] or bumblebees [9, 16] learn to use color information. A mechanism by which this probably improves learning is that color information is available from previous positive and negative reinforced experiences, which applies to CS+ and CS− information, respectively. Thus when a bee is conditioned to the CS+ or the CS−, it retrieves the appropriate memories associated to these particular stimuli, and the newly acquired information is then updated with respect to previously experienced (see Chap. 6.3). Interaction between excitatory memory traces (derived from experiences with the CS+) and inhibitory memory traces (derived from experiences with the CS−) will determine the current performance of individual forager bees.

4.5.5   How Does Color Discrimination Vary with the Perceptual Similarity of Stimuli?

Since honey bees often have to use their color vision in complex natural environments where there may be a wide variety of plant species, each plant potentially having flowers that are more or less different from a given 'target' color, an important question is what type of color discrimination function describes the probability with which individual bees can reliably discriminate a flower? This is an important question because individual bees tend to be flower constant [11], which means that they remain truthful to a flower species as long as its reward (nectar or pollen) remains profitable, and thus the color choices by individual bees are potentially an important driver for what flower colors are successful in nature [14].

It was hypothesized that the function describing how perceptually similar color are discriminated by honey bees would be a soft sigmoidal type function [27], although theoretical analyses using computational modeling of a bee's ability to discriminate flowers have suggested that there is little or no evolutionary pressure for plant flowers to diverge in their color signals [46].

To decide between these two alternative hypotheses, it was possible to provide individual honey bees with differential conditioning and aversive reinforcement to a range of color stimuli that were more, or less similar in color [18]. Individual honey bees were trained on a flat surface to initially discriminate a relatively coarse color difference, which was learnt very quickly, then in small progressive steps the bees were trained to discriminate the same target color from perceptually more similar distractors (Fig. 4.5.4)[18]. Two types of viewing conditions were considered; in the first viewing condition simultaneous color discrimination was tested by making a bee discriminate a five pointed star shape on a background color versus a distractor that only presented the homogeneous background color (Fig. 4.5.4; insert). The reason why this can be described as simultaneous color discrimination is that to

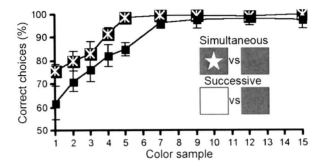

Fig. 4.5.4 The frequency of bees correctly choosing a target color depends upon the color distance between the target color (specified by color number) and the distractor color. There is strong correlation between color sample steps and their perceptual color distance in color space for bee vision. Error bars show ±1.0 s.e.m. Insert shows an example of simultaneous and successive discrimination tasks (Modified from [18]; perceptual color distances in a variety of color spaces for the respective color samples available in the original manuscript)

discriminate the star pattern is has to be seen at exactly the same point in time as the background color. Whilst it is not currently known what the initial temporal decay of color memory is for honey bees when a stimulus is removed from the visual field; for human vision the time course for deterioration in color discrimination ability considering successive viewing conditions is as short as 60 ms after a stimulus has been removed from the visual field [43]. Thus, for human vision colors must be seen side by side for a simultaneous viewing condition [43]. In a second viewing condition bees, were trained using successive color discrimination simply by using two types of different homogeneous color stimuli where no pattern information was available (Fig. 4.5.4 insert) [18]. Thus, unless a bee is presented with a color edge or pattern for enabling simultaneous color discrimination [18], only successive color discrimination is tested. This includes Y-maze experiments where bees often make small saccadic turns to alternatively look at homogeneous color stimuli in the respective arms of the apparatus [1].

The shape of the discrimination function for both simultaneous and successive viewing conditions shows that for more similar colors, an increasing number of errors are made by honey bees choosing the correct color stimuli (Fig. 4.5.4). The data thus clearly supports the hypothesis of a soft sigmoidal function describing honey bee color discrimination [27]. This finding is also supported by earlier studies on honey bees that only used appetitive conditioning [24, 33], and so is likely to be a general principle describing how honey bees use color vision to find similar flower colors in nature. These studies thus explain why there has been significant evolutionary pressure on plants to evolve saliently different flower colors. Salient colors minimize perceptual errors by important pollinators like honey bees, so that pollen can be reliably transferred between conspecific flowering plants [14]. Two other important points can be drawn from this study (i) simultaneous color discrimination is significantly better than successive color discrimination in honey bees (Fig. 4.5.4), which is consistent with studies on human color vision [32, 43]; and (ii) the level of color resolution that the honey bee visual system can resolve for bees trained with differential conditioning and aversive reinforcement is close to the level of color discrimination possible by the color processing of trichromatic primates, including humans [18]. Specifically, for a region of the visual spectrum where both human and bee vision sample light, bees discriminated between color stimuli equivalent to one just-noticeable-difference [35] when the stimuli were quantified according to color models for human vision [18].

4.5.6   Speed-Accuracy and Color Choices in Honey Bees

Since honey bees make an increasing rate of errors for perceptually more similar colors (Fig. 4.5.4), and fine color resolution may take a long time to learn as it requires differential conditioning [21, 23], an important question is what level of color discrimination accuracy might bees use in a complex type situation where there are both rewarding and non-rewarding flowers of a similar color appearance. This question can be formulated in terms of a speed-accuracy trade-off [41], where the frequency of correct choices is evaluated relative to the amount of time required to make a decision. In honey bee foragers that are exclusively collecting nectar (rather than involved in recruitment), this question has been addressed by training individual honey bees with differential conditioning to perceptually similar colors, and then in test conditions varying the ratio of targets and distractors available for a bee to land on [6]. In condition one there was a high target frequency (ratio 1 target: 1 distractor), in condition two there was a low target frequency (ratio 1 target: 2 distractors). Bees in both test conditions that made faster decisions, also made more errors. But why do some bees choose to perform at a low level of accuracy [5]? When there are many targets then making fast decisions with a low level of accuracy can be optimal for collecting the most nutrition per unit time, however, when there are relatively few targets, then sacrificing some decision time to make accurate choices actually collects more nutrition per unit time by avoiding the time cost of landing on incorrect distractor colors [6]. This shows that to understand the decisions made by honey bees when using color cues, it is often necessary to consider the context within which the decisions actually are made [6, 38]. Interestingly, speed-accuracy trade-offs have also been recently reported for several other insect species including bumblebees [9, 15, 28] and ants [22, 42], so this may be a common aspect of decision making in invertebrates and other animals [10].

For experimenters conducting psychophysics testing to understand the building blocks of honey bee visual behavior, speed-accuracy tradeoffs are a potential confounding factor that has to be controlled for. One realistic way to control for behavioral variability is to use aversive reinforcement to make sure that bees participating in behavioral experiments are highly motivated and 'pay attention' to the task under investigation [1, 19, 20].

4.5.7   Outlook

The recent work on honey bee color learning shows that for fine color discrimination tasks there is a large degree of behavioral flexibility depending on individual experience of each bee. This suggests some interesting, necessary and important avenues of research to better understand the basis of different color learning in bees. Future work could approach the selective attention hypothesis for fine color learning in honey bees [1, 23], or other potential hypothesis that fine color learning may involve tuning of different neural circuits in the bee brain [21]. Currently it is also not clear if the reported fine color learning described above is dependent upon the presence of contrast to the long wavelength sensitive (green) photoreceptor, or if differential conditioning could work on stimuli presented at a large visual angle that only modulates the UV-sensitive and blue-sensitive photoreceptors (see Chap. 4.4). Indeed the behavioral flexibility recently observed in honey bees for color learning that is reviewed in this chapter was not anticipated in the development of previous models of honey bee color processing, and understanding the neurobiological basis of this complex color learning behavior and how it should be modeled promises to be an exciting challenge for science over the next 50 years.

References

1. Avarguès-Weber A, de Brito Sanchez MG, Giurfa M, Dyer AG (2010) Aversive reinforcement improves visual discrimination learning in free flying honeybees. PLoS One 5(10):e15370
2. Avarguès-Weber A, Deisig N, Giurfa M (2010) Visual cognition in social insects. Annu Rev Entomol 56:423–443
3. Backhaus W (1991) Color opponent coding in the visual system of the honeybee. Vision Res 31(7/8):1381–1397
4. Bowmaker JK, Dartnall HJ (1980) Visual pigments of rods and cones in a human retina. J Physiol 298(1):501–511
5. Burns JG (2005) Impulsive bees forage better: the advantage of quick, sometimes inaccurate foraging decisions. Anim Behav 70:e1–e5
6. Burns JG, Dyer AG (2008) Diversity of speed accuracy strategies benefits social insects. Curr Biol 18:R953–R954
7. Çamlitepe Y, Aksoy V (2010) First evidence of fine colour discrimination ability in ants (Hymenoptera, Formicidae). J Exp Biol 213(1):72–77
8. Chittka L (1996) Does bee colour vision predate the evolution of flower colour? Naturwissenschaften 83(3):136–138
9. Chittka L, Dyer AG, Bock F, Dornhaus A (2003) Bees trade off foraging speed for accuracy. Nature 424:388
10. Chittka L, Skorupski P, Raine NE (2009) Speed-accuracy tradeoffs in animal decision making. Trends Ecol Evol 24:400–407

11. Chittka L, Thomson JD, Waser NM (1999) Flower constancy, insect psychology, and plant evolution. Naturwissenschaften 86:361–377
12. Chittka L, Spaethe J, Schmidt A, Hickelsberger A (2001) Adaptation, constraint, and chance in the evolution of flower color and pollinator color vision. In: Chittka L, Thomson JD (eds) Cognitive ecology of pollination. University Press, Cambridge, pp 106–126
13. Dyer AG (2001) Ocular filtering of ultraviolet radiation and the spectral spacing of photoreceptors benefit von Kries colour constancy. J Exp Biol 204:2391–2399
14. Dyer AG, Chittka L (2004) Biological significance of discriminating between similar colours in spectrally variable illumination: bumblebees as a study case. J Comp Physiol A 190:105–114
15. Dyer AG, Chittka L (2004) Bumblebees (*Bombus terrestris*) sacrifice foraging speed to solve difficult colour discrimination tasks. J Comp Physiol A 190:759–763
16. Dyer AG, Chittka L (2004) Fine colour discrimination requires differential conditioning in bumblebees. Naturwissenschaften 91(5):224–227
17. Dyer AG, Murphy AH (2009) Honeybees choose "incorrect" colors that are similar to target flowers in preference to novel colors. Isr J Plant Sci 57(3):203–210
18. Dyer AG, Neumeyer C (2005) Simultaneous and successive colour discrimination in the honeybee (*Apis mellifera*). J Comp Physiol A 191(6):547–557
19. Dyer AG, Neumeyer C, Chittka L (2005) Honeybee (*Apis mellifera*) vision can discriminate between and recognise images of human faces. J Exp Biol 208:4709–4714
20. Dyer AG, Spaethe J, Prack S (2008) Comparative psychophysics of bumblebee and honeybee colour discrimination and object detection. J Comp Physiol A 194:617–627
21. Dyer AG, Paulk AC, Reser DH (2011) Colour processing in complex environments: insights from the visual system of bees. Proc R Soc B 278:952–959
22. Franks NR, Dornhaus A, Fitzsimmons JP, Stevens M (2003) Speed versus accuracy in collective decision making. Proc R Soc B 270:2457–2463
23. Giurfa M (2004) Conditioning procedure and color discrimination in the honeybee *Apis mellifera*. Naturwissenschaften 91(5):228–231
24. Giurfa M, Núñez J, Backhaus W (1994) Odour and colour information in the foraging choice behaviour of the honeybee. J Comp Physiol A 175:773–779
25. Giurfa M, Núñez J, Chittka L, Menzel R (1995) Colour preferences of flower-naive honeybees. J Comp Physiol A 177:247–259
26. Gumbert A (2000) Color choices by bumble bees (*Bombus terrestris*): innate preferences and generalization after learning. Behav Ecol Sociobiol 48:36–43
27. Helverson Ov (1972) The relationship between difference in stimuli and choice frequency in training experiments with the honeybee. In: Wehner R (ed) Information processing in the visual system of anthropods. Springer, Berlin, pp 323–334
28. Ings TC, Chittka L (2008) Speed accuracy tradeoffs and false alarms in bee responses to cryptic predators. Curr Biol 18:1520–1524
29. Kelber A (2010) What a hawkmoth remembers after hibernation depends on innate preferences and conditioning situation. Behav Ecol 21:1093–1097
30. Kevan PG, Chittka L, Dyer AG (2001) Limits to the salience of ultraviolet: lessons from colour vision in bees and birds. J Exp Biol 204:2571–2580
31. Kien J, Menzel R (1977) Chromatic properties of interneurons in the optic lobes of the bee. II. Narrow band and colour opponent neurons. J Comp Physiol A 113:35–53
32. Kulikowski JJ, Walsh V (1991) On the limits of colour detection and discrimination. In: Kulikowski JJ, Walsh V, Murray JJ, Cronly-Dillion JR (eds) Vision and visual dysfunction: limits of vision, vol 5. Macmillian, London, pp 202–220
33. Lehrer M (1999) Dorsoventral asymmetry of colour discrimination in bees. J Comp Physiol A 184:195–206
34. Lennie P (2000) Color Vision. In: Kandel ER, Schwartz JH, Jessel TM (eds) Principles of neural science, 4th edn. McGraw Hill, New York, p 583
35. MacAdam DL (1986) Color measurement. Theme and variations, vol 27. Springer series in optical sciences. Springer, Berlin

36. Menzel R (1967) Untersuchungen zum Erlernen von Spektralfarben durch die Honigbiene (*Apis mellifica*). Z vergl Physiol 56:22–62
37. Menzel R (1985) Color pathways and colour vision in the honeybee. In: Ottoson D, Zeki S (eds) Central and peripheral mechanisms of color vision. MacMillan Press, London, pp 211–233
38. Muller H, Chittka L (2008) Animal personalities: the advantage of diversity. Curr Biol 18:R961–R963
39. Peitsch D, Fietz A, Hertel H, de Souza J, Ventura DF et al (1992) The spectral input systems of hymenopteran insects and their receptor-based colour vision. J Comp Physiol A 170:23–40
40. Peter CI, Johnson SD (2008) Mimics and magnets: the importance of color and ecological facilitation in floral deception. Ecology 89:1583–1595
41. Rival C, Oliver I, Ceyte H (2003) Effects of temporal and/or spatial instructions on the speed-accuracy trade-off of pointing movements in children. Neurosci Lett 336:65–69
42. Stroeymeyt N, Giurfa M, Franks NR (2010) Improving decision speed, accuracy and group cohesion through early information gathering in house-hunting ants. PLoS One 5:e13059
43. Uchikawa K, Ikeda M (1981) Temporal deterioration of wavelength discrimination with successive comparison method. Vision Res 21:591–595
44. Vergoz V, Roussel E, Sandoz JC, Giurfa M (2007) Aversive learning in honeybees revealed by the olfactory conditioning of the sting extension reflex. PLoS One 2(3):1–10
45. von Frisch K (1914) Der Farbensinn und Formensinn der Biene. Z Jb Abt allg Zool Physiol 35:1–188
46. Vorobyev M, Menzel R (1999) Flower advertisement for insects: bees, a case study. In: Archer SNea (ed) Adaptive mechanisms in the ecology of vision. Kluwer Academic Publishers, Great Britain, pp 537–553
47. Wehner R (1971) The generalization of directional visual stimuli in the honey bee, *Apis mellifera*. J Insect Physiol 17:1579–1591
48. Yang EC, Lin HC, Hung YS (2004) Patterns of chromatic information processing in the lobula of the honeybee, *Apis mellifera* L. J Insect Physiol 50(10):913–925

Chapter 4.6
Sensory Systems: Commentary

Randolf Menzel

Sensory physiology in general has gained enormously from studies in honeybees during the last century. New senses were discovered in bees (e.g. polarized light, UV and magnetic sensitivity), and well designed experiments fostered by the traditions developed in Karl von Frisch's lab opened avenues into quantitative studies of perception. The key to this success was and still is the potential to train bees to respond to stimuli of different sensory modalities in such an efficient way hardly met by any other animal. The rich knowledge about senses in bees allows now to search for their neural correlates at all levels of integration.

The olfactory pathway is the best studied neural circuit in the bee brain. No other sensory modality has been examined so well from stimulus conditions to behavioral output, and still we are far from understanding even the basics of the gross organization of this neural circuit, its anatomical structure, its coding properties and plasticity. Asking what should be studied next requires consideration of the advantages of working with bees. These are, in my view, robust olfactory learning while recording neurons, and identifiable neurons at least in the more peripheral part of the olfactory pathway. What may be the outcome of such endeavors? We may be able to answer questions like the following.

1. Is there an anatomical and functional separation between coding of odors controlling innate and stereotypical behavior as compared to coding of odors for adaptive behavior? The separation I am dwelling on relates to that between pheromones and general odors. All attempts to trace dedicated pheromone coding in the bee brain were unsuccessful, a surprising result given the large number of well-characterized pheromones controlling social interactions. What are we missing? Should we search for pheromone- dedicated neurons and antennal-lobe glomeruli, or do we need a new concept that bridges the apparent dichotomy

R. Menzel (✉)
Institut für Biologie, Neurobiologie, Freie Universität Berlin, Berlin, Germany
e-mail: menzel@neurobiologie.fu-berlin.de

C.G. Galizia et al. (eds.), *Honeybee Neurobiology and Behavior: A Tribute to Randolf Menzel*, DOI 10.1007/978-94-007-2099-2_24,
© Springer Science+Business Media B.V. 2012

between pheromones and general odors? There is no other way than to make sure that all glomeruli and all neurons projecting out from the antennal lobe are recorded under suitable test conditions. The suspicion is that neural processing in the lateral horn may somehow be connected to the control of stereotypical behavior. The lack of knowledge about the lateral horn both with respect of its anatomical organization and its neural processing is a great deficiency that should be overcome.

2. Three major tracts leave the antennal lobe. Although a few differences in their coding properties have been found, the overlap in their odor profiles makes one think that we may have missed important features of their coding properties. This suspicion is generated by the fact that basic properties of olfactory perception cannot yet be traced to neural characteristics, like intensity coding without loss of odor identity, representation of both time invariant and time dependent stimulus conditions, simultaneous mixture unique and component coding, just to mention a few of the mysteries.

3. The impressive size of the mushroom body calyces and the large number of Kenyon cells hides a secret. A considerable proportion of Kenyon cells is devoted to olfactory processing. We may get closer to the secret of these densely packed neurons if we are able to understand the divergent pattern of projection neurons on to them and relate their individual properties to their structure. Although a number of ambitious graduate students in my and other labs have collected intracellular recordings from projection neurons and marked them afterwards, we still miss something important here. Since the mushroom body does not provide intrinsic structural elements for accurate spatial relations to subgroups of Kenyon cells we have to embed the data from intracellular markings into the atlas of the bee brain, and I bed the secret will be uncovered.

4. The question whether odor coding follows predominantly a spatial or a temporal scheme is still unresolved. Other systems for which synchrony of spikes in parallel pathways was postulated [7] share the same task, namely to detect, discriminate, learn and recognize odors in a highly dynamic world. Ideally one would like to manipulate temporal coding independently of spatial coding, but unfortunately there is simply no way of selectively interfering with the temporal code leaving the spatial code untouched. Maybe the solution lies, as often in science, in the combination of both apparent alternatives, but then we still need to understand how these coding schemes interact. The solution may come from combined physiological and behavioral studies, a requirement that bees can offer more than any other insect.

What kind of sensation might be created by the multiple inputs via the antennae? Inside of the hive bees use the antennae for measuring the wax cells with high precision, probe the larvae, control trophallaxis, inspect their hive mates, and directly or indirectly the queen. Outside of the hive young bees police the traffic of inbound comrades, qualify their nectar load and exchange multiple chemical signals. In flight they determine flight speed (together with the visual input), wind direction and odor plumes. After landing on a flower they register the distribution of olfactory cues and surface structures und use these for quick handling of flowers. As scout bees of a swarm they probe the suitability of a cave with respect to humidity and airflow. In dance communication multiple forms of signals are received despite the darkness surrounding dancer and recruited bees. All these inputs are received by the antennae, a highly mobile device indeed which appears to evaluate its sensory input by its actively probing behavior. Processing of primary afferents from the antennae occurs in the antennal lobe, dorsal lobe, suboesophageal ganglion and lateral protocerebrum (see Chap. 4.3). The anatomical distribution of these multimodal inputs may suggest separation into different sensory modalities (olfaction, mechanosensory via external hair cells and proprioceptive organs) but it could well be that higher order integration combines these inputs and creates a unique sensation, the antennal sense, a view put forward already by Karl von Frisch in his proposal of a topochemical sense created by the antennae [14]. Training experiments like those performed so elegantly in Jochen Erber's lab could be used to ask whether and how the antennal sense is represented in an integrated way, e.g. in the dedicated parts of the mushroom body. It is worth remembering in this context that the functional organization of the basal ring in the mushroom body calyces is practically unknown. The combined input from olfactory and mechanosensory organs may be taken as a hint for such integration. A valuable study case could be contact chemoreception, a highly important antennal (and tibial sense) for social organization inside the hive as well as a guiding sense in pollination and potential nest site inspection outside of the hive. I expect the study of contact chemoreception will uncover a range of novel insight into the neural processing of the antennal sense in particular and across modality integration in general.

Gabriela de Brito-Sanchez stresses the need of more studies on the neural basis of contact chemoreception and taste. Besides our ignorance about the number and classes of receptors for taste we even do not understand neural integration of contact chemoreception and its relation to taste. One way is to study receptors, their structure, location, intracellular signaling cascades, and so forth. Another and possibly even more informative way relates to central processing. Are inputs from contact chemoreceptors, e.g. for long C-chain hydrocarbons, processed in different neuropils than those of taste receptors for small molecules like sugars, amino acids, water, bitter substances? Such studies require the electro- and optophysiological probing of the SOG, a formidable task because of the hidden location of the SOG. Martin Hammer and Ulrike Schröter, who managed to record ventral unpaired neurons in the midline of the SOG, accessed the SOG by stretching the head way down when the animal was fixed to its dorsal side [5, 12]. Bees do not survive well under these conditions, and cutting the buckle of the head capsule surrounding the neck unavoidably damages nerves controlling proboscis movements. In spite of these problems the SOG should be moved more in the focus of our attention.

Higher order processing of contact chemoreception and taste may bring the mushroom body in focus again. Ulrike Schröter's finding of gustatory neurons projecting from the SOG to a special region of the calyx between lip and collar may provide hints for optophysiological recordings. As already said above we should remember how little we know about neural processing going on in the basal ring of the calyx.

It could well be that the accessibility of the calyx may offer more opportunities than the second order processing regions in the SOG.

Intracellular recordings and staining of central neurons in the bee brain are demanding enterprises. Those who take up the daily battle with blocking electrodes, unstable preparations, short lasting recordings and faint or no staining need to be praised for their patience and endurance. Nothing is simple in research but intracellular electrophysiology of the bee brain is known to be tough, indeed. Aiming for a particular neuron deep in the brain calls for additional efforts. Infrared microscopy combined with several other optical tricks has been tried. Unfortunately the trachea contrast sharply and obscure neurons and axon tracts. We hoped to pick-up action potentials from somata using sharp or patch electrodes, but unfortunately without success. Isolated brains and even thick slices of the bee brain have been tried, and have worked so far only with slices of the compound eyes of drones. Thus the community of bee researchers is waiting for a methodological breakthrough in intracellular electrophysiology. It will be necessary to work towards an improvement of this approach because ultimately net-work analysis requires intracellular signal detection and the measurement of synaptic potentials. Furthermore, as long as it is possible to express functional dyes in gene targeted neurons, intracellular filling of neurons provides us with the only valuable data. The time of improvement will come, but meanwhile the patience and endurance of hard working electrophysiologists is asked for.

In 1914 Karl von Frisch published his sensational discovery about color discrimination [13] disproving the commonly accepted understanding at this time that invertebrates are color blind. Hess [6], the proponent of this view was in fact not incorrect with respect to bees because he studied phototaxis, and indeed bees are color blind in their phototactic responses [10]. Today color vision in bees is well understood as Misha Vorobyev and Natalie Hempel de Ibarra show in (see Chap. 4.4). Well founded models allow making predictions about color discrimination, and still a range of questions need to be addressed in the future. Is the general color opponency model based on the noise properties of photoreceptors [16, 18] or the specific color opponent model, Backhaus` COC model [1], more appropriate? The COC model receives additional support from color discrimination data, the interpretation of the perceptual dimensions by multidimensional scaling and by recordings from higher order visual interneurons [9]. Taking all these data together I consider the COC model as the most appropriate way of a quantitative representation of color vision in bees (and most likely other Hymenopteran pollinators with trichromatic color vision), and want to urge the community to use it for calculations of color differences. In my view another model, the color hexagon [2, 17] cannot be used for predictions of color similarity or differences because its formal basis is incorrect, it does not respect the analyses leading to the noise-based model and the COC model as well as the neural data, and it makes incorrect predictions [15]. An appropriate color opponent model is extremely important for any attempt to interpret co-evolutionary adaptations between colored objects like flowers and color vision, e.g. of pollinators like the bee, and the application of an incorrect color vision model will lead astray. Deviations from pure sensory models like those addressed in Chap. 4.5 on modulation of color discrimination by higher level cognitive factors leading to e.g. selective attention require a reference that originates from low level models, and therefore such studies are of great importance.

It is still a mystery why Hymenopteran pollinators in general lack red receptors. The added long wave discrimination (around 560 nm) would provide a rich additional object – background contrast range, and floral pigments in the yellow/red region are metabolically easily accessible. Beetles, butterflies and vertebrate pollinators benefit from this color range, and since beetles are considered to be the most ancient insect pollinators, the lack of red receptors in most Hymenopteran pollinators is likely to reflect a loss of the long wave receptor. Two lines of research are interesting to follow in this respect, on the ultimate level (the ecological-evolutionary argument) and the proximate level (the neural integration argument). In the first case one may assume that flowers specialized for groups of pollinators according to their predominant handling procedures and dependence on reward conditions. This could have lead to reduced competition between pollinators, and pollinators would gain from the fact that they become more specifically guided by color signals. The neural integration argument assumes that a general color coding device as in the case of trichromatic Hymenoptera evaluates colors at the neural level most economically, combining wavelength information with spatial, temporal and e-vector information. A tetrachromatic neural coding system may simply require too many of neural resources. These two lines of arguments are not mutually exclusive, and may be studied side by side.

Colors of flowers are meaningful signals for the pollinator. Bees are equipped with innate search images for floral colors, weighting human-blue colors highest [3, 4], and colors differ in salience as appetitive cues [8]. Thus other than the peripheral sensory system, the central integration of color is highly biased toward particular floral colors, and reflects a mutual adaptation between the predominantly visited flowers by median to large sized Hymenopteran pollinators [11]. But what is the neural basis of the meaning of color signals? The future will hopefully uncover neural mechanisms of high-order visual coding including those in the collar of the mushroom body calyx where visual afferences can be found.

References

1. Backhaus W, Menzel R (1987) Color distance derived from a receptor model of color-vision in the honeybee. Biol Cybern 55(5):321–331
2. Chittka L (1992) The color hexagon – a chromaticity diagram based on photoreceptor excitations as a generalized representation of color opponency. J Comp Physiol A 170(5):533–543
3. Chittka L, Menzel R (1992) The evolutionary adaptation of flower colors and the insect pollinators color-vision. J Comp Physiol A 171(2):171–181
4. Giurfa M, Núñez J, Chittka L, Menzel R (1995) Color preferences of flower-naive honeybees. J Comp Physiol A 177(3):247–259
5. Hammer M (1993) An identified neuron mediates the unconditioned stimulus in associative olfactory learning in honeybees. Nature 366(6450):59–63

6. Hess CV (1913) Experimentelle Untersuchungen über den angeblichen Farbensinn von Bienen. Zool Jb 34:81–106
7. Laurent G, Stopfer M, Friedrich RW, Rabinovich MI, Volkovskii A et al (2001) Odor encoding as an active, dynamical process: experiments, computation, and theory. Annu Rev Neurosci 24:263–297
8. Menzel R (1967) Untersuchungen zum Erlernen von Spektralfarben durch die Honigbiene (*Apis mellifica*). Z vergl Physiol 56:22–62
9. Menzel R, Backhaus W (1991) Colour Vision in Insects. In: Gouras P (ed.) Vision and visual dysfunction. The perception of colour. MacMillan Press, London, pp 262–288
10. Menzel R, Greggers U (1985) Natural phototaxis and its relationship to color-vision in honey-bees. J Comp Physiol A 157(3):311–321
11. Menzel R, Shmida A (1993) The ecology of flower colors and the natural color-vision of insect pollinators – the Israeli flora as a study case. Biol Rev 68(1):81–120
12. Schröter U, Menzel R (2003) A new ascending sensory tract to the calyces of the honeybee mushroom, body, the subesophageal-calycal tract. J Comp Neurol 465(2):168–178
13. von Frisch K (1914) Der Farbensinn und Formensinn der Biene. Z Jb Abt allg Zool Physiol 35:1–188
14. von Frisch K (1967) The dance language and orientation of bees. Harvard University Press, Cambridge
15. Vorobyev MV (1999) Evolution of flower colours – a model against experiments: reply to comments by Chittka. Naturwissenschaften 86:598–600
16. Vorobyev M, Osorio D (1998) Receptor noise as a determinant of colour thresholds. Proc R Soc B 265(1394):351–358
17. Vorobyev MV, Hempel de Ibarra N, Brandt R, Giurfa M (1999) Do "white" and "green" look the same to a bee? Naturwissenschaften 86:592–594
18. Vorobyev M, Brandt R, Peitsch D, Laughlin SB, Menzel R (2001) Colour thresholds and receptor noise: behaviour and physiology compared. Vision Res 41(5):639–653

Part V
Genetics and Molecular Biology

Chapter 5.1
Neurogenomic and Neurochemical Dissection of Honey Bee Dance Communication

Andrew B. Barron, Axel Brockmann, Moushumi Sen Sarma, and Gene E. Robinson

Abstract Honey bee dance communication is a classic form of animal behavior, with over 70 years of intense study. In this chapter, we first discuss conceptually how it is possible to dissect dance communication into simpler behavioral modules for neurogenomics analysis, based on information from prior ethological studies of dance behavior and a rapidly advancing functional analysis of the insect brain. We then review recent studies that have used this conceptual approach and new genomic tools to begin to explore neurogenomic and neurochemical aspects of dance communication, highlighting the following findings. Comparative transcriptomic studies of specific brain regions across *Apis* species that differ in dance behavior have implicated genes involved in the geotactic and odometric elements of dance, and genes involved in learning and memory systems and the circadian clock as important modulators of dance output. This research also has identified distinct patterns of gene expression in different brain regions that provide additional hints about the regulation of dance behavior. Pharmacological studies with octopamine and related compounds have demonstrated the role of the reward system in modulating the likelihood that a bee will dance upon returning from a foraging trip. The results of these early studies provide a foundation for a more comprehensive molecular dissection of dance behavior and suggest that the mechanisms regulating dance communication involve evolutionary reuse and adaptation of neuromolecular systems that control elements of solitary behavior.

A.B. Barron
Department of Biology, Macquarie University, North Ryde,
Sydney, NSW 2109, Australia

A. Brockmann • M.S. Sarma • G.E. Robinson (✉)
Neuroscience Program, Institute for Genomic Biology & Department of Entomology,
University of Illinois at Urbana-Champaign, Urbana, IL, USA
e-mail: generobi@illinois.edu

C.G. Galizia et al. (eds.), *Honeybee Neurobiology and Behavior: A Tribute to Randolf Menzel*, DOI 10.1007/978-94-007-2099-2_25,
© Springer Science+Business Media B.V. 2012

Abbreviations

CNS   Central nervous system
CX   Central complex
MB   Mushroom body
OA   Octopamine
OL   Optic lobes

5.1.1   Introduction

Honey bees gather all their food from flowers, a highly ephemeral resource. They have evolved a complex system to effectively exploit floral resources that involves impressive cognitive and perceptual abilities on the part of individual foragers;

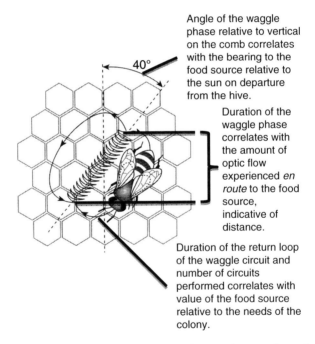

Angle of the waggle phase relative to vertical on the comb correlates with the bearing to the food source relative to the sun on departure from the hive.

Duration of the waggle phase correlates with the amount of optic flow experienced *en route* to the food source, indicative of distance.

Duration of the return loop of the waggle circuit and number of circuits performed correlates with value of the food source relative to the needs of the colony.

Fig. 5.1.1 Waggle dance of the honey bee. Returning foragers dance on the vertical comb inside the hive. The waggle dance encodes information related to the location and relative value of a food source. Bees dance on return to the hive after a successful foraging trip to communicate to their nest mates the location and value of profitable resources needed by the colony [54, 68]. Most dances are performed on a small area of the comb inside the hive close to the entrance, the dance floor [54, 66, 68]. Foragers that are not actively foraging crowd the dance floor and closely follow the movements of the dancers, after which they may be stimulated to leave the hive (Modified from Winston [74])

an intricate system of division of labor among various groups of hivemates to cooperate in the discovery, exploitation and processing of food resources; and a communication system to coordinate all aspects of foraging activity (see Chap. 2.3 for information on the ecology of dance communication). The waggle dance of *Apis mellifera* (Fig. 5.1.1), is an information-rich repeated figure-of-eight movement that represents the distance, direction and relative value of a food source (or for swarming bees; a nest site) from the colony [54, 68]. Discovery of the honey bee's symbolic dance communication system revolutionized perceptions of the behavioral capacities of all animals, but especially of insects [68]. Von Frisch's elegant and methodical work on the subject became pivotal in the establishment of the disciplines of ethology, and later neuroethology, and helped set standards for the rigorous and quantitative analysis of behavior. Dance communication represents a clear and remarkable case of the evolution of a behavioral innovation, and one particularly valuable outcome of studying its molecular basis could be a detailed understanding of how new forms of behavior can evolve.

In this chapter, we first present a conceptual dissection of dance communication into simpler behavioral modules, based on information from prior ethological studies and a rapidly advancing functional analysis of the insect brain. We then review recent studies that have used this 'modular' approach in combination with new genomic tools to begin to explore some elements of dance communication from a molecular perspective. We also present results of neurochemical analyses of dance behavior. We close with an assessment of the prospects and challenges for achieving a comprehensive understanding of the molecular, neural and evolutionary bases of dance communication.

5.1.2 Conceptual Neuroethological Dissection of Dance Behavior

Ethological studies have identified the stimuli dancers use that inform different elements of dance communication, and the stimuli dance followers use to read the dance. From these studies it is possible to conceptually dissect dance communication into component behavioral modules, some that are simpler to investigate experimentally than the entire dance communication system. The concept of modularity is used throughout biology, but modules of any type of complex system can be difficult to precisely define. In developmental biology, an assumption of modularity is that a developing organism can be divided into distinct organizational or functional units and these can be described as modules [11]. Here, we borrow that perspective from developmental biology to describe a behavioral module as a distinct organizational or functional unit in the expression of complex behavior [4]. For example, many complex behaviors include a module of rhythmic behavior; a focus on this module has helped provide molecular dissections for a variety of complex behaviors, i.e., courtship in *Drosophila melanogaster* [23].

We present a scheme for a modular dissection of dance behavior focused on the different stimuli and forms of information communicated by a dancing honey bee to the "follower bees," i.e., those that obtain information from the dancer to motivate and

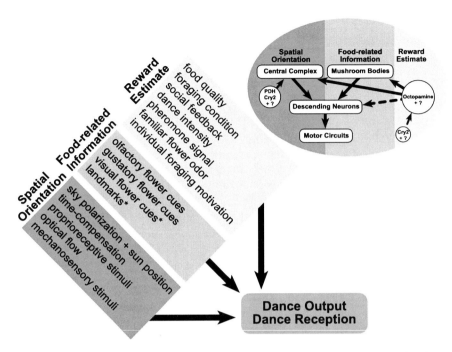

Fig. 5.1.2 Conceptual dissection of dance communication into behavioral modules. This scheme is based on the different types of information communicated by the waggle dancer to the dance follower, the information extracted by the dance follower from the dancer to reconstruct a flight vector for her subsequent foraging flight, and the key stimuli conveying these forms of information. Highlighted here are modules related to: (1) spatial orientation; (2) food-related information; and (3) reward estimation. *Inset* depicts brain regions that can be plausibly linked to these modules, based on experimental evidence (*solid arrows*) or inferences (*dashed arrows*) from honey bees or other insect species discussed in the text. *=Food-related stimuli not directly involved in dance communication, but used during foraging. +?=possibility of additional, as yet unidentified, molecular signals

guide their subsequent foraging behavior. Dancers must signal a spatial vector and value information for a food source, and dance followers must calculate a flight vector from dance movements, each of which involve a variety of sensory and cognitive systems. Dance communication involves behavioral modules associated with: (1) spatial orientation and (2) reward estimation; and (3) food-related information (Fig. 5.1.2). Because most of the modules in the third category are more related to foraging in general, we focus here on the first two. We also present a proposed neuroanatomical analysis of dance communication that is based on these behavioral modules.

5.1.2.1 *Behavioral Modules for Spatial Orientation*

Modules associated with spatial orientation influence both the production and perception of dance information. These modules include: proprioceptive systems; systems to perceive and interpret information about sun azimuth and patterns of polarized light; systems to learn and remember information about landmarks; systems to measure flight distance; and mechanosensory systems. The duration of the waggle phase of the dance signals the distance to the food source [54], which is estimated by the amount of optic flow experienced by the dancer en route to the food source [16]: a visual indicator of distance flown [60, 61]. The dancer represents the direction to the food source as the angle of the waggle phase relative to vertical on the surface of the comb [54, 68], which is referenced against the position of the sun viewed when departing the hive [19, 68] and bees can use either a direct view of the sun, or the polarization pattern in a clear sky and learned landmark information in combination with a time-compensated sun compass to estimate the position of the sun if it is not directly visible [15, 68].

Recruits closely follow the movements of the dancer and must have the ability to back-translate the movements of the dancer into a foraging flight vector (see Chap. 2.5). There is debate over precisely how followers extract information from the dance within the dark hive, but it seems likely airborne vibrations generated by the wings of the dancer [38, 40], vibrations carried by the comb [65] and direct antennal contact with the dancer [50] can all provide information to recruits on the movements of the dancing bee (see Chap. 2.2). One additional possibility is that recruits obtain spatial information from the movements of the dancer by performing the dance movement themselves while following behind the dancer, in which case a bee could use a proprioceptive sense to interpret the dance signal from her own movements (see Chap. 2.1).

5.1.2.2 Behavioral Modules for Reward Estimation

Behavioral modules associated with reward estimation influence the likelihood that a returning forager will dance, and if so, the "vigor" with which she dances [53]. These modules include: innate gustatory abilities to discern differences in nectar sweetness; systems to estimate the energetic efficiency of the foraging trip [54], moderated by an assessment of the extent to which the colony needs the food [53, 54]; and systems that integrate all of this information to modulate the motivation of the bee to engage in foraging and dancing behavior. The higher the food quality, the more likely it is that the bee will initiate dance communication [53]. Once the decision to dance has been made, the higher the quality, the more vigorous is the dance, measured primarily in terms of the number of waggle circuits per unit time [55]. Other modules enable bees to locate the precise patch of flowers once they are directed to the general vicinity by means of dance information. The prodigious learning abilities of honey bees, in both laboratory and field assays, likely reflect the prominence of these behavioral modules in dance communication. For nectar, information on colony hunger is communicated to the returning forager by social interactions, which include the number of bees crowding the dance floor to unload nectar foragers of their nectar load, and their time from entering the hive until they are unloaded [54].

5.1.2.3   Proposed Neuroanatomical Regions and Brain Systems Involved in Dance Communication

Identification of the above behavioral modules and their associated sensory bases in turn suggests parts of the central nervous system (CNS), or neural systems that are likely to be particularly important for dance behavior (Fig. 5.1.2, inset). In the following paragraphs in this section we outline the evidence for the involvement of each of these CNS regions or neural systems.

The central complex (CX) is likely to be particularly important for the calculation of the direction to the food source. In locusts and *Drosophila* the CX receives highly processed input from several different sensory systems: polarization vision, gravity, proprioception, and antennal mechanosensors. Studies of locust polarization vision, and visual memory in *Drosophila melanogaster* have shown that the CX plays a major role in spatial orientation and sun-compass navigation [26, 41]. It seems likely, therefore, that in bees the CX would be involved in the processing of the sun compass information to estimate direction to a food source.

The CX also is involved in the initiation and maintenance of locomotion in other insects, suggesting it might play the role of initiating and controlling the dance movements themselves. However, little is known of how the CX is connected to the descending neurons, which carry motor commands to the motor centers in the thoracic ganglia. The second thoracic ganglion likely controls the abdominal waggles and the sounds produced by precisely patterned wing beats that occur at the same time [15]. Pharmacological and RNAi gene knockdown experiments suggest that tachykinin, GABA and acetylcholine are all important in CX regulation of locomotor activity [71, 75]. In both *Periplaneta americana* and crickets the number of descending neurons is relatively small [42, 62], and assuming a similar organization for honey bees, the motor commands for dancing-related movements or foraging flights most likely exist as a cross-fiber pattern of activated descending neurons.

It is likely that the circadian clock system contributes the time signal necessary to calibrate sun compass navigation. Sun-compass navigation is dependent on a time-compensation mechanism to account for the daily solar movement, and several studies indicate that the circadian clock does this [27, 46]. In *Drosophila* and monarch butterflies (*Danaus plexippus*) the circadian clock system is composed of a central clock including neurons of the pars lateralis and additional peripheral clocks, in the antennae and fat bodies for example [32, 47, 64, 76]. Immunocytochemistry indicates that interneurons expressing two major circadian clock proteins, PDH and Cry2, innervate specific layers of the CX in locusts and monarch butterflies [27, 79], which could provide the time signal information needed for time compensation of the sun-compass. Similar projections from the clock system to the CX seem likely in honey bees [8], but have not yet been identified (see Chap. 1.3).

The mushroom bodies (MBs) likely are involved in the processing of multi-modal sensory information related to the food source. Current knowledge indicates that the MBs are the brain neuropils that receive the most varied forms of sensory input in insects, receiving mechanosensory, olfactory and diverse visual forms of information (color, motion, polarization) [17, 22]. Neuroanatomical tract tracing suggests that the

MBs also receive polarization-sensitive visual input in honey bees [9]. If this is the case, the MBs of honey bees receive all the sensory input necessary to define a food source location as well as the features of the food source, such as floral color, odor, and the landmarks surrounding it. As a key brain region implicated in various forms of learning and memory, the MBs also are likely involved in processing and remembering information on the distance of the food source from the hive [56].

The MBs also are probably important in developing the reward estimate that modulates dance communication. In *Drosophila* and honey bees the MBs have been shown to be essential for learned visual or olfactory information relevant to food reward [37, 52]. The biogenic amine octopamine (OA) has repeatedly been shown to modulate reward learning and reward responses in honey bees, and there are octopaminergic projections in the MBs [59]. In addition, octopaminergic VUM (ventral unpaired medial) neurons in the subesophageal ganglion (SEG) are highly likely to mediate sugar-reward dependent responses in bees, with projections to the MBs, the CX as well as to other parts of the brain [24, 51].

Suggestions of CNS regions that are likely important in dance communication have provided a useful starting point for the neurogenomic studies reviewed below. However, a full explication of dance communication will require detailed neuroanatomical analyses to identify more precisely the roles of these regions and how they connect with each other to regulate this complex behavior. One study [9] attempted to use neuronal tracing to identify specializations in sensory pathways for dance communication in honey bees, but no obvious specializations were found. Staining of sensory neurons and secondary interneurons sensitive to polarization in bees revealed a neural circuit similar in overall complexity to that seen in locusts [9], and did not reveal elaborations that might be reflective of the added complexity of directional signaling in dance. Comparisons of antennal mechanosensory projections between worker, queen, and drone honey bees, as well as between honey bee and bumblebee workers did not reveal any evidence for specific enlargements of these projections in honey bee workers [9]. It was reasonable to assume such enlargements because dance followers likely use complex inter-antennal sensory comparisons to track the body orientation of dancers [39]. Based on these results so far, it appears that the evolution of dance language communication has not involved the evolution of any new sensory pathways, nor extensive specialization of already existing pathways. This might imply that the production of dance behavior involves differences in connectivity or modifications in central brain processing regions instead. This idea awaits experimental verification.

5.1.3   Neurogenomic Dissection of Dance Behavior: Early Results

5.1.3.1   Overview of Honey Bee Neurogenomics

The above conceptual neuroethological dissection of dance behavior identified candidate behavioral modules and brain systems. In this part of the chapter we report on three studies that use transcriptomic analysis of specific modules and brain systems to identify candidate genes for dance communication. These analyses are possible because of the development of gene expression microarrays and the sequencing of the honey bee genome [28]. Transcriptomics measures changes in the expression of many genes that correlate with changes in behavior, and this is proving a powerful tool for identifying genes involved in complex social behavior in a variety of species [49]. There is now strong support for the premise underlying transcriptomic analysis: that differences in transcript abundance reflect a mechanistic link between gene and behavior [49]. The nature of the stimulus often determines the pattern of gene expression changes, with groups of genes upregulated for one condition but downregulated for another [14]. Studies in song birds have shown that when exposed to a behaviorally significant stimulus (novel male song), birds show rapid induction of brain gene expression that often results in long term changes in neural functioning [13], even when stimuli are too short-lived to elicit immediate behavioral responses [33]. Studies such as these present a new paradigm that links genomic activity, neural activity and behaviorally relevant environmental stimuli.

It is already known that there are extensive differences in brain gene expression between foragers and bees that have not yet made the transition from working in the hive to foraging [57, 72, 73]. Alaux et al. [2] have recently shown clear differences in brain gene expression profiles between bees performing a vibratory communication signal and non signalers, and that brain gene expression responds rapidly to pheromone signals [1, 3] demonstrating that in honey bees there is an association between the neurogenomic brain state and communication behavior.

One difficulty in applying a transcriptomics analysis to dance, however, is the speed of dance behavior relative to the time course of RNA abundance changes. Not all forager bees dance, those that do will not dance after every trip, and each dance bout usually lasts less than a minute. It is not clear that there would be meaningful or detectable gene expression changes associated with a single dance communication event. This proved not to be a problem for the study of vibratory communication [2], but in that case there were large and stable differences between individuals in the likelihood of performing the behavior. Some bees perform the vibratory signal intensively for long periods (effectively acting as communication specialists) whereas other bees never perform this behavior [2]. By contrast dancing occurs in short bursts interrupted by foraging flights, and environmental factors strongly modulate dance likelihood. For this reason, some experimental creativity has been needed to allow the application of transcriptomics to the study of dance behavior.

5.1.3.2 Comparative Transcriptomic Analysis of Apis Dancers

While all extant *Apis* species dance, there are informative differences in dance communication behavior between the species [15]. *Apis mellifera, A. dorsata* and *A. florea* differ in how they orient the waggle phases of their dances [43]. In the cavity nesting *A. mellifera* waggle phases are oriented with respect to gravity as the bee dances on a vertical comb. The Asian dwarf honey bee, *A. florea*, is an open-nesting bee that typically builds combs suspended from tree branches. Foragers dance on a horizontal dance floor and do not transpose the solar frame of reference to a gravitational frame of reference for indicating direction. The dances of *A. florea* also lack the sound cues known from the dances of *Apis mellifera* [38]; it is argued that since in the open nesting species the dance movements are well illuminated and clearly visible there is no need to produce sounds to alert recruits to the dancer [43]. The Asian giant honey bee *A. dorsata* (another open-nesting species) dances on a vertical comb using gravity for reference but sometimes uses celestial cues to indicate dance direction [31, 67]. *A. florea* is considered most basal in the genus and it is usually assumed that dancing on a vertical comb and transposing directional information from a solar to a gravitational reference is a derived trait [48, 68]. In addition, while in all these species the waggle phase of the dance is closely correlated with distance to the food source, there are clear differences between these three species in the waggle duration for the same absolute distance to an attractive food source. Such differences in the representation of distance have been called 'dialects' of the dance language [35].

To determine whether the microarray designed for *A. mellifera* could be used effectively for other *Apis* species, Sen Sarma et al. [57] compared expression in the brains of 1-day-old bees and foragers of the three above mentioned species and also *A. cerana* (considered the sister species to *A. mellifera*). This study indicated that the microarray performed reliably for the other *Apis* species. The number of cDNA spots showing hybridization fluorescence intensities above the experimental threshold was reduced by an average of 16% in the Asian species compared to *A. mellifera*, but an average of 71% of genes on the microarray were available for analysis. Brain gene expression profiles between foragers and 1-day-olds showed differences that are consistent with a previous study on *A. mellifera*. The results also were comparable across species, with 1,772 genes showing differences in expression between foragers and 1-day-olds. This result suggests that molecular processes underlying behavioral maturation are conserved across the four *Apis* species. However, there were 218 genes that showed differences in forager/1-day-old expression between species, which in theory could include genes that relate in some way to species differences in dance communication. Among these genes were several homologs to *Drosophila* genes involved in regulating rhythmic activity, and circadian rhythms in *Drosophila*. The *double-time* gene product interacts with PERIOD to modulate clock function [45]. *Slowpoke* is a calcium and voltage activated potassium channel [78] involved in the generation of rhythmic activity [12], *ebony* is involved in both pigmentation and clock function [63], and *dopamine acetyltransferase* is involved in sleep homeostasis in flies [20]. The circadian system can be plausibly linked to dance communication, as discussed above.

A second comparative transcriptomic study focused specifically on dance behavior [58]. Gene expression profiles were generated by microarray for some of the CNS regions expected to be involved in dance communication discussed above, i.e., the optic lobes (OLs), MBs, CX and second thoracic ganglion. The CNS regions were taken from bees sampled while dancing. Sen Sarma et al. [18]

did not compare dancers to non-dancers because this behavioral difference may be too rapid to generate a strong transcriptional signal, and is perhaps more appropriate for quantitative proteomics analysis [10]. Instead Sen Sarma et al. [18] compared brain regions of dancers across *A. mellifera, A. dorsata* and *A. florea*. The assumption was that differences between brain regions that are shared between all species would reflect intrinsic functional specialization within the *Apis* nervous system whereas differences between brain regions not shared by all the species would reflect differences related to inter-species differences in behavior.

Gene expression profiles in the MBs of dancers consistently showed the biggest differences relative to the other CNS regions, as well as the biggest differences between species. These results are consistent with the above suggestion that the MBs, as the integration centre of the CNS, play the biggest role in processing sensory input and dance output. An independent transcriptomic study has also emphasized the importance of the MBs during the performance of dance behavior. Using the immediate early gene, *kakusei*, as a marker of neural activity Kiya et al. [30] reported increased neural activity in the small Kenyon cells of the MBs of dancing and foraging bees (see Chap. 5.2).

Sen Sarma et al. [18] also found surprisingly strong similarities in gene expression between the central brain and the second thoracic ganglion across all three species during dancing, suggesting a coupling of activity during dance output. Many of the similarly expressed genes were involved with energy production and metabolism, likely underlining the energy intensive process of the dance motor output. In addition, *A. mellifera* and *A. florea*, the two species with the biggest differences in dance dialects, also showed the biggest differences in gene expression profiles in the central brain and thoracic ganglion. Some of these genes could be involved in processes that underlie some of the differences in dance communication between these two species.

Species-specific differences in gene expression in selected CNS regions provide particularly attractive candidate genes to explain the differences in dance behavior exhibited by these three honey bee species. Genes identified as possible candidates by this study include *shaggy*, which in *Drosophila* shortens the duration of the circadian clock [36], *cacophony*, which disrupts *Drosophila* courtship song [69], and *CAMKII*, which is involved in learning and memory [21], as well as several genes that relate to synaptic activity and motor control.

5.1.3.3 Distance Responsive Genes

To begin to explore the molecular bases of distance measurement, Sen Sarma et al. [56] used microarray analysis to determine whether there are distance responsive genes in the bee brain, i.e., genes whose expression changes in response to perceived differences in distance. The search was motivated both by findings of gene activation in the brain in response to highly specific naturally occurring environmental stimuli, such as bird song [13], and by evidence for foraging-related effects on brain gene expression in honey bees [73]. Studying molecular representations of distance during natural flight risks confounding effects of an individual's perception of distance and effects of differential energy expenditure resulting from differences in distance flown. To avoid this problem, Sen Sarma et al. [56] used an established method that separates effects of perceived distance from effects of actual distance flown; a tunnel that can manipulate the bee's perception of distance by manipulating the optic flow they experience during flight [61]. Two brain regions were analyzed: the MBs and the OLs. As stated above, the MBs are likely involved in processing and remembering information on the distance of the food source from the hive, and the OLs, as the primary neuropil for visual stimuli, are also likely involved in processing distance information, at least in terms of sensory adaptation to different visual environments.

Regions of the honey bee brain involved in visual processing and learning and memory showed a specific genomic response to distance information. Individuals forced to shift from a short to perceived long distance to reach a feeding site showed differences in expression of 59 genes in either the OLs, MBs, or both, relative to individuals that continued to perceive a short distance, even though they all flew the same distance. Principal component analyses suggested that the expression profiles of the OLs and MBs responded to a change in perceived distance in a similar way. This result was interesting because the two brain regions have different functions; the OLs process visual input from the eyes and the MBs carry out higher order processing of multimodal sensory input from multiple parts of the brain. The similarity of expression profiles is reminiscent of the similarities between the central brain and the second thoracic ganglion detected in dancing bees mentioned above [58] and hints at coordinated activity for the production of dance behavior. In addition, the fact that the MBs showed a genomic response to distance suggests that the effects of distance on the bee brain are not solely related to stimulus perception, but also engage molecular pathways involved in distance-related memories. This hypothesis should be tested in future studies.

Bioinformatic analyses of the differentially expressed genes in this study suggest that the genomic response to distance information involves learning and memory systems associated with well-known signaling pathways, synaptic remodeling, transcription factors and protein metabolism. Some genes seemed to respond to all changes in perceived distance, suggesting a novelty response. Others responded only to specific changes in distance. More detailed neuroanatomical analyses of these genes might provide information on the neural circuitry underlying distance measurement.

5.1.4 Neurochemical Analysis of Dance Communication: The Reward System

Whether a bee dances or not, and how long and vigorously she dances is influenced by the forager's estimate of the relative value of the food resources she collected, which in turn proposes that brain systems involved in reward evaluation should modulate dance output (Fig. 5.1.2). There is now experimental evidence that this hypothesis is correct [5].

In honey bees and other insects associative learning of appetitive rewards is modulated by the biogenic amine OA. Evidence suggests OA may be the neurochemical released by the perception of sucrose that modulates downstream behavioral responses, and 'represents' the sucrose unconditioned stimulus in the brain [25]. OA has a well-established role in reward learning and assessment in solitary insects, but in socially foraging bees OA also modulates dance behavior [5]. Systemic OA treatment modulated the dance behavior of returning forager bees in a manner that was dose-dependent [5]. Dance parameters reporting reward value (dance vigor and circuit number) were by far the most sensitive to OA treatment, while positional information represented in dances was largely unchanged, supporting the hypothesis that OA was involved in the reward assessment module of dance only [5]. The effects of OA treatment were seen in dances for nectar and pollen, and blocked by mianserin, an antagonist with high affinity for OA receptors [44], suggesting that OA modulated dance performance by interaction with OA receptors [5]. Similar effects were seen following cocaine treatment [6], which alters biogenic amine reuptake in the insect brain [5]. That OA specifically modulates dance parameters signaling resource value strongly supports the hypothesis that the assessment and signaling of resource value by the dancer involves general brain reward-assessment pathways. Other biogenic amines need to be studied to determine the precise role played by OA in modulating dance-related reward assessments. Deeper studies of the reward pathways in the honey bee brain have great potential to illuminate the molecular basis of dance communication.

5.1.5 Prospects and Challenges

Mechanistic studies of dance communication are difficult because the behavior itself is so complex and context dependent. Honey bee neurobiology as a discipline flowered with the development by Menzel's research group of simple elegant bioassays for learning that could be performed by bees harnessed in a controlled laboratory setting; most notably proboscis extension reflex (PER) [7, 34]. Having this simple reduced bioassay for learning allowed recording and visualization of neuronal activity in the brains of restrained bees as they learned (see Chap. 4.1) and microinjection of compounds into discrete brain regions to probe the molecular pathways of learning and memory [25].

By contrast, the social nature of dance has made it extremely difficult to extract it into the laboratory. Forager bees will only dance on completion of a successful foraging trip, and only if they have an audience of inactive foragers on the dance floor of the hive to dance to. So far, it has not been possible to stimulate bees to dance away from the social environment of the colony dance floor [5], which has prevented studying dance behavior in the laboratory environment. Since both dancing and following dances involves active movement interspersed with foraging flights it is currently impossible to access the brains of bees for electrophysiological recordings as they are dancing.

These technical issues prevented a traditional neurophysiological investigation of dance behavior, but ethological analyses of dance progressed to the point that there are now few other examples of animal communication understood to the same degree (see Chaps. 2.1, 2.2, and 2.5). This has allowed for the conceptual dissection of dance communication into neuroanatomically grounded behavioral modules, which has provided the foundation for transcriptomic analyses. Initial findings from these analyses have yielded candidate genes that now must be studied more intensively in order to determine what causal roles they play in dance communication.

However, prospects for a comprehensive molecular analysis are bright. Results from the studies reviewed here indicate that molecular analysis of modules of dance communication is a productive approach. Numerous genes were identified with functions that relate to one of the hypothesized behavioral modules and a neurochemical analysis of one module—reward—has already yielded strong findings. One important insight from these early studies is that the evolution of dance behavior involved extensive reuse and adaptation of existing neuromolecular systems. Our findings emphasize the possibility that molecules involved in fundamental neuroethological modules operating in solitary insects have been recruited to regulate dance communication. Understanding better how and why this has occurred should yield important new insights into both the mechanisms and evolution of dance communication.

5.1.6   Outlook

Further development of a neuromolecular and neurogenomic dissection of dance will need to proceed hand-in-hand with both conceptual and technical advances in neurogenetics. Conceptually, this study would benefit from improved methods of analysis of gene expression studies. Transcriptomic studies, like those described here, are clearly informative, but they often propose quite a large number of possible candidate genes for further analysis. Typically the list of candidates is refined by identifying genes showing informative expression differences that are already known to have a plausible or interesting function in the trait in question, but over-reliance on this approach risks missing genes that would not *a priori* be thought of as being involved in this context. New developments in bioinformatics interpret transcriptomic datasets in terms of gene regulatory networks [29], which are pathways of interacting genes whose expression patterns are co-dependent. Adding this level of understanding to transcriptomic data can focus the identification of candidate genes to those most likely to be causal of differences in expression, and hopefully the trait of interest.

Technically, we urgently need better ways to test the function of identified candidate genes in the expression of dance behavior. Current methods of manipulation—RNAi [70] and pharmacology are useful, but need to be improved to target specific brain regions, or better circuits, in order to be able to test candidates identified from transcriptomic studies, and gain insights into neural mechanisms of dance. The *Drosophila* community has already developed remarkable genetic tools that allow rapid and reversible activation or silencing of specific circuits of the fly brain. The challenge to honey bee researchers is to work out how to translate such genetic technologies into the bee system. This effort would have to be developed in parallel with a capacity to maintain lines of mutant and genetically manipulated bees, which is far from easy when each genetically selected queen needs an entire colony to support her.

These challenges are not trivial by any means, but they are solvable. Bioinformatic understanding of the analysis of gene regulatory networks is developing rapidly. Viral-vector systems [77] are one possible solution to producing transgenic honey bees, and new successes with artificial queen rearing may make breeding and maintaining these lines easier. As the capacity for genetic manipulation of bees increases the potential benefits for neurogenomic analyses of the process of social evolution, and the development and evolution of complex behavioral traits such as the dance language will be enormous.

References

1. Alaux C, Robinson GE (2007) Alarm pheromone induces immediate-early gene expression and slow behavioral response in honey bees. J Chem Ecol 33:1346–1350
2. Alaux C, Duong N, Schneider SS, Southey BR, Rodriguez-Zas S et al (2009) Modulatory communication signal performance is associated with a distinct neurogenomic state in honey bees. PLoS One 4:e6694
3. Alaux C, Sinha S, Hasadsri L, Hunt GJ, Guzamán-Novoa E et al (2009) Honey bee aggression supports a link between gene regulation and behavioral evolution. Proc Natl Acad Sci USA 106(36):15400–15405
4. Barron AB, Robinson GE (2008) The utility of behavioral models and modules in molecular analyses of social behavior. Genes Brain Behav 7:257–265
5. Barron AB, Maleszka R, Van Der Meer RK, Robinson GE (2007) Octopamine modulates honey bee dance behavior. Proc Natl Acad Sci USA 104:1703–1707
6. Barron AB, Maleszka R, Helliwell PG, Robinson GE (2009) Effects of cocaine on honey bee dance behaviour. J Exp Biol 212:163–168
7. Bitterman ME, Menzel R, Fietz A, Schäfer S (1983) Classical conditioning of proboscis extension in honeybees Apis mellifera. J Comp Physiol 97:107–119
8. Bloch G, Solomon SM, Robinson GE, Fahrbach SE (2003) Patterns of PERIOD and pigment-dispersing hormone immunoreactivity in the brain of the European honeybee (Apis mellifera): age- and time-related plasticity. J Comp Neurol 464:269–284
9. Brockmann A, Robinson GE (2007) Central projections of sensory systems involved in honey bee dance language communication. Brain Behav Evol 70:125–136
10. Brockmann A, Annangudi SP, Richmond TA, Ament SA, Xie F et al (2009) Quantitative peptidomics reveal brain peptide signatures of behavior. Proc Natl Acad Sci USA 106:2383–2388
11. Carroll SB (2005) Endless forms most beautiful: the new science of evo devo. W.W. Norton, New York
12. Ceriani MF, Hogenesch JB, Yanovsky M, Panda S, Straume M et al (2002) Genome-wide expression analysis in Drosophila reveals genes controlling circadian behavior. J Neurosci 22(21):9305–9319
13. Clayton DF (2000) The genomic action potential. Neurobiol Learn Mem 74:185–216

14. Cummings ME, Larkins-Ford J, Reilly CRL, Wong RY, Ramsey M et al (2008) Sexual and social stimuli elicit rapid and contrasting genomic responses. Proc R Soc B 275:393–402
15. Dyer FC (2002) The biology of the dance language. Annu Rev Entomol 47:917–949
16. Esch H, Zhang S, Srinivasan M, Tautz J (2001) Honeybee dances communicate distance by optic flow. Nature 411:581–583
17. Fahrbach SE (2006) Structure of the mushroom bodies of the insect brain. Annu Rev Entomol 51:209–232
18. Gillman LN, Keeling DJ, Ross HA, Wright SD (2009) Latitude, elevation and the tempo of molecular evolution in mammals. Proc R Soc B 276:3353–3359
19. Gould JL, Gould CG (1988) The honey bee. W.H. Freeman & Company, New York
20. Greenspan RJ, Tononi G, Cirelli C, Shaw PJ (2001) Sleep and the fruit fly. Trends Neurosci 24:142–145
21. Griffith LC, Verselis LM, Aitken KM, Kyriacou CP, Greenspan RJ (1993) Inhibition of calcium/calmodulin-dependent protein kinase in *Drosophila* disrupts behavioral plasticity. Neuron 10:501–509
22. Gronenberg W, Lopez-Riquelme GO (2004) Multisensory convergence in the mushroom bodies of ants and bees. Acta Biol Hung 55:31–37
23. Hall JC (1998) Genetics of biological rhythms in *Drosophila*. Adv Genet 38:135–184
24. Hammer M (1993) An identified neuron mediates the unconditioned stimulus in associative olfactory learning in honeybees. Nature 366:59–63
25. Hammer M, Menzel R (1998) Multiple sites of associative odor learning as revealed by local brain microinjections of octopamine in honeybees. Learn Mem 5:146–156
26. Heinze S, Homberg U (2007) Maplike representation of celestial E-vector orientations in the brain of an insect. Science 315:995–997
27. Homberg U (2004) In search of the sky compass in the insect brain. Naturwissenschaften 91:199–208
28. Honey Bee Genome Consortium (2006) Insights into social insects from the genome of the honeybee *Apis mellifera*. Nature 443:931–949
29. Hyduke DR, Palsson BØ (2010) Towards genome-scale signalling network reconstructions. Nat Rev Genet 11:297–307
30. Kiya T, Kunieda T, Kubo T (2007) Increased neural activity of a mushroom body neuron subtype in the brains of forager honeybees. PLoS One 2:e371
31. Koeniger N, Koeniger G (1980) Observations and experiments on migration and dance communication of *Apis dorsata* in Sri Lanka. J Apicult Res 19:21–34
32. Kowalska E, Brown S (2007) Peripheral clocks: keeping up with the master clock. Cold Spring Harb Symp Quant Biol 72:301–305
33. Kruse AA, Stripling R, Clayton DF (2000) Minimal experience required for immediate-early gene induction in zebra finch neostriatum. Neurobiol Learn Mem 74:179–184
34. Kuwabara M (1957) Bildung des bedingten Reflexes von Pavlovs Typus bei der Honigbiene *Apis mellifica*. J Fac Sci Hokkaido Univ (Ser 6) 13:458–464
35. Lindauer M (1956) Über die Verstandigung bei Indischen Bienen. Z vergl Physiol 38:521–557
36. Martinek S, Inonog S, Manoukian AS, Young MW (2001) A role for the segment polarity gene shaggy/GSK-3 in the *Drosophila* circadian clock. Cell 105(6):769–779
37. Menzel R (2001) Searching for the memory trace in a mini-brain, the honeybee. Learn Mem 8:53–62
38. Michelsen A (1993) The transfer of information in the dance language of honeybees: progress and problems. J Comp Physiol A 173:135–141
39. Michelsen A (2003) Karl von Frisch lecture. Signals and flexibility in the dance communication of honeybees. J Comp Physiol A 189:165–174
40. Michelsen A, Andersen BB, Storm J, Kirchner WH, Lindauer M (1992) How honeybees perceive communication dances studied by means of a mechanical model. Behav Ecol Sociobiol 30:143–150
41. Neuser K, Triphan T, Mronz M, Poeck B, Strauss R (2008) Analysis of a spatial orientation memory in *Drosophila*. Nature 453:1244–1247

42. Okada R, Sakura M, Mizunami M (2003) Distribution of dendrites of descending neurons and its implications for the basic organization of the cockroach brain. J Comp Neurol 459:158–174

43. Oldroyd BP, Wongsiri S (2006) Asian honey bees biology, conservation and human interactions. Harvard University Press, Cambridge

44. Orr N, Orr GL, Hollingworth RM (1991) Characterization of a potent agonist of the insect octopamine-receptor-coupled adenylate-cyclase. Insect Biochem 21:335–340

45. Price J, Blau J, Rothenfluh A, Abodeely M, Kloss B et al (1998) double-time is a novel *Drosophila* clock gene that regulates PERIOD protein accumulation. Cell 94:83–95

46. Reppert S (2007) The ancestral circadian clock of monarch butterflies: role in time-compensated sun compass orientation. Cold Spring Harb Symp Quant Biol 72:113–118

47. Reppert SM, Gegear RJ, Merlin C (2010) Navigational mechanisms of migrating monarch butterflies. Trends Neurosci 33:399–406

48. Rinderer TE, Oldroyd BP, Sylvester HA, Wongsiri S, de Guzman LI (1992) Evolution of bee dances. Nature 360:305

49. Robinson GE, Fernald RD, Clayton DF (2008) Genes and social behavior. Science 322:896–900

50. Rohrseitz K, Tautz J (1999) Honey bee dance communication: waggle run direction coded in antennal contacts. J Comp Physiol A 184:463–470

51. Schröter U, Malun D, Menzel R (2007) Innervation pattern of suboesophageal ventral unpaired median neurones in the honeybee brain. Cell Tissue Res 327:647–667

52. Schwärzel M, Monastirioti M, Scholz H, Friggi-Grelin F, Birman S et al (2003) Dopamine and octopamine differentiate between aversive and appetitive olfactory memories in *Drosophila*. J Neurosci 23:10495–10502

53. Seeley TD (1994) Honey-bee foragers as sensory units of their colonies. Behav Ecol Sociobiol 34:51–62

54. Seeley TD (1995) The wisdom of the hive. Harvard University Press, Cambridge

55. Seeley TD, Mikheyev AS, Pagano GJ (2000) Dancing bees tune both duration and rate of waggle-run production in relation to nectar-source profitability. J Comp Physiol A 186:813–819

56. Sen Sarma M, Rodriguez-Zas SL, Gernat T, Nguyen T, Newman T et al (2010) Distance-responsive genes found in dancing honey bees. Genes Brain Behav 9:825–830

57. Sen Sarma M, Whitfield CW, Robinson GE (2007) Species differences in brain gene expression profiles associated with adult behavioral maturation in honey bees. BMC Genomics 8:202

58. Sen Sarma M, Rodriguez-Zas SL, Hong F, Zhong S, Robinson GE (2009) Transcriptomic profiling of central nervous system regions in three species of honey bee during dance communication behavior. PLoS One 4:e6408

59. Sinakevitch I, Niwa M, Strausfeld NJ (2005) Octopamine-like immunoreactivity in the honey bee and cockroach: Comparable organization in the brain and subesophageal ganglion. J Comp Neurol 488:233–254

60. Srinivasan MV, Zhang SW, Lehrer M, Collett TS (1996) Honeybee navigation en route to the goal: visual flight control and odometry. J Exp Biol 199:237–244

61. Srinivasan MV, Zhang S, Altwein M, Tautz J (2000) Honeybee navigation: nature and calibration of the "odometer." Science 287:851–853

62. Staudacher E (1998) Distribution and morphology of descending brain neurons in the cricket *Gryllus bimaculatus*. Cell Tissue Res 294:187–202

63. Suh J, Jackson FR (2007) *Drosophila* ebony activity is required in glia for the circadian regulation of locomotor activity. Neuron 55:435–447

64. Tanoue S, Krishnan P, Krishnan B, Dryer S, Hardin P (2004) Circadian clocks in antennal neurons are necessary and sufficient for olfaction rhythms in *Drosophila*. Curr Biol 14:638–649

65. Tautz J (1996) Honeybee waggle dance: recruitment success depends on the dance floor. J Exp Biol 199:1375–1381

66. Tautz J, Rohrseitz K (1998) What attracts honeybees to a waggle dancer? J Comp Physiol A 183:661–667
67. Towne WF (1985) Acoustic and visual cues in the dances of four honey bee species. Behav Ecol Sociobiol 16:185–187
68. von Frisch K (1967) The dance language and orientation of honeybees. Harvard University Press, Cambridge
69. von Schilcher F (1976) The behavior of cacophony, a courtship song mutant in *Drosophila melanogaster*. Behav Biol 17:187–196
70. Wang Y, Mutti NS, Ihle KE, Siegel A, Dolezal AG et al (2010) Down-Regulation of honey bee IRS gene biases behavior toward food rich in protein. PLoS Genet 6:11
71. Wenzel B, Hedwig B (1999) Neurochemical control of cricket stridulation revealed by pharmacological microinjections into the brain. J Exp Biol 202:2203–2216
72. Whitfield CW, Cziko A-M, Robinson GE (2003) Gene expression profiles in the brain predict behavior in individual honey bees. Science 302:296–299
73. Whitfield CW, Ben-Shahar Y, Brillet C, Leoncini I, Crauser D et al (2006) Genomic dissection of behavioral maturation in the honey bee. Proc Natl Acad Sci USA 103:16068–16075
74. Winston ML (1987) The biology of the honey bee. Harvard University Press, Cambridge
75. Winther AME, Acebes A, Ferrús A (2006) Tachykinin-related peptides modulate odor perception and locomotor activity in *Drosophila*. Mol Cell Neurosci 31:399–406
76. Xu K, Zheng X, Sehgal A (2008) Regulation of feeding and metabolism by neuronal and peripheral clocks in *Drosophila*. Cell Metab 8:289–300
77. Young LJ, Wang ZX (2004) The neurobiology of pair bonding. Nat Neurosci 7:1048–1054
78. Zhou Y, Schopperle WM, Murrey H, Jaramillo A, Dagan D et al (1999) A dynamically regulated 14-3-3, Slob, and Slowpoke potassium channel complex in *Drosophila* presynaptic nerve terminals. Neuron 22:809–818
79. Zhu H, Sauman I, Yuan Q, Casselman A, Emery-Le M et al (2008) Cryptochromes define a novel circadian clock mechanism in monarch butterflies that may underlie sun compass navigation. PLoS Biol 6:e4

Chapter 5.2
Neuroanatomical Dissection of the Honey Bee Brain Based on Temporal and Regional Gene Expression Patterns

Takeo Kubo

Abstract To identify the molecular and neural bases of honey bee social behavior and dance communication, we performed a neuroanatomical dissection of the honey bee brain based on molecular techniques. We systemically searched for genes, peptides, and proteins that are expressed in a region-preferential manner or whose expression differs depending on the behavior of an individual honey bee. Large- and small-type Kenyon cells (KCs) that comprise the honey bee mushroom bodies (MBs) have distinct gene expression patterns. Based on their temporal and regional expression profiles, the large- and small-type KCs are assumed to play a major part in calcium-signaling-mediated learning and memory, and ecdysteroid-signaling-mediated division of labor of workers, respectively. In addition, analysis of the neural activity in forager brains using a novel immediate early gene indicated that the small-type KCs are active in forager brains, suggesting that the small-type KCs are involved in processing information during the foraging flight. Furthermore, we identified two genes expressed preferentially in the monopolar cells of the optic lobes (OLs), the visual center in insect brains, and a novel gene expressed preferentially in a neural subpopulation located in the anterior to posterior dorsal OL region. Based on these findings, we propose that advanced 'module-functionalization' based on differential gene expression patterns could be a prominent feature of the honey bee brain.

T. Kubo (✉)
Department of Biological Sciences, Graduate School of Science, The University of Tokyo,
Bunkyo-ku, Tokyo 113-0033, Japan
e-mail: stkubo@biol.s.u-tokyo.ac.jp

C.G. Galizia et al. (eds.), *Honeybee Neurobiology and Behavior: A Tribute to Randolf Menzel*, DOI 10.1007/978-94-007-2099-2_26,
© Springer Science+Business Media B.V. 2012

Abbreviations (except gene and protein names)

AL	Antennal lobe
JH	Juvenile hormone
KC	Kenyon cell
MB	Mushroom body
ncRNA	non-coding RNA
OL	Optic lobe

5.2.1   The Theory of Localization of Brain Function in Mammals

As the theory of localization of brain function advocates, many functional areas, including the visual cortex, the auditory cortex and Broca's and Wernicke's language areas, are mapped onto specific regions of the cerebral cortex in human brain (see e.g. review: [17]). One possible explanation for the localization of brain function in mammals is that, during the evolution of brain function, a novel brain function (e.g., language areas) might have been assigned to a new brain region. To gain a better understanding of the molecular and neural bases of higher-order brain functions, it would be very helpful to identify genes that are expressed in a region-preferential manner in the higher-order brain structures: such genes could be used to visualize projections by expressing the green fluorescence protein gene (*gfp*) downstream of the promoter, and to determine the function of the brain region by expressing toxin genes downstream of the promoter and provide indication for the molecular basis of the evolution of the brain areas by analyzing the gene promoters themselves. Genome-wide atlases of gene expression in the mouse and human brains were recently reported (http://www.brain-map.org/). Some genes are also expressed in a region-preferential manner in the cerebral cortex in monkeys (e.g., [44]). The number of genes that are expressed in a region preferential manner, however, is estimated to be small in mammals, and thus the molecular bases of the function of the cerebral cortex regions still remain obscure.

In contrast, in this chapter, I propose that advanced 'module-functionalization' based on the distinct gene expression patterns could be a prominent feature of the brain of the honey bee, a social insect. We expect that this feature could be useful, not only for understanding the mechanism of the honey bee brain structure that regulates their social behaviors, but also to provide clues to understand the mechanism of the higher-order brain functions in mammals.

5.2.2   Social Behaviors and Brain Structure of the Honey Bee

The European honey bee (*Apis mellifera* L.) is a eusocial insect living in large colonies. Despite their comparatively small brains, honey bees exhibit advanced learning ability as well as complex social behaviors (reviews: [26, 47], see also Chaps. 2.5 and 6.6). Female honey bees differentiate into two castes: queens and workers,

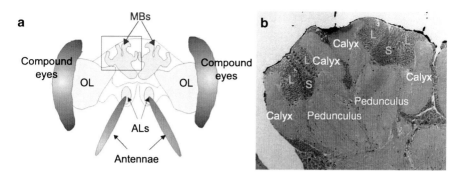

Fig. 5.2.1 Compartments of the honey bee brain and MBs. (**a**) Schematic drawing of the honey bee brain. (**b**) Hematoxylin-eosin staining of the *left* MB, which is boxed in panel (**a**). The somata of large (*L*) and small (*S*) types of KCs are located at the both edges and inner core of the inside of the calyces, respectively. *MB* mushroom body; *OL* optic lobe; *AL* antennal lobe

and workers shift from nursing their brood to foraging for nectar and pollen according to their age (age-polyethism) (review: [49]). In addition, foragers returning to the hives inform their nestmates of the location of a food source using the well-known dance communication (review: [47]), which is an abstract communication that is rare in the animal kingdom, except for human language. In 2006, the whole honey bee genome has been determined. Thus, the honey bee is a useful model animal for studies aimed at understanding the molecular and neural bases underling animal social behaviors and abstract communication.

Honey bee brains comprise several major regions, such as the mushroom bodies (MBs, a higher center), the OLs (a visual center), and the antennal lobes (ALs, an olfactory center) (Fig. 5.2.1).

Visual and olfactory information received at the compound eyes and antennae are first projected to the OLs and ALs, respectively, and then to other brain areas, such as the MBs [27, 35]. The MBs play important roles in sensory integration, learning, and memory (review: [26], see also Chaps. 2.5 and 3.1). The honey bee MBs are well developed compared with those of most other insects, and each of the paired MBs has two cup-like structures, called calyces (Fig. 5.2.1) (see Chap. 3.2). There are three types of Kenyon cells (KCs) that comprise the honey bee MBs: the large-type KCs and class I small-type KCs, whose somata are located at both edges and at the inner core inside of the calyces, respectively, and the class II small-type KCs, whose somata are located at the bottom of the MBs [27] (Fig. 5.2.1). Recently, more detailed subdivision of the Kenyon cells that project different parts of the MB calyx and pedunculus, and a nomenclature for them, have also been proposed ([35], review: [8]). The calyces and pedunculi of the MBs comprise the dendrites and axons of the KCs, respectively. Hereafter, I discuss only the large-type and class I small-type KCs, because gene expression patterns in class II small-type KCs are relatively difficult to analyze due to the restricted localization of the cells.

There are some structural characteristics in the honey bee brain that might reflect the importance of visual ability in the honey bee. The MB structure is altered depending on the age-polyethism of workers, and correlates with the foraging experience of the foragers (review: [8]). In addition, in Hymenopteran insects, visual information processed in the OLs projects directly to the MBs, whereas in many other insect species, the MBs are important for olfactory processing and there are few or no direct neural connections between the OLs and MBs (e.g., [36]).

5.2.3 Our Hypotheses and Neuroanatomical Dissection of the Honey Bee Brain Based on Molecular Techniques

Our methodology to analyze the brain functions that underlie the dance communication ability and social behavior of the honey bee was originally based on two principle hypotheses (for previous review, see [21]). First, we assumed that the honey bee might have newly acquired a higher-order brain function(s) that is responsible for dance communication during evolution, because among insects only bees possess the ability to use dance communication. Second, because each honey bee colony member expresses distinct and stereotyped social behaviors, we assumed that they might have distinct neural networks, and thus gene expression patterns, to be engaged in their distinct tasks. The second hypothesis has now been shown in numerous studies, from several labs (e.g., [2, 23], see also Chaps. 5.1 and 5.4). Therefore, we started to systematically search for and identify genes that are expressed in a MB-preferential manner in the honey bee brain, and those whose expression differs depending on the behavior of the honey bees. Chittka and colleagues have mapped the major events in the evolution of bee dances on the phylogenetic tree of eusocial bees [5]. Brockman and Robinson [4] showed that there is no 'dance-specific' sensory projection in the honey bee brain (see Chap. 5.1). Our hypotheses, however, do not seem to contradict these reports in that all support the notion that the honey bee has acquired higher-order brain functions during evolution.

We initially used the differential display method to search for differentially expressed genes. This allowed us to identify genes whose expression levels are quite low, as compared to cDNA microarray studies. Although the majority of our findings do not overlap with those of other groups (see Chaps. 5.1 and 5.4), possibly due to differences in the molecular techniques used by each group or the honey bee strains used, there is also overlap between our results and those of other groups, as described in the following.

5.2.4 Gene Expression Profiles Characteristic of Each of the KC Subtypes

5.2.4.1 Genes for Proteins Involved in Calcium Signaling

In our differential display screening, we selected bands whose intensities were stronger for MB samples than for OL samples, the latter of which were used as a control neural tissue. Our criterion for 'MB-preferential gene expression' is that expression of the identified gene was stronger in MBs than in most other brain regions, when analyzed by *in situ* hybridization. Table 5.2.1 summarizes the genes that have been identified in our laboratory, and those described in this chapter.

In the MBs, most of these genes are expressed in either a large- or small-type KC preferential manner, but a few are expressed throughout the MBs (Table 5.2.2).

We first identified the gene for 1,4,5-inositol trisphosphate receptor (IP$_3$R), which is involved in calcium-signaling that plays an important role in learning and memory in various animals. We showed that *IP$_3$R* is expressed preferentially in the large-type KCs in the honey bee brain [15].

Based on the assumption that the expression of genes for other proteins involved in calcium-signaling might also be expressed preferentially in the MBs, we analyzed the expression of the genes for protein kinase C (PKC) and Ca^{2+}/calmodulin-dependent protein kinase II (CaMKII), both of which are involved in calcium-signaling, and found that they were also expressed in a MB-preferential manner. *PKC* is expressed throughout the MBs, whereas *CaMKII* is expressed preferentially in the large-type KCs [16]. Subsequent search for genes with MB-preferential expression using cDNA microarray identified *IP$_3$ phosphatase* (*IP$_3$P*), which is also involved in calcium-signaling [40]. The *IP$_3$R*- and *IP$_3$P*-expressions are enriched 5.2- and 2.6-folds in the MBs compared to OLs, respectively. Taken together, these findings demonstrate that gene expression for some proteins involved in calcium-signaling is enhanced in the MBs, especially in the large-type KCs, suggesting that synaptic plasticity based on calcium-signaling is enhanced in the large type-KCs in the honey bee brain (Table 5.2.2).

Recent genome-wide transcriptomic comparisons of different brain regions in the honey bee showed that the expression levels of genes involved in signaling and synaptic remodeling are upregulated in the MBs, and corresponded with our findings that *IP$_3$R* and *CaMKII* are more highly expressed in the MBs [34]. In contrast, Kucharski and Maleszka [23] reported that one of three isoforms of the *IP$_3$ 3-kinase* (*IP$_3$K*) transcript is more enriched rather in OLs than in the other parts of the worker brain, implying that not all of the genes for proteins involved in calcium-signaling are upregulated in the large-type KCs.

5.2.4.2 Mushroom Body Large-Type KC-Specific Protein-1 (Mblk-1)

We next identified *Mushroom Body Large-Type KC-Specific Protein-1* (*Mblk-1*), which encodes a novel transcription factor, as a gene expressed preferentially in the large-type KCs [39] (Table 5.2.2). Although its *Drosophila* homologue, *E93,* is involved in metamorphosis downstream of the ecdysteroid-signaling, its molecular function remained unknown [25]. We showed that Mblk-1 functions as a DNA-sequence specific transcription activator, and its transcriptional activity is enhanced through phosphorylation [29].

Mblk-1 homologues are conserved across animal species, from insects to mammals [14]. To investigate the function of Mblk-1 homologues in the nervous system, we used

Table 5.2.1 Summary of genes expressed in a brain region-preferential manner and other reference genes described in this chapter

Name	Function of the product	Brain region where expressed	Bees analyzed	Reference (publication year)
Calcium-signaling				
IP_3R	Calcium channel on the endoplasmic reticulum	l-KC	W=Q=D	[15] (1998), [16] (2000), [34] (2007)
$CaMKII$	Calcium/calmodulin-dependent protein kinase	l-KC	W=Q=D	[16] (2000), [34] (2007)
PKC	Calcium/lipid-dependent protein kinase	l- and s-KC	W	[16] (2000)
IP_3P	Enzyme that hydrolyzes IP_3	l-KCs	Wᵃ	[40] (2002)
IP_3K	Enzyme that phosphorylates IP_3 (type A/B-transcript)	Whole brain/A-type OL/B-type	0-1 h<48 h<96 h-old W / 0-1 h>48 h=96 h-old W	[23] (2002)ᵇ
Ecdysteroid-signaling				
$Mblk$-$1/E93$	Ecdysone-regulated gene/transcription factor	l-KC	W, W pupae	[39] (2001)
$E74$	Ecdysone regulated genes/transcription factor	s-KC	W	[30] (2005)
BR-C	Ecdysone-regulated gene/transcription factor	l-KC	W	[31] (2006)
$E75$	Ecdysone-regulated gene/transcription factor	l- and s-KC	W	[31] (2006)
$HR38$	Hormone receptor-like 38 (orphan receptor)	s-KC	Q=N<F	[50] (2006)
USP	Cofactor that binds to EcR	s-KC	N>F	[46] (2006)
EcR	Ecdysone-receptor/transcription factor	s-KC	W=Q	[43] (2007)
Dopamine receptors				
$Dop1$	Dopamine D1 receptor	Whole brain	'Adult honey bee'	[3] (1998)
$Dop2$	Dopamine D2-like receptor	s-KC (constitutive) l-KC (increased with age)	neW=N=F, neD=mtD neW<N<F, neD<mtD	[13] (2003)
Other signaling				
RJP-3	A royal jelly protein	A defined population of KCs	'Worker bee'	[22] (1998)
PKA	cAMP-dependent protein kinase	l- and s-KC, weakly in AL/OL	'Adult honey bee'	[6] (2001), [34] (2007)
$For (PKG)$	cGMP-dependent protein kinase	s-KC and lamina in OL	N<F, saN<preF	[2] (2002)

Peptide

Trp	Tachykinin-related peptide/neuromodulator	s-KC and outer part of l-KCs, and some OL and AL neurons	N=F=Q=D	[42] (2004)
JHDK	Enzyme that phosphorylates JH diol	s-KC and outer part of l-KCs	W[c]	[45] (2007)
ncRNAs				
kakusei	Immediate early gene	Mainly s-KC	da=F, n.d. in N, fo	[19] (2007)
Ks-1	Function unknown	s-KC and large somata brain neurons	N=F=Q=D	[32] (2002)
AncR-1	Function unknown	Whole brain	W[d]	[33] (2004)
Nb-1	Function unknown	Octopamine-positive neurons	N>F=Q	[38] (2009)
mir-276	miRNA	s-KC and OL	N=F=Q=D	[12] (2011)
Genes with OL-preferential expression				
Futsch	Microtubule-associated protein	Monopolar cells in OL	N=F=Q=D	[18] (2010)
Tau	Microtubule-associated protein	Monopolar cells in OL	N=F=Q=D	[18] (2010)
MESK2	Implicated in Ras/MAPK-signaling	Horizontal zone in ventral OL	N=F=Q=D	[18] (2010)

Abbreviations: *l-KC* large-type KC, *s-KC* small-type KC, *W* worker with unidentified age/task. *N* nurse bee, *F* forager, *Q* queen, *D* drone, *ne* newly emerged, *mt* mature (older), *sa* same aged, *pre* precocious, *da* dancer, *fo* follower, = means similar expression patterns and/or expression levels, < means higher expression in right than in left, *n.d.* not detected

Notes: [a]Labor-dependent change in expression was not detected by cDNA microarray analysis in the study

[b]For age/sex/task-dependent differential expression of each isoform, see Ref. [23]

[c]Labor-dependent change in expression was not detected by proteomics in the study

[d]For tissue-specific expression, see ref. [33]

Table 5.2.2 Summary of the neuroanatomical dissection of the honey bee MBs based on distinct gene expression patterns

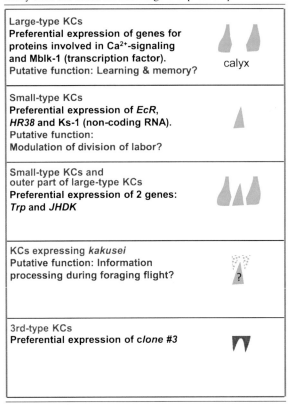

Each line indicates KC types (blue and pink letters), genes preferentially expressed therein (black letters), and putative function of the KC types (red letters). Figure at the right of each line indicates schematically the location of the somata of the corresponding KC type inside of a single MB calyx: green regions indicate large- and small-type KCs (1st to 3rd lines), and KCs active in the forager brain (4th line). Green dots mean that it is possible that *kakusei* is expressed not only in the small-type KCs, as originally reported [19], but also in some KCs located around the small-type KCs, in the forager brains. Pink regions indicate putative '3rd type-KCs'

Caenorhabditis elegans, a model animal with a simple body organization. We found that *Mblk-1-related factor-1* (*MBR-1*, a nematode homologue of *Mblk-1*) is expressed mainly in the neurons and is involved in the pruning of excessive neurites during development, which occurs even in *C. elegans* [14]. Furthermore, Hayashi et al. [10] revealed that the *mbr-1* mutant shows a defect in olfactory adaptation. Assuming that honey bee Mblk-1 also has a similar function, and considering that *Mblk-1* is expressed preferentially in the large-type KCs, this strengthens our hypothesis that large-type KCs are involved in neural plasticity, e.g. learning and memory (Table 5.2.2).

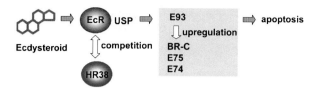

Fig. 5.2.2 Ecdysteroid-signaling that functions in the apoptosis of salivary gland during metamorphosis in *Drosophila melanogaster*. The triangular complex of ecdysterids, ecdysone receptor (*EcR*) and Ultraspiracle (*USP*) activates expression of the ecdysteroid-regulated genes: *E93* (*Drosophila* homologue of honey bee *Mblk-1*), *broad-complex* (*BR-C*), *E75* and *E74*, all of which encode transcription factors. *E93* upregulates *BR-C*, *E75* and *E74*. EcR competes with hormone receptor like-38 (HR38) to bind to their common co-factor USP

Both MBs and ALs are involved in olfactory learning in the honey bee, and cAMP-dependent signaling is implicated in the olfactory learning [28]. The gene for cAMP-dependent protein kinase (PKA), which plays a crucial role in cAMP-dependent signaling, is expressed predominantly in the MBs, and moderately in the OLs and ALs in the honey bee brain [6]. Therefore, it is plausible that both the calcium- and cAMP-dependent signaling play roles in learning and memory in the honey bee MBs.

5.2.4.3   Ecdysone-Regulated Genes and Genes Involved in Ecdysteroid-Signaling

Because *E93* functions downstream of ecdysteroid-signaling during metamorphosis in *Drosophila*, [25] we hypothesized that the expression of genes for other proteins involved in ecdysone-signaling (Fig. 5.2.2) is also enriched in the MBs in the honey bee brain.

Indeed, *Broad-Complex (BR-C)* is expressed preferentially in the large-type KCs [31]. Unexpectedly, however, *E74* and *Ecdysone receptor (EcR)* [48] are expressed preferentially in the small-type KCs [30, 43], whereas *E75* is expressed in both large- and small-type KCs [31]. These findings suggest that the function of distinct parts of ecdysteroid-signaling is enhanced in each of the KC type.

In addition, during the cDNA microarray search for genes whose expression in the brain differs depending on the task of a honey bee, we found that *HR38*-expression is higher in forager brains than in nurse bee brains, and is enriched in the small-type KCs [50] (Table 5.2.2). In *Drosophila*, *HR38* encodes a nuclear hormone receptor that is similar to, but distinct from, EcR. HR38 competes with EcR for the cofactor ultraspiracle (USP) and, by responding to different kinds of ecdysteroids, regulates the expression of a different set of target genes from EcR. Based on the fact that both *EcR* and *HR38* are expressed preferentially in the small-type KCs, and that *HR38*-expression is enhanced in the forager brain, we hypothesize that the mode of ecdysteroid-signaling in the small-type KCs changes from being EcR- to being HR38-dependent [50]. This modal change might result in modification of MB neural circuits that could be involved in the division of labor of workers. Similarly, Velarde et al. [46] reported that *USP* is expressed preferentially in small type-KCs, and its expression decreases with the division of labor of workers. In contrast to juvenile hormone (JH), which is involved in regulating the division of labor of the workers (e.g., [37], see also Chaps. 1.2 and 5.1), little is known about the role of ecdysone in modulating honey bee behaviors. Hartfelder et al. [9] reported that 20-hydroxyecdysone titers are higher in queens than in workers. In addition, genes for enzymes involved in ecdysteroid biosynthesis are expressed in the brain, fat body and ovary of the worker honey bees [51]. These findings raise the possibility that ecdysteroids might also be involved in regulating honey bee social behaviors by affecting brain function.

Distinct gene expression patterns between the large- and small-type KCs are also reported for other genes. *Dopamine D2-like receptor* (*Dop2*) is also expressed preferentially in the MBs [22] (see Chap. 3.6): its expression in the small-type KCs is constitutive, whereas that in the large-type KCs increases with age in both workers and drones, suggesting its role in differential regulation of the KC subtypes [13]. In contrast, the widespread expression of *dopamine D1 receptor* (*Dop1*) in the honey bee brain suggests its role in both sensory and higher-order information processing [3]. The foraging gene (*for*), which encodes a cGMP-dependent protein kinase (PKG), is related to insect foraging behavior. Ben-Shahar et al. [2] reported that *for* is involved in the division of labor of workers in the honey bee. The expression of *for* is enriched in the small-type KCs as well as in the OLs in the worker brain, and increases in the brain with division of labor of workers [2]. The functional relationship among the genes expressed preferentially in the large- or small-type KCs, respectively, requires further investigation.

5.2.4.4 Peptides and Proteins Identified by MALDI-TOF MS and Proteomics

We also searched for peptides and proteins with MB-preferential expression. Takeuchi & Nakajima et al. [41] applied direct MALDI-TOF MS to search for peptides expressed preferentially in the MBs, and identified tachykinin-related peptides (Trp) (see Chap. 3.7), which have neuromodulatory functions in the central nervous system in insects. In the honey bee brain, *Trp* is expressed preferentially in the MBs and in some neurons located in the OLs and ALs [42]. To our knowledge, there is no report of *Trp*-expression in the MB in other insects, implying that the Trp-mediated neuromodulation in the MB is unique to the honey bee. In addition, *Trp* exhibited a unique expression pattern in the MBs:

strong expression in the small-type KCs, moderate expression in the outer part of the large-type KCs, and almost no expression in the inner large-type KCs, suggesting that the large-type KCs are divided into two subtypes based on *Trp*-expression (Table 5.2.2).

We also used proteomics to show that the protein/gene for JH diol kinase (JHDK), which is thought to be involved in JH metabolism, is expressed preferentially in the MBs [45]. JH is involved in regulating the division of labor of workers (e.g., [37]). Therefore, we speculated that JH might modulate MB neural activity by binding to its unknown receptor, and is then later inactivated by JHDK. Interestingly, in the MBs, *JHDK* exhibited an expression pattern similar to that of *Trp* described above: *JHDK* is expressed strongly in the outer part of the large-type KCs, moderately expressed in the small-type KCs, and scarcely expressed in the inner large-type KCs [45] (Table 5.2.2). We recently used a cDNA microarray to identify a novel gene (tentatively termed *Clone #3*), whose expression pattern in the MB is compensatory to those of *Trp* and *JHDK*: *Clone #3* is expressed preferentially in a KC subpopulation (tentatively termed '3rd-type KCs'), in which neither *Trp* nor *JHDK* are expressed. This suggests that the honey bee MBs actually comprise three types of KCs (Table 5.2.2) (Kaneko et al., unpublished). Taken together, our findings strongly suggest that the honey bee MBs have a 'module-like structure' based on the distinct gene expression patterns.

5.2.5 Neural Activity Mapping in the Forager Brain Using an Immediate Early Gene

Then what brain region(s) are important for the highly advanced behaviors of the honey bee? Kiya intended to use an immediate early gene whose expression is increased transiently after the neuron is activated to map active brain regions in dancing workers. We used differential display to identify a novel immediate early gene, termed *kakusei* (which means 'awakening' in Japanese), whose expression is transiently induced in the brains of workers after neural excitation [19]. *In situ* hybridization revealed *kakusei*-expression is detectable only in workers that were dancing in their hives (dancers: just after the foraging flight), but not those that were following dancers (followers: possibly, before the foraging flight) and nurse bees. In the forager brains, *kakusei*-expression was detected mainly in the small-type KCs (Table 5.2.2). Subsequent analysis revealed that the small-type KC preferential *kakusei*-expression is observed in the brains of all foragers with pollen load, suggesting that *kakusei*-expression, and thus the neural activity, is not due to dancing behavior, but rather to the foraging experience. To our knowledge, this is the first study to demonstrate the brain regions active in forager/dancer honey bees. *kakusei* does not encode any significant open reading frames, and the *kakusei*-transcripts are located in the nuclei, indicating that *kakusei* transcripts function as non-coding RNA (ncRNA) [19].

5.2.6 Novel Non-coding RNAs (ncRNA) Derived from the Honey Bee Brain

In addition to kakusei, we have identified three novel ncRNAs: Kenyon cell/small-type preferential gene-1 (Ks-1) [32], Apis non-coding RNA-1 (AncR-1) [33], and Nurse bee-preferential gene-1 (Nb-1) [38]. *In situ* hybridization revealed that Ks-1 is expressed predominantly in the small-type KCs (Table 5.2.2). Ks-1-transcripts (17.5 kb in size) do not encode any significant open reading frames and are exclusively located in the nuclei of Ks-1-expressing cells [32]. Although the sex-specific nuclear ncRNAs that function in chromosome dosage compensation were identified in mammals and *Drosophila*, this was the first report of the expression of a long-type ncRNA that was independent of the sex.

In contrast to Ks-1, whose gene contains no intron, AncR-1 is categorized as an mRNA-type nuclear ncRNA [33]. AncR-1 was expressed in the whole honey bee brain as well as in some other tissues in a tissue-selective manner: AncR-1 is predominantly expressed not only in the whole brain, but also in queen ovaries, drone testes, and worker hypopharyngeal glands (HGs) that synthesize major royal jelly proteins, implying that the function of AncR-1 is related to the physiology of the organs that are specific to queens (ovaries), drones (testes), and workers (HGs), respectively [33].

More recently, Tadano and Takeuchi et al. [38] identified the gene for a novel ncRNA, Nb-1 (approximately 600b in size), as a gene whose expression is higher in the nurse bee brains than in forager or queen brains. Nb-1-expression in the worker brain is restricted to the octopamine-positive neurons that project axons to the corpora allata, which secrete JH [38]. Considering that the hemolymph JH titer depends on social task performance, and JH influences the pace of division of labor of workers [37], it might be that Nb-1-expression level in octopamine-positive neurons is somehow related to hemolymph JH titers (see Chap. 1.2).

Finally, Hori & Kaneko et al. [12] identified two microRNAs (miRNAs: *ame*-mir-276 and -1000) whose expression is enriched in the honey bee brain. In general, miRNAs are 20–25 base long ncRNAs that usually repress eukaryotic gene expression. Because *ame*-mir-276 is expressed in a brain region-preferential manner, it might well be that this miRNA regulates the brain region-preferential expression of target genes at posttranscriptional steps.

5.2.7 Identification of Genes Expressed Preferentially in the OLs

Recent findings have revealed that foragers gauge the flight distance to the food source based on the 'optic flow' perceived during their foraging flight (e.g., [7], see also Chap. 2.1). Assuming that some OL neural populations could be involved in detecting optic flow, we first examined whether there are 'module-like structure(s)' in the OLs. We used a cDNA microarray to search for genes expressed preferentially

a   **b**

futsch and *tau*   *mesk2*

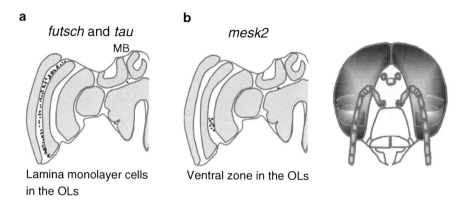

Lamina monolayer cells   Ventral zone in the OLs
in the OLs

Fig. 5.2.3 Schematic drawing of the expression of two types of genes in the OLs. (**a**) Distribution of *Futsch*- and *Tau*-expressing cells (*red dots*) in the left worker brain hemisphere. (*b left panel*) Distribution of *MESK2*-expressing cells (*red dots*) in the left worker brain hemisphere. (*b right panel*) Schematic drawing of the position of the *MESK2*-expressing cells, which are located from anterior to posterior in the dorsal OL region, in the drone head (*pink*). *MB* mushroom body; *OL* optic lobe

in the OLs in the honey bee brain. We identified three genes: *Futsch* and *Tau*, both of which encode microtubule-associated protein family, and *Misexpression Suppressor of Dominant-negative Kinase Suppressor of Ras 2* (*MESK2*), which might be involved in Ras/MAPK-signaling in *Drosophila*.

Futsch and *Tau* are expressed preferentially in the monopolar cells, which may be involved in detecting contrast between visual objects, and are located at the outermost OL layer (lamina) in the honey bee brain (Fig. 5.2.3) [18]. In contrast, *MESK2*-expressing cells form a zone that spans the anterior to posterior dorsal regions of the OLs (Fig. 5.2.3). Considering the location of the *MESK2*-expressing cells in the OLs, it might be that these cells are important for detecting visual cues, e.g., optic flow, present on the ground rather than in the air. Our findings suggest that not only the MBs, but also the OLs of the honey bee brain have a 'module-like structure' based on the distinct gene expression patterns.

5.2.8   Outlook

One strength of neuroanatomical dissection of the brain based on distinct gene expression patterns is that it can lead to the identification of new brain areas that have not been discriminated morphologically (see Chap. 3.7). Another is, that it provides information about the function of the brain area based on the function of the identified genes. Our present hypothesis is that advanced 'module-functionalization (functional specification of brain regions based on distinct gene expression patterns)' could be a prominent feature of the honey bee brain. Most of the genes that we identified as being preferentially expressed in the MBs have not been reported as MB-preferential genes in *Drosophila* [20]. In addition, although mosaic analysis with a repressible cell marker revealed that Kenyon cell subtypes also exist in the *Dsosophila* MBs (review: [8]), endogenous genes expressed in a region preferential manner in the *Drosophila* MBs have scarcely been identified. A comparative study is thus needed to examine whether the 'module-like structure' observed in the honey bee brain coincides with that previously proposed based on anatomical analysis ([35], review: [8]), and whether it is specific to certain Hymenopteran insects or conserved across animal species.

Clarifying the links between the function of each brain region and honey bee social behaviors could be a promising approach to solve the mysteries of the dance communication and social behaviors in the honey bee. For example, to test the possibility that neurons expressing *MESK2* are involved in gauging optic flow, it will be effective to knockdown *MESK2*-expression in these neurons and/or to knockdown the function of the neurons expressing *MESK2*, and analyze the phenotypes of the transgenic honey bees. Although we established two methods to analyze gene functions *in vivo*: electroporation [24] and recombinant baculovirus infection [1], these methods have not yet been used in *in vivo* experiments, and require further refinement. Gene manipulation techniques that can be used routinely for the analysis of *in vivo* gene function are indispensable for further development of honey bee molecular neuroethology.

In addition to the gene manipulation techniques, behavioral assay system using harnessed bees could also be useful to access honey bee brain functions involved in detecting optic flow. Recently, Hori and Takeuchi et al. [11] established a novel conditioning paradigm to associate the proboscis extension reflex with a motion cue, which is generated by projecting videos onto a screen that encompasses the visual field of the harnessed honey bees. These paradigms could be useful for the future analysis of honey bee visual abilities, by combining their use with imaging and/or electrophysiologic methodologies.

Finally, it is our impression that nc RNAs play important roles in regulating honey bee brain function and physiology. Compared to protein-coding genes, the genes for ncRNAs are less conserved among animal species, which makes it difficult to analyze the function of ncRNAs in other model animals. To analyze the function of ncRNA in the honey bee, the transgenic method is also needed for future research.

References

1. Ando T, Fujiyuki T, Kawashima T, Morioka M, Kubo T et al (2007) In vivo gene transfer into the honeybee using a nucleopolyhedrovirus vector. Biochem Biophys Res Commun 352(2):335–340
2. Ben-Shahar Y, Robichon A, Sokolowski MB, Robinson GE (2002) Influence of gene action across different time scales on behavior. Science 296(5568):741–744
3. Blenau W, Erber J, Baumann A (1998) Characterization of a dopamine D1 receptor from *Apis mellifera*: cloning, functional expression, pharmacology, and mRNA localization in the brain. J Neurochem 70(1):15–23

4. Brockmann A, Robinson GE (2007) Central projections of sensory systems involved in honey bee dance language communication. Brain Behav Evol 70(2):125–136
5. Chittka L, Dornhaus A (1999) Comparisons in physiology and evolution, and why bees can do the things they do. Ciencia al Dia International 2:1–17
6. Eisenhardt D, Fiala A, Braun P, Rosenboom H, Kress H et al (2001) Cloning of a catalytic subunit of cAMP-dependent protein kinase from the honeybee (*Apis mellifera*) and its localization in the brain. Insect Mol Biol 10(2):173–181
7. Esch HE, Zhang S, Srinivasan MV, Tautz J (2001) Honeybee dances communicate distances measured by optic flow. Nature 411(6837):581–583
8. Fahrbach SE (2006) Structure of the mushroom bodies of the insect brain. Annu Rev Entomol 51:209–232
9. Hartfelder K, Bitondi MM, Santana WC, Simoes ZL (2002) Ecdysteroid titer and reproduction in queens and workers of the honey bee and of a stingless bee: loss of ecdysteroid function at increasing levels of sociality? Insect Biochem Mol Biol 32(2):211–216
10. Hayashi Y, Hirotsu T, Iwata R, Kage-Nakadai E, Kunitomo H et al (2009) A trophic role for Wnt-Ror kinase signaling during developmental pruning in *Caenorhabditis elegans*. Nat Neurosci 12(8):981–987
11. Hori S, Takeuchi H, Kubo T (2007) Associative learning and discrimination of motion cues in the harnessed honeybee *Apis mellifera* L. J Comp Physiol A 193(8):825–833
12. Hori S, Kaneko K, Saito T, Takeuchi H, Kubo T (2011) Expression of two microRNAs, ame-mir-276 and -1000, in the adult honeybee (*Apis mellifera* L.) brain. Apidologie 42(1):89–102
13. Humphries MA, Mustard JA, Hunter SJ, Mercer A, Ward V et al (2003) Invertebrate D2 type dopamine receptor exhibits age-based plasticity of expression in the mushroom bodies of the honeybee brain. J Neurobiol 55(3):315–330
14. Kage E, Hayashi Y, Takeuchi H, Hirotsu T, Kunitomo H et al (2005) MBR-1, a novel helix-turn-helix transcription factor, is required for pruning excessive neurites in *Caenorhabditis elegans*. Curr Biol 15(17):1554–1559
15. Kamikouchi A, Takeuchi H, Sawata M, Ohashi K, Natori S et al (1998) Preferential expression of the gene for a putative inositol 1,4,5-trisphosphate receptor homologue in the mushroom bodies of the brain of the worker honeybee *Apis mellifera* L. Biochem Biophys Res Commun 242(1):181–186
16. Kamikouchi A, Takeuchi H, Sawata M, Natori S, Kubo T (2000) Concentrated expression of Ca^{2+}/calmodulin-dependent protein kinase II and protein kinase C in the mushroom bodies of the brain of the honeybee *Apis mellifera* L. J Comp Neurol 417(4):501–510
17. Kandel ER, Schwartz JH, Jessell TM (2000) Principles of neural science, 4th edn. McGraw-Hill, Health Professions Division, New York
18. Kaneko K, Hori S, Morimoto MM, Nakaoka T, Paul RK et al (2010) *In situ* hybridization analysis of the expression of *Futsch*, *Tau*, and *MESK2* homologues in the brain of the European honeybee (*Apis mellifera* L.). PLoS One 5(2):e9213
19. Kiya T, Kunieda T, Kubo T (2007) Increased neural activity of a mushroom body neuron subtype in the brains of forager honeybees. PLoS One 2(4):e371
20. Kobayashi M, Michaut L, Ino A, Honjo K, Nakajima T et al (2006) Differential microarray analysis of *Drosophila* mushroom body transcripts using chemical ablation. Proc Natl Acad Sci USA 103(39):14417–14422
21. Kubo T, Ohashi K, Takeuchi H, Sawata M, Kamikouchi A et al (1996) Molecular sociobiology of the honey bee. Chem Biol 34:793–798 (Kagaku to Seibutsu, in Japanese)
22. Kucharski R, Maleszka R (1998) A royal jelly protein is expressed in a subset of Kenyon cells in the mushroom bodies of the honey bee brain. Naturewissenshaften 85:343–346
23. Kucharski R, Maleszka R (2002) Molecular profiling of behavioural development: differential expression of mRNAs for inositol 1,4,5-trisphosphate 3-kinase isoforms in naive and experienced honeybees (*Apis mellifera*). Mol Brain Res 99(2):92–101
24. Kunieda T, Kubo T (2004) In vivo gene transfer into the adult honeybee brain by using electroporation. Biochem Biophys Res Commun 318(1):25–31

25. Lee CY, Wendel DP, Reid P, Lam G, Thummel CS et al (2000) E93 directs steroid-triggered programmed cell death in *Drosophila*. Mol Cell 6(2):433–443
26. Menzel R, Leboulle G, Eisenhardt D (2006) Small brains, bright minds. Cell 124(2):237–239
27. Mobbs PG (1982) The brain of the honeybee *Apis mellifera*. 1. The connections and spatial-organization of the mushroom bodies. Philos Trans R Soc B 298(1091):309–354
28. Müller U (2000) Prolonged activation of cAMP-dependent protein kinase during conditioning induces long-term memory in honeybees. Neuron 27(1):159–168
29. Park JM, Kunieda T, Kubo T (2003) The activity of Mblk-1, a mushroom body-selective transcription factor from the honeybee, is modulated by the ras/MAPK pathway. J Biol Chem 278(20):18689–18694
30. Paul RK, Takeuchi H, Matsuo Y, Kubo T (2005) Gene expression of ecdysteroid-regulated gene *E74* of the honeybee in ovary and brain. Insect Mol Biol 14(1):9–15
31. Paul RK, Takeuchi H, Kubo T (2006) Expression of two ecdysteroid-regulated genes, *Broad-Complex* and *E75*, in the brain and ovary of the honeybee (*Apis mellifera* L.). Zool Sci 23(12):1085–1092
32. Sawata M, Yoshino D, Takeuchi H, Kamikouchi A, Ohashi K et al (2002) Identification and punctate nuclear localization of a novel noncoding RNA, Ks-1, from the honeybee brain. RNA 8(6):772–785
33. Sawata M, Takeuchi H, Kubo T (2004) Identification and analysis of the minimal promoter activity of a novel noncoding nuclear RNA gene, *AncR-1*, from the honeybee (*Apis mellifera* L.). RNA 10(7):1047–1058
34. Sen Sarma M, Whitfield CW, Robinson GE (2007) Species differences in brain gene expression profiles associated with adult behavioral maturation in honey bees. BMC Genomics 8:202
35. Strausfeld NJ (2002) Organization of the honey bee mushroom body: representation of the calyx within the vertical and gamma lobes. J Comp Neurol 450(1):4–33
36. Strausfeld NJ, Hansen L, Li Y, Gomez RS, Ito K (1998) Evolution, discovery, and interpretations of arthropod mushroom bodies. Learn Mem 5(1–2):11–37
37. Sullivan JP, Fahrbach SE, Robinson GE (2000) Juvenile hormone paces behavioral development in the adult worker honey bee. Horm Behav 37(1):1–14
38. Tadano H, Yamazaki Y, Takeuchi H, Kubo T (2009) Age- and division-of-labour-dependent differential expression of a novel non-coding RNA, Nb-1, in the brain of worker honeybees, *Apis mellifera* L. Insect Mol Biol 18(6):715–726
39. Takeuchi H, Kage E, Sawata M, Kamikouchi A, Ohashi K et al (2001) Identification of a novel gene, *Mblk-1*, that encodes a putative transcription factor expressed preferentially in the large-type Kenyon cells of the honeybee brain. Insect Mol Biol 10(5):487–494
40. Takeuchi H, Fujiyuki T, Shirai K, Matsuo Y, Kamikouchi A et al (2002) Identification of genes expressed preferentially in the honeybee mushroom bodies by combination of differential display and cDNA microarray. FEBS Lett 513(2):230–234
41. Takeuchi H, Yasuda A, Yasuda-Kamatani Y, Kubo T, Nakajima T (2003) Identification of a tachykinin-related neuropeptide from the honeybee brain using direct MALDI-TOF MS and its gene expression in worker, queen and drone heads. Insect Mol Biol 12(3):291–298
42. Takeuchi H, Yasuda A, Yasuda-Kamatani Y, Sawata M, Matsuo Y et al (2004) Prepro-tachykinin gene expression in the brain of the honeybee *Apis mellifera*. Cell Tissue Res 316(2):281–293
43. Takeuchi H, Paul RK, Matsuzaka E, Kubo T (2007) *EcR-A* expression in the brain and ovary of the honeybee (*Apis mellifera* L.). Zool Sci 24(6):596–603
44. Tochitani S, Liang F, Watakabe A, Hashikawa T, Yamamori T (2001) The occ1 gene is preferentially expressed in the primary visual cortex in an activity-dependent manner: a pattern of gene expression related to the cytoarchitectonic area in adult macaque neocortex. Eur J Neurosci 13(2):297–307
45. Uno Y, Fujiyuki T, Morioka M, Takeuchi H, Kubo T (2007) Identification of proteins whose expression is up- or down-regulated in the mushroom bodies in the honeybee brain using proteomics. FEBS Lett 581(1):97–101

46. Velarde RA, Robinson GE, Fahrbach SE (2006) Nuclear receptors of the honey bee: annotation and expression in the adult brain. Insect Mol Biol 15(5):583–595
47. von Frisch K (1967) The dance language and orientation of bees. Belknap Press of Harvard University Press, Cambridge
48. Watanabe T, Takeuchi H, Kubo T (2010) Structural diversity and evolution of the N-terminal isoform-specific region of ecdysone receptor-A and -B1 isoforms in insects. BMC Evol Biol 10:40
49. Winston ML (1987) The biology of the honey bee. Harvard University Press, Cambridge
50. Yamazaki Y, Shirai K, Paul RK, Fujiyuki T, Wakamoto A et al (2006) Differential expression of *HR38* in the mushroom bodies of the honeybee brain depends on the caste and division of labor. FEBS Lett 580(11):2667–2670
51. Yamazaki Y, Kiuchi M, Takeuchi H, Kubo T (2011) Ecdysteroid biosynthesis in workers of the European honeybee *Apis mellifera* L. Insect Biochem Mol Biol 41(5):283–293

Chapter 5.3
Molecular Insights into Honey Bee Brain Plasticity

Judith Reinhard and Charles Claudianos

Abstract The honey bee worker experiences changing sensory environments throughout her adult life as she progresses from a young nurse bee living inside the hive to a forager bee that navigates the outdoors. Honey bees continually process and learn new sensory information, and their brain changes accordingly. Numerous studies have demonstrated age- and experience-dependent variations in neuropil volume and synaptic density of the honey bee antennal lobes (ALs) and the mushroom bodies (MBs), in particular linked to foraging and odor learning. Changes in antennal sensitivity and AL neural activity after olfactory learning have also been documented. Here, we present evidence for molecular changes occurring in the adult honey bee brain. We discuss how sensory experience and learning affect expression patterns of olfactory receptor genes in the antennae and synaptic adhesion molecules in higher brain centres. Our studies indicate the molecular basis of sensory processing is highly plastic throughout life, and that it is regulated by sensory input. We discuss how sensory regulated expression of olfactory receptors and synaptic molecules may provide a basis for understanding anatomical and physiological plasticity of the honey bee brain.

Abbreviations

AL   Antennal lobe
CS   Conditioned stimulus
MB   Mushroom body
US   Unconditioned stimulus

J. Reinhard • C. Claudianos (✉)
Queensland Brain Institute, The University of Queensland,
Brisbane, QLD 4072, Australia
e-mail: c.claudianos@uq.edu.au

C.G. Galizia et al. (eds.), *Honeybee Neurobiology and Behavior: A Tribute to Randolf Menzel*, DOI 10.1007/978-94-007-2099-2_27,
© Springer Science+Business Media B.V. 2012

5.3.1   Introduction

It has long been known that the brain is a plastic organ, changing throughout life [30]. Brain plasticity has been well studied in vertebrates, and includes variations in overall brain volume, number of neurons, number and size of synapses, neural wiring and neural activity [8]. Importantly, these changes involve not only increase of neuropil volume and synapse numbers, but also synaptic and neural pruning, which are important apoptotic processes associated with normal postnatal brain maturation [27]. What are the major triggers for changes in the brain to occur? Growing evidence suggests that processing new sensory information and learning associations between such information and specific outcomes are the main driver for brain plasticity. Mammalian studies have shown that an enriched environment leads to dramatic increases in brain size and weight, whereas sensory deprivation can result in the opposite [28, 30]. In a natural environment exposure to new sensory information occurs continually, and the brain needs to remain plastic throughout life to respond to new experiences and environmental changes in a timely fashion. A number of studies have shown that brain plasticity is not an exclusive vertebrate trait, but also occurs in invertebrates [23, 36, 49]. While these studies focus on only a few insect species, it is likely that it occurs in all insects that have a comparatively long life-span, undergo different life stages as an adult, and experience changing environments, such as the honey bee. In this chapter we will review anatomical and physiological evidence for brain plasticity in honey bees that has been compiled over the past decades. We will then focus on recent findings that suggest mechanisms underlying brain plasticity which are found in experience-dependent expression of sensory and synaptic molecules.

5.3.2   Sensory Environment and Honey Bee Brain Plasticity

On average, a honey bee worker lives 40–50 days after emergence, during which she has to perform a variety of age-related tasks [48]. During the first weeks of adulthood, worker bees perform tasks inside the hive as nurse bees, that is they experience the smells, sounds, and tactile stimulations of the social environment of a bee colony. Once they start foraging around 14–21 days of age, the bees enter the outside world [48]. There they experience a completely new environment with novel sensory information. In particular the bees' visual sense, which was never exposed to strong light inside the darkness of the hive, has to process massive amounts of new input. It is during this time, when learning of visual and olfactory stimuli associated with food sources and navigational routes become central to a bee's life. This is reflected in changes in the bee's brain.

Age- and experience-dependent changes in the honey bee brain have been reported in particular for the ALs (discussed in the next section) [42, 47] and the MBs [14, 31]. The latter are a paired structure in the insect brain, which receives sensory input from both the eyes and antennae, as well as mechanosensory and gustatory input. The MBs are a brain region important for multisensory integration, as well as learning, memory, and 'cognitive' function [21, 22, 35].

The transition from life inside the hive as nurse bee to activities outside the hive as forager bee is associated with a distinct increase in MB size [17, 34]. Forager bees have a larger mushroom body calyx than nurse bees of the same age that have not yet left the hive [17, 49]. The increased calyx size in forager bees is due to larger and more complex dendritic arborisations, which may be linked to the numerous associations a foraging bee needs to learn while navigating to and from food sources [17]. Neurogenesis was of course discussed as a potential contributor to the increase in MB size in forager bees, however there is no evidence that experience-dependent neurogenesis actually occurs in the adult honey bee brain [16].

Different environmental rearing conditions have also been shown to affect MB size in the first week of a bee's life [34], equivalent to what has been documented for mammals [28, 30]. MBs were significantly larger in bees that lived in the sensory rich environment of the hive, compared to bees that were kept in sensory and social isolation in an incubator. These findings suggest that the honey bee brain responds with great sensitivity to both social and sensory stimuli throughout adult life.

5.3.3 Plasticity of the Honey Bee Olfactory System

Honey bees need to cope with a constantly changing floral environment to find food. They have therefore evolved an amazing ability to learn different floral odors, to extinguish existing odor memories and replace them with new ones as different plants come into flower. The honey bee sense of smell originates in their antennae, and after odor detection, information is processed in the first olfactory neuropil, the AL. Information is then transferred via projection neurons to the MBs, where the information is integrated with other sensory input and where associative memories are formed, and to the lateral horn [21, 22] (see also Chap. 4.1).

A number of physiological studies have shown learning-induced plasticity of the olfactory periphery using the electroantennogram (EAG) technique. The results varied; some studies showed that antennal sensitivity generally increased with odor learning [11, 45], while others identified no such increase in EAG levels or even a decrease after odor learning, the latter possibly caused by neuronal adaptation to continuous odor presentation [3, 40]. These contradicting findings might be due to the fact that some of the studies used odor mixtures while the others used single odorants. We should also be mindful that all of these studies used slightly different short-term odor conditioning procedures, which could contribute to variability of the results. Furthermore, the effect of long-term odor conditioning on antennal sensitivity has not yet been formally investigated.

5.3.3.1 Plasticity of Honey Bee Antennal Lobes

The ALs, as first centre of olfactory processing, have been a focus of research into olfactory plasticity. The availability of sophisticated calcium imaging techniques, first developed in the late 1990s for insects, made it possible to record neural activity in the ALs during odor processing and learning in the live animal in real time [19]. Using this technique, several studies have reported that odor learning can lead to changes in the neural activity patterns in the ALs of insects, i.e. odors have a different neural representation after learning [15, 18], although other studies did not support this physiological plasticity [25, 37] (see also Chap. 6.1). It has also been reported that the volume of antennal lobe glomeruli changes with shift to foraging duties and odor learning [25, 42, 47]. This suggests that olfactory experience induces synaptogenesis also in peripheral processing centres such as the ALs. Indeed, two studies have shown for selected AL glomeruli that an increase in volume was accompanied by a significant increase in the number of synapses [6, 7]. Furthermore, it has been shown that prolonged exposure to the honey bee alarm pheromone component isopentyl acetate, as well as to the plant odor hexanal, induces changes in gene expression in the ALs [1]. Exposure to these odors increased expression of the immediate early gene and transcription factor, *IEG c-Jun*, which is involved in synaptic plasticity and to link experience to further changes in gene expression.

5.3.3.2   Plasticity of Honey Bee Olfactory Receptor Expression

A question that comes to mind is whether olfactory plasticity also occurs in the very periphery of the olfactory system, that is in the antennae themselves where the neurons first interface with odors. Odor molecules enter the antennae through the minute pores of olfactory sensilla, which house the dendrites of approximately 65,000 olfactory sensory neurons. Located on the dendrites of these olfactory neurons are the olfactory receptors, which are 7-transmembrane proteins. The honey bee has approximately 170 olfactory receptor genes [39]. Olfactory receptors bind incoming odorants, which triggers a signalling cascade, opening of ion channels and an electric message being sent from the olfactory sensory neuron to the brain. Each olfactory receptor binds a select range of odorants, with specificities varying from narrowly tuned pheromone receptors to broadly tuned floral receptors [20]. Each olfactory neuron carries only one type of specific olfactory receptor plus the generic olfactory receptor (*AmOr2*), which is required for signal transduction in insects [41].

Honey bee olfactory receptors have been characterized as part of the honey bee genome project, and the phylogenetic studies show that they form a diverged group, distinct from olfactory receptors in moths, mosquitoes and flies, with the exception of the generic *AmOr2* [39]. This evolutionary divergence is likely due to the special selective pressures associated with the life history of the honey bee, such as the honey bee's diet, which requires a focus on floral scent detection. Interestingly, when olfactory receptor expression patterns for honey bee workers and drones were compared, distinct differences became obvious [46]. The comparative microarray study showed that the olfactory receptor for the queen pheromone (*Or11*) was significantly up-regulated in drones [46]. Given that detection of queen pheromone in order to find and mate with a queen is the main objective in a drone's life, this result was not surprising. However, the study also showed that a number of other receptors (Or63, Or81, Or109, Or150, Or151, Or152) were expressed higher in worker bees than in drones. What is the reason for this differential expression in worker bees? With floral nectar and pollen being the honey bee's main food source, and worker bees constantly processing the odors emanating from these foods, it is highly likely that these six olfactory receptors bind floral odorants.

Assuming that Or63, Or81, Or109, Or150, Or151, Or152 are indeed floral scent receptors, we wondered whether they are subject to plasticity in expression levels depending on the floral scent environment a bee experiences. The reasoning behind this hypothesis was based on the chance observation that during scent-conditioning using the proboscis-extension-reflex (PER) assay, Australian honey bees showed a very poor learning performance for linalool compared to Northern hemisphere honey bees. However, this poor learning performance for linalool in Australian honey bees was only evident in winter (during which honey bees still forage in Australia due to the mild climate). In Australian spring and summer, the bees showed a linalool learning performance which was as good as the one reported for Northern hemisphere honey bees [32]. What had changed? We had used bees from the same hive with the same queen, sitting in the same location in both winter and spring, so it is unlikely that the genetic makeup of the bees had changed within 3–4 months. What was different however, was the type of food source available. In Australian winter, specifically in south-east Queensland where the experiments were conducted, honey bees forage on eucalypts, which are the predominant flowering plant at that time. Eucalypts, however, produce very little or no linalool in their floral bouquet and nectar; in contrast, in spring and summer, a large variety of plants are in flower, among them many exotic varieties rich in linalool [29]. Clearly, the bees were experiencing very different olfactory environments in winter and summer, and this was reflected in a plasticity of olfactory learning performance. We wondered whether this observed behavioral plasticity would be mirrored by a plasticity of olfactory receptor expression.

We therefore conducted a semi-quantitative RT-PCR analysis of the above six olfactory receptors comparing their expression pattern in Australian honey bees from Brisbane, Queensland versus American honey bee workers from Urbana, Illinois. The results revealed that indeed there were differences in expression pattern, with Or109, Or150 and Or151 having a lower expression in Brisbane bees than in Urbana bees, while Or152 had a slightly higher expression in Brisbane bees (Table 5.3.1). This was the first indication that olfactory receptor expression within the same species and sex may be plastic and could possibly depend on the sensory environment. One could speculate that the observed behavioral difference between Australian and Northern hemisphere bees in linalool learning might have its basis in differential olfactory receptor expression. Of course, these are preliminary findings and we need to extend this study by using hives from a range of different environments, cities, and countries. This will allow us to test how different factors including genetics, region, global hemisphere and sensory environment contribute to variability of Or expression.

Table 5.3.1 Relative expression of selected honey bee olfactory receptors (Or) from antenna tissue, comparing foragers from Urbana (USA) with foragers from Brisbane (Australia)

	Urbana IL USA[a]	Brisbane QLD Australia[b]
Or63	++	++
Or81	+	+
Or109	+	+/−
Or150	+++	+
Or151	++	+/−
Or152	+/−	+

+++ very high expression, ++ high expression, + medium expression, +/− low expression
[a]Expression levels based on data from [46]
[b]Expression levels assessed by semi-qt PCR amplification and compared to data from [46]

To test for seasonal effects, we conducted a second *Or* expression analysis using Australian bees from the same hive (to reduce genetic variability) comparing winter foragers with summer foragers. These groups of bees clearly experience different olfactory environments, as outlined above. Again, we found differences in expression patterns of olfactory receptors (Claudianos & Reinhard, unpublished data). These data lend further support to the hypothesis that *Or* expression is plastic and may be regulated by the local olfactory environment honey bees experience.

What is the biological significance of the plasticity in *Or* expression? One could imagine that a plastic and reversible up- and down-regulation of olfactory receptors is an adaptation to the constantly changing floral environment a bee has to manage. Detection of a familiar odor may not require the relevant olfactory receptors to be expressed at high levels, because a whiff of the odor will be sufficient to trigger memory recall of the associated food source [38]. On the other hand, bees may need to express olfactory receptors for novel odors at high levels to maximize the chances of detecting new food sources.

At this stage, the above hypothesis is of course rather speculative. A large body of work investigating plasticity of individual olfactory receptors under controlled scent conditions, linked to their specific odor ligands will be required to provide experimental support for our hypothesis. But importantly, our study has given first insight into molecular plasticity of the honey bee's sense of smell. It suggests that the molecular basis of odor processing is not static, and that changes occur at the molecular level even at the very periphery. The fact that olfactory plasticity seems to be linked to scent environment could explain why sometimes behavioral studies based on olfaction carried out in different laboratories in different parts of the world, do not always produce the exact same results. We must consider that such data might only represent a "snapshot in time", similar to the varying results on linalool learning performance in winter and summer bees in our laboratory in Australia.

5.3.4   Plasticity of Synaptic Molecules in Higher Brain Centres

There is clear anatomical evidence that the higher brain centres of honey bees, in particular the MBs, change after sensory stimulation. Neuropil volume and dendritic outgrowth increase, and the number of microglomerular complexes in the MBs (see Chap. 3.2), as well as their density changes with sensory experience and long-term memory formation [26, 31]. This suggests that the synaptic architecture of the MBs is highly plastic, responding with formation of new synapses when the bee encounters and learns novel sensory stimuli during foraging outside the hive. If indeed sensory stimulation leads to synaptogenesis, we should be able to show that the expression levels of synaptic molecules, which form the physical connection between pre- and postsynaptic neurons, also changes. Two adhesive synaptic molecules have recently been identified in playing a crucial role in synapse formation and maturation in vertebrates, namely the presynaptic *neurexins* and their postsynaptic binding partners, the *neuroligins* (Fig. 5.3.1) [13, 33]. We therefore postulated that *neurexins* and *neuroligins* could similarly be used to investigate brain plasticity

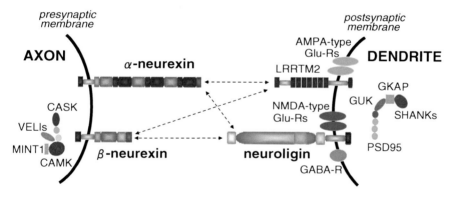

Fig. 5.3.1 Schematic representation of neurexin, neuroligin, and leucine-rich repeat transmembrane (*LRRTM2*) proteins in the synapse, including selected presynaptic and postsynaptic binding proteins involved in scaffolding and signaling. Shown are conserved α-neurexin and β-neurexin isoform proteins that differentially bind with neuroligins 1–4 and LRRTM2 and specify the development of excitatory and inhibitory synapses. Interacting cytoplasmic proteins of neurexin such as calcium/calmodulin-dependent protein kinases (*CASK and CAMK*) associate with a cell polarity-related protein (*VELI*) and neuronal adaptor protein (*MINT1*) to form a multidomain protein scaffold that anchors presynaptic receptors involved in ion channel trafficking. Similarly shown are interacting cytoplasmic proteins of neuroligin such as postsynaptic density protein (*PSD95*), related to CAMK, that forms a scaffold complex with SH3 and ankyrin repeat domain protein (*SHANK*), guanylate kinase (*GUK*) and guanylate kinase-associated protein (*GKAP*) to create a postsynaptic density that connects neurotransmitter N-methyl D-aspartate (*NMDA*), α-amino-3-hydroxy-5-methyl-4-isoxazolepropionic acid (*AMPA*) and gamma-aminobutyric acid (*GABA*) receptors, ion channels, and other membrane proteins to the actin cytoskeleton and G-protein-coupled signaling pathways of the neuron. The neurexin-neuroligin/LRRTM complex is highly conserved between vertebrates and invertebrates [4] many of the proteins occur as reciprocal orthologues between humans and honey bees. This conservation reflects key functional roles associated to cognitive processing [44]

in the honey bee. Indeed, our laboratory has demonstrated for the first time, that these key synaptic molecules also exist in bees, and that they constitute an ancient, highly conserved synaptic adhesion complex [4].

5.3.4.1 Expression of Neuroligins and Neurexin in the Honey Bee Brain

We characterized five *neuroligins* (*neuroligin 1–5*) and one *neurexin* (*neurexin I*) in honey bees [4]. Using whole brains from honey bee larvae, pupae, newly emerged bees (24 h old), 7-day-old and 21-day-old adults we showed that *neuroligins* and *neurexin I* were expressed in the bee brain throughout development from larvae to adult life stages [4]. Expression levels of *neurexin I* and *neuroligins 2–5* significantly increased through development, with particularly pronounced up-regulation after emergence and during adult development from nurse bee to forager bee. An increase in expression of these synaptic molecules could be due to both experience-dependent factors, such as novel and enriched sensory stimulation during adulthood, and experience-independent factors, that are part of the normal developmental path. Interestingly, *neuroligin 1* was the only molecule that did not show any obvious increase in expression during development. However, this could have been due to the generally very low expression of *neuroligin 1*, which would make any increase appear small in comparison to the high expression levels of the other *neuroligins*.

In situ hybridisation was used to investigate the exact distribution of *neuroligins* and *neurexin* in different brain structures. In particular, we investigated neuroligin 3 and *neurexin I*. Our data showed that *neuroligin 3* is predominantly found in the MBs of the adult honey bee brain, with some expression also in the cell bodies of the optic lobes, ALs and central body [4]. A broadly similar distribution pattern was found for its presynaptic binding partner, *neurexin I*. The distinct expression in the MBs as centre for learning, memory, and sensory integration, suggested a role for these molecules in synapse formation associated with sensory processing and learning, a hypothesis which we investigated in the studies described below.

5.3.4.2 Functional Role of Honey Bee Neuroligins and Neurexin

Localisation and developmental profile of these synaptic molecules clearly suggests their involvement in experience- and age-dependent honey bee brain plasticity. Indeed, a role for *neurexin I* in synaptogenesis and learning had already been demonstrated in *Drosophila*, where *neurexin I* null mutants were found to exhibit decreased synapse number as well as learning defects in larvae [50]. We therefore embarked on a behavioral study, investigating the functional role of *neuroligins* and *neurexin I* in sensory processing and learning in adult honey bees [5].

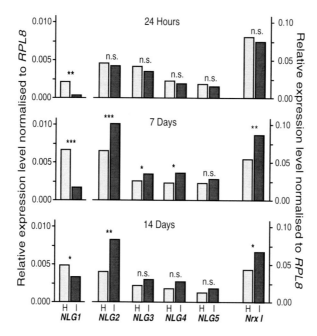

Fig. 5.3.2 Relative expression of *neuroligins* (*NLG1-5*) and *neurexin I* (*NrxI*) in honeybee brain tissue, comparing bees that lived since emergence in a normal hive environment (*H grey bars*) with bees that lived since emergence in isolation in a dark incubator (*I black bars*); n = 10 bees per group. Expression was assessed at 24 h, 7 and 14 days by quantitative real time PCR amplification relative to RPL8; detailed methodologies described in [5]. *** p<0.001, ** p<0.01, * p<0.05, n.s. no significant difference (*t*-test). Figure modified from Biswas et al. [5], with permission

We first isolated a group of honey bees at emergence and kept them in cages in a dark incubator for 14 days. These isolated honey bees had food ad libitum, but limited social interaction and limited sensory experience being kept in the dark and without the tactile and olfactory stimuli from the hive environment. At 24 h, 7 and 14 days of age, we analysed expression levels of *neuroligins 1–5* and *neurexin I* in the brains of these isolated bees, and compared them to bees of the same age from the same hive, which had been reared in the sensory enriched, natural environment of the hive. Intriguingly, only expression of *neuroligin 1*, which has a generally low level of expression, was significantly increased in bees of all three age groups that had been exposed to the enriched hive environment compared to the isolated bees (Fig. 5.3.2) [5]. In contrast to *neuroligin 1*, experience dependent expression of *neuroligins 2–5* and *neurexin I* varied over time. Expression levels were similar at 24 h, but at 7 and 14 days most of them were increased in isolated bees as compared to hive bees.

Clearly, sensory deprivation and sensory enrichment, respectively, had a marked effect on *neurexin* and *neuroligin* expression in the brain, suggesting a role for these molecules in sensory induced brain plasticity during adulthood. Our study also showed that *neuroligin 1* has a different role than the other *neuroligins*. A newly emerged bee constantly receives new sensory input from the hive environment, which influences brain connectivity. The increased *neuroligin 1* expression in sensory stimulated bees may reflect establishment of these new connections. Given *neuroligin 1* has been shown to specify excitatory glutamatergic synapses involved in regulating afferent neuronal circuits in vertebrates [43] and neuromuscular motor control in *Drosophila* [2], it is likely that it has the same or a similar role in honey bees. Arguably, a lack of sensory stimulation as experienced by the isolated bees reduces the requirement for excitatory synapses, which is reflected in the lower *neuroligin 1* expression level in isolated bees [5].

What role, then, do the other *neuroligins* and *neurexin* play? These molecules generally show a higher expression in isolated bees, which seems counter-intuitive. One explanation may be that in absence of sensory input, the brain remains in the state it was at emergence, foregoing synaptic or neuronal pruning associated with normal postnatal brain development. That is, the brain of an isolated bee is kept in a pre-adapted state. The phenomenon of synaptic elimination has been reported in vertebrate models [27], and has also been associated with changes in the *Drosophila* brain [24]. Our data on decreased *neuroligin 2–5* and *neurexin 1* expression in sensory stimulated adult bees suggest similar processes may occur during adult brain development in the honey bee.

The functional role of *neuroligin 1* in sensory induced brain plasticity in honey bees was confirmed in a second study [5]. We used the well-established proboscis-extension-reflex conditioning paradigm (see Chap. 6.2) to train 21 day-old bees to associate lemon scent (CS, conditioned stimulus) with a sugar reward (US, unconditioned stimulus). Nine trials over 2 days were used, resulting in long-term memory (LTM) of the association (Fig. 5.3.3).

The control group of bees was subjected to unpaired conditioning, separating US and CS by a 15 min inter-stimulus interval; thus the bees received the same amount of sensory exposure, but were prevented from associative learning. After 2 days of conditioning, the brains of trained and control bees were analysed for *neuroligin 1* expression, revealing a *3.3-fold* higher expression in trained bees (Fig. 5.3.3) [5]. This suggests that *neuroligin 1* is not only involved in synaptogenesis linked to general sensory processing, but also plays a significant role in synapse formation during associative learning and/or long-term memory formation. We also found an equivalent increase in the expression of *neurexin 1* after conditioning (*3.6-fold* increase), which likely reflects the required interaction of these pre- and postsynaptic binding partners in the synapse.

The up-regulation of synaptic connectivity is a well-documented consequence of intensive learning and memory in vertebrates [8], and our study provides molecular evidence that it also occurs in honey bees. However, we are mindful that the forager bees we used in the above study are likely to have had natural foraging experience prior to our conditioning experiment, and would have already learnt a range of olfactory associations outside the hive. The additional scent training received in our experiment might have led to reinforcement or rewiring of existing foraging-related memories, as reflected by the increased *neuroligin* and *neurexin* levels in trained bees. Alternatively, the lower expression levels in control bees could be interpreted

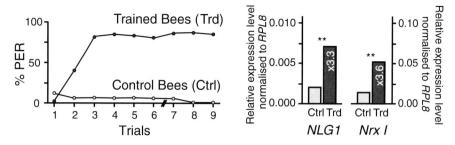

Fig. 5.3.3 Relative expression of *neuroligin 1 (NLG1)* and *neurexin I (NrxI)* in honeybee brain tissue, comparing bees that received associative scent conditioning to control bees. *Left:* Acquisition of the Proboscis Extension Response (*PER*) by bees conditioned to associate lemon scent (*CS*) with a sugar reward (*US*) (*Trained Bees*), and bees that received the unpaired control (US presented 15 min before CS) (*Control Bees*); nine trials were conducted over 2 days (six trials on day 1 and three trials on day 2) using 20 bees per group. *Right:* Expression of *NLG1* and *NrxI* in trained bees (*Trd*) and control bees (*Ctrl*); n = 10 bees per group. Expression levels were assessed by quantitative real time PCR amplification relative to *RPL8*; detailed methodologies described in [5]. ** p < 0.01, * p < 0.05, n.s. no significant difference (*t*-test). Figure modified from Biswas et al. [5], with permission

as a decrease in *neuroligin 1* and *neurexin I* due to fading and extinction of prior foraging memories linked with synaptic pruning.

Many questions remain to be answered with respect to *neuroligin* and *neurexin* function, especially considering the extensive number of splice variants [4]. Also, we need to investigate the role of receptors linked to *neuroligins* and other postsynaptic molecules such as leucine-rich repeat transmembrane proteins (LRRTM2), which is a second binding partner for *neurexin* (Fig. 5.3.1) [12]. Nevertheless, our work highlights that synaptic molecules, and in particular *neuroligin 1* are important players in the mechanisms underlying sensory-dependent honey bee brain plasticity.

5.3.5   Future Research into Molecular Plasticity of the Honey Bee Brain

Publication of the honey bee genome has accelerated the field of honey bee molecular neuroscience. We are now in a position to identify genes and gene products involved in honey bee brain plasticity, and through clever integrated approaches we are able to assign functional roles to these genes. The emphasis in future research clearly needs to be on integrated science: molecular genetics, bioinformatics, electro- and optophysiology, and behavioral methods all need to come together to answer the big question of how the bee brain works and which mechanisms underlie brain plasticity.

Molecular brain plasticity is a novel field of research, and of course, some painstaking and time-consuming basic technology needs to be developed first. For example we need to design specific mono-clonal antibodies for *neuroligins*, *neurexin* and other synaptic molecules, including their splice variants, to untangle their specific functions within the brain. *Neuroligins* and *neurexins* are of particular interest, because they have been linked to human mental disorders such as autism and schizophrenia [44]. The honey bee might thus prove to be an invaluable model for investigating the molecular basis of human psychiatric disorders [9].

In the field of olfaction, the most burning issue is identification of odor ligands for olfactory receptors. Only a few have been identified to date, and progress is slow due to the technical limitations of the available *in vitro* assays, which are transient expression assays using small-field sampling techniques. We need to develop a new, reliable, high-throughput method for identifying olfactory receptor ligands. The way forward may be to use stably transformed insect (Sf9) or vertebrate (HEK) cells that can be screened against thousands of odors in an automated cell sorting system using flow cytometry analysis. Combining molecular experiments with physiological approaches will be a main focus of future research in olfactory plasticity. If we can assign specific olfactory receptors and their odor ligands to individual glomeruli in the AL, it will significantly advance our understanding of how scents are processed within the complex neural network of the ALs.

Furthermore, bioinformatic analyses will become increasingly essential to develop system networks based on protein-protein and coordinated regulatory gene interactions. This approach has already been successfully used to identify networks for genes involved in honey bee caste development and reproduction [10]. Feeding data from existing microarray and mass spectrometric analyses into these complex network models might be a rapid method to identify molecules crucial for specific brain functions and various honey bee behaviors including the dance language.

Finally, due to the honey bee being a social insect, genetic manipulations have always been difficult in bees. However, honey bee studies similar to the ones described above could be combined with targeted gene knockdown studies and genetic manipulation in *Drosophila* to confirm the functional role of molecules of interest. The combined use of these two insect models represents a real option for furthering honey bee molecular neuroscience.

References

1. Alaux C, Robinson G (2007) Alarm pheromone induces immediate–early gene expression and slow behavioral response in honey bees. J Chem Ecol 33(7):1346–1350
2. Banovic D, Khorramshahi O, Owald D, Wichmann C, Riedt T et al (2010) *Drosophila* neuroligin 1 promotes growth and postsynaptic differentiation at glutamatergic neuromuscular junctions. Neuron 66(5):724–738
3. Bhagavan S, Smith BH (1997) Olfactory conditioning in the honey bee, *Apis mellifera*: effects of odor intensity. Physiol Behav 61:107–117
4. Biswas S, Russell RJ, Jackson CJ, Vidovic M, Ganeshina O et al (2008) Bridging the synaptic gap: *neuroligins* and *neurexin I* in *Apis mellifera*. PLoS One 3(10):e3542

5. Biswas S, Reinhard J, Oakeshott J, Russell R, Srinivasan MV et al (2010) Sensory regulation of neuroligins and neurexin I in the honeybee brain. PLoS One 5(2):e9133
6. Brown SM, Napper RM, Thompson CM, Mercer AR (2002) Stereological analysis reveals striking differences in the structural plasticity of two readily identifiable glomeruli in the antennal lobes of the adult worker honeybee. J Neurosci 22(19):8514–8522
7. Brown SM, Napper RM, Mercer AR (2004) Foraging experience, glomerulus volume, and synapse number: a stereological study of the honey bee antennal lobe. J Neurobiol 60(1):40–50
8. Bruel-Jungerman E, Davis S, Laroche S (2007) Brain plasticity mechanisms and memory: a party of four. Neuroscientist 13(5):492–505
9. Burne T, Scott E, van Swinderen B, Hilliard M, Reinhard J et al (2011) Big ideas for small brains: what can psychiatry learn from worms, flies, bees and fish? Mol Psychiatry 16:7–16
10. Cristino AS, Núñez FM, Lobo CH, Bitondi MM, Simoes ZL et al (2006) Caste development and reproduction: a genome-wide analysis of hallmarks of insect eusociality. Insect Mol Biol 15(5):703–714
11. de Jong R, Pham-Delègue MH (1991) Electroantennogram responses related to olfactory conditioning in the honey bee (*Apis mellifera ligustica*). J Insect Physiol 37(4):319–324
12. de Wit J, Sylwestrak E, O'Sullivan M, Otto S, Tiglio K et al (2009) LRRTM2 interacts with neurexin1 and regulates excitatory synapse formation. Neuron 64:799–806
13. Dean C, Dresbach T (2006) Neuroligins and neurexins: linking cell adhesion, synapse formation and cognitive function. Trends Neurosci 29(1):21–29
14. Durst C, Eichmüller S, Menzel R (1994) Development and experience lead to increased volume of subcompartments of the honeybee mushroom body. Behav Neural Biol 62(3):259–263
15. Faber T, Joerges J, Menzel R (1999) Associative learning modifies neural representations of odors in the insect brain. Nat Neurosci 2:74–78
16. Fahrbach SE, Strande JL, Robinson GE (1995) Neurogenesis is absent in the brains of adult honey bees and does not explain behavioral neuroplasticity. Neurosci Lett 197:145–148
17. Farris SM, Robinson GE, Fahrbach SE (2001) Experience- and age-related outgrowth of intrinsic neurons in the mushroom bodies of the adult worker honeybee. J Neurosci 21(16):6395–6404
18. Fernandez PC, Locatelli FF, Person-Rennell N, Deleo G, Smith BH (2009) Associative conditioning tunes transient dynamics of early olfactory processing. J Neurosci 29(33):10191–10202
19. Galizia CG, Joerges J, Kuttner A, Faber T, Menzel R (1997) A semi-in-vivo preparation for optical recording of the insect brain. J Neurosci Methods 76:61–69
20. Getz W, Akers R (1993) Olfactory response characteristics and tuning structure of placodes in the honeybee, *Apis mellifera* L. Apidologie 24:195–217
21. Heisenberg M (1998) What do the mushroom bodies do for the insect brain? An introduction. Learn Mem 5(1–2):1–10
22. Heisenberg M (2003) Mushroom body memoir: from maps to models. Nat Rev Neurosci 4(4):266–275
23. Heisenberg M, Heusipp M, Wanke C (1995) Structural plasticity in the *Drosophila* brain. J Neurosci 15:1951–1960
24. Hiesinger P, Zhai R, Zhou Y, Koh T, Mehta S et al (2006) Activity-independent prespecification of synaptic partners in the visual map of *Drosophila*. Curr Biol 16:1835–1843
25. Hourcade B, Perisse E, Devaud J-M, Sandoz J-C (2009) Long-term memory shapes the primary olfactory center of an insect brain. Learn Mem 16(10):607–615
26. Hourcade B, Muenz TS, Sandoz J-C, Rössler W, Devaud J-M (2010) Long-term memory leads to synaptic reorganization in the mushroom bodies: a memory trace in the insect brain? J Neurosci 30(18):6461–6465
27. Kano M, Hashimoto K (2009) Synapse elimination in the central nervous system. Curr Opin Neurobiol 19(2):154–161
28. Kim J, Diamond D (2002) The stressed hippocampus, synaptic plasticity and lost memories. Nat Rev Neurosci 3:453–462
29. Knudsen JT, Eriksson R, Gershenzon J, Stahl B (2006) Diversity and distribution of floral scent. Bot Rev 72:1–120

30. Kolb B, Whishaw I (1998) Brain plasticity and behaviour. Annu Rev Psychol 49:43–64
31. Krofczik S, Khojasteh U, de Ibarra NH, Menzel R (2008) Adaptation of microglomerular complexes in the honeybee mushroom body lip to manipulations of behavioral maturation and sensory experience. Dev Neurobiol 68(8):1007–1017
32. Laloi D, Bailez O, Blight M, Roger B, Pham-Delegue M et al (2000) Recognition of complex odors by restrained and free-flying honeybees, *Apis mellifera*. J Chem Ecol 26:2307–2319
33. Lise MF, El-Husseini A (2006) The neuroligin and neurexin families: from structure to function at the synapse. Cell Mol Life Sci 63(16):1833–1849
34. Maleszka J, Barron A, Helliwell P, Maleszka R (2009) Effect of age, behaviour and social environment on honey bee brain plasticity. J Comp Physiol A 195(8):733–740
35. Menzel R, Giurfa M (2001) Cognitive architecture of a mini-brain: the honeybee. Trends Cogn Sci 5:62–71
36. Murphey R (1986) The myth of the inflexible invertebrate: competition and synaptic remodelling in the development of invertebrate nervous systems. J Neurobiol 17:585–591
37. Peele P, Ditzen M, Menzel R, Galizia CG (2006) Appetitive odor learning does not change olfactory coding in a subpopulation of honeybee antennal lobe neurons. J Comp Physiol A 192(10):1083–1103
38. Reinhard J, Srinivasan MV, Zhang S (2004) Scent-triggered navigation in honeybees. Nature 427(6973):411
39. Robertson HM, Wanner KW (2006) The chemoreceptor superfamily in the honey bee, *Apis mellifera*: Expansion of the odorant, but not gustatory, receptor family. Genome Res 16(11):1395–1403
40. Sandoz JC, Pham-Delegue MH, Renou M, Wadhams LJ (2001) Asymmetrical generalisation between pheromonal and floral odours in appetitive olfactory conditioning of the honey bee (*Apis mellifera* L.). J Comp Physiol A 187:559–568
41. Sato K, Pellegrino M, Nakagawa T, Nakagawa T, Vosshall LB et al (2008) Insect olfactory receptors are heteromeric ligand-gated ion channels. Nature 452(7190):1002–1006
42. Sigg D, Thompson CM, Mercer AR (1997) Activity-dependent changes to the brain and behavior of the honey bee, *Apis mellifera* (L.). J Neurosci 17(18):7148–7156
43. Song JY, Ichtchenko K, Sudhof TC, Brose N (1999) Neuroligin 1 is a postsynaptic cell-adhesion molecule of excitatory synapses. Proc Natl Acad Sci USA 96(3):1100–1105
44. Südhof T (2008) Neuroligins and neurexins link synaptic function to cognitive disease. Nature 455(7215):903–911
45. Wadhams LJ, Blight MM, Kerguelen V, Le Metayer M, Marion-Poll F, Masson C, Pham-Delegue MH, Woodcock CM (1994) Discrimination of oilseed rape volatiles by honey bee: novel combined gas chromatographic-electrophysiological behavioral assay. J Chem Ecol 20:3221–3231
46. Wanner KW, Nichols AS, Walden KKO, Brockmann A, Luetje CW et al (2007) A honey bee odorant receptor for the queen substance 9-oxo-2-decenoic acid. Proc Natl Acad Sci USA 104(36):14383–14388
47. Winnington AP, Napper RM, Mercer AR (1996) Structural plasticity of identified glomeruli in the antennal lobes of the adult worker honey bee. J Comp Neurol 365(3):479–490
48. Winston ML (1987) The biology of the honeybee. Harvard University Press, Cambridge
49. Withers GS, Fahrbach SE, Robinson GE (1993) Selective neuroanatomical plasticity and division of labour in the honeybee. Nature 364(6434):238–240
50. Zeng X, Sun M, Liu L, Chen F, Wei L et al (2007) Neurexin-1 is required for synapse formation and larvae associative learning in *Drosophila*. FEBS Lett 581(13):2509–2516

Chapter 5.4
Elucidating the Path from Genotype to Behavior in Honey Bees: Insights from Epigenomics

Ryszard Maleszka

Abstract One of the key unresolved issues in biology is the relationship between a limited number of genes and virtually unlimited behavioral and phenotypic complexity of organisms belonging to different phyla. Recent advances in epigenetics suggest that genomic modifications via DNA methylation provide the level of flexibility that is important for generating morphological and behavioral diversity from the same genome that might be of particular importance for post-mitotic neurons. This robust and reversible chemical modification has the capacity of creating cell-specific epigenetic signatures that can persist even in the absence of the original stimulus because of the self-perpetuating properties of the DNA methylation system. These long-lasting effects are essential to maintaining cellular memory of context-dependent patterns of transcriptional activity. The critical contribution of DNA methylation to development and brain plasticity has already been demonstrated in mammals and in honey bees. Like humans, the honey bees utilize a conserved family of enzymes called DNA methyltransferases (DNMTs) to mark their genes with methyl tags and are capable of producing highly plastic outcomes from a static genome. The honey bee offers an easily manageable and ecologically applicable model for studying the role of epigenetic mechanisms in development and behavior. The incorporation of epigenomic technologies into behavioral studies in honey bees is likely to accelerate the lingering process of translating the raw genomic sequences into a relevant neurobiological knowledge.

Abbreviations

CpG   Cytosine-phosphate-Guanine
DNMTs   DNA methyltransferases

R. Maleszka (✉)
Research School of Biology, The Australian National University,
Canberra, ACT 200, Australia
e-mail: ryszard.maleszka@anu.edu.au

C.G. Galizia et al. (eds.), *Honeybee Neurobiology and Behavior: A Tribute to Randolf Menzel*, DOI 10.1007/978-94-007-2099-2_28,
© Springer Science+Business Media B.V. 2012

LTM   Long-term memory
ncRNAs   non-protein-coding RNAs
RNAi   RNA interference
STM   Short-term memory

5.4.1   Introduction

The honey bee *Apis mellifera* gained prominence in the era of genomics and is often described as a useful system to facilitate the bridging of biological levels from molecules to behavior. How realistic are these aspirations? The conversion of raw DNA sequences into knowledge represents a great challenge for future brain research in honey bees. The first draft of the *Apis* genome and its predicted neural proteome has already reenergised honey bee researchers and redirected the community's attention to what genomic information means for behavioral complexity and organismal function. Now, with the availability of a massive amount of genomic data, how do we go about making rapid progress in understanding the bee's nervous system at any given level and then of understanding phenomena between levels? What technologies are best suited for uncovering the molecular intricacies of learning, memory and complex emergent behaviors? What are the most promising ways forward for honey bee behavioral research? Here I discuss some of the initial benefits for neurobiological studies in honey bees flowing from the genome-inspired projects and draw attention to the recent advances in epigenomics that are likely to reinforce the value of this organism in comparative neuroscience.

5.4.2   The Genome, the Transcriptome and the Predicted Brain Proteome of Honey Bees

Not so long ago, gene number was considered "a pragmatic measure of biological complexity" and a widely held view was that evolution of vertebrates was accompanied by a dramatic increase in protein-coding capacity of the genome [35]. Estimates of more than 100,000 genes for humans prevailed in the 'gene sweep pool' during the 2000 Cold Spring Harbor Genome meeting [33]. Another common expectation in the field of genomics was that functional knowledge gained in powerful model systems like the fly *Drosophila melanogaster*, would fuel a wide range of comparative cross-species studies. This notion, based on the Rosetta stone analogy, in which the decipherment of three ancient scripts has been extended to the decipherment of biological systems, carried the proviso that the meaning of a gene sequence from one organism is directly transferable to another [6, 32]. In recent years, however, it became clear that phenotypic prediction in complex Metazoa cannot be automatically derived from the putative protein function typically inferred from sequence homologies. Problems

Table 5.4.1 Comparison of genome sizes, gene numbers and neuron numbers in organisms with different levels of behavioral complexity

Organism	Genome size (Mb)	Gene number	Neuron number	Behavioral complexity
Human	~3,000	25,000 or less	>85,000 millions	HIGH
Mouse			40,000 millions	
Whale/elephant			200,000 millions	
Zebra fish	1.700		16,000 millions	
Miniature salamander	>25,000		400,000	
Honey bee	260	13,000 or less	1 million	
Vinegar fly	180		250,000	
Miniature wasp	~180		5,000	LOW
Nematode	100	20,000	302	
Cnidarian	1,000	18,000	5,000	
Sponges/Placozoa	950		0	
Arabidopsis	160	30,000	–	–

Sources: Animal Genome Size Database (www.genomesize.com); NCBI (www.ncbi.nlm.nih.gov); [27, 28]

arise not only because of the context-dependent nature of protein function and the indirectness of the path from genes to phenotypes, but also because organismal complexity does not appear to correlate with an increased number of genes [12, 28, 29].

The finding that complex Metazoa including humans have fewer genes than *Arabidopsis* and other plants (Table 5.4.1) came as a surprise to many experts who believed that evolutionarily more advanced organisms with sophisticated sensory systems and complex brains evolved by increasing the number of genes. The preliminary gene count of around 10,000 for *Apis mellifera* is likely to be attuned after the completion of an upgraded genome assembly. However, even if the supplementary analyses uncover 3,000 additional genes, this highly social insect is unlikely to have more than the 13,000 protein-coding genes found in solitary Diptera, or the 20,000 genes encoded by the genome of *C. elegans*, a eutelic species with only 302 neurons and no centralized brain. As these and other examples in Table 5.4.1 plainly illustrate there is no obvious correlation between the gene number, neuron number and the apparent behavioral complexities of diverse organisms. Honey bees with their complex social organization and unique communication skills have significantly less genes than nematodes whose behavioral repertoires are very limited. On the other hand the bee has fewer genes, but more neurons than a miniature salamander, an organism with an impressive behavioral repertoire [27].

It is clear that massively expanded mammalian brains are constructed using information from genomes encoding less than 25,000 genes, whereas the development of a plant, *Arabidopsis*, requires 30,000 genes. Interestingly, this observed reduction in the protein coding capacity of mammalian genomes, known as the 'G-value paradox', is accompanied by a significant increase in the total genome size and its capacity to transcribe a variety of both short and long non-protein-coding RNAs (ncRNAs).

There is increasing evidence that these ncRNAs are vital components of the intricate molecular machinery driving brain plasticity. One possibility is that they regulate the epigenome by providing the necessary specificity for protein complexes involved in DNA methylation and histone modifications that have to be directed to their sites of action [24, 25, 34]. Consequently, ncRNAs could be involved in mediating cell identity and generating the enormous array of regional neuronal and glial cell subtypes that are present in the brain. Another salient feature of nervous system evolution is that brain complexity occurred via a modular construction strategy by utilizing a limited repertoire of pre-existing protein domains and an unknown number of regulatory elements [27, 29]. As a result, not only are most proteins in all lineages orthologous, but a huge overlap exists between genes expressed in the brain and in other tissues.

Evidence based on both molecular and genetic approaches indicates that 60–70% of protein-coding genes in higher animals are expressed in the brain [37]. For example, most of the mouse genome is activated in the building of the brain. Expression of nearly 80% of the 21,500 mouse genes was detected in the brain by a massive *in situ* hybridization effort [23]. Individual neurons, however, have been found to express a much smaller fraction of genes (about 15%) that still allows for huge molecular diversity of neurons, synaptic connections and brain regions. The honey bee is no different. Our estimates of the honey bee transcriptome size in different tissues suggest that at least 60–70% of its genes are used in the nervous system [8]. Like in other species, only a small proportion of these genes appear to be brain-unique with most genes also expressed in other tissues. It is now widely accepted that the pathways used to induce and determine neuronal cell types have been co-opted from those employed at earlier developmental stages to control the differentiation of other cells and tissues. Many of the proteins involved in the specification of neuronal subtype identity, such as the Hedgehog FGF and TGFbeta gene families, have parallel functions in controlling cell fate in other non-neuronal tissues. Even recently evolved lineage-specific genes, for example those encoding Royal Jelly proteins in *Apis*, are also expressed in the brain and appear to be involved in activity-dependent functions [13, 17]. Other examples of important neuronal genes ubiquitously expressed in honey bees are shown in Table 5.4.2.

With regard to the predicted neuronal proteome the differences between *Apis* and *Drosophila* are relatively minor, but nonetheless might be crucial to the apparent behavioral disparity between bees and flies. For example, the bee genome encodes three NMDA receptor subtypes, instead of two in *Drosophila*, and there is also one extra nicotinic acetylcholine receptor subunit. A further three ligand-gated ion channel subunits of the *cys*-loop superfamily remain to have their functions defined and may well represent novel key components of signalling in the honey bee nervous system. Compared with *Drosophila*, there are several extra metabotropic glutamate-like receptors in the bee as well as three extra glutamate transporters. Although the classes and numbers of ligand-gated ion channels are largely similar between *Drosophila* and honey bees, a few *Apis* neuronal gene families are smaller than in other invertebrates. In contrast to the around 50 two-pore (TWIK) potassium channels found in *C. elegans*, the honey bee genome

Table 5.4.2 Examples of ubiquitously expressed methylated honey bee genes involved in neuronal functions. The percentage of similarity to both flies and humans underlies their high level of conservation

| Gene name | GENE ID | % similarity to a closest relative in | | Predicted function |
		Fly	Human	
Dynactin p62	XP001121083	57	55	The binding of dynactin to dynein is critical for neuronal function
Myotubularin myopathy related	GB19180	72	69	Multiple cellular functions. In humans, brain protein linked to neuropathies
Histone methyltransferase	GB13959	53	48	Histone modifications, has been implicated in learning and memory (Wolf-Hirschhorn syndrome)
Nadrin	GB16176	57	62	A novel GTPase-activating protein expressed in neurons
Receptor of activated protein kinase C	GB12499	55	60	Linked to PKC that is involved in learning, such as spatial learning in rats
TATA-box binding protein	GB19036	77	76	Broadly expressed. A general transcription factor for RNA polymerase II
Casein kinase II beta	GB12504	95	95	Involved in circadian rhythm and regulation of mushroom body development
Neuropathy target esterase	GB10208	71	65	Involved in neuronal development and regulation of interactions between neurons and glia

encodes only 10. A similar contraction of channel number occurs in the degenerin/amiloride-sensitive sodium channel family, where honey bees have only 8 genes compared to the 24 in *Drosophila*.

In terms of synaptic signalling pathways, both *Drosophila* and *Apis* lack conserved Brain-derived Neurotrophic Factor (BDNF) signalling machinery that is important for the growth and differentiation of mammalian neurons. However, the bee has all the components of the agrin synapse formation pathway, several of which, including agrin itself, are missing in flies. The core synaptic vesicle trafficking machinery is largely conserved between the currently sequenced invertebrate genomes, although the bee synaptotagmin family is more similar to mice and humans than to *Drosophila* or mosquito. Likewise, the bee huntingtin homolog displays more similarity to the vertebrate huntingtin members. Similar to vertebrates, bees contain a single NSF gene (a vesicle-fusing ATPase), in contrast to the duplication found in *Drosophila*. Several synaptic assembly protein families have also expanded in honey bees, as there is a second PSD95 isoform and a second neurexin isoform compared to *Drosophila*. Like in flies, there are no bassoon/piccolo active zone proteins encoded in the bee genome, suggesting that vertebrates and invertebrates are likely to have distinct active zone components.

5.4.3 DNA Methylation in the Bee Brain: An Ancient Molecular Device by Which Life Experiences Are Encoded?

Comparative genomic analyses revealed that honey bees retain more than 600 genes found in other organisms such as nematodes, yeasts or mammals, which fast evolving dipteran insects have lost. For example, the bee has genes coding for Amyloid beta A4 precursor binding protein, Hydrocephalus protein and a synaptic protein Agrin all of which are missing in flies. Many other proteins such as the Huntingtin homolog, Synaptotagmins, several components of the circadian rhythm and RNAi interference system display more similarity to their vertebrate relatives than to *Drosophila* or mosquitoes. As shown in Fig. 5.4.1 *Apis* has a full complement of all three functional DNA methyltransferases (DNMTs); two orthologues of Dnmt1, one orthologue of Dnmt2 and one of Dnmt3, and similarly to mammals methylates cytosines predominantly in the CpG nucleotide context [8, 39]. Assuming that *Apis* enzymes have the same catalytic capabilities as their mammalian equivalents, DNMT3 is capable of adding new methyl tags to a gene's DNA backbone, whereas DNMT1 maintains the fidelity of the established patterns of modified cytosines. The third, most conserved and ancient protein DNTM2 does not use the DNA template *in vitro* and is believed to be a tRNA methylating enzyme [9]. However, the existence of a second paralog of DNMT1 in *Apis* suggests that there might be some important differences in the functioning of DNA methylation machinery between insects and mammals. Similarly to vertebrates, DNA methylation in *Apis* and other insects is specific to CpG dinucleotides, but in contrast to vertebrates, DNA methylation in invertebrates is almost exclusively intragenic, and transposons and other repetitive elements are not methylated [20]. Finally, the number of methylated CpGs in insects is two to three orders of magnitude lower than in mammals.

The honey bee, nonetheless, possesses a mechanism for storing epigenetic information that controls states of gene expression similar to that found in vertebrate species. Although it is still unclear how epigenetic mechanisms might be linked to gene regulatory networks, it has been proposed that DNA methylation together with changes in the histone code has the capacity to adjust DNA accessibility to cellular machinery by changing chromatin density [1, 4]. Recent findings in honey bees support this notion and suggest that this mechanism provides an additional level of transcriptional control to fine tune the levels of messenger RNAs, including differentially spliced variants, encoded by the methylated genes [8, 20]. It also has been established that honey bees utilize methyl tags to mark a core of 'ubiquitously' expressed critical genes whose activities cannot be switched off in most of the tissues (Table 5.4.2). Thus, rather than switching the genes on and off by promoter methylation in a vertebrate-like manner, the honey bee system operates as a modulator of gene activities.

Epigenetic regulation of transcription is crucial for normal development, providing a mechanism for cellular memory and the inheritance of gene expression pattern information during mitosis. DNMTs are highly expressed in developing tissues, but decline during differentiation and are expressed only at low levels in mature non-dividing tissues. However, brain expression of DNMTs is not only high throughout adulthood, but its pattern is dynamically regulated. This process is generally set as part of the standard genetic program, but recent studies have shown that methyl tags

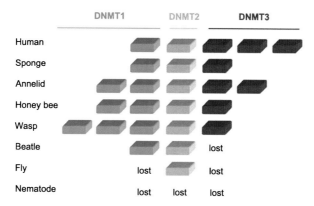

Fig. 5.4.1 Phylogenetic distribution of DNA methyltransferases in selected animal lineages. The number of gene copies encoding DNMTs in a given species is indicated by colored bars

are sensitive to external influences and can be reset by a variety of factors including viral infection, drug treatment or even simple components of a regular diet [16].

Commonly studied epigenetic 'settings' include DNA methylation and histone modifications. Until recently, it wasn't clear whether invertebrates possessed DNA methylation enzymology similar to that found in vertebrates and consequently this line of research was restricted to mammalian models. Recent work in a number of invertebrates has shown that DNA methylation is widespread across the animal king-dom, albeit with several interesting differences between vertebrates and invertebrates [20, 21, 30]. As can be seen in Fig. 5.4.1 the distribution of DNMTs in animals is intriguing. In some species DNA methylation is absent or minimal while in others it is as common as in vertebrates. In particular, the lack of a full DNA methylation toolkit in *Drosophila* and *C. elegans* suggests that DNA methylation is either not essential for basic organismal functions, or can be replaced by other regulatory mechanisms. In contrast, the honey bee uses a full complement of DNMTs to mark its DNA in a vertebrate-like manner [8, 39]. Most recently, whole-genome bisulfite sequencing was used to map global methylation patterns in several invertebrates including four insects, the honey bee, silkworm, fly and flour beetle [20, 40]. The results confirmed the very low levels of methylation in insects with only 0.2% of cytosines methylated in the silkworm and 0.7% in honey bees compared to more than 4% in vertebrates. In addition, whole-genome methylomics provided more evidence that DNA methylation in insects is CpG specific and enriched in gene bodies.

5.4.4   How Is Memory Storage in the Brain Impervious to Molecular Turnover?

Memory can be defined as an animal's capacity to store information about life experiences that can later be retrieved with high distinctiveness to direct behavioral functions in accordance with the new information (acquired by learning). Memory consolidates in distinct but increasingly stable phases with each phase overlapping the preceding one. Short-term memory (STM) gives rise to mid-term memory (MTM) that under certain conditions becomes consolidated into a persistent long-term memory (LTM). Studies in various animals, including the honey bee [26], have shown that this sequence of events is conserved across species [38]. There are fundamental differences between STM and LTM. STM lasts for seconds to hours and is susceptible to interferences of various kinds, whereas LTM lasts for the better part of a life-cycle and is remarkably resistant to disruption (see Chaps. 6.2 and 6.3). The persistence and fidelity of the LTM trace is perhaps one of the most puzzling aspects of brain function. Given the high level of molecular turn-over of cellular machinery the faithfulness of memory is remarkably resistant to a constant flux and renewal. Pioneering experiments conducted in the 1960s by Dingman and Sporn [7] clearly established that formation of long-term memory requires de novo protein and RNA synthesis and therefore gene expression. Since then specific sets of genes involved in different phases of memory consolidation were identified in both *Drosophila* and mice [38]. However, until recently the remarkable immunity of memory storage to molecular flux over a long period of time remained puzzling.

In 1968 Hyden [15] put forward an idea that "a brain mechanism (...) could exist in which the genes and the DNA-code not only permit learning, but also direct the operations of memory". Several years later Crick proposed a molecular mechanism for persistent memory storage involving self-perpetuating alterations of specific proteins [5]. His model, inspired by epigenetic modification of DNA by methylation, was later refined by Holliday who argued that particular sites in the DNA of neurons involved in memory might exist in alternative methylated or non-methylated states [14]. Thus, a gene with just 10 methylation targets could have 2^{10} (1024) epigenotypes and potential phenotypes suggesting that methylation states of thousands of genes would be capable of providing the needed complexity to control overall patterns of gene transcription in non-dividing neurons with fixed genetic hardware. Since DNA methylation is both stable and long-lasting and involves the macromolecule that needs to be preserved intact throughout the animal's lifespan, such models were intuitively attractive, but until recently, lacked direct experimental support.

Seminal work published in 2007 by Sweatt's lab provided strong support for the idea that DNA methylation does in fact play an important role in learning and memory [31]. Using the rat model they uncovered that contextual fear conditioning upregulates DNA methyltransferase 3 (DNMT3) and *de novo* methylation in the adult hippocampus. They also showed that fear conditioning is associated with rapid methylation and transcriptional silencing of the memory suppressor gene *pp1* and demethylation and downregulation of the synaptic plasticity gene *reelin*. These findings not only established a clear link between DNA methylation and memory storage, but also provided evidence that DNA methylation, once thought to be a static process after cellular differentiation, is dynamically regulated in the adult brain. Rapidly accumulating evidence suggests that the nervous system has co-opted these epigenetic mechanisms utilized during development for the generation of long-term behavioral memories in adulthood [36]. Although both DNA methylation and chromatin remodelling have been implicated in these processes the specific biological mechanisms underlying such adaptations remain largely unknown.

Importantly, the involvement of epigenetic modifications of DNA in memory processing appears to be conserved in honey bees. A study employing associative olfactory learning has shown that DNA methyltransferase inhibition significantly reduces extinction memory, or within-session extinction, depending on when DNMT function is inhibited. In addition, extinction was found to be more DNA methyltransferase-dependent than acquisition in agreement with the idea that these two processes represent two competing memory traces rather than two subsequent stages of the same memory trace [19]; (see Chap. 6.3).

5.4.5 The Honey Bee Genome Does Not Explicitly Encode One Behavioral System

Social insects offer special advantages in the analysis of epigenetic phenomena because they are renowned for their astonishing morphological and behavioral polymorphisms [21]. Indeed, the nutritionally-controlled queen/worker developmental divide in the honey bee is one of the best known examples. Despite their identical nature at the DNA level, the queen bee and her workers are strongly differentiated by anatomy, behavior and physiology and by the longevity of the queen. This is undoubtedly the most striking example of developmental flexibility in any phylum. The high level of intake of royal jelly during larval development in honey bees markedly influences the epigenetic status of cells of an individual without altering any of the hardwired characteristics of the genome. As a result, two contrasting organismal outputs, namely fertile queens and non-reproductive workers are generated from the same genome. In addition to profound physiological and anatomical differences including brain architecture (Fig. 5.4.3) these two types of female bees also show remarkable behavioral divergence. However, diet is not the only modulator of developmental trajectories in *Apis*. The same output is accomplished by a very different input, namely gene silencing of components of the DNA methyltransferase system. When the DNMT3 transcript is down regulated by injection of larvae with a small interfering RNA, the majority of the treated individuals emerging from the pupal stage are queens [18, 21]. These newly emergent RNAi-induced queens already have the behavioral characteristics of naturally-reared queens, such as the inclination to kill a weaker queen. Thus the simple perturbation of the DNA methyltransferase system during larval development leads to exactly the same developmental trajectory as a diet of royal jelly (Fig. 5.4.2). Importantly, both royal jelly feeding and DNMT3 silencing lead to alterations in genome-wide DNA methylation patterns and changes in the expression of physiometabolic genes [2, 18, 20] suggesting that such conditional phenotypes are created by epigenetic reprogramming of global regulatory networks.

Although the honey bee data support the key role of environmental factors in controlling developmental trajectories, it needs to be remembered that the genotype–phenotype interaction is a two-way communication. For example, the construction of neural connections is controlled by genes, but the precise instructions as to when and where these connections should be built arrive from the phenotype, which receives

Queen   Workers

Fig. 5.4.2 Silencing of DNMT3 expression in *Apis* by RNA interference (RNAi). Newly-hatched larvae were injected with small RNA molecules complementary to the DNMT3 transcript. The majority of treated larvae emerged as queens, whereas the control larvae emerged as workers [18]

them through its own processing from the environment. When the phenotype receives external input, various genes are activated in order to make the physical changes in the brain such as those corresponding to learning. Gottlieb coined a term 'probabilistic epigenesis' to contrast these bidirectional interactions with a deterministic view on behavior [10, 11]. He argued that complex phenotypic traits, including behaviors, are not predetermined. Instead, they are creations of "reciprocal influences within and between levels of an organism's developmental manifold (genetic activity, neural activity, behavior, and the physical, social, and cultural influences of the external environment) and the ubiquity of gene–environment interaction." In other words, genes depend on input from the phenotype and influence behavior in a probabilistic rather than deterministic manner. The gene and its current transcriptional state are only part of a complex picture that involves both present and previous states of the genotype/phenotype interactions. These dynamic processes can now be studied in honey bees by applying epigenomic approaches, for example by measuring methylation profiles of populations of neurons in brains of individuals subjected to external stimuli or learning paradigms.

5.4.6   Epigenomics of Brain and Behavior in Honey Bee Queens and Workers

The queen/worker developmental divide in honey bees offers an unparalleled experimental system in which the epigenomic modifications of the same genome can be analysed in the context of two distinct phenotypic and behavioral outcomes. As shown in Fig. 5.4.3, the organization of brains in queens, workers and drones shows significant differences in spite of being produced from the same genome. This architectural diversity correlates with the functional specialization of each caste. For example, drones have massively expanded optic lobes (OL) that help them to locate virgin queens during the mating flights, whereas workers performing more complex and diverse tasks have the largest antennal lobes (ALs) and mushroom bodies (MBs). The queen brain also has a distinct organization with both AL and MB smaller than those in workers, but bigger than in drones. Interestingly, these diverse brain structures of behaviorally distinct castes correlate with unique brain epigenomes. The distribution of the methyl-cytosine in the brains of queens and workers has recently been determined at single-base-pair resolution using shotgun bisulfite sequencing technology [20]. This whole-genome sequencing approach revealed nearly 600 differentially methylated genes that are epigenetically fine-tuned in the brains of workers and queens to produce their extraordinarily different behaviors. In addition, the study also confirmed the uniqueness of the brain methylome in drones and found strong correlation between methylation patterns and splicing sites including those that have the potential to produce alternative exons. The capacity of DNA methylation to influence alternative splicing of ubiquitously expressed housekeeping genes suggests that gene body methylation can regulate the relative quantity of alternate transcript variants in a context-dependent manner. Such modulation of alternative splicing is one mechanism by which epigenetic gene regulation in honey bees can produce protein diversity from a limited number

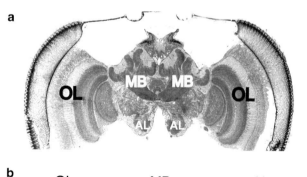

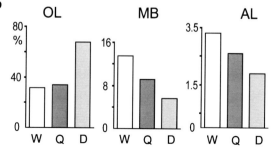

Fig. 5.4.3 (**a**) Frontal section of the brain of a worker honeybee at a depth of 310 μm stained with toluidine blue [22], OL: optic lobes, MB: mushroom bodies, AL: antennal lobes. (**b**) Relative sizes of parts of the brains. The Y axis indicates the percentage of the whole brain occupies by each brain region, OL, MB and AL in workers (W), queens (Q) and drones (D). Measurements taken from Bullock and Horridge [3]

of genes. This previously unknown mechanism might be important for generating phenotypic flexibility not only during development, but also in the adult post-mitotic brain.

5.4.7   Outlook

We already know nearly the entire DNA sequence of the *Apis* genome and most of its neural proteome, but the predicted behaviors of the organism do not emerge from this knowledge. The trajectories from genotype to complex phenotypes are indirect, multi-level and virtually unknown. Most types of behaviors depend on interplay between environmental factors and multiple genes operating in highly interconnected, frequently overlapping networks. The discovery of epigenomic mechanisms in *Apis* brings a fresh perspective to behavioral studies in this organism. In particular, the honey bee system is poised to allow a transition from static molecular data to flexible epigenomes to neural circuitry and to sophisticated behaviors, all under completely natural environmental conditions. The low and manageable number of methylated sites in the *Apis* genome combined with powerful next-generation sequencing technologies provides an excellent opportunity to study the dynamics of activity-related changes in brain-specific epigenomes. In this context, the honey bee is perfectly placed to become a truly innovative model for functional neuroepigenomics.

References

1. Ball MP et al (2009) Targeted and genome-scale strategies reveal gene-body methylation signatures in human cells. Nat Biotechnol 27(4):361–368
2. Barchuk AR, Cristino AS, Kucharski R, Costa LF, Simoes ZL et al (2007) Molecular determinants of caste differentiation in the highly eusocial honeybee *Apis mellifera*. BMC Dev Biol 7:70
3. Bullock TH, Horridge GA (1965) Structure and function in the nervous systems of invertebrates. A Series of books in biology. W. H. Freeman, San Francisco/London
4. Cedar H, Bergman Y (2009) Linking DNA methylation and histone modification: patterns and paradigms. Nat Rev Genet 10(5):295–304
5. Crick F (1984) Memory and molecular turnover. Nature 312(5990):101
6. Das S, Yu LH, Gaitatzes C, Rogers R, Freeman J et al (1997) Biology's new Rosetta stone. Nature 385(6611):29–30
7. Dingman W, Sporn MB (1964) Molecular theories of memory – any theory of memory in nervous system must consider structure + function in entire neuron. Science 144(361):26–29
8. Foret S, Kucharski R, Pittelkow Y, Lockett GA, Maleszka R (2009) Epigenetic regulation of the honey bee transcriptome: unravelling the nature of methylated genes. BMC Genomics 10:472
9. Goll MG, Kirpekar F, Maggert KA, Yoder JA, Hsieh CL et al (2006) Methylation of tRNA(AsP) by the DNA methyltransferase homolog Dnmt2. Science 311(5759):395–398

10. Gottlieb G (1998) Normally occurring environmental and behavioral influences on gene activity: from central dogma to probabilistic epigenesis. Psychol Rev 105(4):792–802
11. Gottlieb G (2007) Probabilistic epigenesis. Dev Sci 10(1):1–11
12. Greenspan RJ (2001) The flexible genome. Nat Rev Genet 2(5):383–387
13. Hojo M, Kagami T, Sasaki T, Nakamura J, Sasaki M (2010) Reduced expression of major royal jelly protein 1 gene in the mushroom bodies of worker honeybees with reduced learning ability. Apidologie 41(2):194–202
14. Holliday R (1999) Is there an epigenetic component in long-term memory? J Theor Biol 200(3):339–341
15. Hyden H (1968) Biochemical approaches to learning and memory. In: Koestler A, Smythies JR (eds) The Alpbach Symposium 1968. Beyond reductionism, New perspectives in the life sciences. Hutchinson, London
16. Jaenisch R, Bird A (2003) Epigenetic regulation of gene expression: how the genome integrates intrinsic and environmental signals. Nat Genet 33(Suppl):245–254
17. Kucharski R, Maleszka R, Hayward DC, Ball EE (1998) A royal jelly protein is expressed in a subset of Kenyon cells in the mushroom bodies of the honey bee brain. Naturwissenschaften 85(7):343–346
18. Kucharski R, Maleszka J, Foret S, Maleszka R (2008) Nutritional control of reproductive status in honeybees via DNA methylation. Science 319(5871):1827–1830
19. Lockett GA, Wilkes F, Maleszka R (2010) Brain plasticity, memory and neurological disorders: an epigenetic perspective. Neuroreport 21(14):909–913
20. Lyko F, Foret S, Kucharski R, Wolf S, Falckenhayn C et al (2010) The honey bee epigenomes: differential methylation of brain DNA in queens and workers. PLoS Biol 8(11):e1000506
21. Maleszka R (2008) Epigenetic integration of environmental and genomic signals in honey bees. Epigenetics 3(4):188–192
22. Maleszka R, Maleszka J, Barron AB, Helliwell PG (2009) Effect of age, behaviour and social environment on honey bee brain plasticity. J Comp Physiol A 195(8):733–740
23. Markram H (2007) Bioinformatics - Industrializing neuroscience. Nature 445(7124):160–161
24. Mattick JS, Amaral PP, Dinger ME, Mercer TR, Mehler MF (2009) RNA regulation of epigenetic processes. Bioessays 31(1):51–59
25. Mattick JS, Taft RJ, Faulkner GJ (2010) A global view of genomic information - moving beyond the gene and the master regulator. Trends Genet 26(1):21–28
26. Menzel R (2001) Searching for the memory trace in a mini-brain, the honeybee. Learn Mem 8(2):53–62
27. Miklos GLG (1993) Molecules and cognition - the latterday lessons of levels, language, and Lac - evolutionary overview of brain structure and function in some vertebrates and invertebrates. J Neurobiol 24(6):842–890
28. Miklos GL, Maleszka R (2000) Deus ex genomix. Nat Neurosci 3(5):424–425
29. Miklos GL, Maleszka R (2001) Protein functions and biological contexts. Proteomics 1(2):169–178
30. Miklos GLG, Maleszka R (2011) Epigenomic communication systems in humans and honey bees: from molecules to behavior. Horm Behav 59(3):399–406
31. Miller CA, Sweatt JD (2007) Covalent modification of DNA regulates memory formation. Neuron 53(6):857–869
32. Nadeau JH, Grant PL, Mankala S, Reiner AH, Richardson JE et al (1995) A Rosetta stone of mammalian genetics. Nature 373(6512):363–365
33. Pennisi E (2003) Human genome. A low number wins the GeneSweep Pool. Science 300(5625):1484
34. Qureshi IA, Mattick JS, Mehler MF (2010) Long non-coding RNAs in nervous system function and disease. Brain Res 1338:20–35
35. Simmen MW, Leitgeb S, Clark VH, Jones SJ, Bird A (1998) Gene number in an invertebrate chordate, Ciona intestinalis. Proc Natl Acad Sci USA 95(8):4437–4440

36. Sweatt JD (2009) Experience-dependent epigenetic modifications in the central nervous system. Biol Psychiatry 65(3):191–197
37. Thaker HM, Kankel DR (1992) Mosaic analysis gives an estimate of the extent of genomic involvement in the development of the visual system in *Drosophila melanogaster*. Genetics 131(4):883–894
38. Tully T, Preat T, Boynton SC, Del Vecchio M (1994) Genetic dissection of consolidated memory in *Drosophila*. Cell 79(1):35–47
39. Wang Y, Jorda M, Jones PL, Maleszka R, Ling X et al (2006) Functional CpG methylation system in a social insect. Science 314(5799):645–647
40. Zemach A, McDaniel IE, Silva P, Zilberman D (2010) Genome-wide evolutionary analysis of eukaryotic DNA methylation. Science 328(5980):916–919

Chapter 5.5
Genetics and Molecular Biology: Commentary

Randolf Menzel

A major step in understanding the working of a brain is to relate structures to functions. As long as functions are characterized by sensory or motor components, structure-function relations can be established rather easily in all brains, from worms to primates. Localization of brain function becomes problematic when cognitive components of behavioral organization are concerned. This is because we do not know how behaviorally-defined cognitive functions relate to brain functions. Learning and memory are not suitable categories of isolated brain functions because they are characterized by many kinds of neural process at many – probably all - levels of neural integration. The mushroom body of the insect brain should not – in my view – be addressed as a general learning and memory device. When years ago we asked whether and how the bee mushroom body is involved in learning and memory we posed this question in such a way that a particular and specific component (consolidation from short-term to cooling resistant mid-term memory) in a particular sensory domain (olfaction) and a specific form of learning (appetitive classical conditioning) were addressed [2]. In my view it is important to well define the behavioral components under study if the goal is to search for the neural structures that are necessary and/or sufficient to account for the behavior observed. The ongoing debate around this topic in *Drosophila* studies is highly relevant in this context [1]. Although the mushroom body of *Drosophila* is less complex than that of the honeybee (for instance, in terms of the number of constitutive Kenyon cells: a few thousands vs. more than hundred thousand). *Drosophila* researchers would not believe the mushroom body to be a general learning and memory device. It contributes differently and specifically to some but not to all forms of learning (e.g. visual learning), and it may play different roles during consolidation phases of specific memory forms.

R. Menzel (✉)
Institüt fur Biologie, Neurobiologie, Freie Universität Berlin, Berlin, Germany
e-mail: menzel@neurobiologie.fu-berlin.de

C.G. Galizia et al. (eds.), *Honeybee Neurobiology and Behavior: A Tribute to Randolf Menzel*, DOI 10.1007/978-94-007-2099-2_29,
© Springer Science+Business Media B.V. 2012

The strength of molecular data lies in the fact that the identified molecules can be used as tools to manipulate the system if the respective genes can be switched on and off, or the pathways of protein synthesis can be manipulated e.g. by RNAi, RNAsi, antisense nucleotide technology, etc. Work on *Drosophila* and the mouse will guide the way. If neuron specific promoters are identified, genes are switched on and off and dyes are expressed for light control of neural activity, hypotheses can be tested rather directly, an exciting avenue of research, indeed. How should we proceed?

I am skeptical whether one should aim for transgenic bees now. Such an approach requires highly sophisticated and laborious treatments to produce, control and maintain the respective genetic lines. The haplo-diploid cycle (males are haploids while females are diploids) and the social life of the bee is a big hurdle for such attempts. I rather believe that emphasis should be put on somatic transfection techniques focusing on single animals in a controllable environment. Well designed viruses and other gene transfection vehicles can be used to treat single bees for specific purposes avoiding the problems with whole colonies and germ line manipulations. In addition, the variability in rate and specificity of transfected animals will be a highly valuable source of information. I envisage that molecular research will lead us to develop specifically designed transfection agents for subtypes of neurons allowing to apply some of the powerful techniques developed for *Drosophila* and the mouse, i.e. conditional knock-outs, expression of proteins for the up and down regulation of excitation by light, or fluorescence probes for monitoring neural excitation patterns. Only after understanding the complex molecular networks of gene regulation in bees we may be able to alter the germ line and produce transgenes that if escaping into the environment will not become a hazard.

How and where is the content of the memory trace stored? Memories are about something, they have content, called the engram. The engram is stored in the nervous system such that it can be retrieved on demand and will then influence behavior. Our molecular, physiological and anatomical tools help us to characterize processes that lead to an engram, but they are not the engram. What we learn from the new discoveries on epigenetic processes in the bee brain summarized by Ryszard Maleszka is that gene expression is not only regulated by transcription factors and co-factors interacting together and with regulatory elements but also by chemically altering the molecular backbone of the DNA. DNA does not store memory content as it stores the information for building an organism but takes part in multiple processes that provide the mechanisms for the storage of memory content. In other words the engram is written into the neural network by multiple letters, epigenetic being one of them. So where is the engram? The broadly accepted concept to-day envisages the engram as the total number of changes in synaptic communication as a consequence of changes of synapses themselves or alterations of neural excitability. Will it ever be possible to directly "see" the engram? It will certainly be possible to visualize processes leading to the engram, but how about the engram itself?

An allegory might help to understand the problem. The text of a book, its content, can be stored in multiple ways (hand written or printed on paper, stored digitally on disc, etc.) but extracting the content requires a knowledgeable reader. The knowledgeable reader of memory content is the whole brain, or at least a large part of it. We shall need to dive deep into the functional structure of the brain to enable our own brain to read an engram of another brain. Insects with their reduced brain complexity and the possibility of tracking functions to single identifiable neurons may help in this respect. Identifying neurons that house the processes for forming the engram will be an important first step. Like reading the text of a book you need to open the book or switch on the computer, localize the area of interest and feed the knowledgeable reader. Thus tracking synaptic changes to single neurons, as wonderful as it may be, does not offer more than reading a word or a sentence in a whole book without understanding it. The knowledgeable readers of even such a word or a sentence are the postsynaptic neurons and many others to follow. It is hard to imagine whether such a task will ever be possible. But we should not give up. May be the way back into the nervous system from behavior to motor programs and on to premotor control circuits will help. In any case precise knowledge about the neuroanatomy on many levels of spatial resolution will be essential, and methods will be necessary to image the changes of neuronal excitability and synaptic transmission on to the neural structures. Are there any more suitable brains than those of insects?

References

1. Heisenberg M, Gerber B (2008) Behavioral analysis of learning and memory in *Drosophila*. In: Menzel R (ed) Learning theory and behavior, vol 1. Elsevier, Amsterdam, pp 549–560
2. Menzel R, Erber J, Masuhr T (1974) Learning and memory in the honeybee. In: Barton-Browne L (ed) Experimental analysis of insect behaviour. Springer, Berlin, pp 195–217

Part VI
Learning and Memory

Chapter 6.1
Distributed Plasticity for Olfactory Learning and Memory in the Honey Bee Brain

Brian H. Smith, Ramón Huerta, Maxim Bazhenov, and Irina Sinakevitch

Abstract Honey bees have a number of sophisticated learning abilities to track rapidly changing distributions of nectar and pollen resources the colony needs for survival. The honey bee is an excellent animal in which to study learning because these abilities can be evaluated both in the field and under laboratory conditions that permit use of physiological analyses. Our focus is on how the neural bases for these learning abilities can be tracked into different levels of processing in the CNS (central nervous system). We specifically review two kinds of conditioning protocols to show first how behavior changes over conditioning and second how plasticity can be tracked into the antennal lobe (AL) as well as in the mushroom body (MB). We begin by pointing out that, particularly when the learning problem becomes difficult, the behavioral response to conditioned stimuli proceeds in a nonlinear manner. Honey bees may have difficulty in making an appropriate response until some point when a precipitous change in their behavior occurs. We then discuss how plasticity related to behavioral conditioning has been reported at subsequent processing levels in the AL as well as in the MB, which receives input from the AL. We point out that this distributed plasticity in the CNS for any kind of learning raises an important conceptual issue, which regards how changes at a higher level

B.H. Smith (✉) • I. Sinakevitch
School of Life Sciences, Arizona State University, PO Box 874501,
Tempe, AZ 85287, USA
e-mail: brianhsmith@asu.edu

R. Huerta
Biocircuits Institute, University of California, San Diego, CA 92093, USA
e-mail: rhuerta@ucsd.edu

M. Bazhenov
Department of Cell Biology and Neuroscience, University of California, Riverside,
CA 92521, USA
e-mail: maksim.bazhenov@ucr.edu

C.G. Galizia et al. (eds.), *Honeybee Neurobiology and Behavior: A Tribute to Randolf Menzel*, DOI 10.1007/978-94-007-2099-2_30,
© Springer Science+Business Media B.V. 2012

of processing (the MB) can be adapted to track and perhaps be augmented by changes at an earlier level (the AL). We show by example how coupled behavioral and physiological analyses combined with computation modeling can begin to address these important issues.

Abbreviations

AL   Antennal lobe
CA   Mushroom body calyces
CS   Conditioned stimulus
KC   Kenyon cell
MB   Mushroom body
PN   Projection neuron

6.1.1   Learning and Memory in the Honey Bee Olfactory System

Honey bees must learn about the associations between floral odors and the nectar and pollen resources the colony needs for survival [16, 30, 43]. How reliably a particular flower's odor is associated with nectar or pollen can change hour-to-hour and day-to-day, potentially many times within a foraging honey bee's lifetime. Therefore, honey bees must be prepared to quickly learn a new association, or, as we will show, learn about the lack of an association, in order to maximize its ability to collect resources. Much of the honey bee's impressive learning ability has probably evolved because of the instability of information about nectar and pollen relative to a honey bee's foraging lifetime.

Honey bees have a rich repertoire of olfactory learning behaviors that range from nonassociative through associative and operant conditioning [30, 43]. These behaviors can be studied under controlled laboratory conditions [2], in which neurophysiological, bioimaging and molecular techniques can be employed in parallel to, or simultaneous with, behavioral experiments. Furthermore, the laboratory procedures allow for use of pharmacological and molecular genetic techniques to disrupt targeted neurotransmitter and neuromodulatory pathways in conjunction with behavioral [11], bioimaging [12, 22, 39] and electrophysiologial recordings [18, 29, 32, 44]. Therefore, in the honey bee it is now possible to study correlations, and in many cases causal relationships, across several different levels of analysis from molecular though neural systems and behavior.

6.1.2   Learning to Pay 'Attention' to, or to 'Ignore', an Odor[1]

When describing how a honey bee learns about floral reward, what normally comes to mind is how a honey bee associates floral cues with nectar and pollen. That is, a honey bee learns to seek out and approach flowers with cues that have been successful in predicting rewards. There have been many studies of this kind of learning in honey bees, and those studies have revealed many simpler as well as more complex forms of learning (see Chap. 6.6). However, various forms of 'inhibitory' learning have only begun to be investigated in detail. Here, we will focus on two very different forms of learning in an attempt to demonstrate how they can begin to be encoded in early processing in the brain.

Honey bees can be easily conditioned to discriminate an odor associated with sucrose reward (CS+) from an odor explicitly not associated with reward (CS−) [12, 30]. This kind of protocol has been effectively used in many studies to investigate whether different odors are detectable and discriminable to honey bees [17, 41]. Honey bees typically solve discrimination tasks easily. However, the problem can be made more difficult by reducing the intensity (concentration) and/or exposure time of each odor [50] or systematically changing the ratio of two odors in a binary mixture [12] (Fig. 6.1.1). At higher intensities, honey bees can take as few as two or three trials with each of two odors to show discrimination of the CS+ from the CS−. However, when faced with lower intensities, honey bees take longer to successfully discriminate two odors [12]. Moreover, especially at lower intensities the response to both odors initially increases. The heightened responses to both odors continue over several trials until there is a rapid increase in response to the CS+ and a decrease in response to the CS−. The behavior is as though they have finally attained an ability to differentiate one odor from the other, which leads to a precipitous change in behavior (an 'aha!' effect). This rapid separation between a CS+ and CS− depends on the ease of the task as well as on the outcome associated with the CS− (see Chap. 23). When associated with punishment instead of nothing, the effect occurs much more quickly [42]. This rapid change in behavior is frequently evident both in population means responses (Figs. 6.1.1 and 6.1.2) as well as in the response patterns of individual honey bees [42].

The presence of unrewarding flowers has an important influence on choice behavior in freely flying honey bees, and learning about the *lack* of an association of an odor with nectar or pollen is also an important form of learning [9]. One specific form is called Latent Inhibition [27]. When confronted with a series of exposures to an odor not followed by reinforcement, honey bees learn to ignore that odor.

---

[1] We use the words 'attention', 'ignore' and the 'aha effect' in this section in a very descriptive sense, and we do not necessarily mean more cognitive interpretations frequently implied by these terms. The precise mechanisms of each form of learning and decision making still remain to be thoroughly investigated.

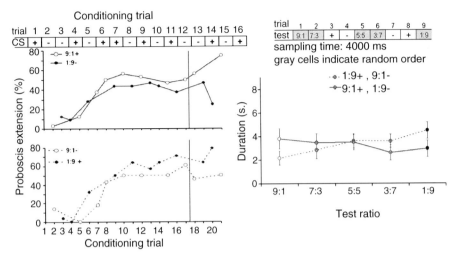

Fig. 6.1.1 Differential conditioning of two groups of honey bee workers to ratios of binary mixtures (From Fernandez et al. [12]). *Top row left*: schematic of PER (proboscis extension reflex) conditioning protocol for bees trained either CS + = 9:1, CS− = 1:9 or vice versa. The odors were 1-hexanol and 2-octanone. The mixtures were conditioned at low (0.02 M) concentration to make it more difficult for the bees to make the discrimination. *Top row right*: After conditioning a series of unreinforced test trials were performed to test responses to the conditioned mixtures generalization to a range of intermediate mixtures. White boxes intermingled with test trials indicate retraining. *Bottom*: Graphs (*left*) represent percent proboscis extension during training, showing that under these conditions honey bees had difficulty discriminating the odor mixtures until the final few trials. Graphs (*right*) show changes in slope of the generalization gradients depending on whether 1:9 or 9:1 was the odor reinforced with sucrose (+) (Reprinted with permission from Fernandez et al. [12])

This can be easily demonstrated by associating that odor, after the unrewarded exposure phase, with sucrose in a way that normally produces robust excitatory conditioning (Fig. 6.1.2; [4]). After a number of unrewarded exposures honey bees learn the association much more slowly than they normally would. Moreover, the same 'aha!' effect occurs with this kind of learning. At first, there is a slight drop in response levels with only a one to five unrewarded trials. Then as the number of exposures initially increases, there is little additional effect on excitatory conditioning. However, there is a precipitous decline in responding after 15–25 unrewarded trials.

In summary, these two forms of learning – learning about odor association with sucrose reward and learning about the lack of an association – represent in a sense opposite forms of learning. They are only two of several forms of learning identified in honey bees (for example, see discussion of extinction learning in Chap. 6.3), but we will use them to illustrate how their neural bases can be tracked into plasticity represented in the central nervous system. Finally, the computational modeling below will provide insights into how abrupt changes in behavior – a decision, in a

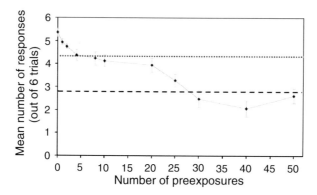

Fig. 6.1.2 Retardation of acquisition is a function of the number of preexposure trials. This figure shows the mean (±SE) number responses to odor during a test phase as a function of the number of preexposures (different groups of honey bees received zero through 50 unreinforced presentations with odor). In this figure the total number of responses each subject made to odor over six trials in the test period was summed. Fewer responses means stronger retardation of acquisition (latent inhibition) to the preexposed odor. Subjects were randomly assigned to 11 groups that differed on the basis of the number of stimulus preexposures but which were equated for exposure to the conditioning context. The means below the *dotted line* (8, 10, 20, 25, 30, 40 & 50) indicate that they are different from the mean of the zero preexposure group. Means below the *dashed line* (30, 40 and 50 trials) indicate differences from all groups with fewer preexposures. Sample sizes for each group *left* to *right*, resp., were: n=17, 15, 16, 18, 17, 16, 16, 21, 19, 20, 21 (Reprinted with permission from Chandra et al. [4])

sense – might be implemented in the neural networks of the brain. In particular, these models may eventually yield insights into how graded plasticity-related changes at one level (the AL) can translate into abrupt plasticity-related changes at the next level (the MB).

6.1.3   Convergent Evolution of Neural Solutions to Olfactory Encoding, Pattern Recognition and Plasticity

The honey bee brain contains just under 10^6 neurons. Odor ligands interact with approximately 60,000 sensory cells distributed along each antenna (see Chap. 19). There are approximately 170 odorant receptors encoded in the honey bee genome (see Chap. 5.3). Sensory axons project to the Antennal Lobe (AL) at the base of the brain, where axons from sensory cells converge onto approximately 160 glomeruli. In the fruit fly, sensory axons from cells that express the same receptor converge to the same glomerulus [49]. These glomeruli are interconnected via a network of GABAergic and histaminergic inhibitory local interneurons [39]. Dendrites from three to five projection neurons (PN) innervate each glomerulus. PN axons on each side of the brain project onto dendritic fields of 170,000 Kenyon Cells (KCs) that are intrinsic to the mushroom body (MB), where olfaction, vision, mechanosensory and taste modalities converge ([38], see also Chap. 3.2). KC axons project ventrally through the MB and branches interact with different types of extrinsic neurons. One type of extrinsic neuron referred to as 'EN' below (approx 400); [15, 38] provides output to premotor centers. Another type of extrinsic neuron (PCT) provides GABAergic inhibitory feedback to the MB calyces (CA) [15].

Much of the circuitry in the honey bee brain is analogous to circuits that perform similar functions in the mammalian brain. Neural networks in the AL and mammalian olfactory bulb, for example, are anatomically very similar [19]. Both encode sensory inputs for odors as spatiotemporal patterns [24]. These similarities most likely evolved independently [45]. Therefore, these network architectures likely reflect a convergence on important, fundamental solutions to encoding sensory patterns at different levels of processing.

6.1.4   The Problem of Distributed Plasticity

Both reinforced (e.g. Fig. 6.1.1) and unreinforced (e.g. Fig 6.1.2) forms of plasticity, correlated to different forms of behavioral plasticity reviewed above, have been identified in the honey bee [7, 10, 12, 20, 29, 47] and moth [6] AL and MB. These have distinct parallels in olfactory processing in the mammalian brain [26]. Therefore, it seems unlikely that there is a single locus that is both necessary and sufficient to account for learned behavior. As of yet we have very little understanding of the mechanism and roles of plasticity at either level. Nor do we yet have a means for understanding how plasticity at an early stage of processing (the AL) interacts with plasticity at a later stage (the MB).

Distributed plasticity raises a fundamental computational problem. *How can a neuropil (e.g. the MB) that receives a reinforcement (teaching) signal encode a memory for a pattern of input when that pattern may be changing as a result of the same reinforcement signal operating at an earlier stage of processing (e.g. the AL)?* We suspect that the plasticity in serial stages of processing might be tuned to interact in a way that enhances pattern recognition. It may be, for example, that plasticity in the antennal lobe helps to make more distinct the patterns of sensory input necessary for solving a current problem. This plasticity might be dismantled later when the problem changes. We also suspect that the increased separation of two sensory patterns as an animal learns is more like a gradual than a step-like process. The graded nature of the change in the AL may allow a later neuropil (e.g. the MBs) to more effectively track, and perhaps anticipate, the direction of the change. Finally, the tracking problem may be solved in part by two pathways emanating from the AL [14]. One pathway may be subject to plasticity while the other remains static.

However, this is currently only speculation. More substantial insights into the solution to this problem will require coupling an analysis of different types of plasticity at two different stages of processing in the brain with computational models that involve realistic representations of neural activity patterns and plasticity at each of those stages. Unfortunately, the solution to this problem is still out of reach. Nevertheless, we need to begin by understanding how sensory information is encoded in the olfactory system, and how it is influenced by plasticity.

6.1.5 Transient Dynamics in Early Olfactory Processing in the AL

Olfactory coding in several animals including the honey bee is a dynamic process. Calcium-based imaging has been recently employed to study transient dynamics and plasticity encoded by the antennal lobe network [12, 13] (Fig. 6.1.3). Odor stimulation sets off a sequence of activity patterns (a transient) that reaches maximal separation from transients to other odors after approx 400–500 ms. The sequence can be visualized in Fig. 6.1.3a as glomeruli are activated to differing degrees through time. Each step along the transient corresponds to a specific pattern of activation of PNs across glomeruli that helps to separate the odors (Fig. 6.1.3b, c). The sequence of steps, and not just the final state, is important for perceptual separation of odors [36].

The time-dependent nature of encoding in the AL implies that longer exposure times should lead to improved behavioral discrimination of odors. In regard to this time-accuracy trade-off, there has been controversy in research with honey bees [8, 12, 50] and with mammals [37, 48]. When animals are allowed time to sample, and specifically when there is little or no cost to making an error, they make choices very fast. However, under conditions when stimuli become much more difficult to discriminate, for example with restricted sampling times and/or low stimulus intensities, accuracy in choices of alternative odors can be compromised. For example, with short sampling times honey bees recognize an odor better if the test stimulus has the same duration as the conditioning stimulus, now sorted in a memory template [12]. Above 800 ms the need for a match is not evident. So there is a transition between time-dependent memory and time-independent memory at approximately the time needed for the transients to evolve maximal separation in the AL. Furthermore, freely flying honey bees hover above an odor source for 700 ms before landing [8]. This response latency might correspond to the sampling time required for the AL to reach a reliable classification implied in the imaging studies cited above.

Finally, much of the debate about response latencies in regard to stimulus dynamics in the brain make an implicit assumption that honey bees stop evaluating odors once they have released a response. That is, when a honey bee responds after 400 ms, then it does not use information in the evolving spatiotemporal dynamics of the AL patterns beyond that point. This assumption is likely to be incorrect. Many times honey bees will show a pattern of multiple extensions and retractions in quick succession following odor stimulation, almost as thought they are 'indecisive' about the identity of the odor [40]. Thus it is likely that evaluation of odor quality continues well after initial extension of the proboscis.

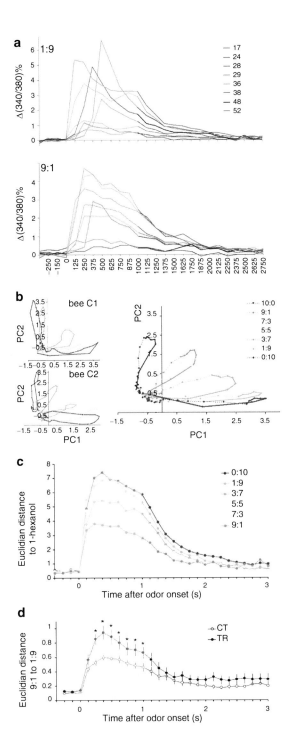

6.1.6   Plasticity in Transient Dynamics in the AL

Differential conditioning of two odors increases the distances between the reinforced and unreinforced transients in the AL (Fig. 6.1.3d). Presumably the increase in separation of the neural patterns represents an increase in discriminability of the CS+ and CS− in this case. This is consistent with a model in which a role for plasticity is to tune the AL to increase a honey bee's ability to detect and discriminate the odors important for solving a specific problem at hand [43]. This tuning would amount to a slight reconfiguration, and thus more efficient use, of the large perceptual space set up for encoding odors by the 170 or so sensory receptors. Most of that space would never be used because the number of odors relevant to a honey bee in its lifetime, and certainly at any particular point in its lifetime, is very small relative to the number of odors than can be detected.

The reinforcement signal for plasticity in the AL, as well as in the MB, is mediated at least in part by a set of cells (includes VUM_{mx1}) that receive input from sucrose-sensitive taste receptors on the honey bee's mouthparts (see Chap. 4.2). These cells project outputs to many, if not all, glomeruli in the AL and throughout the calyces of the MB. When stimulated they are thought to release a biogenic amine – octopamine (OA) – broadly in the AL and MB [23]. Several studies using electrophysiological recordings [18] and molecular disruption of an OA receptor [11] have identified OA released by these neurons as a key component of the reinforcement pathway in the honey bee AL and MB.

Fig. 6.1.3 Spatiotemporal response patterns show a smooth transition along ratios from one single component to the other. (**a**) Mean responses in 6 selected glomeruli to ratios 9:1 (*top*) and 1:9 (*bottom*) over 125 ms time steps from just before odor delivery through 2,750 ms. Line colors correspond to different glomeruli (see legend). Stimulus pulse (1,000 ms) is marked by the *shaded area*. (**b**) (*Left*) Odor specific trajectories for 2 control bees; 7 out of 9 bees follow the same general pattern. To generate this figure the original 17-dimensional space has been projected onto the first two principal components for each bee. Under these conditions, 86.2–97.5% of the variance is explained for each bee. All ratios were presented to each bee and Ca^{++} transients were recorded at fixed time intervals (125 ms). Accordingly, the distance between different color data points represents the divergence of the odor representations over time. Trajectories depart rapidly from baseline and slow down when they approach odor-specific regions. (*Right*) PCA of odor-evoked activity patterns for a control "average" bee obtained by averaging 9 control bees by the activity of 17 glomeruli along 27 125 ms-time intervals. The response of each glomerulus was used as a dimension for the analysis. (**c**) Euclidean distances between ratio 10:0 (pure 1-hexanol) and all the other test ratios based on a 17-dimensional space over 125 ms time steps from just before odor delivery through 3,000 ms after odor onset. Stimulus pulse (1,000 ms) is marked by the shaded area. (**d**) Differential conditioning increases the distance between spatiotemporal patterns. Worker honey bees were differentially trained to 9:1+ and 1:9−. Nine hours after conditioning, brains were treated as above with fura-2 and 8–12 h later were imaged. These bees were evaluated in parallel to the untrained "control" bees. The Euclidean distances between ratios 9:1 and 1:9 (i.e. CS+ vs. CS−) are based on a 17-dimensional space over 125 ms time steps from just before odor delivery through 3,000 ms after odor onset. Stimulus pulse (1,000 ms) is marked by the shaded area. Close and open symbols indicate respectively trained and untrained bees. Asterisks indicate significant. All figures reprinted with permission from Fernandez et al. [12]

6.1.7   Processing of PN Inputs in the Mushroom Bodies

Axons that carry PN outputs project to the dendritic fields of KCs, which are the intrinsic cells of the MBs (see Chap. 3.2). The axons of a given PN will fan out and make contact with many different KCs, and each Kenyon cell receives input from several different PNs. Therefore KCs are in a position to act as coincidence detectors for input from several different PNs [34] (and in the honey bee possibly synchronized inputs from other sensory modalities). The activity of the KCs is sparse and distributed under odor stimulation [33, 46]. The combination of the connectivity just described, a high firing threshold and the need of coincident inputs from several PNs make KCs fire in a sparse and reliable manner (see Chap. 4.1). Remarkably, this sparseness is maintained across entire ranges of odor concentrations [1]. From the theoretical point of view, sparse activity improves the capacity for forming associative memories [28], and areas of the brain that display sparse activity are likely to be important in formation of memories. In fact, VUM_{mx1} that likely drives associative memory in the AL also projects broadly into the Kenyon cell dendritic fields [18, 23]. This information is very consistent with behavioral analyses that implicate the MBs as a very important area of the brain for formation of memory.

In locusts, spike-timing dependent plasticity (STDP) regulates the connections from the KCs to the ß-Lobe neurons [3]. One of the functions of STDP could be to allow ß-lobe neurons in the locust to act as coincidence detectors for KC inputs. In addition, given the existence of sparse coding in the KCs, another function could be to contribute to memory formation [21].

In summary, like the AL the MB also performs a transformation of olfactory processing. The spatiotemporal patterns set up by the AL networks are transformed into a sparse distributed pattern by the networks in the MB, which is theoretically ideal for storing a large number of associative memories. Studies have begun to examine how transformations at this level may enhance pattern recognition and memory [33]. But we need a computational framework in which to place much of this information if we are to understand how plasticity at each level contributes to olfactory pattern recognition and memory in the honey bee.

6.1.8   Computational Modeling of Neural Networks
         in the Mushroom Body

To illustrate how computational modeling can help to generate hypotheses about the function of the neural networks described above, we propose a model for how learning progresses in studies of conditioning represented in Figs. 6.1.1 and 6.1.2. We showed how conditioning progresses in a nonlinear manner, particularly under conditions in which the detection or discrimination is difficult. Our model implemented a simplified form of Hebbian learning at the output of the MB by enhancing those connections that follow reinforcement while reducing those connections that do not.

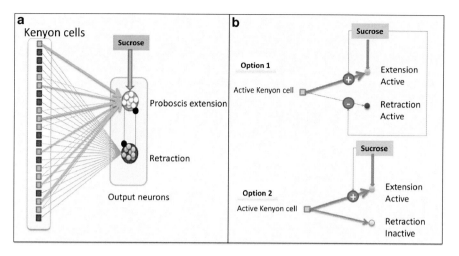

Fig. 6.1.4 (**a**) Model description for decision making in the Mushroom Body. An odor activated specific set of Kenyon cells which projected randomly to the output neurons (ENs in **b**). One group of output neurons was responsible for proboscis extension and another one for retraction. Sucrose conditioning increased excitability of extension group. (**b**) The two types of conditional learning implemented for a model of decision making in the MBs: *Option 1* represents the situation when the retraction neurons are active but they should not because sucrose has been released. In that case connection to retraction neurons was decreased and connection to extension group was increased. *Option 2* is applied when the retraction neuron is not active. Thus nothing is to be done with that connection but the connection to the extension group is increased

The implementation of such a rule was not deterministic; the synapses were modified with some given probability. This is likely the form of plasticity occurring at the synaptic level [3, 31, 32]. In this case, we implement sparse activity, Hebbian plasticity and competition via mutual inhibition to create a model of decision making in the insect brain. A very similar approach was proposed in [21] where a computational model mimicking the MB olfactory circuit was used to successfully solve a classical example of pattern recognition of hand written digit identification [25]. In this particular implementation we chose to use the same database for simplicity reasons, but with digits representing putative odorants. Therefore, below we will refer to different inputs to the model as odors.

We built two groups of output neurons that receive input from the KCs in the calyx (CA) (Fig. 6.1.4a). Activation of one group of neurons led to proboscis extension while the other was responsible for proboscis retraction. The basic principle for 'decision making' between these two groups of neurons that represent different behavioral responses is inhibition [35]. Initially, before any training, we assumed that the dominant behavioral response is retraction. For that purpose we assume that the majority of the connections from the KCs are projecting into the retraction group, while the number of connections into the extension group was initially only 10%. A specific odor (specific pattern of KC activation) was used for training. Release of sucrose was modeled as increase of excitability of the group of output neurons responsible for proboscis extension. Training odor was modified from one presentation to another to account for odor fluctuation found in real life. During training phase, whenever this odor paired with sucrose was presented, Hebbian learning modified synapses between the activated KCs and the output neurons. If the KCs and the output neurons corresponding to proboscis extension are coactivated (correct decision in a presence of sucrose) the connections were increased by probability p+. On the other hand, if the KCs and the output for retraction neurons are coactivated (incorrect decision when sucrose was released) the connections were reduced with probability p−. The two options are described in Fig. 6.1.4b where two different rules were applied depending whether the neurons of the retraction group are active or not.

The results of these rules applied to the test data [21] can be seen in Fig. 6.1.5. In these simulations a specific odorant input (odor+) was paired with sucrose. The probability of synaptic changes p+ and p− was small to track the evolution of the decision making process. To determine the generalization ability of the learning process, one odor from the data set was used for training and the entire set of odors was used for testing. The test set presentation was alternated with training odor presentation to track how the decision making model is performing throughout the training session. Thus, the model changed the connections only if the odorant belongs to the training set. During the testing phase, PE (proboscis extension) activity during odor+ presentation was classified as correct decision (Fig. 6.1.5). At presentation number 13–14, the system suddenly starts to associate the training odorant with the sucrose (Fig. 6.1.5, top). This corresponds to the "aha!" moment which is similar to behavioral experiments performed with honey bees. If one tracks in parallel the evolution of the percentage of the connections to the proboscis extension neurons, it can be seen that a sufficient number of connections is achieved after 13–14 trials such that the two groups of neurons can compete for decision making purposes (Fig. 6.1.5, bottom).

The mechanism underlying decision making in this model is relatively simple. Initially the majority of the connections were pointing into the retraction group of neurons. Therefore, retraction group became active during majority of trials including those with odor+ presentation. As the training procedure starts to operate, specific (corresponding to the odor+) connections from the KCs to the extension (output) neurons were enhanced, while the connections pointing to retraction were eliminated. Note that other odors which have not been paired with sucrose did not activate the extension group, although there always was some small probability of having false positives. It might be the case that some false positives in the model might be actual generalizations in the real environment of the insect. Once a critical mass of connections into the extension group was achieved, the inhibition between output neurons provided the mechanism to shut down the retraction neurons during odor+ presentation. The inhibitory mechanism and the modification of the balance of the synaptic weight provided the nonlinearity of the "Aha!" moment. In our model the speed of decision making depended on the probability of connection increase p+ or decrease p− and also on how much each connection was modified after single event. In the example in Fig. 6.1.5 the probabilities of reinforcing

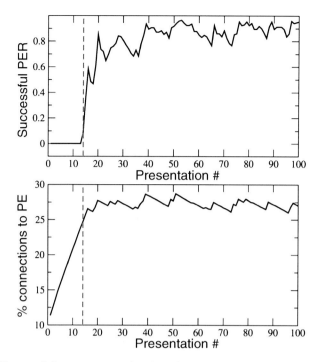

Fig. 6.1.5 (*Top panel*) Success rate as a function of the number of presentations of the training odorant. To determine the performance depending on the # of CS + presentations one can use a completely different data set to be able to determine the real ability to generalize. Note that the time scale have been slowed down to be able to find the sharp change in the decision making process. (*Bottom panel*) Evolution of the percentage of connections into the proboscis extension group

connections (p+) was rather low, probably lower than in real-life honey bees, which explains the relatively slow process of learning.

This model illustrates how some of the basic principles in the neural networks of the MB can be used to explain abrupt changes in behavior found in conditioning studies with honey bees. We implemented known sparse activity patterns across KCs of the MB, and combined these patterns with plasticity mechanisms described in the locust [3] and the honey bee [29, 32]. A basic mechanism that still remains to be observed and, in this case it is a prediction that can be empirically tested, is mutual inhibition between outputs from the MB. This model not only explains a decision-like process in conditioning, but can also account for a recent report of an interesting behavior called 'olfactory interference' [5]. When fed a droplet of sucrose, a honey bee will extend its proboscis and make feeding movements well beyond the time required to consume the sucrose-water droplet. When presented with an odor during this extended portion of the response, many times honey bees will abruptly retract their proboscis. That is, odor 'interferes' with the response in a way that would be predicted by our model. This model does not attempt to provide a final mechanism of how olfactory stimuli are learned in the insect brain, but to illustrate a possibility of testing with a simple model whether a specific hypothesis may lead to solving a pattern recognition problem in olfaction. Note that there is no unique way to learn; there can be multiple neural architectures to provide learning. Nevertheless, some basic principles must remain. Those are sparse coding to provide large storage space, local synaptic plasticity, inhibition as a tool to provide competitions, and global reward signals to guide the system into the proper set of answers.

6.1.9 Conclusions and Outlook

We feel that the data we have presented support a model for distributed plasticity that is sufficient in accounting for olfactory learning and memory in the honey bee. Known components of plasticity in the AL interact with likely components of plasticity in the MB. Moreover, we have shown how gradual changes in odor representations in the AL, which are associated with appetitive reinforcement, can interact with plasticity in the MB to produce rapid and precipitous changes in MB output. There are undoubtedly also components of plasticity yet to be described. We advocate integration of these components into computational models to begin to understand and formulate testable hypotheses about how these distributed components of plasticity interact to produce observed patterns of conditioning. In summary, the honey bee is an excellent model animal species for revealing interactions of distributed plasticity that are bound to be general principles for understanding learning and memory in all animals. Furthermore, these same principles will provide inspiration to other fields in computer science and engineering, particularly in regard to development of computationally efficient biomimetic algorithms for pattern recognition and classification.

References

1. Assisi C, Stopfer M, Laurent G, Bazhenov M (2007) Adaptive regulation of sparseness by feedforward inhibition. Nat Neurosci 10(9):1176–1184
2. Bitterman ME, Menzel R, Fietz A, Schäfer S (1983) Classical conditioning of proboscis extension in honeybees (*Apis mellifera*). J Comp Psychol 97(2):107–119
3. Cassenaer S, Laurent G (2007) Hebbian STDP in mushroom bodies facilitates the synchronous flow of olfactory information in locusts. Nature 448(7154):709–713
4. Chandra SB, Wright GA, Smith BH (2010) Latent inhibition in the honey bee, *Apis mellifera*: is it a unitary phenomenon? Anim Cogn 13(6):805–815
5. Dacher M, Smith BH (2008) Olfactory interference during inhibitory backward pairing in honey bees. PLoS One 3(10):e3513
6. Daly KC, Christensen TA, Lei H, Smith BH, Hildebrand JG (2004) Learning modulates the ensemble representations for odors in primary olfactory networks. Proc Natl Acad Sci USA 101(28):10476–10481

7. Denker M, Finke R, Schaupp F, Grun S, Menzel R (2010) Neural correlates of odor learning in the honeybee antennal lobe. Eur J Neurosci 31(1):119–133

8. Ditzen M, Evers JF, Galizia CG (2003) Odor similarity does not influence the time needed for odor processing. Chem Senses 28(9):781–789

9. Drezner-Levy T, Shafir S (2007) Parameters of variable reward distributions that affect risk sensitivity of honey bees. J Exp Biol 210(Pt 2):269–277

10. Faber T, Joerges J, Menzel R (1999) Associative learning modifies neural representations of odors in the insect brain. Nat Neurosci 2(1):74–78

11. Farooqui T, Robinson K, Vaessin H, Smith BH (2003) Modulation of early olfactory processing by an octopaminergic reinforcement pathway in the honeybee. J Neurosci 23:5370–5380

12. Fernandez PC, Locatelli FF, Person-Rennell N, Deleo G, Smith BH (2009) Associative conditioning tunes transient dynamics of early olfactory processing. J Neurosci 29(33):10191–10202

13. Galan RF, Weidert M, Menzel R, Herz AV, Galizia CG (2006) Sensory memory for odors is encoded in spontaneous correlated activity between olfactory glomeruli. Neural Comput 18(1):10–25

14. Galizia CG, Rössler W (2010) Parallel olfactory systems in insects: anatomy and function. Annu Rev Entomol 55:399–420

15. Ganeshina O, Menzel R (2001) GABA-immunoreactive neurons in the mushroom bodies of the honeybee: an electron microscopic study. J Comp Neurol 437(3):335–349

16. Giurfa M (2007) Behavioral and neural analysis of associative learning in the honeybee: a taste from the magic well. J Comp Physiol A 193(8):801–824

17. Guerrieri F, Schubert M, Sandoz JC, Giurfa M (2005) Perceptual and neural olfactory similarity in honeybees. PLoS Biol 3(4):e60

18. Hammer M (1993) An identified neuron mediates the unconditioned stimulus in associative olfactory learning in honeybees. Nature 366:59–63

19. Hildebrand JG, Shepherd GM (1997) Mechanisms of olfactory discrimination: converging evidence for common principles across phyla. Annu Rev Neurosci 20:595–631

20. Hourcade B, Perisse E, Devaud JM, Sandoz JC (2009) Long-term memory shapes the primary olfactory center of an insect brain. Learn Mem 16(10):607–615

21. Huerta R, Nowotny T (2009) Fast and robust learning by reinforcement signals: explorations in the insect brain. Neural Comput 21(8):2123–2151

22. Joerges J, Kuttnet A, Galizia G, Menzel R (1997) Representations of odours and odour mixtures visualized in the honeybee brain. Nature 387:285–288

23. Kreissl S, Eichmüller S, Bicker G, Rapus J, Eckert M (1994) Octopamine-like immunoreactivity in the brain and subesophageal ganglion of the honeybee. J Comp Neurol 348(4):583–595

24. Laurent G (2002) Olfactory network dynamics and the coding of multidimensional signals. Nat Rev Neurosci 3(11):884–895

25. LeCun Y, Cortes Chylcem (1998) MNIST database. http://yannlecuncom/exdb/mnist/

26. Linster C, Menon AV, Singh CY, Wilson DA (2009) Odor-specific habituation arises from interaction of afferent synaptic adaptation and intrinsic synaptic potentiation in olfactory cortex. Learn Mem 16(7):452–459

27. Lubow RE (1973) Latent inhibition. Psychol Bull 79:398–407

28. Marr D (1969) A theory of cerebellar cortex. J Physiol 202(2):437–470

29. Mauelshagen J (1993) Neural correlates of olfactory learning paradigms in an identified neuron in the honeybee brain. J Neurophysiol 69(2):609–625

30. Menzel R (1990) Learning, memory, and 'cognition' in honeybees. In: Kesner RP, Olton DS (eds) Neurobiology of comparative cognition. Lawrence Erlbaum, Hillsdale, pp 237–292

31. Menzel R, Manz G (2005) Neural plasticity of mushroom body-extrinsic neurons in the honeybee brain. J Exp Biol 208(Pt 22):4317–4332

32. Okada R, Rybak J, Manz G, Menzel R (2007) Learning-related plasticity in PE1 and other mushroom body-extrinsic neurons in the honeybee brain. J Neurosci 27(43):11736–11747

33. Perez-Orive J, Mazor O, Turner GC, Cassenaer S, Wilson RI et al (2002) Oscillations and sparsening of odor representations in the mushroom body. Science 297(5580):359–365

34. Perez-Orive J, Bazhenov M, Laurent G (2004) Intrinsic and circuit properties favor coincidence detection for decoding oscillatory input. J Neurosci 24(26):6037–6047
35. Quinn M, Smith L, Mayley G, Husbands P (2003) Evolving controllers for a homogeneous system of physical robots: structured cooperation with minimal sensors. Philos Transact A Math Phys Eng Sci 361(1811):2321–2343
36. Rabinovich M, Huerta R, Laurent G (2008) Neuroscience. Transient dynamics for neural processing. Science 321(5885):48–50
37. Rinberg D, Koulakov A, Gelperin A (2006) Speed-accuracy tradeoff in olfaction. Neuron 51(3):351–358
38. Rybak J, Menzel R (1993) Anatomy of the mushroom bodies in the honey bee brain: the neuronal connections of the alpha-lobe. J Comp Neurol 334(3):444–465
39. Sachse S, Galizia CG (2003) The coding of odour-intensity in the honeybee antennal lobe: local computation optimizes odour representation. Eur J Neurosci 18(8):2119–2132
40. Smith BH, Menzel R (1989) An analysis of variability in the feeding motor program of the honey bee; the role of learning in releasing a modal action pattern. Ethology 82:68–81
41. Smith BH, Menzel R (1989) The use of electromyogram recordings to quantify odorant discrimination in the honey bee, *Apis mellifera*. J Insect Physiol 35:369–375
42. Smith BH, Abramson CI, Tobin TR (1991) Conditional withholding of proboscis extension in honeybees (*Apis mellifera*) during discriminative punishment. J Comp Psychol 105(4):345–356
43. Smith BH, Wright GA, Daly KS (2006) Learning-based recognition and discrimination of floral odors. In: Dudareva N, Pichersky E (eds) The biology of floral scents. CRC Press, Boca Raton, pp 263–295
44. Stopfer M, Bhagavan S, Smith BH, Laurent G (1997) Impaired odour discrimination on desynchronization of odour-encoding neural assemblies. Nature 390(6655):70–74
45. Strausfeld NJ, Hildebrand JG (1999) Olfactory systems: common design, uncommon origins? Curr Opin Neurobiol 9(5):634–639
46. Szyszka P, Ditzen M, Galkin A, Galizia CG, Menzel R (2005) Sparsening and temporal sharpening of olfactory representations in the honeybee mushroom bodies. J Neurophysiol 94(5):3303–3313
47. Szyszka P, Galkin A, Menzel R (2008) Associative and non-associative plasticity in kenyon cells of the honeybee mushroom body. Front Syst Neurosci 2:3
48. Uchida N, Mainen ZF (2003) Speed and accuracy of olfactory discrimination in the rat. Nat Neurosci 6(11):1224–1229
49. Vosshall LB, Wong AM, Axel R (2000) An olfactory sensory map in the fly brain. Cell 102(2):147–159
50. Wright GA, Carlton M, Smith BH (2009) A honeybee's ability to learn, recognize, and discriminate odors depends upon odor sampling time and concentration. Behav Neurosci 123(1):36–43

Chapter 6.2
The Molecular Biology of Learning and Memory – Memory Phases and Signaling Cascades

Uli Müller

Abstract In species as diverse as mollusks, insects, birds, and mammals memories are highly dynamic and cover phases from seconds to a lifetime. In honey bees as in other species, the induction of distinct memory phases depends on parameters like the number and succession of the training trials. Employing techniques developed to monitor and manipulate the activity of signaling cascades in intact honey bees, training parameters could be linked to temporal modulations of signaling cascades that contribute to distinct memory phases. This analysis uncovered a dynamic network of signaling events in the antennal lobes (ALs) and the mushroom bodies (MBs) that are required for defined aspects of both, the induction and the maintenance of distinct memory phases.

Abbreviations (excluding gene/protein names)

eLTM	Early phase LTM
lLTM	Late phase LTM
LTM	Long-term memory
MTM	Mid-term memory
NO	Nitric oxide
OA	Octopamine
CS	Conditioned stimulus
US	Unconditioned stimulus

U. Müller (✉)
Natural Sciences and Technology III, Dept. 8.3 – Biosciences
(Zoology and Physiology/Neurobiology), Saarland University, Saarbrücken, Germany
e-mail: uli.mueller@mx.uni-saarland.de

C.G. Galizia et al. (eds.), *Honeybee Neurobiology and Behavior: A Tribute to Randolf Menzel*, DOI 10.1007/978-94-007-2099-2_31,
© Springer Science+Business Media B.V. 2012

6.2.1 Introduction

Numerous studies in the recent years provided compelling evidence that the molecular basis of learning and memory is highly conserved in mammals and invertebrate species as *Aplysia*, *Drosophila* and honey bees [4, 18, 23, 28]. Molecular processes that modulate the strength of the synaptic connections between neurons are the main substrates of both, short- and long-term plasticity. Although the molecular machinery of long-lasting plasticity and long-term memory (LTM) is located in the synapses, it still requires molecular processes in the nucleus and thus signaling processes between synapse and nucleus [6, 23]. The requirement of transcription processes mechanistically distinguishes the robust LTM from the transient short-term memory (STM) and mid-term memory (MTM). Usually, the energy demanding processes underlying LTM or long-lasting plasticity requires repeated training sessions or repeated stimulation protocols for their induction [5, 38].

A series of studies demonstrated a critical role of cAMP mediated signaling processes in learning and memory formation in both, mammals and invertebrates. While initial studies implicated the cAMP-activated signaling cascade in learning and short-term plasticity, the further characterization disclosed the conserved and critical role of cAMP mediated transcription processes in the formation of LTM and long-lasting neuronal plasticity [1, 23, 43]. Especially studies in *Aplysia* and honey bee identified the cAMP-cascade as a central player that links the training pattern and the processes required for the induction of long-lasting plasticity and LTM [33, 35].

However, in addition to the cAMP cascade other processes that regulate protein-protein interaction, intracellular transport, translation, transcription, and many more are critically involved in learning and memory formation [23, 39, 44].

Blocking of distinct molecular processes often causes very specific impairments and is thus suited to uncover characteristic features of memory formation such as its organization in distinct phases. The identification of single memory phases however can be difficult. Distinct memory phases can overlap or can be temporally separated from training and thus may become evident hours to days after training. Moreover, the induction and expression of distinct memory phases may be localized in different neuronal networks.

Although the connection between distinct features of learning and memory formation and the underlying molecular processes has been demonstrated in numerous studies, it is mostly unclear how these processes are regulated by learning and how this contributes to the dynamic process of memory formation at the behavioral level. In this respect, especially studies in honey bees that aimed to unravel the link between the behavioral and the molecular levels provided essential information to understand the dynamic events at the molecular level induced as a consequence of learning and memory formation [27, 34, 42].

6.2.2   Olfactory Learning in Honey Bees: Connecting Training Parameters to Memory Phases

The robust learning in honey bees is an absolute requirement to identify and characterize the dynamic molecular processes responsible for the behavioral changes [16, 28]. In the honey bee, as in other species at least three memory phases can be identified at the behavioral level: a short-term memory in the range of minutes, a MTM in the range of hours, and a stable LTM which lasts for days and weeks.

A single associative conditioning trial consists of the pairing of odor and reward (Fig. 6.2.1). An odor stimulus (CS, conditioned stimulus, 3 s) is immediately followed by a partially overlapping sucrose reward (US, unconditioned stimulus, 3 s). Although this pairing lasts only a few seconds it induces a memory that decays over several days and is independent of translation and transcription. Interestingly, in the range of days the memory induced by a single-trial conditioning is resistant to amnesic treatment like cooling [30], can not be erased by known pharmacological tools, and shows striking parallels to amnesia resistant memory observed in *Drosophila* [22].

Translation and transcription dependent LTM is induced already by three successive conditioning trials applied within a time window of a few minutes [14, 27, 28]. This LTM is mechanistically separable into an early and a late phase (Fig. 6.2.2).

The early phase (eLTM, 1–2 days) requires translation-processes during associative training that can be blocked by translation blockers like emetine and anisomycine. The late phase (lLTM, ≥3 days) depends on transcription processes during and up to several hours after conditioning as demonstrated by the transcription blocker actinomycin [27, 34, 46]. However, all these blockers do not totally erase the memory: they reduce the performance to the level observed after a single-trial conditioning.

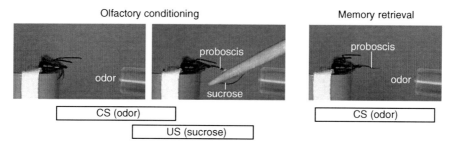

Fig. 6.2.1 Associative olfactory conditioning of the proboscis extension reflex (*PER*). For associative olfactory conditioning honey bees were secured in tubes. The head protruded fully to allow free movement of proboscis and antennae. An olfactory conditioning trial is composed of the overlapping pairing of an odor stimulus (*CS* conditioned stimulus) and a sucrose stimulus (*US* unconditioned stimulus) to the antennae and the proboscis. The scheme illustrates the stimulation procedure, which lasts a few seconds only. For memory retrieval at distinct times after conditioning, the animals were stimulated with the CS alone

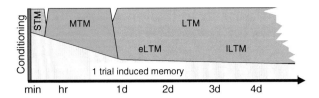

Fig. 6.2.2 Memory phases in honey bees. Memory formation strongly depends on training parameters like the number of training trials. A single-trial conditioning induces a memory that starts at a high level, decays within 1 day, and remains at a basal level for several days. This single trial induced memory (*one trial induced memory*) is insensitive to blockers of translation and transcription and reveals properties similar to the amnesia resistant memory in *Drosophila*. Multiple-trial conditioning induces a memory that stays at a high level for many days and consist of mechanistically distinguishable phases: the short-term memory (*STM*) up to 1 h, the mid-term memory (*MTM*) that occurs in a time window of 1 h to about 1 day, and the long-term memory (*LTM*). While STM and MTM are independent of translation and transcription, the early phase LTM (*eLTM*) is sensitive to translation blockers and the late phase LTM (*lLTM*) requires translation and transcription processes

Thus, in the honey bee our present knowledge concerning molecular processes focuses on memory phases induced by multiple-trial conditioning and here especially on processes implicated in the induction of distinct memory phases.

The overall very short conditioning sessions in honey bees allow a clear distinction between acquisition and consolidation phase and provide the basis to monitor transient *in vivo* induced activities of signaling cascades [20, 21]. By rapid termination (<0.5 s) of signaling processes freezing the whole honey bee in liquid nitrogen, tissue dissection, and subsequent biochemical assays, it is possible to monitor learning induced changes of signaling cascades in defined brain areas at any given time. This analysis concentrates on the ALs and the MBs, two areas in the insect brain known to play a critical role in associative learning and memory formation [8, 15, 16, 19].

6.2.3   Molecular Processes Underlying the Mid-term Memory: Interaction Between Ca²⁺ Regulated Kinases and Proteases

Calcium controls different aspects of cellular physiology and calcium-regulated enzymes play a fundamental role in the regulation of cellular processes including synaptic plasticity. In addition to other pathways, phosphorylation of target protein by Ca^{2+}/phospholipid-dependent protein kinase C (PKC) plays a major role in synaptic plasticity and memory formation in mammals [41].

Initiated by the observation that odors induce characteristic glomerular changes in Ca^{2+} levels in the honey bee [12], measurements in the AL reveal a transient activation of the Ca^{2+}-regulated PKC in the range of a few minutes by both the US and the CS [14]. Surprisingly, odor stimulation, sucrose stimulation, CS-US, and US-CS

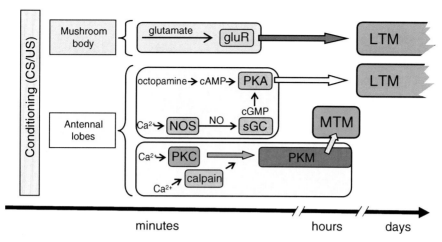

Fig. 6.2.3 Molecular processes localized in different brain areas contribute to induction and maintenance of associative olfactory memory. In the ALs, multiple-trial conditioning activates the Ca^{2+}-dependent protease calpain and the PKC. Calpain cleaves the activated PKC and leads to the formation of the constitutively active PKM. PKM is critically required for the maintenance of MTM. In a parallel and independent process also located in the ALs, multiple-trial conditioning induces a prolonged activation of the PKA that is essential for the induction of LTM. The prolonged activation is mediated by the synergistic action of cyclicAMP (*cAMP*) and cyclicGMP (*cGMP*) on PKA that triggers yet unknown subsequent events in LTM formation. cGMP is formed by the soluble guanylate cyclase (*sGC*) after its activation by nitric oxide (*NO*) formed by the nitric oxide synthase (*NOS*). In addition to these processes located in the ALs, glutamate mediated signaling events in the MBs contribute to LTM formation. Photolytic release of glutamate in the MBs immediately after conditioning facilitates LTM formation. However, the molecular targets triggered by these early events of LTM formation in the ALs and the MBs remain to be identified. In contrast to the pictured processes underlying multiple-trial induced memory the mechanism responsible for memory induced by single-trial conditioning are still unclear

stimulation all induce a similar transient activation of PKC in the ALs. Since blocking this transient PKC activation during the conditioning phase neither affects learning nor memory formation, the immediate stimulus induced PKC activation in the ALs is not critically implicated in processes of learning and memory formation. However, measurements at later time points after conditioning reveal a link between the training procedure and the temporal pattern of PKC activation in the ALs. While multiple-trial conditioning cause an elevation of PKC activity beginning 1 h after conditioning and lasting up to 3 days, a single-trial conditioning does not induce such a prolonged PKC activation [14]

The detailed characterization uncovered two independent and mechanistically distinct processes underlying this PKC activation triggered by associative conditioning. From 1 h until 16 h after conditioning the elevated PKC activity is due to the constitutive active protein kinase M (PKM) (Fig. 6.2.3). The PKM is a shortened form of the PKC, cleaved by the Ca^{2+}-dependent protease calpain. The cleavage of PKC to PKM by calpain only occurs if PKC is activated. Inhibition of calpain during training prevents formation of PKM. This suggests that multiple-trial conditioning activates both Ca^{2+}-dependent enzymes, the PKC and calpain, which enables the interaction of these enzymes and consequently the formation of PKM. Since inhibition of calpain also impairs memory in a time-window of 1–16 h after repeated training, formation of PKM during training is involved in maintaining mid-term memory (MTM) (Fig. 6.2.3). Acquisition, the early memory phase tested at 30 min, and memory after 1 day is not affected by blocking calpain.

The long-lasting increase in PKC activity 1–3 days after multiple-trial conditioning is insensitive to inhibition of calpain but is erased by blockers of translation and transcription during conditioning [14]. This late phase of training-induced PKC activation in the ALs is not affected by inhibition of the early phase and vice versa. The latter suggests that the early and late phase of training induced elevation of PKC activity in the ALs are parallel processes. While the mechanism and contribution of the elevated PKC activity in the late phase is unclear yet, findings in *Aplysia*, *Drosophila* and mammals point to a conserved function of PKM in memory formation [7, 45].

6.2.4 CyclicAMP-Dependent Processes Mediate the Induction of LTM

Blocking cAMP-dependent processes during the training period impairs the formation of LTM without affecting memory in the range of hours [9, 33]. The restriction of this critical time window to the training phase suggested a critical implication of a fast and transient modulation of cAMP-dependent processes during conditioning. This hypothesis was verified by a technique that enabled the determination of *in vivo* induced changes in protein kinase A (PKA) activity [20, 21]. At defined times after associative conditioning the honey bees were shock frozen in liquid nitrogen, the brain areas of interest dissected, and subjected to a special assay to determine the status of PKA activity. As potential brain areas involved in associative olfactory learning [15, 17], the ALs and the MBs were tested. Interestingly, learning related changes in PKA activity could only be detected in the ALs but not the calyces of the MBs.

Stimulation of the antenna with sucrose, the US stimulus in olfactory conditioning, induces a fast activation of PKA in the ALs [20, 21]. PKA activity increases within less than 1 s and is back to baseline within 3 s. Odor stimulation, the CS used in olfactory conditioning, as well as mechanical stimulation of the antennae does not affect PKA activity in the ALs [21]. The sucrose induced PKA activation in the ALs is mediated by octopamine (OA) [20] that is most likely released by the VUM_{mx1} neuron that plays a crucial role in US processing [15, 17]. The dense innervation of the ALs by the VUM_{mx1} neuron together with the predominant localization of the PKA in the local interneurons [31] point to a sucrose (US) induced modulation of PKA-dependent processes throughout the ALs.

In contrast to this, *in vivo* stimulation with sucrose (US) does not lead to detectable changes in PKA activity in the mushroom body calyces that are also densely innervated by the VUM$_{mx1}$ neuron. However, since biogenic amines including OA can activate PKA in cultured Kenyon cells [31] the OA receptors receiving input from VUM$_{mx1}$ in the MBs are rather coupled to Ca^{2+} regulated pathways than to the cAMP pathway [2, 13].

The temporal dynamic of PKA activity induced by a sucrose stimulus to the antennae differs from US stimulation in olfactory conditioning (sucrose stimulation of the antenna immediately followed by proboscis stimulation). A single-trial conditioning (CS-US pairing) induces a prolonged PKA activation in the ALs. After a fast increase, PKA activity returns to baseline within 60 s after the conditioning trial [33]. Latter elevation of PKA activity in the ALs is prolonged up to more than 3 min after the third conditioning trial (2 min interval). In contrast to the duration, the amplitude of PKA activation induced by single and multiple-trial conditioning does not differ. Single or repeated backward pairing (US-CS pairing) induce a short lasting PKA activation similar to that induced by a single-trial conditioning. This suggests that a prolonged PKA activation may be implicated in the induction of molecular processes that lead to LTM formation.

This idea was tested by photolytic release of caged cAMP that enables a locally- and temporally-defined PKA activation in the ALs. Mimicking this prolonged PKA activation by photolytic release of caged cAMP during a single-trial conditioning is sufficient to induce a long-lasting memory (Fig. 6.2.3) [33]. This demonstrates that learning induced PKA activation within a few minutes during conditioning is critical for LTM formation. The upstream signaling pathways and the target networks of this learning triggered transient PKA activation in the ALs are presently unknown.

6.2.5 Nitric Oxide/cGMP-System Acts as an Integrator in LTM Induction

After the detection that nitric oxide (NO) acts as a signaling molecule in the nervous system, the NO system has been implicated in functions like neuronal development and plasticity. The characterization of the Ca^{2+}-dependent NO synthase (NOS) and its major target molecule the soluble guanylate cyclase (sGC) revealed a high conservation between mammals and invertebrates [29, 32]. NOS is abundant in the ALs of the honey bee and is implicated in integrative processing of chemosensory stimuli [36, 37]. As a radical, NO can freely diffuse to target molecules like the sGC and thus induce the formation of the second messenger cGMP independent of synaptic transmission.

Inhibition of NO synthase (NOS) affects both, the prolonged PKA activation induced by multiple-trial conditioning and the formation of LTM [30, 33]. Biochemical evidence shows that the honey bee PKA is synergistically activated by cAMP and cGMP at low cAMP concentrations [24]. The link between both processes is supported by the fact that inhibition of the sGC impairs LTM and erases the prolonged PKA activation triggered by multiple-trial conditioning. Moreover, like uncaging cAMP, photo release of cGMP during a single-trial conditioning leads to the formation of a long-lasting memory. All these findings support the idea that the NO/cGMP system mediates the multiple-trial induced prolonged PKA activation in the ALs, an early and transient event required for the formation of LTM [33] (Fig. 6.2.3).

In contrast to the improvement of memory by uncaging cAMP or cGMP during conditioning, uncaging NO totally impairs memory formation [33]. This suggests that release of NO during conditioning interferes with yet unknown functions of NO in the ALs that are critical during associative odor learning.

In addition to the critical role of the NO/cGMP system in integrative processes in olfactory conditioning, NO/cGMP is also involved in the integrative processing of appetitive signals (sucrose) during habituation [36]. Although the implication in processes concerning signal integration is a common feature in these different paradigms, the detailed comparison revealed differences with respect to the temporal parameters between NO/cGMP function in associative and non-associative learning [37]. Most likely, the NO/cGMP affects different neuronal networks contributing to the different forms of learning.

6.2.6 Glutamate Transmission in the Mushroom Bodies Is Implicated in LTM Induction

In contrast to the well-characterized function of glutamate in the mammalian brain, its role as neurotransmitter in the insect brain is poorly understood [40]. Characterization of components of the glutamate transmission machinery in honey bees [11] together with findings that knockdown of NMDA-type glutamate receptors in *Drosophila* brain impair aversive olfactory learning and long-term memory [47] points to an implication of glutamate in synaptic plasticity as known from mammals. This notion is supported by pharmacological studies demonstrating that drugs characterized for their action on mammalian glutamate receptors and glutamate reuptake machinery results in behavioral changes in honey bees [26]. However, due to the unknown pharmacological profile of these drugs on the insect glutamate receptor and re-uptake machinery, the interpretation of the results with respect to glutamate transmission is difficult.

To avoid these problems, glutamate was directly released within defined brain areas by photolytic uncaging *in vivo* [25]. Release of glutamate in the MBs but not in the ALs affects processes of memory formation. Interestingly only the release immediately (≈ 3 s) after conditioning improves memory formation: it elevates memory tested 2 days after training to a level observed after repeated conditioning-trials (Fig. 6.2.3). Uncaging glutamate 1 min before conditioning has no effect, the memory does not differ from that induced by a single-trial conditioning. This demonstrates that the uncaging technique enables the dissociation of glutamate actions separated by 1 min and allows identifying a very narrow glutamate sensitive time window in associative learning. It is possible that glutamate accelerates the acquisition processes itself or that additional glutamate triggers parallel molecular processes that, together with the single-trial conditioning leads to a long-lasting form of memory. Although, this long-lasting memory has not been characterized as transcription and translation dependent LTM yet, these findings support the idea that glutamatergic neurotransmission in honey bees is implicated in the induction of long-lasting neuronal plasticity as known from mammals [40] (see Chaps. 3.3–3.5).

6.2.7 Parallel cAMP-Mediated Processes Contribute to LTM Formation

Studies in different systems provide convincing evidence that memory formation is a dynamic and continuous process [1, 4, 5, 39, 42]. In the honey bee, LTM induced by multiple-trial conditioning can be dissected into a translation-dependent early phase LTM, the eLTM (1–2 days) and a transcription-dependent late phase LTM, the lLTM ($\geq$3 days) (Fig. 6.2.2) [14, 34, 46]. As in other systems [1, 3], the induction of LTM (eLTM+lLTM) requires cAMP/PKA-triggered processes [9, 33, 34]. This suggests that eLTM and lLTM occur sequentially and are triggered by the same molecular events. Studies on the impact of the satiation level on appetitive conditioning in the honey bee however, provide evidence that eLTM and lLTM are triggered by different processes [10].

In appetitive conditioning paradigms, formation of reliable memories require conditioning of hungry or thirsty animals. The induction of MTM and the two LTM phases (eLTM and lLTM) in the honey bee requires a multiple-trial conditioning of animals starved for many hours. Irrespective of the number of trials, appetitive conditioning of fed honey bees leads to an impairment of acquisition and memory formation [10]. Since this is presently the only treatment that affects memory induced by a single-trial training this is especially interesting with regard to yet unknown processes triggered by single-trial conditioning.

The fact that induction of LTM depends on a prolonged PKA activity [33] points to a potential connection between the satiation level and the cAMP/PKA signaling pathway. Determination of the basic PKA activity in the brain of honey bees at different satiation levels supported this idea. As compared to satiated animals, the basal PKA activity is higher in the brains of hungry honey bees [10]. Conditioning of satiated honey bees with an artificially elevated PKA activity that corresponds to that of hungry animals leads to a normal lLTM with still impaired acquisition, MTM and eLTM. This difference between eLTM and lLTM and the fact that both phases depend on PKA activity during conditioning [33] argues that distinct pathways induce these phases already during conditioning. Although the details have not worked out yet, it is feasible that conditioning directly induces cAMP/PKA-mediated transcription processes mediated via the cAMP response element binding protein (CREB) [43]. However, since acquisition, MTM, and eLTM is not restored by elevation of basic PKA activity during conditioning, it is evident that other yet unknown molecular processes are implicated in the network of molecular interactions between satiation and learning processes.

6.2.8 Contribution of Antennal Lobe and Mushroom Body to Associative Learning

Cooling distinct brain areas immediately after olfactory conditioning in honey bees were the first experiments demonstrating a contribution of the ALs and the MBs in associative learning [8, 28]. In these studies, differences in the sensitive time windows pointed to different contributions of the ALs and the MBs. These ideas have been supported by studies on the role of OA and the octopaminergic VUM_{mx1} neuron in olfactory learning [15–17]. The analysis of the molecular events underlying memory formation in honey bee disclosed a molecular network of parallel and independent acting processes localized in the ALs and the MBs [34, 42].

Both, the ALs and the MBs are implicated in events of LTM formation (Fig. 6.2.3). While the cAMP/PKA system required for LTM induction is located in the ALs [30, 33], cascades activated by glutamate are localized in the MBs [25]. These are most likely only two components of a molecular network of molecular processes necessary for LTM formation. The finding that LTM formation requires at least two different cAMP-dependent processes has supported this; one process specifically contributes to lLTM formation [10]. In all these cases however it remains unclear yet in which neuronal network of the honey bee brain and by which molecular mechanisms LTM is maintained.

Independent of the mechanisms implicated in LTM induction, the ALs contribute to formation and maintenance of MTM [14]. Thus, different signaling cascades located in the ALs contribute to distinct aspects of memory formation (Fig. 6.2.3). This is of special interest, since the contribution of the ALs to associative learning has so far only been studied in the honey bee while the work in *Drosophila* explicitly focused on the MBs. In *Drosophila* it was demonstrated that the MBs are sites where aversive olfactory memories are formed and stored for at least a few hours [4, 19]. The findings in the honey bee reveal a critical contribution of the ALs and provide clear evidence for a parallel organization of the molecular network contributing to memory formation (see Chap. 6.1), similar to the distributed processing in mammals.

6.2.9 Outlook

The identification of transient signaling events triggered by learning, clearly demonstrates the necessity of techniques that allow both, fast monitoring and fast manipulation of signaling cascades *in vivo*. To date, the only technique that allows such a precise temporal resolution is the release and thus activation of caged molecules by short light pulses. However, small molecules such as caged second messengers or caged transmitters require injection into the tissue resulting in diffusion and thus limited spatial resolution. To overcome this problem, a future goal is to modify the caged forms of these small molecules to enable controlled application and thus activation in subsets of neurons. Extending this approach to chemically manufactured light-activated kinases, phosphatases, proteases, etc. would allow the direct manipulation of defined components of signaling cascades *in vivo*.

Bees are highly developed insects, their social organization and their metabolism are both driven by environmental aspects. Environmental changes result in altered behavior, gene expression and physiology. Also learning and memory are environmentally affected. To date the responsible signaling cascades have not been worked out. Linking the developmental changes in signaling components to the molecular network underlying learning and memory will certainly provide a better understanding of the relation between developmental and activity dependent plasticity.

References

1. Abel T, Nguyen PV, Barad M, Deuel TA, Kandel ER et al (1997) Genetic demonstration of a role for PKA in the late phase of LTP and in hippocampus-based long-term memory. Cell 88(5):615–626
2. Balfanz S, Strunker T, Frings S, Baumann A (2005) A family of octopamine [corrected] receptors that specifically induce cyclic AMP production or Ca2+ release in *Drosophila melanogaster*. J Neurochem 93(2):440–451
3. Dash PK, Hochner B, Kandel ER (1990) Injection of the cAMP-responsive element into the nucleus of *Aplysia* sensory neurons blocks long-term facilitation. Nature 345(6277):718–721
4. Davis RL (2005) Olfactory memory formation in *Drosophila*: from molecular to systems neuroscience. Annu Rev Neurosci 28:275–302
5. Davis HP, Squire LR (1984) Protein synthesis and memory: a review. Psychol Bull 96(3):518–559
6. Deisseroth K, Mermelstein PG, Xia H, Tsien RW (2003) Signaling from synapse to nucleus: the logic behind the mechanisms. Curr Opin Neurobiol 13(3):354–365
7. Drier EA, Tello MK, Cowan M, Wu P, Blace N et al (2002) Memory enhancement and formation by atypical PKM activity in *Drosophila melanogaster*. Nat Neurosci 5(4):316–324
8. Erber J, Masuhr T, Menzel R (1980) Localization of short-term memory in the brain of the bee, *Apis mellifera*. Physiol Entomol 5:343–358
9. Fiala A, Müller U, Menzel R (1999) Reversible downregulation of protein kinase A during olfactory learning using antisense technique impairs long-term memory formation in the honeybee, *Apis mellifera*. J Neurosci 19(22):10125–10134
10. Friedrich A, Thomas U, Müller U (2004) Learning at different satiation levels reveals parallel functions for the cAMP-protein kinase A cascade in formation of long-term memory. J Neurosci 24(18):4460–4468
11. Funada M, Yasuo S, Yoshimura T, Ebihara S, Sasagawa H et al (2004) Characterization of the two distinct subtypes of metabotropic glutamate receptors from honeybee, *Apis mellifera*. Neurosci Lett 359(3):190–194
12. Galizia CG, Menzel R (2000) Odour perception in honeybees: coding information in glomerular patterns. Curr Opin Neurobiol 10(4):504–510
13. Grohmann L, Blenau W, Erber J, Ebert PR, Strunker T et al (2003) Molecular and functional characterization of an octopamine receptor from honeybee (*Apis mellifera*) brain. J Neurochem 86(3):725–735
14. Grünbaum L, Müller U (1998) Induction of a specific olfactory memory leads to a long-lasting activation of protein kinase C in the antennal lobe of the honeybee. J Neurosci 18(11):4384–4392

15. Hammer M (1993) An identified neuron mediates the unconditioned stimulus in associative olfactory learning in honeybees. Nature 366:59–63
16. Hammer M, Menzel R (1995) Learning and memory in the honeybee. J Neurosci 15(3 Pt 1):1617–1630
17. Hammer M, Menzel R (1998) Multiple sites of associative odor learning as revealed by local brain microinjections of octopamine in honeybees. Learn Mem 5(1–2):146–156
18. Heisenberg M (2003) Mushroom body memoir: from maps to models. Nat Rev Neurosci 4(4):266–275
19. Heisenberg M, Borst A, Wagner S, Byers D (1985) *Drosophila* mushroom body mutants are deficient in olfactory learning. J Neurogenet 2(1):1–30
20. Hildebrandt H, Müller U (1995) Octopamine mediates rapid stimulation of protein kinase A in the antennal lobe of honeybees. J Neurobiol 27(1):44–50
21. Hildebrandt H, Müller U (1995) PKA activity in the antennal lobe of honeybees is regulated by chemosensory stimulation in vivo. Brain Res 679(2):281–288
22. Isabel G, Pascual A, Preat T (2004) Exclusive consolidated memory phases in *Drosophila*. Science 304(5673):1024–1027
23. Kandel ER (2001) The molecular biology of memory storage: a dialogue between genes and synapses. Science 294(5544):1030–1038
24. Leboulle G, Müller U (2004) Synergistic activation of insect cAMP-dependent protein kinase A (type II) by cyclicAMP and cyclicGMP. FEBS Lett 576(1–2):216–220
25. Locatelli F, Bundrock G, Müller U (2005) Focal and temporal release of glutamate in the mushroom bodies improves olfactory memory in *Apis mellifera*. J Neurosci 25(50):11614–11618
26. Maleszka R, Helliwell P, Kucharski R (2000) Pharmacological interference with glutamate re-uptake impairs long-term memory in the honeybee, *Apis mellifera*. Behav Brain Res 115(1):49–53
27. Menzel R (1999) Memory dynamics in the honeybee. J Comp Physiol A 185:323–340
28. Menzel R, Müller U (1996) Learning and memory in honeybees: from behavior to neural substrates. Annu Rev Neurosci 19:379–404
29. Müller U (1994) Ca2+/calmodulin-dependent nitric oxide synthase in *Apis mellifera* and *Drosophila melanogaster*. Eur J Neurosci 6(8):1362–1370
30. Müller U (1996) Inhibition of nitric oxide synthase impairs a distinct form of long-term memory in the honeybee, *Apis mellifera*. Neuron 16(3):541–549
31. Müller U (1997) Neuronal cAMP-dependent protein kinase type II is concentrated in mushroom bodies of *Drosophila melanogaster* and the honeybee *Apis mellifera*. J Neurobiol 33(1):33–44
32. Müller U (1997) The nitric oxide system in insects. Prog Neurobiol 51(3):363–381
33. Müller U (2000) Prolonged activation of cAMP-dependent protein kinase during conditioning induces long-term memory in honeybees. Neuron 27(1):159–168
34. Müller U (2002) Learning in honeybees: from molecules to behaviour. Zoology (Jena) 105(4):313–320
35. Müller U, Carew TJ (1998) Serotonin induces temporally and mechanistically distinct phases of persistent PKA activity in *Aplysia* sensory neurons. Neuron 21(6):1423–1434
36. Müller U, Hildebrandt H (1995) The nitric oxide/cGMP system in the antennal lobe of *Apis mellifera* is implicated in integrative processing of chemosensory stimuli. Eur J Neurosci 7(11):2240–2248
37. Müller U, Hildebrandt H (2002) Nitric oxide/cGMP-mediated protein kinase A activation in the antennal lobes plays an important role in appetitive reflex habituation in the honeybee. J Neurosci 22(19):8739–8747
38. Nguyen PV, Abel T, Kandel ER (1994) Requirement of a critical period of transcription for induction of a late phase of LTP. Science 265(5175):1104–1107
39. Reissner KJ, Shobe JL, Carew TJ (2006) Molecular nodes in memory processing: insights from *Aplysia*. Cell Mol Life Sci 63(9):963–974
40. Riedel G, Platt B, Micheau J (2003) Glutamate receptor function in learning and memory. Behav Brain Res 140(1–2):1–47

41. Sacktor TC (2008) PKMzeta, LTP maintenance, and the dynamic molecular biology of memory storage. Prog Brain Res 169:27–40

42. Schwärzel M, Müller U (2006) Dynamic memory networks: dissecting molecular mechanisms underlying associative memory in the temporal domain. Cell Mol Life Sci 63(9):989–998

43. Silva AJ, Kogan JH, Frankland PW, Kida S (1998) CREB and memory. Annu Rev Neurosci 21:127–148

44. Skoulakis EM, Grammenoudi S (2006) Dunces and da Vincis: the genetics of learning and memory in *Drosophila*. Cell Mol Life Sci 63(9):975–988

45. Sutton MA, Bagnall MW, Sharma SK, Shobe J, Carew TJ (2004) Intermediate-term memory for site-specific sensitization in *Aplysia* is maintained by persistent activation of protein kinase C. J Neurosci 24(14):3600–3609

46. Wüstenberg D, Gerber B, Menzel R (1998) Short communication: long- but not medium-term retention of olfactory memories in honeybees is impaired by actinomycin D and anisomycin. Eur J Neurosci 10(8):2742–2745

47. Xia S, Miyashita T, Fu TF, Lin WY, Wu CL et al (2005) NMDA receptors mediate olfactory learning and memory in *Drosophila*. Curr Biol 15(7):603–615

Chapter 6.3
Extinction Learning in Honey Bees

Dorothea Eisenhardt

Abstract Extinction describes the decrease of a conditioned behavior after reinforcement has failed. This paper discusses studies on extinction in harnessed honey bees with the aim of understanding the relevance of this learning phenomenon for the natural behavior of free-flying honey bees. It has been demonstrated that the reward memory is crucial to the extinction outcome and that the memory phase during which the reward memory is extinguished is critical. Based on these considerations we suggest that extinction plays a role in the adaptive behavior of foraging honey bees to variable food sources.

Abbreviations

PER   Proboscis extension reflex
CR   Conditioned response
CS   Conditioned stimulus
US   Unconditioned stimulus

6.3.1   Adaptation of Foraging to Changing Rewards Is Based on Different Memories

Forager bees collect nectar and pollen during the summer season to provide food for their offspring and to ensure adequate food supply during hibernation of the colony (Fig. 6.3.1a). Accordingly, the survival of a honey bee colony

D. Eisenhardt (✉)
Department of Biology, Chemistry, Pharmacy, Institute for Biology, Neurobiology,
Freie Universität Berlin, Königin-Luise-Strasse 28/30, 14195 Berlin, Germany
e-mail: dorothea.eisenhardt@fu-berlin.de

C.G. Galizia et al. (eds.), *Honeybee Neurobiology and Behavior: A Tribute to Randolf Menzel*, DOI 10.1007/978-94-007-2099-2_32,
© Springer Science+Business Media B.V. 2012

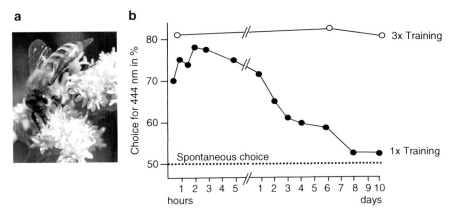

Fig. 6.3.1 The stability of reward memories correlates with the number of learning trials in foraging honey bees (**a**) Foraging honey bee collecting nectar (Photo: Katrin Gehring) (**b**) Retention test at different time points after one (*filled circles*) or three (*empty circles*) learning trials in a visual learning paradigm. Shown is the percentage of free-flying honey bees that choose the previously rewarded color (444 nm) at different time points after training. Each data point depicts a different group of animals (Adapted from [20])

critically depends on the exploitation of profitable food sources. However, honey bees are confronted with high variability of nectar availability, volume and concentration, both within and between flower patches (e.g. [16, 28, 31]). To ensure efficient nectar collection and to optimize their choice of profitable food sources they have to adapt to this changing nectar supply. Depending on nectar flow, different foraging strategies are used. When the nectar flow is high, bees show flower constancy, i.e. they remain faithful to the exploited food source. Hence, animals choose the food source with the best cost/reward ratio, a strategy termed maximization [13]. When the reward is low, animals switch to matching behavior, i.e. they match the frequency of their visits to the reward delivered at the feeder. When the actual reward is greater than that previously experienced, bees stay at the current feeder; if, however, it is lower, they shift to another feeder [13]. Accordingly, bees change their foraging strategy based on previous reward experiences. To be capable of doing so, they need to form a short-term memory about the most recently experienced reward to compare this with the actually occurring reward [22]. The outcome of this comparison is critical for the bee's decision-making: If the reward is as high as the reward experienced shortly before animals stay at their current food source. When the reward is lower than the reward experienced shortly before, retrieval of long-term flower-specific and patch-specific memories might lead to shifting to a different nectar source [12, 13]. Taken together, to be able to adapt to a changing nectar availability honey bees need to form memories with different stabilities about a particular nectar source and its reward.

6.3.2 The Stability of Reward Memories in Free-Flying Honey Bees Depends on the Number of Reward Experiences

Experiments using free-flying bees trained to color stimuli (see Chap. 6.6) showed that one training trial leads to a transient memory that is stable for a period of minutes to 1 day, whereas three training trials lead to a stable memory lasting for several days [20] (Fig. 6.3.1b). This dependency matches with the properties of the short- and long-term memories that are proposed to be involved in adapting to variable food availability: a short-term memory has to be formed after a single reward experience but doesn't need to be long lasting, because it is retrieved as soon as the honey bee visits the next nectar source. In contrast, a long-term memory that is formed after three trials indicates that the reward is reliably associated with a nectar source. This reliability is a precondition for an effective decision towards a nectar source.

However, it is not sufficient to know that a flower is associated with a reward. In addition, information about the reward magnitude and the reward variability of a certain food source has to be at hand to make an effective decision towards this food source. Indeed, honey bees form long-lasting memories about the magnitude and the variability of a reward associated with a certain food source [10, 11]. Most likely these memories are utilized when honey bees shift from a less profitable to a profitable food source, enabling the animals to decide in favor of the most profitable food source. Consequentially, when the reward provided by a nectar source fails to appear, honey bees should learn and memorize this lack of reward.

6.3.3 Adaptation to Failing Reward Resembles Extinction

Indeed, if a food source fails to provide a previously experienced reward, honey bees decrease their visits to this particular food source [4, 5, 13, 20], (Fig. 6.3.2). This decline in foraging behavior resembles the behavioral phenomenon of extinction. The term extinction stems from studies on associative learning and describes the declining frequency of a conditioned behavior after withdrawal of the reinforcement.

Interestingly, two studies on free-flying bees demonstrate a critical role of training and reward for the occurrence of extinction [4, 5, 20]: They show that the number of training trials and the reward duration affect an animal's resistance to extinction (Fig. 6.3.2). Extinction is thought to be based on learning about the absence of a previously experienced reinforcement [26]. Therefore, the extinction time course resembles the acquisition curve for extinction learning. Hence, the fact that there is an effect of the number of training trials and the reward duration on extinction learning hints towards an interplay between two learning processes: learning about the reward provided by a food source and learning about the failure of this previously experienced reward.

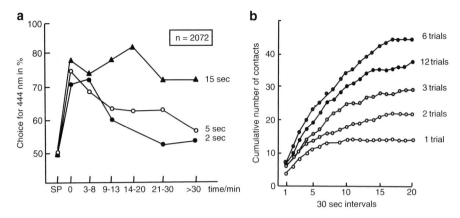

Fig. 6.3.2 Reward duration and the number of reward-learning trials affect extinction in free-flying honey bees (**a**) Extinction and reward duration. The percentage of animals that choose the learned color decreases over time when the color is presented without the previously experienced reward. This decrease is termed extinction and depends on the reward duration during the learning phase (2, 5, 15 s reward duration). Shown is the percentage of free flying honey bees that choose the previously rewarded color (444 nm) at different time intervals after training. Every data point depicts the same group of animals. *SP* spontaneous choice (Adapted from [20]) (**b**) Extinction and number of training trials. The number of contacts with a feeder that fails to deliver the previously experienced reward was studied depending on the number of training trials (1-, 2-, 3-, 6-, and 12-trials). Shown is the cumulative number of contacts with the feeder over five consecutive 30 s intervals. The resistance to extinction is affected by the number of training trials (Adapted from [4])

6.3.4 Formation of Extinction Memories Is Studied in Harnessed Honey Bees

Memories about the reward magnitude and its variability over time are proposed to be the basis for an optimal decision towards a profitable food source. Hence, it is important to memorize food sources that failed to provide nectar. In free-flying honey bees the formation of extinction memories has not been systematically examined. Yet, another experimental preparation, the olfactory conditioning of the proboscis extension reflex (PER), has been used to this end. In this preparation, single, harnessed bees are individually trained to establish a classical (Pavlovian) association between an odorant (the conditioned stimulus or CS) and sucrose reward (the unconditioned stimulus or US) [1]. Hungry honey bees reflexively extend their proboscis (PER) when their antennae are touched with a drop of sucrose solution. During the acquisition phase, bees learn to associate CS and US. Once the association has been formed, the odor alone elicits the PER. This reaction to the odor is the conditioned response (CR). The CR can be elicited by the learned odor (CS alone) immediately after acquisition and up to several days later, indicating the formation of short (STM) as well as long-term memories (LTM) [21, 22]. Comparable to the situation in free-flying honey bees (Fig. 6.3.1b), in harnessed honey bees the stability of the appetitive memory also depends on the number of training trials, i.e. the number of CS-US pairings: One CS-US pairing leads to a transient memory that is stable for a period of minutes to 1 day, whereas three training trials lead to a stable memory lasting for several days [21] (Fig. 6.3.3b).

6.3.5 Spontaneous Recovery Hints Towards Two Contrasting Memory Traces After Extinction

Several studies have examined extinction of short-term memory upon olfactory PER conditioning. This is comparable to the situation described above for free-flying bees and resembles the situation where a forager learns that the present food source fails to deliver the expected reward. Extinction can be observed in harnessed honey bees when the odor used as CS is presented alone multiple times after the animals were successfully conditioned. These multiple CS presentations result in a successive decrease of the CR. This decrease of the CR is termed extinction and can be depicted as an extinction learning curve [1, 26, 33, 37].

Bitterman et al. [1] demonstrated extinction after five CS presentations, 1 min after olfactory conditioning. A subsequent CS presentation 35 min after extinction revealed the reappearance of the CR: the decreased CR spontaneously recovered within this time interval from only 10% at the end of extinction (i.e. 10% of the animals reacted with a PER to the unrewarded odor) to up to 70% [1]. This phenomenon of a reappearing CR after extinction is well-known from many other extinction studies. It is termed spontaneous recovery [2, 26] (Fig. 6.3.4a).

Besides spontaneous recovery, two additional phenomena are described in the literature that relate to extinction. In renewal, the CR reappears when the CS alone is presented in the context in which it was trained but not extinguished; in reinstatement, the CR reappears when the US is presented several times during the interval between the extinction session and a subsequent retention test, but only when US presentation and retention test take place in the same context [3, 26]. Common to these three behavioral phenomena is the time- and context-dependent reappearance of the CR after extinction. These phenomena illustrate that the CS-US memory still exists after extinction although its retrieval is not always possible. Hence, several authors argue that extinction does not comprise the destruction of the CS-US association but rather a new learning about the CS-noUS association. Extinction learning results in an extinction memory (i.e. a memory about the CS-noUS association) that is stored in parallel to the previously formed CS-US memory. Thus, two conflicting memories about the CS would coexist [2, 30]. If this holds true, the CS acquires a second meaning during extinction learning, so that it becomes an ambiguous stimulus for the animal [2].

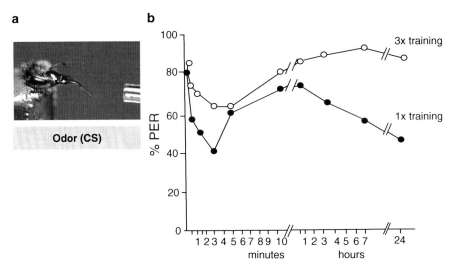

Fig. 6.3.3 The stability of reward memories correlates with the number of learning trials in harnessed honey bees (**a**) A harnessed honey bee showing the PER when a conditioned odor is presented after olfactory conditioning (Photo: Uli Müller) (**b**) Retention test at different time points after one (*filled circles*) or three (*empty circles*) training trials in classical conditioning of the PER. Shown is the percentage of harnessed honey bees that show the PER when memory retention is tested at the indicated time point. Each data point depicts a different group of animals (Adapted from [21])

6.3.6   Extinction of Short-Term Memories Reveals the Interaction of Two Memories

The occurrence of spontaneous recovery in honey bee olfactory conditioning suggests that after extinction two memories about the odor (the CS) are formed: One memory about the association of the odor with the reward (the US), which we will here refer to as the *reward memory*, and one memory about the absence of the reward, the *extinction memory*. To examine the interaction between the reward memory and the extinction memory Sandoz and Pham-Delègue [33] studied absolute spontaneous recovery. They defined absolute spontaneous recovery as the percentage of honey bees that did not react to the last extinction trial but to a CS presentation 1 h later. Absolute spontaneous recovery correlates with the number of CS-US pairings [33]: The more CS-US pairings honey bees experienced, the more absolute spontaneous recovery was observable (Fig. 6.3.4b). Absolute spontaneous recovery also correlates with the number of extinction trials: After five extinction trials absolute spontaneous recovery is higher than after ten extinction trials (Fig. 6.3.4c) [33]. Taken together, these data confirm that after reward learning and extinction learning two memories are formed each of which gains a certain weight depending on the number of CS-US pairings during reward learning or the number of CS-noUS pairings during extinction learning. According to their weight these two memories contribute to behavior after extinction.

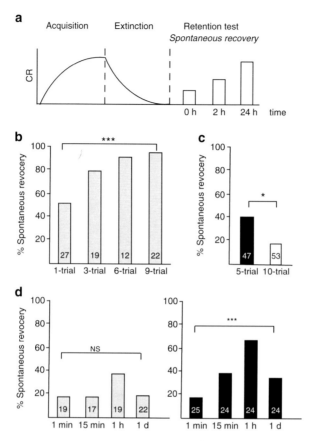

Fig. 6.3.4 Effect of the number of conditioning trials, extinction trials, and the time interval between training and extinction on spontaneous recovery (**a**) Spontaneous recovery is defined as the reappearance of the previously extinguished conditioned response (*CR*) during the time interval after extinction (here exemplified by 0, 2, 24 h) (Adapted from [26]) (**b**) Absolute spontaneous recovery correlates with the number of training trials. The number of acquisition trials is varied (from one to nine trials) whereas the number of extinction trials is kept constant (five extinction trials). Shown is the proportion of bees that do not react with a PER at the last extinction trials but that respond in the recovery test 1 h after the last extinction trial (% spontaneous recovery) (**c**) The number of extinction trials affects absolute spontaneous recovery. Honey bees are trained with one acquisition trial and extinguished with different numbers of extinction trials (5- or 10-trials). Shown is the proportions of bees that do not react with a PER at the last extinction trial but that respond in the recovery test 1 h after the last extinction trial (% spontaneous recovery) (**d**) Time point of extinction affects absolute spontaneous recovery. Proportion of bees that do not react with a PER at the last extinction trial but that respond in the recovery tests (% spontaneous recovery) at 1, 15 min, 1 h, 1 day after extinction. *Left*: 1 min interval between acquisition and extinction. *Right*: 10 min interval between acquisition and extinction ((**b–d**) Adapted from [33])

6.3.7   The Time Interval Between Reward Learning and Extinction

A third variable that is critical for the occurrence of spontaneous recovery is the time interval between acquisition and extinction. Sandoz and Pham-Delègue [33] presented five extinction trials after they trained honey bees with one CS-US pairing. When they presented five extinction trials 10 min after one CS-US pairing spontaneous recovery was higher than 1 min after training [33] (Fig. 6.3.4d). How can these results be interpreted? In olfactory conditioning of the PER four different memory phases have been identified that are defined by the time interval after learning during which they control behavior and by the biochemical processes that are necessary for their formation [22, 25] (see also Chap. 6.2): (1) a short-term memory (STM) which lasts from minutes to hours and is formed after one CS-US pairing; (2) a mid-term memory (MTM) which lasts several hours to 1 day, is formed after three CS-US pairings, and depends on protein kinase C activity [14]; (3) an early long-term memory (eLTM) which depends on translation and can be retrieved 24–48 h after three CS-US pairings [9], (4) a late long-term memory (lLTM) which depends on transcription and translation and can be retrieved 72 and 96 h after three and five CS-US pairings [18, 23, 38]. The existence of these four memory phases suggests that, depending on the time interval between learning and extinction, different memory phases are retrieved during extinction. Based on this consideration, Sandoz and Pham-Delègue [33] hypothesized that extinction provokes a decay of the particular memory phase that controls behavior when extinction is applied. In this case the behavioral effect of extinction resembles the destroyed CS-US memory phase and spontaneous recovery presents the memory phase that is not inhibited by the extinction procedure. Interestingly, this hypothesis is partially in line with a study in *Drosophila melanogaster* where it was demonstrated that extinction antagonizes intracellular signalling cascades underlying the CS-US memory [35]. If this holds true, the phase of the reward memory at which extinction takes place would be crucial.

Memory phases differ in their stability (see above). Sandoz and Pham-Delègue [33] studied extinction of short-term reward memories. These memories are short-lasting and are therefore regarded as labile and easy to disturb. In contrast, long-lasting memories that can no longer be disturbed by amnesic treatment, i.e. inhibition of protein synthesis, are thought to be stable, because it is assumed that these memories have undergone a stabilization process. This stabilization process is termed memory consolidation and the respective memories are defined as consolidated memories [6]. These long-lasting, stable memories might be more resistant to extinction than the unstable short-term memories extinguished in the study of Sandoz and Pham-Delègue [33].

However, extinction also takes place when these consolidated memories are retrieved [37]. Comparable to extinction of a short-term memory, extinction of a consolidated memory depends on the number of extinction trials: One extinction trial does not lead to extinction whereas two extinction trials and five extinction trials result in extinction [37]. Thus, the more extinction trials are applied, the more extinction can be observed.

This suggests that extinction of consolidated, stable memories is comparable to extinction of labile short-term memories. Sandoz and Pham-Delègue [33] propose that extinction provokes a decay of the particular memory phase that controls behavior. However, a decay of a consolidated memory that controls behavior when extinction is applied is only conceivable if one assumes that extinction destabilizes consolidated memories; an assumption that contradicts the definition of a consolidated, i.e. stable, memory.

6.3.8 Spontaneous Recovery After Extinction of a Consolidated Memory Depends on Protein Synthesis

However, it might well be that extinction destabilizes consolidated memories in the honey bee. When a consolidated reward memory is extinguished with five extinction trials, spontaneous recovery is observed 1, 2, 4, 24 and 48 h after extinction (Fig. 6.3.5b, above). Interestingly, spontaneous recovery observed 24 h after five-trial extinction is inhibited by the application of the protein synthesis inhibitor emetine shortly before extinction [37] (Fig. 6.3.5b, above) but not if emetine is injected at the same time point without applying extinction trials [37]. Thus, only when the animals experience extinction does the reward memory become susceptible for protein synthesis-inhibition. This result might be interpreted as follows: By the application of extinction trials the consolidated reward memory becomes labile and undergoes a second round of consolidation depending on protein synthesis. When the reward memory is destabilized this way it does not contribute to the control of behavior, which can be seen in the decline of the CR during extinction. Interestingly, this interpretation resembles reconsolidation [27, 34]. Reconsolidation describes the assumed process of memory stabilization after a consolidated memory has been reactivated [27, 34]. This reactivation of the consolidated memory induces a state of plasticity, which requires a reconsolidation process to gain stability. Thus it might be hypothesized that extinction reactivates the consolidated reward memory, which then undergoes a re-consolidation process that underlies spontaneous recovery. This interpretation is in line with the hypothesis of Sandoz and Pham-Delègue [33] explaining the transient decline of the conditioned behavior after extinction.

However, there is at least one alternative explanation: the so-called internal reinforcement hypothesis [8], that is based on the properties of the reinforcement system of the honey bee, namely the Vum_{mx1} neuron [17]. It proposes that with each extinction trial the CS-noUS association gains strength resulting in a decline of the conditioned behavior. However, during extinction an additional new learning process takes place, the reminder learning. During reminder learning the CS (that constitutes the extinction trial) is associated with an internal reinforcement signal and a reminder memory about this association is formed and consolidated. This reminder memory replaces the reward memory and now controls behavior together with the

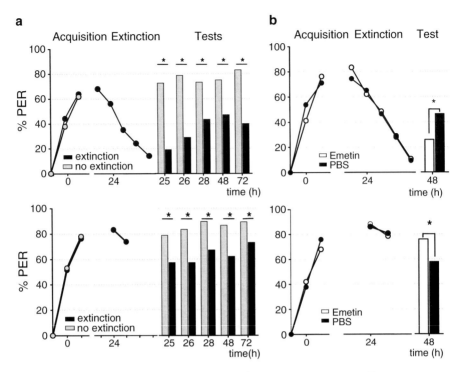

Fig. 6.3.5 Spontaneous recovery and memory consolidation after extinction of long-term memories (**a**) Spontaneous recovery occurs after extinction with five extinction trials (*above*) but not after extinction with two extinction trials (*below*) when an olfactory memory is extinguished 24 h after training. Proportion of harnessed honey bees showing the CR (*%PER*) when the conditioned stimulus (*CS*) is presented during acquisition, extinction, memory retention tests (**b**) Memory retention 24 h after extinction is blocked when the protein synthesis inhibitor emetine is injected 30 min before extinction depending on the number of extinction trials. *Above*: After extinction with five extinction trials, spontaneous recovery is blocked, thus the re-consolidation of the reward memory depends on protein synthesis. *Below*: After extinction with two extinction trials retention of the extinction memory is blocked, thus the consolidation of the extinction memory depends on protein synthesis (Adapted from [37])

extinction memory. During the consolidation time course the strength of this memory increases. This compensates the behavioral decline after extinction. Hence, the increasing strength of the reminder memory underlies spontaneous recovery. When the consolidation of the reminder memory is blocked, spontaneous recovery is inhibited [8].

Conceptually, the reconsolidation hypothesis and the internal reinforcement hypothesis are very different. However, both are in line with the hypothesis of Sandoz and Pham-Delègue [33] that extinction interferes with the reward memory. Whether this reward memory is restabilized thereafter or whether its failure is compensated by a third, newly formed reminder memory remains to be shown.

6.3.9   A Consolidating Extinction Memory

Spontaneous recovery after five extinction trials is, however, not complete: memory retention of the extinguished group and the non-extinguished control group is significantly different 24 h after extinction, and a significant decrease between the first extinction trial and memory retention 24 h later can be observed (Fig. 6.3.5a, below). Therefore, in addition to the spontaneous recovery after 24 h, a 1-day extinction memory is observed [37]. Stollhoff et al. [37] therefore conclude that two memory traces are formed after extinction of a consolidated reward memory, the re-consolidated reward memory and a long-lasting extinction memory. It remains unclear whether this extinction memory is also protein synthesis-dependent because protein synthesis inhibition by emetine has no visible effect on the extinction memory formed after extinction with five extinction trials. There are two possible explanations: either this extinction memory does not depend on protein synthesis, or both the extinction memory and the spontaneous recovery of the reward memory depend on protein synthesis and the behavioral effect observed is the net output of these two inhibited memories. However, as it has been demonstrated that extinction memories also undergo consolidation, the latter seems more likely.

Indeed, when a consolidated reward memory is extinguished with two extinction trials, an extinction memory, but no spontaneous recovery, is visible 24 h later. Systemic injection of the protein synthesis inhibitor emetine shortly before extinction inhibits retention of this extinction memory [37], (Fig. 6.3.5b, below). Thus, the formation of a long-lasting extinction memory after two extinction trials depends on protein synthesis [19, 37].

Interestingly, shaking the animals for 15 h after two-trial extinction also blocks the formation of this extinction memory [19]. Shaking results in sleep deprivation because all shaken animals show a sleep rebound during the subsequent night [19]. Therefore, it has been concluded that sleep deprivation is the reason for the disturbed formation of a long-term extinction memory [19]. Both the susceptibility to protein synthesis inhibition and the sensitivity to an amnesic treatment like shaking the animals suggest that the extinction memory is progressively stabilized after the last extinction trial. As noted above, this progressive stabilization of a memory trace is termed memory consolidation [6, 7]. Therefore it can be concluded that, in addition to the reward memory, an extinction memory in honey bees is formed upon CS-noUS experiences, which also undergoes a consolidation process.

6.3.10   The Size of the Prediction Error Correlates with the Induction of a Consolidation Process

However, consolidation of the extinction memory is only observed when the reward duration (i.e. the length of the US) during reward learning exceeds 2 s [36]. When the reward duration is only 2 s long, the long-term extinction memory is not susceptible to protein synthesis inhibition. How can these results be explained? In classical conditioning several authors have proposed that the difference between the US presented during acquisition and the absence of US during extinction is the cause for extinction and extinction learning [3, 32]. Regarding the US duration, the absolute value of this difference is smaller when animals are rewarded with a short US (i.e. 2 s US presentation) than with a long US (i.e. >2 s US presentation). These absolute values can be interpreted as different magnitudes of the mismatch between the previously and the currently experienced US. It seems that the magnitude of this mismatch has to exceed a certain threshold to trigger consolidation of an extinction memory. When animals received a short-lasting reward during reward learning a small mismatch between the previously experienced reward and the current failure of the reward is detected and a consolidation process is not induced. In contrast, animals that receive a long-lasting reward during reward learning experience a bigger mismatch and consolidation of the extinction memory is induced. Taken together, honey bees only form a stable and long-lasting memory about the reward's failure if a substantial mismatch between the previously memorized reward and its failure is detected. This might be a mechanism to ensure that only meaningful changes in the reward magnitude are memorized, preventing irrelevant fluctuations of reward magnitude from influencing a honey bee's decision-making in a foraging context.

6.3.11 How Can the Results on Extinction in Harnessed Honey Bees Be Reconciled?

Based on extinction studies in harnessed honey bees, two basic assumptions have been put forward: (1) studies on extinction of labile short-term memories led to the hypothesis that extinction provokes a decay of the memory phases that control behavior when extinction is applied [33], (see above) and (2) the occurrence of spontaneous recovery as well as studies on extinction of consolidated long-term memories indicated that at least two parallel memories, the extinction memory and the reward memory, are responsible for controlling behavior after extinction [37]; in doing so, the contribution of each memory depends on the current strength of each memory gained during reward learning or during extinction learning.

Work on memory extinction in vertebrates demonstrates that extinction memory formation results from a reorganization of the initial CS-US memory rather than from erasure of this initial memory [29]. This is realized by changing networks. Part of changing networks seems to be the reversal of synaptic alterations that have been induced by conditioning [29]. Interestingly, also the decay of the extinguished memory as proposed by Sandoz and Pham-Delègue [33] as well as the reconsolidation after extinction [37] can only take place if one assumes the reversal of synaptic alterations. Hence, if extinction in honey bees leads to a reversal of synaptic alterations, both the decay of the extinguished reward memory and the reconsolidation of the reward memory could take place. How can this be reconciled? It can be hypothesised that depending on the memory phase that is extinguished either one or the other mechanism occurs: If the extinguished reward memory is not yet consolidated the particular memory phases during which extinction takes place is decaying. If a consolidated reward memory is extinguished, reconsolidation takes place and leads to the stabilization of the extinguished reward memory. Hence, the decay of a short-term reward memory and the reconsolidation of a long-term reward memory after extinction might both take place due to the same mechanism namely the activation and reorganisation of the reward memory by the reversal of synaptic alterations. However, because it has been shown several times that extinction learning leads to the formation of an extinction memory, it has to be proposed that in parallel to both mechanisms an extinction memory is formed.

6.3.12   Extinction and Adaptation to Variable Food Sources

Foraging honey bees have to adapt to variable food sources to ensure their colony's survival. When they experience a reward that is lower than expected, they switch towards a more profitable food source. Given the high variability in nectar concentration, volume and availability, it is nevertheless important to take into account not only flowers that are currently rewarding, but also those that were rewarding before and that have been afterwards experienced as non-rewarding. These flowers may still constitute an important although highly variable nectar source. Therefore, it is important to weigh up the information about a previously rewarding flower or flower patch against its failure to provide a reward. How is this contrasting information balanced during foraging?

Two memories are proposed to be required for adapting to variable food sources: a short-term memory about the previously experienced reward that is the basis of a forager's decision to change its strategy, and a long-term memory about the previously experienced reward that is the basis for a decision towards a new food source [13]. Interestingly, the results we have described above on extinction and the formation of extinction memories in harnessed honey bees mirror these requirements.

Each extinction trial simultaneously constitutes a retrieval trial. Therefore, the rapid decrease of the CR at every consecutive extinction trial resembles retention of a short-term extinction memory. This short-term extinction memory meets the requirements during foraging within a foraging bout, where reward experiences take place at short inter-trial intervals and only the short-lasting memory about the last reward experience is relevant for the decision to shift to a new food source [22]. However, within minutes after extinction spontaneous recovery can be observed, demonstrating that the reward memory is not extinguished when honey bees learn about the reward's failure. Rather two parallel memories are formed. These memories gain different weights depending on the memory phase that is extinguished and the number of extinction trials. Together they drive behavior during the decision towards a new food source according to their previously acquired weight. It has to be emphasized that the extinguished reward memory most likely plays a crucial role in the balance between the two contrasting memory traces. On the one hand it is important what is memorized about the reward, i.e. the reward magnitude [36]. On the other hand the memory phase at which extinction takes place is crucial [33, 37]. Finally, it might well be that additional factors like the time of year, weather, food source availability, colony resources, etc. play a role in shifting the balance towards the reward or the extinction memory [15, 24].

6.3.13 Outlook

Here studies on extinction in harnessed honey bees were discussed with the aim of understanding the relevance of this learning phenomenon for the natural behavior of free-flying honey bees. We suggest that extinction plays a role in a foraging context, in which bees have to exhibit an adaptive behavior towards variable food sources. To confirm this notion a thorough study of extinction in free-flying honey bees would be necessary. It is especially important to study spontaneous recovery in free-flying bees to learn about the interplay between reward and extinction memories. Our considerations are based on different hypotheses about the mechanisms of extinction which need to be corroborated. In particular, it will be necessary to dissociate memories that have been proposed to form and consolidate after extinction (reward memory, extinction memory). This would elucidate the mechanisms underlying extinction and spontaneous recovery.

References

1. Bitterman ME, Menzel R, Fietz A, Schäfer S (1983) Classical conditioning of proboscis extension in honeybees (*Apis mellifera*). J Comp Psychol 97(2):107–119
2. Bouton ME (2002) Context, ambiguity, and unlearning: sources of relapse after behavioral extinction. Biol Psychiatry 52(10):976–986
3. Bouton ME, Moody EW (2004) Memory processes in classical conditioning. Neurosci Biobehav Rev 28(7):663–674
4. Couvillon PA, Bitterman ME (1980) Some phenomena of associative learning in honeybees. J Comp Physiol Psychol 94:878–885
5. Couvillon PA, Bitterman ME (1984) The overlearning-extinction effect and successive negative contrast in honeybees (*Apis mellifera*). J Comp Psychol 98(1):100–109
6. Dudai Y (2004) The neurobiology of consolidations, or, how stable is the engram? Annu Rev Psychol 55:51–86
7. Dudai Y, Eisenberg M (2004) Rites of passage of the engram: reconsolidation and the lingering consolidation hypothesis. Neuron 44(1):93–100
8. Eisenhardt D, Menzel R (2007) Extinction learning, reconsolidation and the internal reinforcement hypothesis. Neurobiol Learn Mem 87(2):167–173
9. Friedrich A, Thomas U, Müller U (2004) Learning at different satiation levels reveals parallel functions for the cAMP-protein kinase A cascade in formation of long-term memory. J Neurosci 24(18):4460–4468
10. Gil M, De Marco RJ (2009) Honeybees learn the sign and magnitude of reward variations. J Exp Biol 212(17):2830–2834

11. Gil M, De Marco RJ, Menzel R (2007) Learning reward expectations in honeybees. Learn Mem 14(7):491–496
12. Greggers U, Mauelshagen J (1997) Matching behavior of honeybees in a multiple-choice situation: the differential effect of environmental stimuli on the choice process. Anim Learn Behav 25(4):458–472
13. Greggers U, Menzel R (1993) Memory dynamics and foraging strategies of honeybees. Behav Ecol Sociobiol 32(1):17–29
14. Grünbaum L, Müller U (1998) Induction of a specific olfactory memory leads to a long-lasting activation of protein kinase C in the antennal lobe of the honeybee. J Neurosci 18(11):4384–4392
15. Hadar R, Menzel R (2010) Memory formation in reversal learning of the honeybee. Front Behav Neurosci 4:186
16. Hammer O (1949) Investigations on the nectar-flow of red clover. Oikos 1(1):34–47
17. Hammer M (1993) An identified neuron mediates the unconditioned stimulus in associative olfactory learning in honeybees. Nature 366(6450):59–63
18. Hourcade B, Muenz TS, Sandoz JC, Rössler W, Devaud JM (2010) Long-term memory leads to synaptic reorganization in the mushroom bodies: a memory trace in the insect brain? J Neurosci 30(18):6461–6465
19. Hussaini SA, Bogusch L, Landgraf T, Menzel R (2009) Sleep deprivation affects extinction but not acquisition memory in honeybees. Learn Mem 16(11):698–705
20. Menzel R (1968) Das Gedächtnis der Honigbiene für Spektralfarben I. Kurzzeitiges und langzeitiges Behalten. Z vergl Physiol 60:82–102
21. Menzel R (1990) Learning, memory, and "cognition" in honey bees. In: Kesner RP, Olton DS (eds) Neurobiology of comparative cognition. Lawrence Erlbaum Associates, Inc., Hillsdale, pp 237–292
22. Menzel R (1999) Memory dynamics in the honeybee. J Comp Physiol A 185(4):323–340
23. Menzel R, Manz G, Menzel R, Greggers U (2001) Massed and spaced learning in honeybees: the role of CS, US, the intertrial interval, and the test interval. Learn Mem 8(4):198–208
24. Moore D, Van Nest BN, Seier E (2011) Diminishing returns: the influence of experience and environment on time-memory extinction in honey bee foragers. J Comp Physiol A 197(6):641–651
25. Müller U (2002) Learning in honeybees: from molecules to behaviour. Zoology (Jena) 105(4):313–320
26. Myers KM, Davis M (2002) Behavioral and neural analysis of extinction. Neuron 36(4):567–584
27. Nader K (2003) Memory traces unbound. Trends Neurosci 26(2):65–72
28. Núñez J (1977) Nectar flow by melliferous flora and gathering flow by *Apis mellifera ligustica*. J Insect Physiol 23(2):265–275
29. Pape HC, Pare D (2010) Plastic synaptic networks of the amygdala for the acquisition, expression, and extinction of conditioned fear. Physiol Rev 90(2):419–463
30. Pavlov IP (1927) Conditioned reflexes: an investigation of the activity of the cerebral cortex. Oxford University Press, Oxford
31. Percival MS (1946) Observations on the flowering and nectar secretion of *Rubus fruticosus* (Agg.). New Phytol 45(1):111–123
32. Rescorla R (1972) A theory of classical conditioning: variations in the effectiveness of reinforcement and non-reinforcement. In: Black P (ed) Classical conditioning II: current research and theory. Appleton, New York, pp 64–99
33. Sandoz JC, Pham-Delègue MH (2004) Spontaneous recovery after extinction of the conditioned proboscis extension response in the honeybee. Learn Mem 11(5):586–597
34. Sara SJ (2000) Retrieval and reconsolidation: toward a neurobiology of remembering. Learn Mem 7(2):73–84
35. Schwärzel M, Heisenberg M, Zars T (2002) Extinction antagonizes olfactory memory at the subcellular level. Neuron 35(5):951–960
36. Stollhoff N, Eisenhardt D (2009) Consolidation of an extinction memory depends on the unconditioned stimulus magnitude previously experienced during training. J Neurosci 29(30):9644–9650

37. Stollhoff N, Menzel R, Eisenhardt D (2005) Spontaneous recovery from extinction depends on the reconsolidation of the acquisition memory in an appetitive learning paradigm in the honeybee (*Apis mellifera*). J Neurosci 25(18):4485–4492
38. Wüstenberg D, Gerber B, Menzel R (1998) Short communication: long- but not medium-term retention of olfactory memories in honeybees is impaired by actinomycin D and anisomycin. Eur J Neurosci 10(8):2742–2745

Chapter 6.4
Tactile Antennal Learning in the Honey Bee

Joachim Erber

Abstract The different forms of tactile antennal learning in the honey bee are based on operant activity of the antennae. Flexible motor programs of the antennae are used for monitoring multimodal signals in the space around the head.

Bees can learn the three-dimensional location of an object within the reach of the antennae by touching it frequently. During operant conditioning bees learn that antennal contacts with an object lead to a sucrose reward. Operant antennal conditioning is side specific and bees learn to discriminate between different objects. Operant antennal conditioning can be reduced to conditioning of the activity of the fast flagellum flexor muscle (FFF muscle) which is innervated by a single motoneuron.

Using the proboscis extension reflex (PER) bees can be conditioned to discriminate between different surface structures, forms, sizes and locations of objects. The characteristics of PER conditioning are similar to those of olfactory PER conditioning under laboratory conditions. Mechanoreceptors on the antennal tip are used for surface discrimination. Bees that discriminate between different surface structures show characteristic antennal scanning movements.

Abbreviations

AMMC	Antennal-mechanosensory motor center
MB	Mushroom bodies
PER	Proboscis extension reflex

J. Erber (✉)
Institut für Ökologie, Technische Universität Berlin, Berlin, Germany
e-mail: joachim.erber@tu-berlin.de

C.G. Galizia et al. (eds.), *Honeybee Neurobiology and Behavior: A Tribute to Randolf Menzel*, DOI 10.1007/978-94-007-2099-2_33,
© Springer Science+Business Media B.V. 2012

6.4.1   Introduction

The antennae of honey bees are multimodal sense organs for chemical and physical signals. A bee can localize olfactory, chemical and mechanical stimuli with active movements of both antennae. A number of behavioral contexts, like trophallaxis, depend on the interactions between reception of antennal information and transmission of antennal signals. As honey bees spend most of their life inside the dark hive, the antennae are extremely important for communication and the coordination of labor and social activities within a colony. Antennal movements can be used as monitors for sensory signal processing and for different forms of learning [4].

The capacity of the honey bee to learn rapidly a variety of sensory cues and to remember these signals for a long time is a major reason why this insect became so attractive for behavioral, sensory-physiological, and molecular studies. Most experiments analysing discrimination of sensory signals or orientation in the environment are based on the complex learning abilities of the honey bee (see Chap. 6.6). For the last 30 years olfactory conditioning under laboratory conditions has been in the centre of interest for physiological analyses of learning in the honey bee [10, 19]. The olfactory conditioning protocol is very attractive due to rapid learning, excellent discrimination of sensory cues, long-term memory, accessibility for electrophysiological recordings and *in vivo* imaging analysis ([9, 19], see also Chaps. 6.1–6.3 and 6.6).

The discovery of antennal fine-scale textural learning by Kevan and Lane [12] in restrained honey bees has been the starting point for a number of studies investigating different forms of tactile learning under laboratory conditions. Active antennal scanning movements which are the basis for different forms of tactile antennal learning are a powerful monitor for information processing in the nervous system.

6.4.2   The Antennal Motor System

The annulated antenna of the honey bee consists of three functional segments, the scapus, the pedicellus and the long, flexible flagellum [38]. The scapus, the rigid base of the antenna, is connected to the head by a socket joint which allows three-dimensional rotatory movements. A hinge joint between the scapus and the pedicellus enables two-dimensional movements of the flagellum which is rigidly connected to the pedicellus and carries most of the sensory receptors (see Chap. 4.3). Rotatory movements of the scapus are controlled by four muscles which are innervated by nine motoneurons in the antennal-mechanosensory motor center (AMMC) [15]. In the scapus two antagonistic muscle systems which are innervated by six motoneurons control the two-dimensional movements of the pedicellus with the attached flagellum [15]. The fast flagellum flexor muscle (FFF muscle) in the scapus is used for very rapid movements of the flagellum; it is innervated by

Fig. 6.4.1 Stroboscopic photos of antennal scanning movements in bees. The flashes were delivered at a frequency of 12/s. Positions of the scapus-pedicellus joints and the tips of both antennae are indicated by *asterisks* (**a**) A bee responding with PER and rapid antennal scanning movements to a drop of water (**b**) and (**c**) A bee scanning a wire with the *right* antenna in the picture

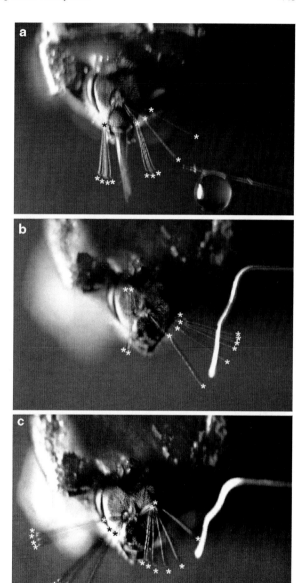

a single motoneuron [7]. The rapid flicking movements of the flagellum are used to localize chemical signals (Fig. 6.4.1a), to examine objects (Fig. 6.4.1b, c) or to probe the texture of a surface.

Antennal scanning activity varies with the behavioral context and is modulated by biogenic amines and sensory stimulation [25]. Antennal scanning of an object is characterized by the frequency of contacts and the contact duration. Contact frequencies with an object have a range between few contacts/min to over 300 contacts/min.

Contact durations with an object are usually shorter than 10 ms. Bees show rapid antennal scanning movements even when the antennal nerve in the flagellum is cut or when the flagellum is replaced by a silver wire [2]. In many situations tactile scanning is controlled by the frequency of the motoneuron innervating the FFF muscle. Activity of this motoneuron can be enhanced by compound sucrose stimuli applied to the antenna and the proboscis. Injection of the biogenic amine octopamine (OA) into the AMMC has a similar effect as the compound sucrose stimulus, while injection of the biogenic amine serotonin reduces the activity of the FFF motoneuron. Projections of neurons with immunoreactivity for modulatory transmitters like FMRFamide, serotonin, octopamine and dopamine (DA) were found in the AMMC [26, 27, 35–37]. These findings suggest that neuromodulators control antennal activity.

6.4.3   Antennal Motor Learning

Harnessed bees whose eyes are occluded show antennal motor learning that functions without an external reinforcer (Fig. 6.4.2a, b; Table 6.4.1). After scanning the edges and the surface of a small object with the antennae for several minutes, the bee continues to search with the antennae for >10 min the area where the surface of the object was located before [5]. Covering the mechanoreceptors on the antennal tips does not impair antennal motor learning, but learning is abolished after blocking the joints between pedicellus and scapus.

Compared to other antennal learning protocols, antennal motor learning is a slow process that takes relatively long to develop. Significant changes of behavior can be observed after a bee has scanned an object for >10 min or after >3 presentations each lasting 5 min (Fig. 6.4.3a). After approximately 650 contacts a bee continues to search the area where the surface of an object was located before. Areas where the edges of the object were located are avoided after approximately 2,500 contacts. The experiments demonstrate the plasticity of goal directed antennal motor activity without an external reinforcer.

6.4.3.1   Operant Conditioning of Antennal Movements

The experiments on antennal motor learning led to the hypothesis that antennal movements could also be used in an operant paradigm in which bees learn that specific antennal movements lead to a reward. Such a protocol is similar to operant conditioning of a vertebrate in a Skinner box. Instead of pressing a lever in order to receive a reward, a bee has to make antennal contact with a plate located near the head. Several experimental protocols using harnessed bees were developed

Tactile antennal motor learning

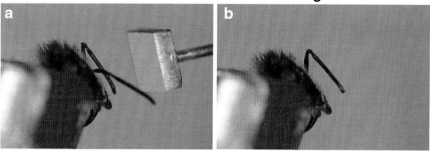

scanning   test

Tactile antennal operant conditioning

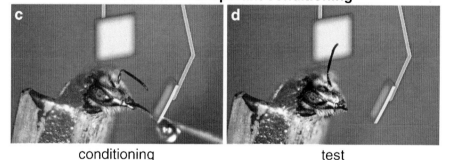

conditioning   test

Tactile antennal PER conditioning

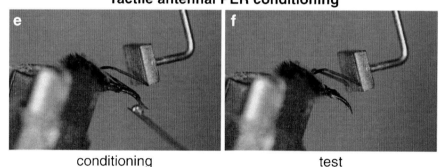

conditioning   test

Fig. 6.4.2 Three different antennal learning protocols; the *left column* shows bees during the learning phase, the *right column* during the test phase (**a**) During antennal motor learning a bee scans an object with the antennae (**b**) After several minutes of scanning, the object is removed and the bee continues to search with the antennae the area where the object was positioned before (**c**) During antennal operant conditioning the bee is rewarded when the contact frequency of one antenna with one of the two plates exceeds a defined threshold (**d**) After several conditioning trials the contact frequencies for both plates are measured (**e**) During antennal PER conditioning the bee is rewarded with a drop of sucrose after scanning a plate with vertical grooves for ≈2–3 s (**f**) After several conditioning trials the bee shows the PER during contact with the conditioning plate

Table 6.4.1 Different forms of tactile learning in harnessed bees

Type of tactile learning	Characteristics of the experimental protocol	Measured parameter	Characteristics of the learning process	What is learned?
Tactile antennal motor learning	Bee scans a small plate with both antennae; bee is not rewarded with sucrose	Antennal scanning behavior before and after presentation of the object	**Slow learning**; significant change of antennal scanning behavior after ≥10 min of exposure; effect lasts >10 min after removal of the object	Antennal scanning preference for areas where the surface of the object was presented during exposure
Operant conditioning of antennal movements	Bee touches an object with one antenna and receives a sucrose reward when the frequency of contacts exceeds a defined threshold	Antennal contact frequency with one or two plates within the reach of the antenna	**Fast learning**; significant increase of contact frequency after 10 sucrose rewards; effect lasts >30 min after the end of conditioning	Discrimination of two plates; reversal learning with two plates; side specific learning; independent conditioning of both antennae
Operant conditioning of antennal fast flagellum flexor muscle activity	Activity of the fast flagellum flexor muscle is recorded, bee receives a sucrose reward when the frequency of muscle potentials exceeds a defined threshold	Frequency of potentials in the fast flagellum flexor muscle	**Fast learning**; significant increase of muscle potential frequency after 10 sucrose rewards; antennal joints must be freely movable; effect lasts >30 min after the end of conditioning	Increase of action potential frequency in a single antennal motoneuron
Tactile PER conditioning	Bee scans a plate with both antennae and receives a sucrose reward after ≈2–3 s of scanning	Proboscis extension reflex (PER)	**Fast learning**; ≈80% conditioned PER after 3–4 rewards; retention >2 days	Discrimination of size and form; discrimination of angle and wavelength of surface texture; side specific learning

[13] (see Table 6.4.1). Contacts of one antenna with usually two plates at different positions relative to the head were registered electronically (Fig. 6.4.2c, d). After measuring baseline antennal contact frequencies for the two plates, a reward criterion for the plate with the lower frequency was defined. The reward criterion was one or two standard deviations above the mean baseline activity for this plate. Antennal contacts were measured on-line and whenever the reward criterion was reached, the experimental bee was rewarded with a drop of sucrose (Fig. 6.4.2c). Bees learn fast that antennal contacts with a plate lead to a sucrose reward. After 10 rewards they significantly increase the frequency of antennal contacts for the rewarded plate and significantly reduce the frequency for the other plate (Fig. 6.4.3b). Yoked control bees do not show this behavioral change. In these control bees the temporal sequence of the sucrose rewards is the same as in an experimental animal but the stimulation is not contingent on a defined antennal contact frequency with a plate.

Reversal learning is apparent when the bee is first rewarded for frequently touching one plate and then is rewarded for frequent contacts with a second plate. Operant conditioning of antennal movements is side-specific. Conditioning of one antenna does not affect movements of the contralateral antenna and conditioned movements on one side are maintained after operant conditioning of the contralateral antenna [14]. A number of experiments suggest that the neural structures underlying side specific operant conditioning of antennal movements are localized in the AMMC where the dendritic projections of antennal mechanoreceptors, antennal sucrose receptors and antennal motoneurons converge [11, 15, 17].

6.4.3.2 Operant Conditioning of Antennal Muscle Activity

We tested the hypothesis that side specific operant conditioning can be reduced to the conditioning of the activity of a single antennal motoneuron (see Table 6.4.1). The motoneuron of the FFF muscle which controls rapid scanning movements of the flagellum (see Fig. 6.4.1a, c), is a good candidate for such an experiment. Activity of this muscle can be recorded extracellularly from the scapus without impeding movements of the antenna. The large potentials of the FFF muscle can be easily identified and they are correlated 1:1 with action potentials of an identified antennal motoneuron in the AMMC [7, 15]. Activity of the FFF motoneuron can be enhanced by stimulating with sucrose first the antenna and then the proboscis, demonstrating convergence of gustatory input with motor output in the antennal system [25].

Analogous to the protocol for operant conditioning of antennal movements, the bee was rewarded with sucrose when the instantaneous frequency of muscle potentials was one standard deviation higher than the mean baseline activity. In several respects the results of the operant electrophysiological experiments are similar to the behavioral experiments. The frequency of muscle potentials is significantly enhanced for at least 30 min after 10 operant conditioning trials compared to yoked

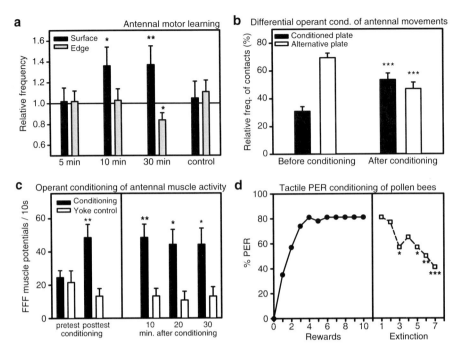

Fig. 6.4.3 Four different antennal learning protocols (**a**) Antennal motor learning for an object presented to both antennae (Data taken from Erber et al. [5]). Mean relative frequencies and SEM of antennal positions for different experiments are shown. The relative frequencies on the ordinate give the number of antennal positions after presentation of the object divided by the number of positions before presentation. Data are shown for the surface and the edges of the object. Measurement duration was 5 min. The *times* indicate how long the bee scanned the object. In the control group antennal activity was measured twice with an interval of 30 min without presenting an object. Significances: 2-sided t-test; *p<0.05; **p<0.01 (**b**) Differential operant conditioning of antennal movements (Data taken from Kisch and Erber [13]). Mean relative contact frequencies with SEM for two plates before and after ten conditioning trials are shown on the ordinate. Before conditioning the bees touched the conditioned plate less frequently than the alternative plate. Operant conditioning leads to a significant increase of the contact frequency at the conditioned plate and to a significant decrease at the alternative plate. Significances: 2-sided t-test ***p<0.001 (**c**) Operant conditioning of the FFF antennal muscle activity (Data taken from Erber et al. [7]). Mean frequencies of muscle potentials and SEM are shown on the ordinate for a conditioning and a yoked control group. The *left* diagram shows the frequencies measured for 10 min before and 10 min after 10 conditioning trials. The *right* diagram shows the frequencies at different times after ten conditioning trials. Significances: 2-sided t-test: *p<0.05; **p<0.01 (**d**) Tactile PER conditioning and extinction for pollen foraging bees (Data taken from Erber et al. [6]). The ordinate gives the percentages of conditioned PER; the bees were conditioned to a plate with vertical grooves. During acquisition the responses differ significantly from spontaneous behaviour already after the first reward (p<0.001, Fisher exact probability test). Significances: Fisher exact probability test *p<0.05; **p<0.01; ***p<0.001

controls (Fig. 6.4.3c). The ten conditioning trials take approximately 25 min, which is similar to the behavioral experiments. Operant conditioning is reduced or abolished if the antennal joints are fixed. Obviously, signals from mechanoreceptors on the antennal joints are necessary for operant conditioning even if the paradigm is reduced to a single motoneuron.

In the AMMC the projections of mechanical and gustatory receptors from the antennal tip overlap with the dendritic projection of the FFF motoneuron. Muscle potentials which are correlated 1:1 with the activity of the FFF motoneuron show responses with very short latencies (<10 ms) when mechanical and gustatory hairs of the ipsilateral antennal tip are stimulated. These experiments suggest direct projections from gustatory and mechanical antennal receptors to the FFF motoneuron [11]. When FFF muscle activity above the reward criterion is paired with ipsilateral antennal sucrose stimulation, operant conditioning is successful and as effective as compound sucrose stimulation of the ipsilateral antenna and the proboscis. It is remarkable that a correlate for the complex behavioral operant paradigm exists at the level of a single neuron which is part of the antennal sensorimotor network in the AMMC. So far it is not known whether other antennal motoneurons of the network also receive gustatory antennal input and whether they show the same operant plasticity as the FFF motoneuron.

6.4.4 Tactile PER Conditioning

For over 35 years the olfactory conditioning protocol of the proboscis extension response (PER) has been a tool for the analysis of learning in the honey bee ([20], see also Chap. 6.2). Bees can be conditioned very efficiently to tactile cues by using a modification of the olfactory PER protocol. Similar to operant conditioning, the tactile PER protocol is based on operant antennal scanning of the bee. For conditioning, a small plate (usually 3×4 mm) is positioned so that the bee can scan the plate with both antennae for 2–3 s. PER is elicited by stimulating the antennae with a drop of sucrose and the bee is then rewarded with a small drop of sucrose presented to the proboscis (Fig. 6.4.2e; Table 6.4.1; [6]). After three to four conditioning trials the acquisition curve reaches a stable asymptote and the bee responds in ≈80% of the trials with conditioned PER when the plate is presented to the antennae (Figs. 6.4.2f and 6.4.3d). The maximum rate of conditioned PER is reduced by ≈50% after seven unrewarded extinction trials. After four conditioning trials long-term retention was found to last more than 48 h [33]. Similar to operant antennal conditioning, bees can be conditioned side-specifically by presenting the plate during conditioning only to one antenna [31].

The characteristics of tactile PER conditioning are very similar to those of olfactory PER conditioning when acquisition, extinction and long-term retention are compared. This is astonishing because there is good experimental evidence that different neuropils are involved in the two types of learning. Many experiments with olfactory PER conditioning have demonstrated that the antennal lobes (ALs) and the mushroom bodies (MBs) are involved in olfactory learning ([3, 18, 21], see also Chaps. 6.1 and

6.2). Tactile signals, on the other hand, are processed in the AMMCs of the deutocerebrum. To test a possible contribution of the MB in tactile PER conditioning, we performed an experiment together with Dagmar Malun using the mitotic blocker hydroxyurea (HU) to ablate parts of the MB during larval development [31]. We found that ablations of parts of the MB did not affect side specific tactile antennal learning, while similar experiments with side specific olfactory PER conditioning demonstrated that HU ablation of a median calyx in the MB had significant effects [16]. These experiments suggest that, different from olfactory learning, the MB do not play a major role in antennal tactile PER conditioning. We hypothesize that the AMMC is the neural substrate for tactile learning and memory.

6.4.4.1 Tactile Discrimination

Bees can learn to discriminate surface texture patterns [6]. Best discrimination was found for plates that had a size of 3×4 mm with grooves engraved in the surface. Bees can discriminate surfaces with grooves from smooth surfaces and even the angles of the grooves. If bees are conditioned to a surface with vertical grooves they can discriminate this texture from grooves that deviate in angle by 22.5°. Tactile angle discrimination is similar to visual angle discrimination in free flying bees [39–41]. Also the spatial wavelengths of the grooves can be discriminated if the trained wavelength λ differs from an alternative by approximately $\lambda/2$. Again, tactile discrimination is comparable to visual discrimination [41]. Apparently, neural signal processing in two different neuropils, the visual ganglia and the AMMC, leads to comparable discrimination of stimuli in two different modalities. Discrimination of different forms and sizes in the tactile conditioning protocol is not as impressive as discrimination of surface structures. Coarse differences of form and size can be discriminated. The PER experiments demonstrate that the discrimination of surfaces is the most prominent feature of tactile learning.

6.4.5 Antennal Joints, Mechanoreceptors and Learning

During tactile PER conditioning bees use three-dimensional antennal movements to scan objects within the reach of the antennae. The experiments with operant conditioning of an antennal motoneuron demonstrated that even in this reduced preparation conditioning is only successful if the antenna is freely movable. To test the contribution of the mechanoreceptors and the antennal joints for tactile PER conditioning and discrimination, Susanne Schnitt from our laboratory performed a series of experiments in which she either covered the tips of the antennae or blocked the antennal joints [34].

   If the mechanical and gustatory receptors on the tips of both antennae are blocked by covering them with paint, bees show slightly slower acquisition but the same

acquisition asymptote compared to untreated controls during tactile PER conditioning. When the antennal tips are covered, discrimination of textures is abolished
but discrimination of forms remains intact. When the joints between scapus and
pedicellus are blocked on both antennae and bees cannot move the flagella relative
to the scapus, tactile acquisition is reduced and bees do not show any texture discrimination. After blocking the ball and socket joint at the base of both antennae
tactile PER conditioning is significantly reduced and discrimination of textures is
abolished. These experiments demonstrate that discrimination of surface textures,
the most demanding task in tactile learning, depends on intact mechanoreceptors
[11] on the antennal tips and freely movable antennal joints. At the moment there
is no experimental evidence that the antennal mechanoreceptors which are used
during tactile learning show topological projections in the brain (Maronde, unpublished observations). These findings are in contrast to the topographic organization
of antennal mechanoreceptors from the Johnston organ shown by Ai ([1], see also
Chap. 4.3).

6.4.6   Antennal Scanning and Learning

Similar to olfactory PER conditioning, three types of individuals can be distinguished in a tactile PER conditioning experiment: (1) bees that discriminate
textures by responding with PER only to the conditioned but not to an alternative
pattern (discriminating bees), (2) bees that do not discriminate textures and respond
to the conditioned and to an alternative pattern (non-discriminating bees), and
(3) bees that do not learn (non-learning bees). As operant antennal scanning behavior
is a necessary condition for tactile learning, one can analyse whether there exist
specific antennal scanning strategies that differ between the three groups of bees.
Celia Moebius in our laboratory has analysed a number of behavioral parameters
during tactile PER conditioning by recording the three-dimensional movements of
the antennae. The bees were conditioned to a plate with grooves. Discrimination
was tested after five conditioning trials by presenting the conditioned pattern and
then an alternative that had the same pattern rotated by 90°. The experiments demonstrated that antennal contact activity is the most prominent parameter that differs
between bees of the different groups. During conditioning and testing antennal
activity in non-discriminating bees was always higher than activity in bees which
discriminated, did not learn or served as unpaired controls.
    Also the two-dimensional antennal scanning patterns during the tests differ
between discriminating, non-discriminating and unpaired control bees (Fig. 6.4.4).
Bees scan a pattern before conditioning with wide-ranging antennal movements
that cover large parts of the test plate. Unpaired control bees do not show any
obvious changes of the antennal scanning movements after five unpaired sucrose
stimulations. In contrast to the controls, the antennal scanning patterns change in
bees which show successful PER conditioning. After five conditioning trials these
individuals tap the conditioned pattern frequently without wide-ranging movements

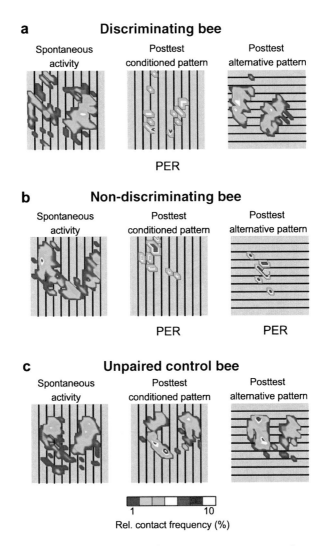

Fig. 6.4.4 Two-dimensional antennal scanning movements across test plates with engraved grooves. The scanning movements for three bees before and after five conditioning trials are shown. The graphs show the relative frequencies of contacts with the test plates in a colour coded scale which is shown at the *bottom* of the graph (**a**) The bee discriminated the two patterns in the posttests and showed conditioned PER after five conditionings only when the conditioned pattern was presented (**b**) The bee did not discriminate the patterns in the posttests and responded with PER when either of the patterns was presented (**c**) Unpaired control which was fed five times sucrose without pairing sucrose feeding with presentation of the plate. PER did not occur during the posttests. Measurement durations were 30 s for spontaneous activity. During the posttests antennal movements were either evaluated until PER occurred or up to 20 s when PER did not occur

and then respond with the conditioned PER. No obvious differences in antennal scanning of the conditioned pattern were found between discriminating and non-discriminating bees. Discriminating bees which do not respond with PER to the alternative pattern show wide-ranging antennal movements when the non rewarded pattern is presented. Non-discriminating bees show frequent antennal tapping movements both for the conditioned and for the alternative pattern. We conclude from these experiments that the antennal scanning movements change during conditioning and that they differ in bees that discriminate compared to bees that do not discriminate.

6.4.7   Significance of the Gustatory Input

Honey bees which perform different foraging tasks differ in their thresholds for sucrose ([23], see also Chap. 1.1). The threshold for sucrose can be measured under laboratory conditions by stimulating bees with different concentrations of sucrose at the antennae. Pollen collecting bees have low sucrose thresholds and, therefore, show the PER already at low sucrose concentrations or even when they are stimulated with water. Nectar collectors, which have higher sucrose thresholds, usually respond to higher sucrose concentrations. In addition to the foraging role also the age and the genotype of the bee affect sucrose sensitivity ([24], see also Chap. 1.1). Under laboratory conditions sucrose thresholds correlate with thresholds for pollen, odor and light [8, 32]. Similar to the findings under laboratory conditions, free flying nectar foragers also show different thresholds for sucrose [22].

Sucrose is the reward in most associative learning protocols with bees. Bees which differ in sucrose thresholds also differ in learning performance when sucrose serves as a reward. Pollen foragers that are very sensitive to sucrose show better acquisition and less extinction during tactile PER conditioning than nectar foragers that are less sensitive to sucrose (Fig. 6.4.3d) [28]. In a series of studies it was shown that the correlations between sucrose sensitivity and acquisition, extinction and retention are valid for olfactory and tactile PER conditioning [29, 30]. Sucrose thresholds also show correlations with the frequency of antennal scanning [33]. The contributions of all these covariants on learning performance have not yet been quantified in detail.

6.4.8   Conclusions

The experiments demonstrate that there exist different levels of plasticity in tactile learning, discrimination and retention. During tactile learning several neural subsystems interact in the brain of the bee. Visual, olfactory, gustatory or tactile sensory inputs elicit stimulus specific three-dimensional antennal movements. Antennal movements lead to sensory feedback that controls three-dimensional antennal motor activity. Antennal sensorimotor feedback is the basis for the acquisition of three-dimensional tactile maps. Gustatory inputs from the antennae and the proboscis can interact with the antennal sensory and motor systems leading to different forms of tactile learning and memory.

Information from the environment controls the motor output of the antennal system at the sensory level. Moving visual patterns, olfactory stimuli, sucrose stimuli or tactile contact with objects induce stimulus specific antennal movements that optimize the sampling of sensory information. Short-term enhancement of motor activity after gustatory stimulation demonstrates that an external stimulus can modify the basic motor programs. The dendritic projections of gustatory antennal sensilla into the antennal motor neuropil are the neural substrate for interactions of gustatory information with the motor system.

Motor learning is the simplest form of associative plasticity at the level of the antennal motor system. In contrast to sucrose dependent learning, antennal motor learning is a slow process that takes many hundred antennal contacts to develop. Operant conditioning of antennal activity with sucrose demonstrates that a sucrose reward can induce rapid and long-lasting modifications of antennal motor patterns. Operant conditioning of muscle activity shows that the basic mechanisms for operant learning can be found even at the level of a single motoneuron. At the same time these experiments also demonstrate that operant conditioning of a single motoneuron depends on information from intact mechanoreceptors at the antennal joints.

Tactile PER conditioning represents the highest level of plasticity in this system. In this conditioning protocol bees learn the position of an object in the three-dimensional space around the head. They also learn to discriminate form, size and surface textures. To solve these complex problems bees adapt the antennal scanning movements to the characteristics of the scanned object. The antennal system of the bee is a very useful model for the study of learning because there exists detailed knowledge about neuroanatomical projections, about stimulus controlled motor output, about plasticity at different system levels, and about the different functions of sucrose stimuli.

6.4.9 Outlook

In the future tactile antennal learning could be used to analyse three basic questions that are of great importance for general behavior not only in insects: (a) how does the nervous system acquire three-dimensional maps of sensory cues in the environment; (b) are the physiological and molecular mechanisms underlying different forms of learning and memory similar for different sensory modalities; (c) how is sensory information from a multimodal environment processed and stored in the nervous system?

Tactile antennal motor learning is a very good example for the acquisition of a three-dimensional map of the environment. On another scale similar tasks are solved by free flying bees using visual cues to acquire maps of the environment (see Chap. 2.5). For the future it will be necessary to analyse the structural and dynamic mechanisms underlying the formation of such maps. At the neuroanatomical level one could analyse in the antennal system whether there exist topographic maps of the three-dimensional tactile environment. The analysis of the physiological mechanisms underlying coincidence detection between antennal motor activity and mechanoceptive feedback will be extremely important for understanding the dynamic aspects of acquiring a three-dimensional map. Such an analysis can be performed under laboratory conditions which is a great advantage compared to experiments in the field.

The obvious similarities between olfactory and tactile PER conditioning can be used to unravel a number of open questions concerning the acquisition and retention of information during associative learning. While the MB are involved in olfactory learning, tactile antennal learning and discrimination apparently do not need intact MB and probably are based on signal processing in the AMMC. For the future it will be most interesting to compare the molecular and physiological mechanisms underlying learning and memory in the AMMC and the MB. One first important step for the future could be the development of a preparation for *in vivo* imaging of neural activity in the AMMC.

Multimodal signal processing, learning, retention and discrimination could be analysed in the future with behavioral methods by studying visual and tactile learning in free flying bees. Can information which was learned in one modality be transferred to the other modality? Is tactile information which is learned in the hive transferred to visual tasks in the field? Answering these questions would also help to develop physiological experiments in which the respective neural mechanisms are analysed under laboratory conditions.

References

1. Ai H, Nishino H, Itoh T (2007) Topographic organization of sensory afferents of Johnston's organ in the honeybee brain. J Comp Neurol 502(6):1030–1046
2. Erber J, Pribbenow B (2000) Antennal movements in the honey bee: how complex tasks are solved by a simple neuronal system. In: Prerational intelligence: adaptive behavior and intelligent systems without symbols and logic. Kluwer, Dordrecht, pp 109–121
3. Erber J, Masuhr T, Menzel R (1980) Localization of short-term-memory in the brain of the bee, *Apis mellifera* 10. August 2011 11:15. Physiol Entomol 5(4):343–358
4. Erber J, Pribbenow B, Bauer A, Kloppenburg P (1993) Antennal reflexes in the honeybee – tools for studying the nervous-system. Apidologie 24(3):283–296
5. Erber J, Pribbenow B, Grandy K, Kierzek S (1997) Tactile motor learning in the antennal system of the honeybee (*Apis mellifera* L.). J Comp Physiol A 181(4):355–365
6. Erber J, Kierzek S, Sander E, Grandy K (1998) Tactile learning in the honeybee. J Comp Physiol A 183(6):737–744
7. Erber J, Pribbenow B, Kisch J, Faensen D (2000) Operant conditioning of antennal muscle activity in the honey bee (*Apis mellifera* L.). J Comp Physiol A 186(6):557–565
8. Erber J, Hoormann J, Scheiner R (2006) Phototactic behaviour correlates with gustatory responsiveness in honey bees (*Apis mellifera* L.). Behav Brain Res 174(1):174–180

 9. Faber T, Joerges J, Menzel R (1999) Associative learning modifies neural representations of odors in the insect brain. Nat Neurosci 2(1):74–78
10. Hammer M (1997) The neural basis of associative reward learning in honeybees. Trends Neurosci 20(6):245–252
11. Haupt SS (2007) Central gustatory projections and side-specificity of operant antennal muscle conditioning in the honeybee. J Comp Physiol A 193(5):523–535
12. Kevan PG, Lane MA (1985) Flower petal microtexture is a tactile cue for bees. Proc Natl Acad Sci USA 82(14):4750–4752
13. Kisch J, Erber J (1999) Operant conditioning of antennal movements in the honey bee. Behav Brain Res 99(1):93–102
14. Kisch J, Haupt SS (2009) Side-specific operant conditioning of antennal movements in the honey bee. Behav Brain Res 196(1):131–133
15. Kloppenburg P (1995) Anatomy of the antennal motoneurons in the brain of the honeybee (*Apis mellifera*). J Comp Neurol 363(2):333–343
16. Komischke B, Sandoz JC, Malun D, Giurfa M (2005) Partial unilateral lesions of the mushroom bodies affect olfactory learning in honeybees *Apis mellifera* L. Eur J Neurosci 21(2):477–485
17. Maronde U (1991) Common projection areas of antennal and visual pathways in the honeybee brain, *Apis mellifera*. J Comp Neurol 309(3):328–340
18. Menzel R, Manz G (2005) Neural plasticity of mushroom body-extrinsic neurons in the honeybee brain. J Exp Biol 208(Pt 22):4317–4332
19. Menzel R, Müller U (1996) Learning and memory in honeybees: from behavior to neural substrates. Annu Rev Neurosci 19:379–404
20. Menzel R, Erber J, Masuhr T (1974) Learning and memory in the honey bee. In: Barton Browne L (ed) Experimental analysis of insect behavior. Springer, Berlin/Heidelberg/New York, pp 195–217
21. Menzel R, Durst C, Erber J, Eichmüller S, Hammer M et al (1994) The mushroom bodies in the honey bee: from molecules to behavior. Fortschr Zool 39:81–102
22. Mujagic S, Erber J (2009) Sucrose acceptance, discrimination and proboscis responses of honey bees (*Apis mellifera* L.) in the field and the laboratory. J Comp Physiol A 195(4):325–339
23. Page RE, Erber J, Fondrk MK (1998) The effect of genotype on response thresholds to sucrose and foraging behavior of honey bees (*Apis mellifera* L.). J Comp Physiol A 182(4):489–500
24. Page RE, Scheiner R, Erber J, Amdam GV (2006) The development and evolution of division of labor and foraging specialization in a social insect (*Apis mellifera* L.). Curr Top Dev Biol 74:253–286
25. Pribbenow B, Erber J (1996) Modulation of antennal scanning in the honeybee by sucrose stimuli, serotonin, and octopamine: behavior and electrophysiology. Neurobiol Learn Mem 66(2):109–120
26. Rehder V, Bicker G, Hammer M (1987) Serotonin-immunoreactive neurons in the antennal lobes and suboesophageal ganglion of the honey bee. Cell Tissue Res 247:59–66
27. Schäfer S, Rehder V (1989) Dopamine-like immunoreactivity in the brain and suboesophageal ganglion of the honeybee. J Comp Neurol 280(1):43–58
28. Scheiner R, Erber J, Page RE Jr (1999) Tactile learning and the individual evaluation of the reward in honey bees (*Apis mellifera* L.). J Comp Physiol A 185(1):1–10
29. Scheiner R, Page RE Jr, Erber J (2001) The effects of genotype, foraging role, and sucrose responsiveness on the tactile learning performance of honey bees (*Apis mellifera* L.). Neurobiol Learn Mem 76(2):138–150
30. Scheiner R, Page RE Jr, Erber J (2001) Responsiveness to sucrose affects tactile and olfactory learning in preforaging honey bees of two genetic strains. Behav Brain Res 120(1):67–73
31. Scheiner R, Weiß A, Malun D, Erber J (2001) Learning in honey bees with brain lesions: how partial mushroom-body ablations affect sucrose responsiveness and tactile antennal learning. Anim Cogn 4:227–235

32. Scheiner R, Page RE, Erber J (2004) Sucrose responsiveness and behavioral plasticity in honey bees (*Apis mellifera*). Apidologie 35(2):133–142
33. Scheiner R, Kuritz-Kaiser A, Menzel R, Erber J (2005) Sensory responsiveness and the effects of equal subjective rewards on tactile learning and memory of honeybees. Learn Mem 12(6):626–635
34. Scheiner R, Schnitt S, Erber J (2005) The functions of antennal mechanoreceptors and antennal joints in tactile discrimination of the honeybee (*Apis mellifera* L.). J Comp Physiol A 191(9):857–864
35. Schürmann FW, Erber J (1990) FMRFamide-like immunoreactivity in the brain of the honeybee (*Apis mellifera*) – a light-microscopy and electron-microscopy study. Neuroscience 38(3):797–807
36. Schürmann FW, Klemm N (1984) Serotonin-immunoreactive neurons in the brain of the honeybee. J Comp Neurol 225(4):570–580
37. Sinakevitch I, Niwa M, Strausfeld NJ (2005) Octopamine-like immunoreactivity in the honey bee and cockroach: comparable organization in the brain and subesophageal ganglion. J Comp Neurol 488(3):233–254
38. Snodgrass RE (1956) Anatomy of the honey bee. Comstock Publishing Associates, Ithaca
39. Srinivasan MV (1994) Pattern-recognition in the honeybee – recent progress. J Insect Physiol 40(3):183–194
40. Wehner R (1967) Pattern recognition in bees. Nature 215(5107):1244
41. Wehner R (1981) Spatial vision in arthropods. In: Autrum H (ed) Vision in invertebrates (Handbook of sensory physiology), vol 7/6 C. Springer, Berlin/Heidelberg/New York, pp 287–616

Chapter 6.5
Testing Mathematical Laws of Behavior in the Honey Bee

Ken Cheng

Abstract Two mathematical laws of behavior derived from work on vertebrate animals were tested in honey bees in my research. One law concerned the ubiquitous phenomenon of generalization in learning. An animal obtaining a reward for a response to one stimulus will often make that response to similar but discriminably different stimuli. Under a suitably ideal characterization, generalization gradients ought to come out exponential in shape (Shepard RN, Science 237:1317–1323, 1987). In spatial generalization in honey bees, this prediction was upheld in a number of different studies. A second law concerned the weighting of different and conflicting evidence. A piece of evidence is supposed to be weighted by its recency, with more recent evidence given higher weight (Devenport L, Hill T, Wilson M, Ogden E, J Exp Psychol Anim Behav Process 23:450–460, 1997). With the passage of time since the last evidence was obtained, overall profitability of a 'patch' rather than recency of profits should dominate. Tests with honey bees failed to uphold this law, instead finding circadian modulation of preferences, with 'patch' preference highest at the circadian time at which reward was obtained on the previous (training) day. I attempted a speculative reformulation in terms of modulation of preferences according to different oscillators.

6.5.1 Introduction

This volume and earlier work show that the honey bee has been a model for the study of learning, memory, navigation, and neurobiology [25, 35, 36]. I have used honey bees to test mathematical laws of behavior derived from the study of other animals.

K. Cheng (✉)
Department of Biological Sciences, Macquarie University, Sydney, NSW 2109, Australia
e-mail: ken.cheng@mq.edu.au

C.G. Galizia et al. (eds.), *Honeybee Neurobiology and Behavior: A Tribute to Randolf Menzel*, DOI 10.1007/978-94-007-2099-2_34,
© Springer Science+Business Media B.V. 2012

I report here two study cases testing what might be universal laws, one of which upheld universality while the other one failed to do so. I end with some reflections concerning universal laws of behavior.

6.5.2   About Universality and Mathematical Laws

The physical sciences often formulate mathematical laws relating different variables, laws that have wide application under certain idealised conditions. Gas laws and Newtonian laws of gravitation are examples. Ideal conditions are typically needed to approach a match to the quantitative predictions. Thus, friction interferes with the accuracy of predictions based on laws of gravitation. Mathematical laws concerning behavior have also been formulated, with the best known laws probably those from psychophysics. *Weber's law* [47] is a well known case, originally formulated by Ernst Weber concerning human discrimination of differences in weights. The law states that the *just noticeable difference* in weight is a constant proportion of the reference weight, the weight that one is comparing a test stimulus to. For example, for a 40 g reference weight, a human could be expected to discriminate a test stimulus at least 1 g different from the target (e.g., 41 g), but if the reference was double, at 80 g, a human would need a ±2 g difference in the test stimulus to make the discrimination. This law has been adapted to search behavior, and found to hold in some instances. Thus in landmark-based search, the spread of search scales linearly with the distance to the nearest landmark in pigeons [3]. The spread is a constant proportion of the target distance being measured. Weber's law also holds for honey bees in searching for a target after flying a certain distance in a uniform environment, a narrow channel with textured walls. The bees were trained to fly a constant distance in the channel to reach a feeder. After sufficient training, the feeder was removed for a test. The spread of search scales linearly with the training distance from the start of the channel to the feeder [9].

Laws such as Weber's law, however, are not expected to be universal, holding for all species and all perceptible stimuli. Life being enormously diverse, one might think that a search for universality in mathematical laws of behavior would be doomed to fail. Nevertheless, all life lives on Earth, and some physical conditions are constant over vast regions of the Earth. The Earth rotates on its axis, producing daily light-dark cycles that drive circadian rhythms in all but the polar regions. We might expect all circadian clocks to have a period near 24 h (see Chap. 1.3).

Functional considerations about statistical or probabilistic properties of the world might also lead us to expect universal laws of behavior. Functional explanations are one of the four types of answers to 'why' questions that Niko Tinbergen famously proposed for explaining behavior [46]. It concerns adaptive function, or what the behavior is currently good for. One variety of functional explanation is to say that the behavior or law of behavior reflects "evolutionary accommodations to universal properties of the world" [42, p. 1319]. To give an example for illustration, all animals searching for hidden food should be expected to search in a 'patch' with a

higher density of food rather than a 'patch' with a lower density of food, even if both
patches contain plenty enough food to satiate the animal. A condition for this 'law'
of preference for higher-density patch is that the animal can tell which patch actually has the higher density (e.g., by learning from past experience). We can say that
searching in the higher-density patch increases the probability of finding food with
each search attempt, and decreases the expected time needed to satiate the animal,
'expected' in the statistical sense. This benefits the animal in that it will spend on
average less time foraging at the higher-density patch before it is satiated. My two
case studies concern more complicated cases of such statistical properties.

6.5.3   Testing a Universal *Law of Generalization*

The first case concerns the ubiquitous learning phenomenon of generalization,
found in most well studied cases of learning, such as pigeons generalizing across
wavelengths of keylight in autoshaping [30] or rats generalizing across frequencies
of tones in fear conditioning [19]. In generalization in both classical conditioning
and operant conditioning, an animal responds to a range of stimuli similar to but
different from the training conditioned stimulus. In the case of fear conditioning in
rats, for example, the training stimulus that predicts foot shock might be a 1,000-Hz
tone, while a range of different frequencies might be presented in tests to examine
generalization [19].

Honey bees show generalization in different learning tasks, including appetitive
classical conditioning of proboscis extension with odor as conditioned stimuli [29]
(see also Chap. 6.1), landmark image matching [2], and generalization of abstract
rules such as matching to sample [26]. I tested *Shepard's universal law of generalization* in honey bees with spatial locations with respect to a landmark. This was a
good dimension to use because we can specify the degree of image-based mismatches between a test location and the training location. But first, a brief explication of Shepard's [42] theory is presented using spatial locations as example.

6.5.4   *Shepard's Law of Generalization*

In generalization in operant conditioning, an animal is trained on one particular
stimulus called the S+ to obtain reward. It is then presented with a range of stimuli, usually including S+. Typically, the animal responds most to S+, and an
orderly monotonic generalization gradient is obtained as a function of how different a test stimulus is from S+ (e.g., [30]). The more different a test stimulus is
from the S+, the less the animal responds. Shepard's [42] law applies only to a
subset of generalization data, meaning that certain conditions must be met for the
law to apply. The two crucial conditions are that (1) test stimuli must not be too
similar to S+, so that the animal has trouble discriminating a test stimulus from

S+, and (2) the *x*-axis along which the test stimuli are scaled must be correctly formulated. The measure of distance or dissimilarity between stimuli needs to be formulated in terms of how the animal's brain codes such differences (a psychological scale), rather than in terms of physical measures such as wavelengths of light (a physical scale).

Discrimination refers to telling two stimuli apart. It is not thought to be an all-or-none process. In the vast majority of perceptual systems in most animals, some stimuli can be easily (almost perfectly) discriminated from a target stimulus, while a range of stimuli can only sometimes be discriminated, the 'performance' of the perceptual system being noisy and variable from one time to another. Dyer in this volume provides some striking examples of discrimination experiments on honey bees (see Chap. 4.5). If an animal cannot tell a test stimulus from the training stimulus, it will respond to both similarly, in this sense 'generalizing' from one to the other. In experiments testing discrimination ability, the experimenter typically provides motivation for the animal to differentiate two similar stimuli, rewarding the target, and not the distractor. Cases of generalization in which a test stimulus is difficult to discriminate from the training stimulus follow a different mathematical relation. The function has been argued to be Gaussian [37, 41], and a good number of generalization gradients including stimuli confusable with S+ show Gaussian forms [20]. Shepard's [42] law, on the other hand, specifies an exponential form (discussed below). Indeed, one sense of generalization is that an animal will accept one stimulus for another because the two are coded in the same way and hence indistinguishable. Much work has been done on how honey bees discriminate visual stimuli or generalize in this sense, with a complex picture emerging, dependent on stimuli and training conditions [18, 24, 45, review: 31] (see also Chap. 4.5).

Even when an animal can tell that a test stimulus differs from S+, however, it still might respond to it. With only a single S+ presented in training, the animal would not be treating a test stimulus in general as a test of discrimination. Rather, it would have evolved to respond or not to respond on the basis of the expected consequences of responding. In this characterization of generalization, the animal is 'betting' on whether the test stimulus would deliver the consequence of interest, this being the reward associated with S+. *Shepard's law* concerns such cases.

Suppose that an animal has been trained with a spatial S+ (Fig. 6.5.1). It has come to a particular location in a particular context, the S+ location, and found a container with palatable food. We can suppose that the animal has learned that a container at S+ has a consequence of interest to it, namely that it contains food. Although it is of interest how the animal has learned and how the brain has encoded the training experiences, the formulation of *Shepard's law* is not dependent on these mechanistic theories. Now on a 'test trial', it finds that the location of the container has changed to X, which is noticeably different from S+. The problem for the animal in this generalization test is to estimate the probability that X also has the consequence of interest. To the extent that this estimated probability is high, the animal should respond to the container at X. To the extent that this estimated probability is low, the animal should not respond to the

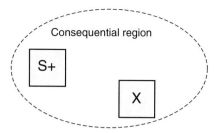

Fig. 6.5.1 Illustration of Shepard's [42] formulation of the problem of generalization, using spatial generalization as an example. An animal has found something of consequence (food) in a container located at S+. This constitutes training in experimental paradigms. Some region around S+ is taken to be the consequential region, the region in which a container has the same consequence of interest (i.e., contains food). The animal has learned that S+ is in the consequential region. If it encounters a container at a different location, X, the problem of generalization can then be asked. The formulation is that given that S+ is in the consequential region, what is the probability that X is also in the consequential region?

container at X. *Shepard's law* relates this probability estimate to the psychological distance between X and S+.

Shepard's formulation says that in some region, called the consequential region, the container has the consequence of interest, in this case the property of containing the food. What the animal has learned from its training with the container at S+ is that S+ is in the consequential region. The question of generalization becomes: given that S+ is in the consequential region, what is the probability that X is also in the consequential region? Given this formulation, Shepard [42] showed that a general law may be formulated that relates the probability estimate to the psychological or subjective distance between a test stimulus and S+. Under many different forms of the possible consequential region, an upwardly concave generalization gradient should be obtained. While the gradients differ mathematically, they look highly similar, and biologically, they are indistinguishable. The function is an exponential:

$$y = e^{-kx}$$

where y is the probability estimate that the test stimulus x is in the consequential region, and x is the subjective distance between stimulus x and S+.

Different methods may be used to obtain the required x axis for testing Shepard's law. It is generally insufficient simply to assume that the physical stimulus dimension represents the subjective scale because animals' brains often recode physical stimuli on a different scale. For example, physically, hue is derived from a one-dimensional linear scale of wavelengths of the electromagnetic spectrum. But the subjective representation of hues can be circular in the sense that the ends join up (not necessarily forming a perfect circle; in humans: [32]; in bees: [1]). Measures of confusability from discrimination experiments may be used to estimate the subjective scale (see Chap. 4.4).

6.5.4.1 Testing Shepard's Law

Shepard [42] presented 12 empirical data sets from humans and pigeons in support of the general law. I extended the data range by testing honey bees on spatial generalization [4–6]. Spatial location was signaled by a nearby landmark in all cases. In fact, this cue was the only reliable cue as the set-up of landmark and target (small dish of sugar water) was moved about the experimental table inside a laboratory from trial to trial. The distance from the landmark was one dimension in which spatial location varied [5]. In one experiment, the dish of sugar water was placed on a long narrow yellow strip of cardboard, near a blue cylinder serving as a landmark (the S+ location). Under the sugar dish, and on top of the yellow cardboard, was a small piece of blue cardboard. The honey bees learned such a simple task quickly. After due training, they were tested occasionally with a dish of tap water replacing the sugar water. The dependent variable was expressed as a proportion of the amount of searching at the target on a test at the S+ location. The prediction was that an exponential generalization gradient would be found on a suitably scaled x-axis.

Two different ways of scaling the x-axis both supported *Shepard's law* by showing excellent exponential fits (Fig. 6.5.2). The x-axis scaling was based on the extent of mismatches in cues that research on landmark-based search in honey bees has suggested. One scale was based on mismatch in the retinal height and the retinal width projected by the landmark [2] (Fig. 6.5.2, top). Figure 6.5.2 shows equal weighting of retinal height and width differences, but the fits were similarly good over a wide range of weighting parameters for retinal height differences vs. retinal width differences. Also of note in Fig. 6.5.2 is the fact that a linear scale of distance does not produce the exponential function. The second scale is based on motion parallax, another cue that honey bees encode [33]. Motion parallax refers to the fact that as the bee (or any other visual perceiver) moves, so does the retinal image. The amount of movement or parallax depends on the distance to the landmark. This cue is crucial for odometry in honey bees, the estimation of the distance flown [44]. Figure 6.5.2 (bottom) shows good fits to an exponential function with equal weighting of parallax differences in height and width dimensions. Again, the weighting for combining height and width parallax did not matter much.

Similarly good exponential generalization gradients were obtained in other experiments. In one, the direction to a single landmark was varied on tests while keeping the distance equal to the training distance [5]. Other experiments combining both angular and retinal size differences between the test locations and the target location also produced generalization gradients fitting an exponential function [4, 6].

Thus, using spatial generalization in honey bees as a method, a variety of different tests all confirmed *Shepard's law* [42]. Confirmation from one more species of course does not add up to universality. All we can say is that the law has been upheld also in an invertebrate animal. It makes an ideal test case for such proposals of universality stemming from work on vertebrate animals. *Shepard's law of generalization* can be probed further in honey bees. We have theoretical scales for

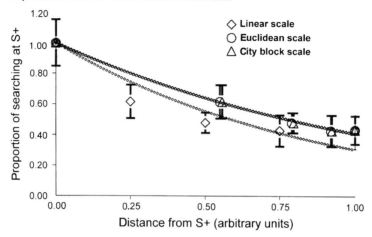

a   Exponential fits for scales based on LM size

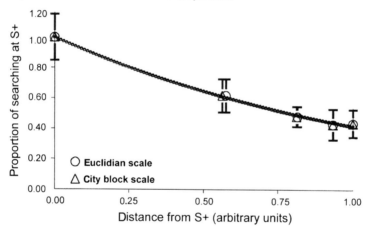

b   Exponential fits for scales based on parallax

Fig. 6.5.2 Best exponential least-square fits for distance generalization in honeybees based on projected landmark (LM) size (**a**) and motion parallax (**b**). The data (*large symbols*) are expressed as proportion of searching relative to searching at S+. Error bars show 95% confidence intervals. The two different metrics (*Euclidean* and *city-block*) refer to different ways of combining mismatches in two variables (projected width and projected height of landmark in (**a**) or amount of parallax in width and parallax in height in (**b**)). The Euclidean metric is the square root of the sums of squares of the differences in the two measures (width and height differences), while the city-block metric is simply the sum of the two measures of differences. In (**a**), the best exponential fit based on a linear scale is shown as well. The data points from corresponding locations in each series (e.g., the second data point from the *left* for all curve fits) do have the same y value. That they appear to differ is a visual illusion (Reprinted from [5] (Fig. 3), with permission)

color perception [1, 10] (see also Chap. 4.4) and odor perception [29] (see also Chap. 4.1), and by now a sizeable set of extant data. We have a good chance of designing suitable experiments for the occasion, or mining existing data. On odor generalization, the research can take on a new angle, as imaging work can link functional aspects such as a *law of generalization* to neurobiology [29] (see also Chap. 6.1). And if a data set fails to provide support for a purported universal law, it can serve to send us back to the drawing board in our considerations of universality, as the next case shows.

6.5.5 Testing the *Temporal Weighting Rule*

Nature being not totally regular and predictable, animals often encounter conflicting evidence and still make decisions. Let us consider a simple situation in which one object can take on one of two properties. Suppose that a small bottle cap on a yellow cardboard on a table could either contain sugar water or tap water. A visiting honey bee would have to decide whether to sample from this bottle cap or sample elsewhere. Intuition would suggest that two factors would matter. One is the amount of evidence in favor of the 'hypothesis' (sugar water vs. tap water), and the other is the recency of this evidence. Presumably, one might rely more on the most recent evidence than on evidence from long ago. These two intuitions were put into one formulation by Lynn Devenport and colleagues, and tested on a range of vertebrate animals [13–16].

The *Temporal Weighting Rule* combines the intuitions regarding amount and recency of evidence by stating that each piece of evidence ought to be weighted by the inverse of elapsed time since the evidence was obtained: $1/t$, where t stands for elapsed time. What this means is that the evaluation of evidence changes dynamically with the passage of time since the last piece of evidence was obtained. With a short delay, recent evidence looms large because $1/t$ is sizeable when t is small. With the passage of time (increase in t), however, pieces of evidence are weighted similarly because for large t values, $1/t$ does not diminish much as t increases further. Thus, when $t = 1$ unit of time, a passage of 1 unit of time halves the weight, from 1 to ½. When $t = 100$ units of time, on the other hand, the passage of 1 unit of time hardly affects the weight.

6.5.5.1 Testing the Temporal Weighting Rule in Mammals

Tests of the *Temporal Weighting Rule* are made in the context of foraging experiments, with animals given two choices or 'patches' A and B. They might be two food containers or two levers to press in the case of rats. One patch, say A, delivers food in the first phase, while patch B does not. After a while, the rules change abruptly in the second phase, and patch B delivers food while patch A is barren. Animals are then offered the choice between A and B after various post-training delays. The *Temporal Weighting Rule* predicts that shortly after the two training phases, animals would prefer B because the recent evidence in favor of B is weighted highly.

Immediate recency should dominate. After some passage of time without further evidence, however, the preference should change to reflect patch averages. This is because the weights accorded to each piece of evidence becomes roughly equal, and preference depends much more on the total amount of evidence in favor of A versus B. Thus, if A and B had delivered about the same amount (for example if the first and second phases were the same in duration), indifference between the patches should be found with delay. If patch A actually had a longer history than B of being profitable, the animal should reverse preferences after delay, and switch from B to A.

6.5.5.2 Testing the Temporal Weighting Rule in Honey Bees: No Empirical Support

Honey bees foraging in the wild need to deal with changing foraging conditions, shifting their area of concentrated search over the course of a season [40]. Not surprisingly, experimental work shows that honey bees and bumblebees learn readily to cope with changes in foraging contingencies, keeping track in memory of the status of multiple feeders [27, 28], and adjusting to the profitability of 'patches' [8, 11, 17].

To test the *Temporal Weighting Rule* in honey bees, Prabhu and I used a conceptually similar approach to that used on mammals [38, 39]. Bees were given some training in one phase with one target that contained food, and then conditions switched abruptly and without announcement in a second phase. After being attracted to the experimental table, the bees were given 20 trials of training in phase 1 followed by 10 trials of training in phase 2. This preponderance of phase-1 training made it easier to see the signature pattern of dynamic changes predicted by the *Temporal Weighting Rule*. Soon after training, the bees should prefer the most recently rewarded target, that associated with phase 2 (phase-2 target). With the passage of time, for example, the next day, the bees should revert to preferring the target that delivered the reward in phase 1 (phase-1 target). With color cues, contrary to the *Temporal Weighting Rule*, a pattern of circadian modulation was found (Fig. 6.5.3). At 0 delay, the bees preferred strongly the phase-2 target. At 22 h delay, corresponding roughly to the start of training in Phase 1 the previous day, the bees had reversed their preferences. But at 24 h delay, the preference had reverted back to the Phase-2 target (Fig. 6.5.3) [38]. Such a pattern was found in a number of variations of the experiment. In fact, with multiple occasions of testing between 0 and 22 h post phase-2 training, an orderly sinusoidal pattern of preferences was found. Preference for the phase-1 target was low on immediate testing (0 h delay), and rose to a peak at 18 and 22 h delay. The midpoint of the rise was at ~11 h, the midpoint between 0 and 22 h (see Fig. 9 in [38]).

When odors formed the stimuli to be discriminated, the honey bees stuck to preferring the phase-2 target odor no matter what the delay after training [39]. The preference for the most recently rewarded odor diminished a bit (statistically significantly) at longer delays (22 and 24 h). A hint of circadian modulation could be seen

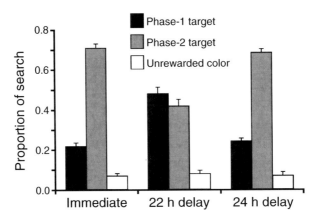

Fig. 6.5.3 Proportion of searching by honey bees on tests done immediately after Phase-2 training and after 22 and 24 h of delay, over the target color for Phase 1, the target color for Phase 2,
and the never rewarded color. The results showed a significant interaction between color (Phase-
1, Phase-2, or never rewarded) and time of testing. While the Phase-2 color was preferred significantly over the Phase-1 color at immediate testing and after 24 h, preference was equivalent
(statistically) between Phase-1 and Phase-2 colors at 22 h delay (Adapted from Experiment 4 of
[38], with permission)

in the data, with the preference for the phase-2 target odor numerically weaker at
22 h than at 24 h, but the difference did not turn out statistically significant. With an
even more lopsided training regime in favor of phase 1, with 28 phase-1 training trials and only 7 phase-2 training trials, circadian modulation was found (unpublished
data). A preference for the task-1 target odor was found at 22 h delay, but again contrary to the *Temporal Weighting Rule*, the preference disappeared at 24 h delay.

It would appear that a combination of circadian modulation and preference for
the most recently rewarded stimulus account for the data with both visual (color)
and olfactory stimuli. The circadian modulation was found more readily with the
colors, but can also be found with olfactory stimuli. In the case of a 2:1 ratio of
phase-1 training to phase-2 training, a recency preference is found for both types of
stimuli. That is, averaging the preferences at 22 and 24 h delay gave a preference in
favor of the phase-2 target. This recency preference might have resulted from a suppression of the earlier acquired memory, a process thought to occur in other contexts
[11, 34]. In the Prabhu and Cheng experiments, when the conditions changed unan-
nounced to those of phase 2, the honey bees could be observed to persist in visiting
the phase-1 target stimulus for a number of visits. To learn to switch to the phase-2
target, they would have had to suppress this tendency. Such a suppressive process
might have led to an overall preference for the phase-2 target, one that was, however, modulated by circadian time. It would be interesting to do these experiments
while eliminating the suppression of the phase-1 target. This can be done by simply
removing the phase-1 target during phase-2 training.

A series of studies on honey bees' reaction to dynamic reward schedules by Gil
and colleagues also failed to support the *Temporal Weighting Rule* [21–23]. A key

contrast was comparing an increasing series of rewards with a constant series of rewards that offered the same average amount of reward per visit. Tests a day after training showed that the increasing series elicited more visit time and inspection time in free flying bees [22] and quicker proboscis extensions to sucrose offerings in harnessed bees [23]. With the long delay, the *Temporal Weighting Rule* would predict that the average reward value should be driving the behavior. This means that the bees experiencing the increasing and constant series with the same average reward rate should have behaved similarly. Gil et al. interpreted the results in terms of expectations formed by the bees. Clear evidence was also found that the effect was modulated by circadian time. Honey bees behaved differently after being trained with the constant versus increasing series at 24 h delay and 48 h delay, but not at ~25 h delay [22].

6.5.6   Discussion: Universality and Its Lack

One way to look at these two cases is that sometimes intuitions about universality of behavioral laws may turn out correct, and other times our intuitions go awry somewhere and predictions turn out wrong. One may conclude that the universality of *Shepard's law* has so far been upheld, but that the *Temporal Weighting Rule* is not universal. Such an argument is coldly logical. But while it is sound, it might miss something about universality in biological laws. In the case of the weighting of evidence for foraging, it seems clear that what appeared intuitively a candidate for a universal law (the *Temporal Weighting Rule*) in fact fractionates at least into a number of different cases according to differences in foraging ecology. The data reviewed certainly suggest that much. Perhaps when we carve the foraging ecologies at the correct joints, we will come up with a set of non-universal laws. But is there perhaps some commonality among the diversity?

One speculation concerning the temporal modulation of weights assigned to evidence is that the weights are modulated in cycles of time rather than in a monotonically decreasing fashion. The cases supporting the *Temporal Weighting Rule* would be thought of as very long cycles of time, beyond the lifetime of an animal (so that the cycle never turns within the life of the animal). Mechanistically, a recent proposal for rats suggests that timing on all scales, from seconds to a day to longer periods, may be computed and represented by oscillators cycling at different periods [12]. Classically, circadian timing and interval timing (of durations typically from seconds to minutes) were thought to be different kinds of mechanisms, but now the distinction has been blurred [7]. Properties thought to characterize the circadian system but not interval timing have been found in interval timing in rats, of chief importance the property of endogenous oscillation [7, 12]. Thus, when a rat has been provided regular periodic food, in intervals ranging from 48 s to 21 h, it shows periodic anticipatory behavior at around the period of the cycle for a number of cycles after food provision is discontinued [7, 12]. Behavior oscillates in periodic cycles for a while without external input, a classic property of circadian timers. In this view, the circadian oscillator is just one (albeit especially important) oscillator among multiple oscillators used for timing. It would be most interesting to examine if multiple endogenous oscillators can be found in other species, including insects such as honey bees. If such a view is correct in broad outline, then the mechanistic basis for different forms of time-based modulation in many animals may well be common. What is universal may be different cycles of time putting their stamp on animal behavior, including long cycles such as yearly cycles.

Universality on functional grounds need not depend on any common mechanistic basis. All that is needed is for some universal condition found on Earth to drive evolution. But natural selection often works in conservative ways, and having a common mechanistic basis may well help to bring universal laws to fruition. Mechanistic analyses should also form a part of this functional and behavioral ecological examination of universalities.

6.5.7 Conclusion and Outlook

I have reviewed two plausible candidates for universal laws derived from work with vertebrate animals and extended to the honey bee in my research. A *law of generalization* formulated by Roger Shepard [42, 43] was upheld by a variety of tests on honey bees. The *Temporal Weighting Rule* [15] for dealing with conflicting evidence, on the other hand, was not supported by analogous experiments on honey bees. The honey bees did not weight evidence as a declining function of elapsed time since the evidence was obtained. Instead, circadian modulation was a better description. I explored whether this diversity might harbor a deeper commonality in time-based modulation of behavior.

Two major lines of research have been raised in the course of the discussion. With regard to the *law of generalization*, an examination of Shepard's *law of generalization* in classical conditioning with odors would be exciting. Not only can a range of controlled data be gotten (and a substantial extant set already awaits), but this line of research presents a link to the neurobiology of generalization because the controlled conditions of experimentation allows brain imaging to be done. With regard to the temporal modulation of weights assigned to evidence, research on interval timing in honey bees would be fruitful for examining the possible role of oscillators of different periods used for timing.

References

1. Backhaus W (1991) Color opponent coding in the visual system of the honeybee. Vision Res 31:1381–1397
2. Cartwright BA, Collett TS (1983) Landmark learning in bees. J Comp Physiol A 151:521–543

3. Cheng K (1992) Three psychophysical principles in the processing of spatial and temporal information. In: Honig WK, Fetterman JG (eds) Cognitive aspects of stimulus control. Lawrence Erlbaum Associates, Hillsdale, pp 69–88
4. Cheng K (1999) Spatial generalization in honeybees confirms Shepard's law. Behav Process 44:309–316
5. Cheng K (2000) Shepard's universal law supported by honeybees in spatial generalization. Psychol Sci 11:403–408
6. Cheng K (2002) Generalisation: mechanistic and functional explanations. Anim Cogn 5:33–40
7. Cheng K, Crystal JD (2008) Learning to time intervals. In: Menzel R (ed) Learning theory and behavior vol 1 of Learning and memory: a comprehensive reference, 4 vols (J Byrne Editor). Elsevier, Oxford, pp 341–364
8. Cheng K, Wignall AE (2006) Honeybees (*Apis mellifera*) holding on to memories: response competition causes retroactive interference effects. Anim Cogn 9:141–150
9. Cheng K, Srinivasan M, Zhang S (1999) Error is proportional to distance measured by honeybees: Weber's law in the odometer. Anim Cogn 2:11–16
10. Chittka L (1992) The color hexagon: a chromaticity diagram based on photoreceptor excitations as a generalized representation of color opponency. J Comp Physiol A 170:533–543
11. Chittka L (1998) Sensorimotor learning in bumblebees: long-term retention and reversal training. J Exp Biol 201:515–524
12. Crystal JD (2006) Time, place, and content. Comp Cogn Behav Rev 1:53–76
13. Devenport JA, Devenport LD (1993) Time-dependent decisions in dogs (*Canis familiaris*). J Comp Psychol 107:169–173
14. Devenport JA, Patterson MR, Devenport LD (2005) Dynamic averaging and foraging decisions in horses (*Equus caballus*). J Comp Psychol 119:352–358
15. Devenport L, Hill T, Wilson M, Ogden E (1997) Tracking and averaging in variable environments: a transition rule. J Exp Psychol Anim Behav Process 23:450–460
16. Devenport LD, Devenport J (1994) Time-dependent averaging of foraging information in least chipmunks and golden-mantled ground squirrels. Anim Behav 47:787–802
17. Dukas R (1995) Transfer and interference in bumblebee learning. Anim Behav 49:1481–1490
18. Dyer AG, Chittka L (2004) Fine colour discrimination requires differential conditioning in bumblebees. Naturwissenschaften 91:547–557
19. Frieman JP, Warner J, Riccio DC (1970) Age differences in conditioning and generalization of fear in young and adult rats. Dev Psychol 3:119–123
20. Ghirlanda S, Enquist M (2003) A century of generalization. Anim Behav 66:15–36
21. Gil M, De Marco RJ (2009) Honeybees learn the sign and magnitude of reward variations. J Exp Biol 212:2830–2834
22. Gil M, De Marco RJ, Menzel R (2007) Learning reward expectations in honeybees. Learn Mem 14:491–496
23. Gil M, Menzel R, De Marco RJ (2008) Does an insect's unconditioned response to sucrose reveal expectations of reward? PLoS One 3:e2810
24. Giurfa M (2004) Conditioning procedure and color discrimination in the honeybee *Apis mellifera*. Naturwissenschaften 91:228–231
25. Giurfa M (2007) Behavioral and neural analysis of associative learning in the honeybee: a taste from the magic well. J Comp Physiol A 193:801–824
26. Giurfa M, Zhang S, Jenett A, Menzel R, Srinivasan MV (2001) The concepts of 'sameness' and 'difference' in an insect. Nature 410:930–933
27. Greggers U, Mauelshagen J (1997) Matching behavior of honeybees in a multiple-choice situation: the differential effect of environmental stimuli on the choice process. Anim Learn Behav 25:458–472
28. Greggers U, Menzel R (1993) Memory dynamics and foraging strategies of honeybees. Behav Ecol Sociobiol 32:17–29
29. Guerrieri F, Schubert M, Sandoz J-C, Giurfa M (2005) Perceptual and neural olfactory similarity in honeybees. PLoS Biol 3:e60

30. Guttman N, Kalish HI (1956) Discriminability and stimulus generalization. J Exp Psychol 51:79–88
31. Horridge A (2009) Generalization in visual recognition by the honeybee (*Apis mellifera*): a review and explanation. J Insect Physiol 55:499–511
32. Hurvich LM, Jameson D (1957) An opponent-process theory of color vision. Psychol Rev 64:384–404
33. Lehrer M, Srinivasan MV, Zhang SW, Horridge GA (1988) Motion cues provide the bee's visual world with a third dimension. Nature 332:356–357
34. Menzel R (1969) Das Gedächtnis der Honigbiene für Spektralfarben. II. Umlernen und Mehrfachlernen [The honey bee's memory of spectral colours. II. Multiple reversal learning]. Z vergl Physiol 63:290–309
35. Menzel R (1999) Memory dynamics in the honeybee. J Comp Physiol A 185:323–340
36. Menzel R (2001) Searching for the memory trace in a mini-brain, the honeybee. Learn Mem 8:53–62
37. Nosofsky RM (1986) Attention, similarity, and the identification-categorization relationship. J Exp Psychol Gen 115:39–57
38. Prabhu C, Cheng K (2008) One day is all it takes: circadian modulation of the retrieval of colour memories in honeybees. Behav Ecol Sociobiol 63:11–22
39. Prabhu C, Cheng K (2008) Recency preference of odour memory retrieval in honeybees. Behav Ecol Sociobiol 63:22–32
40. Seeley TD (1985) Honeybee ecology: a study of adaptation in social life. Princeton University Press, Princeton
41. Shepard RN (1986) Discrimination and generalization in identification and classification: comment on Nosofsky. J Exp Psychol Gen 115:58–61
42. Shepard RN (1987) Toward a universal law of generalization for psychological science. Science 237:1317–1323
43. Shepard RN (2001) Perceptual-cognitive universals as reflections of the world. Behav Brain Sci 24:581–601
44. Srinivasan MV, Zhang SW, Bidwell NJ (1997) Visually mediated odometry in honeybees. J Exp Biol 200:2513–2522
45. Stach S, Benard J, Giurfa M (2004) Local-feature assembling in visual pattern recognition and generalization in honeybees. Nature 429:758–761
46. Tinbergen N (1963) On aims and methods in ethology. Z Tierpsychol 20:410–433
47. Weber EH (1846/1948) The sense of touch and common feeling. In: Dennis W (ed) Readings in the history of psychology. Appleton-Century-Crofts, East Norwalk, pp 194–196

Chapter 6.6
Visual Cognition in Honey Bees: From Elemental Visual Learning to Non-elemental Problem Solving

Martin Giurfa

Abstract Visual learning admits different levels of complexity, from the formation of a simple associative link between a visual stimulus and its outcome, to more sophisticated performances, such as object categorization or rules learning. Not surprisingly, higher-order forms of visual learning have been studied primarily in vertebrates with larger brains, while simple visual learning has been the focus in animals with small brains such as insects. This dichotomy has recently changed as studies on visual learning in free-flying honey bees have shown that these animals can master extremely sophisticated tasks. Here we review a spectrum of visual learning forms in honey bees, from color and pattern learning, visual attention, and top-down image recognition, to category learning, and rule extraction. We discuss the necessity and sufficiency of simple associations to account for complex visual learning in honey bees. We maintain that progresses in understanding the neural bases of visual cognition will be possible through novel protocols – unavailable until now – combining visual performances and simultaneous access to the nervous system.

Abbreviations

DMTS   Delayed matching to sample
DNMTS   Delayed non-matching to sample

M. Giurfa (✉)
Centre de Recherches sur la Cognition Animale, Université de Toulouse; UPS,
118 route de Narbonne, F-31062 Toulouse Cedex 9, France

Centre de Recherches sur la Cognition Animale, CNRS, 118 route de Narbonne,
F-31062 Toulouse Cedex 9, France
e-mail: giurfa@cict.fr

C.G. Galizia et al. (eds.), *Honeybee Neurobiology and Behavior: A Tribute to Randolf Menzel*, DOI 10.1007/978-94-007-2099-2_35,
© Springer Science+Business Media B.V. 2012

6.6.1 Introduction

Visual learning refers to the capacity of acquiring experience-based information pertaining to visual stimuli so that adaptive responses can be produced when viewing such stimuli again. This capacity, which is present in almost all living animals with a functional visual system, intervenes in contexts as diverse as food search, partner recognition, navigation and orientation and defense against potential enemies. It admits different levels of complexity as it may vary from a simple associative link connecting a visual target (e.g. a specific visual pattern) and its outcome (e.g. a reward or a punishment following visual target presentation) to learning abstract rules such as "larger than", "on top of" or "inside of", which allow responding to novel stimuli and classifying them as fulfilling or not the learned rule.

The former situation, i.e. the establishment of univocal, unambiguous links between a visual target and its outcome, constitutes a case of elemental learning (Table 6.6.1). What is learned for a color or a pattern is valid only for that color or that pattern and not for different ones. The latter situation, the learning of relational rules, constitutes a case of non-elemental learning as the individual's response cannot be accounted for by simple links between two stimuli (in the case of Pavlovian conditioning) or between a stimulus and a response (in the case of operant conditioning) (Table 6.6.1). Individuals using rules can transfer their choice to novel stimuli which they have never experienced. The subject's response is, therefore, not based on the particular outcome of a visual target but is flexible enough to generate novel responses that are not purely based on the physical properties of the visual stimuli considered.

Table 6.6.1 Examples of elemental (left) and non-elemental (right) learning paradigms

Elemental learning paradigms	Non-elemental learning paradigms
Absolute conditioning	Negative patterning
A+	A+, B+, AB−
Differential conditioning	Biconditional discrimination
A+, B−	AB+, CD+, AC−, BD−
Feature positive discrimination	Feature neutral discrimination
B−, AB+	AC+, C−, AB−, B+

Simple links between a stimulus (A or B) and reinforcement (+) (or its absence: −) allow solving the three elemental problems on the left: in absolute conditioning, the subject has to learn to respond to A, which is unambiguously associated with reinforcement; in differential conditioning, the subject has to learn to respond to A and not to B; A is unambiguously associated with reinforcement and B is unambiguously associated with the absence of reinforcement; in feature-positive discrimination, the subject has to learn to respond to the compound AB and not to B; although B is ambiguous as it appears twice, once reinforced and once non-reinforced, a simple link between A and reinforcement allows solving the problem. Simple links between a stimulus and reinforcement do not allow solving the three non-elemental problems on the right as in all cases each stimulus appears as often rewarded as non-rewarded. In negative patterning, the subject has to learn to respond to the single stimuli A and B but not to their compound AB; in biconditional discrimination, the subject has to learn to respond to the compounds AB and CD and not to the compounds AC and BD; in feature neutral discrimination, the subject has to learn to respond to B and to the compound AC but not to C and the compound AB

Honey bees are interesting models for the study of visual learning because in a natural context they have to solve a diversity of visual problems of varying complexity. They learn to navigate between the central place of the hive and flower patches (their food sources), and memorize the local cues characterizing these sites [25, 27]; (see also Chap. 2.5). In the case of flowers, learning and memory allow recognizing a profitable flower species that is currently exploited. This capacity is the basis of floral constancy, a behavior exhibited by honey bees, which consists in foraging on a unique floral species as long as it offers profitable nectar and/or pollen reward [5, 17]. Recognition of the species under exploitation is mediated by different sensory cues among which visual ones play a fundamental role [12, 32]. Learning and memorizing visual cues in such appetitive context can be studied in controlled experimental conditions first established by Karl von Frisch [38].

6.6.2 Visual Learning in Honey Bees – A Brief History

Karl von Frisch [38] marked individually free-flying bees using a color-spot code in order to recognize them during training and testing procedures. By indentifying single individuals he aimed at controlling their color experience and demonstrating that they have the capacity of seeing colors, contrarily to what was explicitly postulated by Carl von Hess at the beginning of the last century [39], who made the mistake of considering phototactic responses –which are exclusively mediated by light intensity– as a proof of a general color blindness in bees. Von Frisch paired various color cardboards with sucrose solution and asked, in each case, whether the rewarded color could be discriminated from different achromatic cardboards, some of which shared the same achromatic intensity with the trained color. In the test situation bees had therefore to choose between the learned color and various achromatic cardboards of varying intensity, all without reinforcement. The result is meanwhile well known: bees always chose the color associated with sucrose and never confused that color with an achromatic cardboard. Based on this experiment, von Frisch could show that bees see colors and determined that their visual spectrum spans from 300 (ultraviolet or UV) to 650 nm (orange-red) [38]. At the same time, he provided, without explicitly willing it, one of the first controlled examples of associative learning in bees as his experiment relied on acquisition of the color-sucrose association and testing subsequent discrimination in extinction conditions (no reward provided).

Several years had to pass until a comparable protocol was used to characterize the learning process of color signals *per se*. This was achieved by Randolf Menzel's PhD, who focused on color learning in honey bees and quantified for the first time acquisition and retention of different wavelengths [23, 24]. In this study, individually marked free-flying honey bees were trained with rewarded monochromatic lights, one at a time, and were then tested with the wavelength previously rewarded versus an alternative wavelength. This study showed that, under these experimental conditions, bees learned all wavelengths after few learning trials. Some wavelengths, particularly 413 nm, were learned faster than others, requiring one to three acquisition trials [23]. This result argued in favor of innate biases in color learning, probably reflecting the intrinsic biological relevance of the color signals that are learned faster [25]. Indeed, color-naive honey bees in their first foraging flight prefer those colors that experienced bees learn faster [13] and that label flowers that tend to be highly associated with a profitable nectar reward [13].

Further experiments by Menzel [24] allowed determining that one learning trial leads to a memory trace that fades a few days after learning if the animal is not allowed to learn anything else during this time, while three learning trials lead to a life-long color memory. This was the basis for discovering the existence of different memory phases in honey bees, some of which are short-term memories susceptible to interferences from additional color trials while others are mid-term memories, and long-term memories which are resistant to such interferences [26] (see also Chaps. 6.2 and 6.3).

At the time at which Menzel characterized honey bee color learning [23, 24] (see above) studies on pattern perception by bees were simultaneously performed by Rüdiger Wehner [40, 41]. Wehner did not focus on pattern learning but on the perceptual capabilities of bees confronted with pattern discrimination tasks. Certainly, visual conditioning to patterns was also used in previous and later works on pattern perception (reviews: [22, 31]) but neither pattern memory nor acquisition curves were quantified in these works. This tradition was continued until the 1990s as visual learning was mainly used as a tool to answer questions on visual perception and discrimination close to the goal. These works focused on visual capabilities like visual spatial resolution, shape discrimination, orientation detection, movement perception and parallax, among others, and were not concerned with learning itself.

Thus, during almost three decades, starting with the work on color learning and retention by Randolf Menzel [23], most studies focused on the sensory and central aspects of visual perception in honey bees but only few of them concentrated on learning capabilities *per se*. Visual learning capacities analyzed in this context were mostly elemental as specific visual cues (color, pattern, movement cues, etc.) were unambiguously associated with sucrose reward. Yet, in the last decade, researchers have produced evidence showing that bees are capable of non-elemental forms of learning (reviews: [9, 11, 32]). Here we will discuss new findings on honey bee visual cognition, which in the last years have changed our perspective about the visual abilities of honey bees, the cognitive potential of the miniature brain of insects and the uniqueness of vertebrates in terms of certain cognitive achievements.

6.6.3 Attentional and Experience-Dependent Modulation of Visual Learning

Visual learning, as studied in classical color conditioning experiments, is elemental as bees are just presented with a single color target paired with sucrose solution. It was supposed to be a fast form of learning [23] (see above), compared, for instance, to learning of visual patterns which usually takes longer (20 or more trials). It was long thought that what an animal sees and visually learns is constrained by its perceptual machinery with no or little room for experience-dependent modulations of perception. However, studies on honey bees [10] and on bumblebees [7] have shown that this idea is wrong. In some cases, learning one and the same color may need few trials only but in other cases it may take more than 20 trials. The critical feature to explain this difference is the nature of the learning process. For instance, *absolute conditioning*, in which a subject is trained with a single color rewarded with sugar water, in general yields fast learning. *Differential conditioning*, in which the same subject has to learn to discriminate a rewarded from a non-rewarded color, takes more trials, even if the rewarded color is the same as in absolute conditioning (see Chap. 4.5 and Table 6.6.1). When bees are asked to discriminate colors in a test, their performance differs dramatically. While bees trained in differential conditioning can discriminate colors that are very similar, bees trained in absolute conditioning cannot discriminate the same pair of colors [7, 10].

Comparable results were obtained in a study on pattern learning and discrimination by honey bees [15]. Bees trained to discriminate circular patterns differing in the spatial cues presented in the upper and lower halves behave differently depending on the training procedure. After absolute conditioning, they discriminate patterns using mainly the spatial cues available in the lower half of the patterns. However, after differential conditioning, they discriminate patterns using the cues available in the entire patterns. In other words, bees expanded or restricted their use of spatial cues for pattern recognition depending on the kind of learning. The difference in performance suggests, therefore, that attentional processes are involved. In differential conditioning the bee has to focus on the difference and not on the mere presence of a visual target, thus making learning slower.

Top-down processes affecting visual performances in honey bees have been shown in experiments on pattern recognition [45]. In this case, it was shown that the previous visual experience of a bee can speed up the analysis of the retinal image when a familiar object or scene is encountered. Zhang and Srinivasan [45] first attempted to train bees to distinguish between a ring and a disk when each shape was presented as a textured figure placed a few cm in front of a similarly textured background (Fig. 6.6.1a). The figures were, in principle, detectable through the relative motion that occurred at the figure borders, which were at a different distance than the background when bees fly towards the targets. Despite intensive training, the bees were incapable of learning the difference between the ring and the disk (Fig. 6.6.1a), a discrimination that posed no problems when the bees experienced these stimuli as plain (non-textured) shapes. Zhang and Srinivasan trained then a group of bees to this 'easy' problem which could be solved without problems (Fig. 6.6.1b). Bees were then confronted with the difficult problem of learning the textured disk vs. the ring and this time, they solved immediately the discrimination (Fig. 6.6.1c). Thus, pre-training with plain stimuli primed the pattern recognition system in such a way that it detected shapes that otherwise could not be distinguished. Again, it may be that such pre-training triggers attentional processes that allow better focusing on the targets that have to be discriminated.

This idea has been explicitly studied in honey bees trained to choose a colored disc ('target') among a varying number of differently colored discs ('distractors') [30]. Accuracy and decision time were measured as a function of distractor number and color. For all color combinations, decision time increased and accuracy decreased with increasing distractor

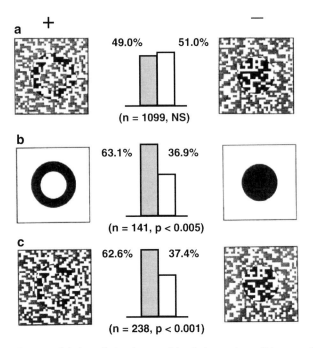

Fig. 6.6.1 Top-down modulation of visual recognition in honey bees. Prior experience enhances pattern discrimination in honey bees (Adapted from [45]). (**a**) Bees were trained in a dual-choice Y-maze to distinguish between a ring (+: rewarded) and a disk (-: non rewarded) when each shape was presented as a textured figure placed a few cm in front of a similarly textured background. Despite intensive training, the bees were incapable of learning the difference between a ring and a disk (n: number of choices; the percentages correspond to the choice of stimuli presented). (**b**) When these stimuli were presented as plain (non-textured) shapes, few cm in front of a *white background*, the bees could, as expected, easily learn the task. (**c**) They were then confronted with the difficult problem of learning the textured disk versus the ring and this time, they solved the discrimination. Pre-training with the plain stimuli may trigger attentional processes that allow better focusing on the targets whose discrimination is difficult

number, whereas performance increased when more targets were present. These findings are characteristic of a serial search in primates, when stimuli are examined sequentially, thus indicating that at the behavioral level, the strategies implemented by bees converge with those of animals in which attention is commonly studied [30].

6.6.4 Categorization of Visual Stimuli

A higher level of complexity is reached when animals respond in an adaptive manner to novel stimuli *that they have never encountered before and that do not predict a specific outcome per se based on the animals' past experience*. Such a positive transfer of learning [29] is therefore different from elemental forms of learning, which link known stimuli or actions to specific rewards (or punishments).

Positive transfer of learning is a distinctive characteristic of categorization. Visual categorization refers to the classification of visual stimuli into defined functional groups [18]. It can be defined as the ability to group distinguishable objects or events on the basis of a common feature or set of features, and therefore to respond similarly to them [43]. A typical categorization experiment trains an animal to extract the basic attributes of a category and then tests it with novel stimuli that were never encountered before and that may or may not present the attributes of the category learned. If the animal chooses the novel stimuli based on these attributes it classifies them as belonging to the category and therefore exhibits positive transfer of learning.

According to several recent studies, free-flying honey bees are indeed able to categorize different patterns and shapes based on specific visual features. For instance, van Hateren et al. [36] trained bees to discriminate two given gratings presented vertically and oriented differently (e.g. 45° versus 135°) by rewarding the choice of only one of these gratings with sucrose solution. Each bee was trained with a changing succession of pairs of different gratings, one of which was always rewarded and the other not. Despite the difference in pattern quality, all the rewarded patterns had the same edge orientation and all the non- rewarded patterns had a common orientation as well (perpendicular to the rewarded one). Under these circumstances, the bees had to extract and learn the orientation common to all rewarded patterns to solve the task. This was the only cue predicting reward delivery. In the tests, bees were presented with novel patterns, which they had never been exposed to before. These patterns were all non-rewarded but had the same stripe orientations as the rewarding and non-rewarding patterns employed during the training. In such transfer tests, bees chose the appropriate orientation despite the novelty of the structural details of the stimuli. Thus, bees could categorize visual stimuli on the basis of their global orientation.

Bees can also categorize visual patterns based on their bilateral symmetry. When trained with a succession of changing patterns to discriminate bilateral symmetry from asymmetry, they learn to extract this information from very different figures and indeed transfer it to novel symmetrical and asymmetrical patterns [14]. Similar conclusions apply to other visual features such as radial symmetry, concentric pattern organization and pattern disruption [3] and even photographs belonging to a given class (e.g. radial flower, landscape, plant stem) [46].

How could bees classify different photographs of radial flowers appropriately [46] if these vary in color, size, outline, etc.? An explanation was provided by Stach et al. [33] who showed that different coexisting orientations can be considered at a time, and can be integrated into a global stimulus representation that is the basis for the category [33]. Honey bees trained with a series of complex patterns sharing a common layout comprising four edge orientations remembered these orientations simultaneously in their appropriate positions, and transferred their response to novel stimuli that preserved the trained layout (Fig. 6.6.2). Thus, bees extract regularities in their visual environment and establish correspondences among correlated features such that they generate a large set of object descriptions from a finite set of elements. The same strategy explains that honey bees learn to recognize visual stimuli that are otherwise fully artificial to them, namely human faces [8]. Using controlled face-like stimuli (two dots in the upper part as the eyes, a vertical line below as the nose, and a horizontal line in the lower part as the mouth), Avargués-Weber et al. [1] showed that bees distinguish between different variants of the face-like stimuli, and that they grouped faces together if trained to do so.

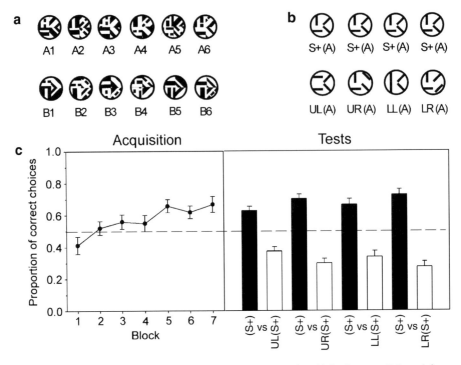

Fig. 6.6.2 Categorization of visual patterns based on sets of multiple features (Adapted from [33]). (**a**) Training stimuli. Bees were trained to discriminate A from B patterns during a random succession of A versus B patterns. A patterns (A1 to A6) differed from each other but shared a common layout of orientations in the four quadrants. B patterns (B1 to B6) shared a common layout perpendicular to that of A patterns. (**b**) Test stimuli used to determine whether bees extract the simplified layout of four bars from the rewarded A patterns. S+, simplified layout of the rewarded A patterns; *UL*, upper-left bar rotated; *UR*, upper-right bar rotated; *LL*, lower-left bar rotated; *LR*, lower-right bar rotated. (**c**) *Left panel*: acquisition curve showing the pooled performance of bees rewarded on A and B patterns. The proportion of correct choices along seven blocks of six consecutive visits is shown. Bees learned to discriminate the rewarding patterns (A or B) and improved significantly their correct choices along training. *Right panel:* proportion of correct choices in the tests with the novel patterns. Bees always preferred the simplified layout of the training patterns previously rewarded (S+) to any variant in which one bar was rotated

6.6.5 Rule Learning in Honey Bees

In rule learning, positive transfer occurs independently of the physical nature of the stimuli considered. The animal learns relations between objects and not the objects themselves. Typical examples are the so-called rules of *sameness* and *difference*. They are demonstrated through the protocols of delayed matching to sample (DMTS) and delayed non-matching to sample (DNMTS), respectively. In DMTS, animals are presented with a sample and then with a set of stimuli, one of which is identical to the sample. Choice of this stimulus is rewarded while choice of the different stimuli is not. Since the sample is regularly changed, animals must learn the sameness rule, i.e. '*always choose what is shown to you (the sample), independent of what else is shown to you'*. In DNMTS, the animal has to learn the opposite, i.e. *'always choose the opposite of what is shown to you (the sample)'*. Honey bees foraging in a Y-maze learn both rules [16]. They were trained in a DMTS experiment in which they were presented with a changing non-rewarded sample (i.e. one of two different color disks or one of two different black-and-white gratings, vertical or horizontal) at the entrance of a maze (Fig. 6.6.3). The bees were rewarded only if they chose the stimulus identical to the sample once within the maze. Bees trained with colors and presented in transfer tests with black-and-white gratings that they had not experienced before solved the problem and chose the grating identical to the sample at the entrance of the maze. Similarly, bees trained with the gratings and tested with colors in transfer tests also solved the problem and chose the novel color corresponding to that of the sample grating at the maze entrance. Transfer was not limited to different types of visual stimuli (pattern vs. color), but could also operate between drastically different sensory modalities such as olfaction and vision [16]. Bees also mastered a DNMTS task, thus showing that they learn a rule of difference between stimuli as well. These results document that bees learn rules relating stimuli in their environment.

The capacity of honey bees to solve a DMTS task has recently been verified and studied with respect to the working memory underlying it [47]. It was found that the working memory for the sample underlying the solving of DMTS lasts for approximately 5 s [47]. This time length coincides with the duration of other visual and olfactory short-term memories characterized in simpler forms of associative learning in honey bees [26]. Moreover, bees trained in a DMTS task can learn to pay attention to one of two different samples presented successively in a flight tunnel (either to the first or to the second) and can transfer the learnt relevance of the sequence to novel samples [47].

A recent study determined furthermore that honey bees learn an above/below relationship between visual stimuli and transfer it to novel stimuli that are perceptually different from those used during the training [2]. In other words bees had to learn that reward was provided whenever visual stimuli of variable nature were above (or below, depending on the group) a defined reference (i.e. a horizontal line or another constant stimulus). Bees learned the task using a conceptual above/below relationship between stimuli, irrespectively of their physical nature. They transferred their choice to novel instances of the trained concept in spite of variations in the distance separating the reference and the target, the spatial location within the visual field, the fact that targets were variable and randomized during training and that novel stimuli were introduced in the tests. None of these manipulations affected the performance of the bees, which learned to choose stimuli based on an abstract above/below relationship [2].

Thus, learning of relational rules such as 'same' or 'different', 'above of' or 'below of', is mastered by honey bees. This kind of problem solving is usually considered as a form of conceptual cognition as it is not based on perceptual similarity but relies on relational rules linking different instances together [43, 44]. The fact that honey bees, with their miniature brains, are capable of such performances raises a fundamental question for which so far answers are scarce: which

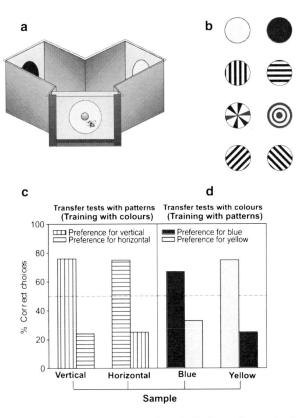

Fig. 6.6.3 Rule learning in honey bees (Adapted from [16]). Honey bees trained to collect sugar solution in a Y-maze (**a**) on a series of different patterns (**b**) learn a rule of sameness. Learning and transfer performance of bees in a delayed matching-to-sample task in which they were trained to colors (Experiment 1) or to *black-and-white*, vertical and horizontal gratings (Experiment 2). (**c, d**) Transfer tests with novel stimuli. (**c**) In Experiment 1, bees trained on the colors were tested on the gratings. (**d**) In Experiment 2, bees trained on the gratings were tested on the colors. In both cases bees chose the novel stimuli corresponding to the sample although they had no experience with such test stimuli

are the mechanisms underlying this level of cognitive processing? Which nervous architecture allows extracting conceptual rules from a learning task?

6.6.6   The Search for the Neural Bases of Visual Cognition in Honey Bees: Overcoming Technical Limitations

Despite the fascination that some of the results mentioned above may produce, there is a technical burden that limits the expansion of research on honey bee visual cognition towards the study of its underlying neural mechanisms. Using free-flying honey bees for studies on visual learning is important because in the experimental conditions described in this chapter, bees reveal the full potential of their cognitive resources when facing problems in the visual domain. However, at the same time, the fact that these bees freely move between the hive and the experimental site precludes the access to the neural bases of visual learning and retention. If honey bees have been so successful for the cellular and molecular study of learning and memory during the last decades [11, 26], it was precisely because researchers were able to translate the basic behavioral components of *olfactory learning* from the field to the laboratory by establishing a conditioning protocol in which harnessed bees exhibit all the features of *olfactory learning* and retention [4, 26, 34]: the olfactory conditioning of the proboscis extension reflex (see Chap. 4.1). In this Pavlovian framework, bees learn to associate an odorant (the conditioned stimulus) and sucrose reward (the unconditioned stimulus) (see also Chap. 6.2). This olfactory conditioning can be combined with a variety of invasive techniques allowing to record neural activity in the bee brain during olfactory acquisition and retention [26] (see also Chap. 6.1).

An equivalent progress has not yet been possible in studies on visual learning and memory in bees, even in their simplest form. Interestingly, the first work on PER conditioning was published by Matsutaro Kuwabara in 1957 [21], years before the olfactory PER conditioning was established in its first version by one of his students, Kimihisa Takeda [34], and used color stimuli as conditioned stimuli associated with sucrose solution delivered to the tarsi of harnessed bees. This work, which showed acquisition for color stimuli [21], could not be reproduced during many years despite many efforts in that direction until recently Hori et al. [19, 20] realized that the critical procedure was to section the antennae, as originally done by Kuwabara [21]. Under these conditions, Hori et al. [19, 20] were able to train harnessed bees to extend its proboscis to colors [19] or motion cues [20] paired with sucrose solution delivered directly to the proboscis. Yet, this protocol has not reached the efficiency and reliability of olfactory PER conditioning. The damage inflicted to the bees by cutting their antennae affects their appetitive motivation [6], thus resulting in poor learning performances. Improving this protocol or, even better, conceiving newer ones in order to combine behavioral and neurobiological access to visual performances is a priority for future research on honey bee visual cognition.

Key structures of the bee brain, the mushroom bodies, should attract the attention of researchers interested by the neural bases of cognitive processing. Mushroom bodies are multimodal structures, which receive input from visual, olfactory, mechanosensory and gustatory pathways. Output neurons are multimodal thus suggesting that cross-talk and information exchange, necessary to higher forms of cognition, could take place within these structures. In the fruit fly *Drosophila melanogaster*, mushroom bodies are required for choice behavior in facing contradictory visual cues [35] and have been related with visual attentional processes [37] despite not having an obvious visual input. They mediate assessment of the relative saliency of conflicting visual cues [35, 42] and are also involved in improving the extraction of visual cues after pre-training in *Drosophila* [28]. The mushroom bodies of honey bees, which receive clear visual input, may play a similar role, favoring attention processes and better problem solving and discrimination in the visual domain.

6.6.7   Conclusion and Overlook

Since Randolf Menzel's first studies on color learning in honey bees we have gained an impressive amount of information about how honey bees see the world and learn about visual cues in their environment. New discoveries in this field have shown that besides simple forms of visual learning that can be easily conceived in the life of a bee (e.g. flower color – reward or pattern – reward association), higher-order forms of visual learning going from conditional discriminations and observational learning to rule learning can also be mastered. Visual learning capabilities allow therefore extracting the logical structure of the world and attain different levels of complexity.

Determining which neural circuitry allows different forms of visual learning is an ineludible task for the immediate future. The challenge is not only to record neural activity in the visual centers of the honey bee brain, which has been done on the basis of a single-neuron approach, but to combine behavioral protocols for the study of visual learning in harnessed bees in the laboratory with this and other kinds of neural recordings. Populational measures of neural activity in visual centers of the bee brain such as the lamina, medulla, lobula or optic tuberculum, among others, are necessary to understand whether there is a specific populational code for visual stimuli and whether it changes with different forms of visual experience. Future research should answer these questions and overcome the historic burden of not having a window open to the neural and molecular basis of visual learning, irrespective of the level of complexity considered.

References

1. Avarguès-Weber A, Portelli G, Benard J, Dyer A, Giurfa M (2010) Configural processing enables discrimination and categorization of face-like stimuli in honeybees. J Exp Biol 213(4):593–601
2. Avarguès-Weber A, Dyer A, Giurfa M (2011) Conceptualization of above and below relationships by an insect. Proc R Soc B 278(1707):898–905
3. Benard J, Stach S, Giurfa M (2006) Categorization of visual stimuli in the honeybee *Apis mellifera*. Anim Cogn 9(4):257–270
4. Bitterman ME, Menzel R, Fietz A, Schäfer S (1983) Classical conditioning of proboscis extension in honeybees (*Apis mellifera*). J Comp Psychol 97(2):107–119
5. Chittka L, Thomson JD, Waser NM (1999) Flower constancy, insect psychology, and plant evolution. Naturwissenschaften 86:361–377
6. de Brito Sanchez MG, Chen C, Li J, Liu F, Gauthier M et al (2008) Behavioral studies on tarsal gustation in honeybees: sucrose responsiveness and sucrose-mediated olfactory conditioning. J Comp Physiol A 194(10):861–869

7. Dyer AG, Chittka L (2004) Fine colour discrimination requires differential conditioning in bumblebees. Naturwissenschaften 91(5):224–227
8. Dyer AG, Neumeyer C, Chittka L (2005) Honeybee (*Apis mellifera*) vision can discriminate between and recognise images of human faces. J Exp Biol 208(24):4709–4714
9. Giurfa M (2003) Cognitive neuroethology: dissecting non-elemental learning in a honeybee brain. Curr Opin Neurobiol 13(6):726–735
10. Giurfa M (2004) Conditioning procedure and color discrimination in the honeybee *Apis mellifera*. Naturwissenschaften 91(5):228–231
11. Giurfa M (2007) Behavioral and neural analysis of associative learning in the honeybee: a taste from the magic well. J Comp Physiol A 193(8):801–824
12. Giurfa M, Menzel R (1997) Insect visual perception: complex abilities of simple nervous systems. Curr Opin Neurobiol 7:505–513
13. Giurfa M, Núñez JA, Chittka L, Menzel R (1995) Colour preferences of flower-naive honeybees. J Comp Physiol A 177:247–259
14. Giurfa M, Eichmann B, Menzel R (1996) Symmetry perception in an insect. Nature 382:458–461
15. Giurfa M, Hammer M, Stach S, Stollhoff N, Muller-deisig N et al (1999) Pattern learning by honeybees: conditioning procedure and recognition strategy. Anim Behav 57(2):315–324
16. Giurfa M, Zhang S, Jenett A, Menzel R, Srinivasan MV (2001) The concepts of 'sameness' and 'difference' in an insect. Nature 410:930–933
17. Grant V (1951) The fertilization of flowers. Sci Am 12:1–6
18. Harnard S (1987) Categorical perception: the groundwork of cognition. Cambridge University Press, Cambridge
19. Hori S, Takeuchi H, Arikawa K, Kinoshita M, Ichikawa N et al (2006) Associative visual learning, color discrimination, and chromatic adaptation in the harnessed honeybee *Apis mellifera* L. J Comp Physiol A 192(7):691–700
20. Hori S, Takeuchi H, Kubo T (2007) Associative learning and discrimination of motion cues in the harnessed honeybee *Apis mellifera* L. J Comp Physiol A 193(8):825–833
21. Kuwabara M (1957) Bildung des bedingten Reflexes von Pavlovs Typus bei der Honigbiene, *Apis mellifica*. J Fac Sci Hokkaido Univ Ser VI Zool 13:458–464
22. Lehrer M (1997) Honeybees' visual spatial orientation at the feeding site. In: Lehrer M (ed) Detection and communication in arthropods. Birkhäuser, Basel, pp 115–144
23. Menzel R (1967) Untersuchungen zum Erlernen von Spektralfarben durch die Honigbiene (*Apis mellifica*). Z vergl Physiol 56:22–62
24. Menzel R (1968) Das Gedächtnis der Honigbiene für Spektralfarben. I. Kurzzeitiges und langzeitiges Behalten. Z vergl Physiol 60:82–102
25. Menzel R (1985) Learning in honey bees in an ecological and behavioral context. In: Hölldobler B, Lindauer M (eds) Experimental behavioral ecology and sociobiology. Gustav Fischer, Stuttgart, pp 55–74
26. Menzel R (1999) Memory dynamics in the honeybee. J Comp Physiol A 185:323–340
27. Menzel R, Greggers U, Hammer M (1993) Functional organization of appetitive learning and memory in a generalist pollinator, the honey bee. In: Papaj DR, Lewis AC (eds) Insect learning: ecology and evolutionary perspectives. Chapman & Hall, New York, pp 79–125
28. Peng YQ, Xi W, Zhang W, Zhang K, Guo AK (2007) Experience improves feature extraction in *Drosophila*. J Neurosci 27(19):5139–5145
29. Robertson SI (2001) Problem solving. Psychology Press, Hove
30. Spaethe J, Tautz J, Chittka L (2006) Do honeybees detect colour targets using serial or parallel visual search? J Exp Biol 209(6):987–993
31. Srinivasan MV (1994) Pattern recognition in the honeybee: recent progress. J Insect Physiol 40(3):183–194
32. Srinivasan MV (2010) Honey bees as a model for vision, perception, and cognition. Annu Rev Entomol 55:267–284
33. Stach S, Benard J, Giurfa M (2004) Local-feature assembling in visual pattern recognition and generalization in honeybees. Nature 429(6993):758–761

34. Takeda K (1961) Classical conditioned response in the honey bee. J Insect Physiol 6:168–179
35. Tang YP (2001) Choice behavior of *Drosophila* facing contradictory visual cues. Science 294(5546):1543–1547
36. van Hateren JH, Srinivasan MV, Wait PB (1990) Pattern recognition in bees: orientation discrimination. J Comp Physiol A 167:649–654
37. van Swinderen B, Greenspan RJ (2003) Salience modulates 20–30 Hz brain activity in *Drosophila*. Nat Neurosci 6(6):579–586
38. von Frisch K (1914) Der Farbensinn und Formensinn der Biene. Z Jb Abt allg Zool Physiol 35:1–188
39. von Hess C (1913) Experimentelle Untersuchungen über den angeblichen Farbensinn der Bienen. Z Jb Abt allg Zool Physiol 34:81–106
40. Wehner R (1967) Pattern recognition in bees. Nature 215:1244–1248
41. Wehner R (1971) The generalization of directional visual stimuli in the honey bee, *Apis mellifera*. J Insect Physiol 17:1579–1591
42. Xi W, Peng YQ, Guo JZ, Ye YZ, Zhang K et al (2008) Mushroom bodies modulate salience-based selective fixation behavior in *Drosophila*. Eur J Neurosci 27(6):1441–1451
43. Zentall TR, Galizio M, Critchfield TS (2002) Categorization, concept learning, and behavior analysis: an introduction. J Exp Anal Behav 78(3):237–248
44. Zentall TR, Wasserman EA, Lazareva OF, Thompson RKR, Rattermann MJ (2008) Concept learning in animals. Comp Cogn Behav Rev 3:13–45
45. Zhang SW, Srinivasan MV (1994) Prior experience enhances pattern discrimination in insect vision. Nature 368:330–332
46. Zhang SW, Srinivasan MV, Zhu H, Wong J (2004) Grouping of visual objects by honeybees. J Exp Biol 207(19):3289–3298
47. Zhang SW, Bock F, Si A, Tautz J, Srinivasan MV (2005) Visual working memory in decision making by honey bees. Proc Natl Acad Sci USA 102(14):5250–5255

Chapter 6.7
Learning and Memory: Commentary

Randolf Menzel

Many of the studies reported in this volume use the olfactory conditioning of the proboscis extension response of confined bees (PER paradigm). We cite Masutaro Kuwabara and acknowledge his discovery [12]. It may be interesting to remember how Kuwabara, a post doc in Karl von Frisch's lab in the 1950s of the last century, noticed that bees can be classically conditioned using color stimuli. Von Frisch (1965) reports in his book (p. 533) that the student Kantner measured an extremely high sensitivity of bees for sucrose (0.0034% = 1/10.000 mol). Von Frisch asked him for more and more control experiments, and finally the student gave up. Kuwabara told me in 1980 at a conference in Kyoto that he observed students repeating these experiments from 1934 and noticed that they were feeding the bee from time to time. He realized that the bee started to respond with the proboscis extension already before the antennae were touched with the test solution. At this time Kuwabara worked on hygroreceptors on the bee antennae [13] and studied the spatial range of antennal probing [11]. He deduced that the animal may have formed an association between the water vapor or visual stimuli and the sucrose reward. In an attempt to eliminate the interfering effects via the antennae and to demonstrate visual reward conditioning he cut off the antennae in his first report and conditioned the bee to visual stimuli. It is interesting to note that von Frisch did not cite Kuwabara's conditioning paper indicating that he may have been skeptical about the learning effect of harnessed bees.

I discovered the PER paradigm during a visit to the Max Planck Institute in Seewiesen in 1969 where I watched a graduate student, Ekkehard Vareschi in Dietrich Schneider's lab training bees for odor discrimination [20]. At that time I had just established my own small research group at the Technical University of Darmstadt with Jochen Erber as a graduate student and Thomas Masuhr as a Diploma and later graduate student. I was searching for a preparation which would allow us to address the question whether the mushroom bodies are involved in

R. Menzel (✉)
Institut für Biologie, Neurobiologie, Freie Universität Berlin, Berlin, Germany
e-mail: menzel@neurobiologie.fu-berlin.de

C.G. Galizia et al. (eds.), *Honeybee Neurobiology and Behavior: A Tribute to Randolf Menzel*, DOI 10.1007/978-94-007-2099-2_36,
© Springer Science+Business Media B.V. 2012

learning. We knew already that bees have short- and long-term memory. I mentioned in my seminar talk in Seewiesen that what I have just seen in Professor Schneider's lab will hopefully help us to address the question where the transition from short- to long-term memory may occur in the bee brain. Thomas Masuhr discovered soon afterwards how robust the PER preparation is. He exposed the brain to cooled needles after removing the cuticle and the air sacks proving that the mushroom bodies are essential structures for the consolidation of an early form of olfactory memory [15]. There are only a few other experiments performed in my lab over the years that I consider equally important. The kind reader may notice from the way we published this work in a low impact factor journal how little I understood about research publication at that time.

A topic coming up frequently in this book relates to the functions of the mushroom body (MB). A computational model as discussed in Chap. 6.1 may indeed help us to compile data, clarify the logic of the arguments and to consider potential neural implementations. The persuasiveness of the model will be enhanced if it captures as much existing data and avoids being inconsistent with data. The proposed model for learning related plasticity in the MB captures some of the data (e.g. the overall anatomy at the input and output sides, sparse coding in Kenyon cells) but it is inconsistent with a range of other data. Solving these inconstancies will help to improve the modeling approach. For example, the reinforcing signal via VUMmx1 targets Kenyon cells at their input sites (calyx) and not at their output sites (lobes), associative plasticity is assumed to depend on pre-postsynaptic coincidence of neural excitation (Hebb plasticity) only at the output side although associative plasticity was found in Kenyon cells at their input sides [19], MB extrinsic neurons do not serve as premotor and do not mimic PER, and MB extrinsic neurons do not fall into two groups (proboscis extension and retraction or non-extension related neurons) [6, 17]. Furthermore, the most consistent result on both Kenyon cells and MB extrinsic neurons is not captured by the model, that is their non-associative response reduction to stimulus repetition. I believe that what a model of the MB should provide us with is an idea why the olfactory pathway including sparse coding appears to represent odors in such a multitude of parallel neurons, the Kenyon cells, and then looses all this information in a small number of output neurons without any odor specificity. Obviously the MB is a recoding device, a system that codes at its input sides sensory information and their combinations highly specifically, and at its output sides behaviorally relevant categories, like novelty, value, salience, learned, innate, memory processed, attention attracting and directing. To this end, different forms of non-associative and associative plasticity may act concurrently at the input and the output sides coordinated by inhibitory (and possibly also excitatory) feedback neurons, the PCT neurons, which change their odor driven activation patterns with learning [5, 6], and which appear to control associative response reduction in the PE1 neuron [6, 17]. Such a view of the MB has not yet been captured in a model and constitutes a task for the future.

Ultimately we want to identify the content of the memory trace e.g. in the MB (see Commentary to Part 5), and incorporate it into a model that may have the structure of an associative net. We expect different patterns of changes for different contents, and we expect different locations of these patterns in different phases of memory consolidation. Such a formidable goal requires at the experimental front much more information about neural identity and connectivity and their plasticity during development and experience. A standard atlas of the living bee brain will therefore be an essential component in these endeavors. New methods will have to be developed in close collaboration between neuroanatomists, molecular biologists, biochemists and imaging researchers. Sidney Brenner, the famous molecular biologist, once said: "Progress in science depends on new techniques, new discoveries, and new ideas, probably in that order." Indeed methods limit progress in tracing the multiple steps of memory formation. We are fortunate to work with a highly cooperative animal, we established already a rich inventory of basic phenomena of a robust form of associative learning in this animal (the PER conditioning) and we know quite a bit about its brain and some of the molecular machinery related to memory formation. Now we need to combine all this knowledge with cellular anatomy. An instructive example is the imaging of cAMP dependent processes in the *Drosophila* brain [2], and with the knowledge reported in Chap. 6.2 we should have chances to visualize key component of the memory trace in the MB.

The process of building the memory trace is known as consolidation, and one form of consolidation includes protein synthesis for establishing a stable trace. A surprising result of our early studies on protein synthesis dependent memory consolidation was its late occurrence [21]. Dorothea Eisenhardt addresses the question of what happens to a consolidated memory trace when new learning interferes with it. Since any kind of learning changes the strength and content of existing memories it needs to be asked whether and how retrieving consolidated memories makes them again labile and subjected to a novel consolidation, the so-called reconsolidation process. Obviously this process depends on how closely the contents of the respective memory traces are related. The current model of a memory trace conceives it as a pattern of multiple locations of neural plasticity (e.g. the synaptic weight of specific neural connections) in an associative matrix. Such patterns are partially overlapping for related memory contents, and depending on this overlap new learning will affect more or less the existing memory trace. As long as memory is stored in short-term memory the ongoing learning process is thought to continuously update associative strength, for example by Ca^{2+} dependent processes leading to the induction of long-term depression (LTD) and long-term potentiation (LTP) (as for example formalized in [1] that provides a good model for these interactions). These ongoing processes create patterns of synaptic weights to be consolidated into a coherent memory trace that stores the accumulated experience during extended periods of acquisition. If, however, the memory trace is already consolidated and translated into structural changes the question arises of whether contradicting experience from extinction learning overwrites the old memory trace by making it labile again and thus transforming it into a labile stage before consolidation. Such a process would convert the consolidated memory trace into a form equivalent to that of short-term memory. The behavioral phenomenon of spontaneous recovery is difficult to reconcile by a reversion into short-term memory. In spontaneous recovery the initially learned and consolidated memory becomes stronger over time after extinction learning. Dorothea Eisenhardt has proven that spontaneous recovery after extinction learning is protein synthesis dependent leading us to propose a different model of "reconsolidation" based on the properties of the reward pathway (the internal reinforcement hypothesis, see below). Since this model is radically different from the usual interpretation of the reconsolidation phenomenon and does not need to assume a transition of the consolidated memory trace into a labile form it would be ideal if one could "see" the memory trace when extinction learning happens. In such a situation one might be able to confirm or reject the transition of consolidated memory and visualize the traces of competing memories. Olfactory learning in the PER paradigm may well offer such an exciting perspective.

The essence of the internal reinforcement hypothesis refers to the mechanism of reminder learning, a strengthening of an existing memory by new learning of the same content. We proposed that reminder learning can also occur during extinction learning, thus in the absence of an external reinforcing stimulus. The neuron VUMmx1 not only represents appetitive reinforcement but also learns about the CS+ [7, 14] leading to an activation of VUMmx1 in response to CS+ alone, a situation occurring in extinction learning. Thus whenever the CS+ occurs, the reward system will be activated even in absence of the physical reward providing an internal reinforcement signal. The internal reinforcement hypothesis offers an experimental approach to at least some components of the assumed interactions between memory traces. It will be necessary to manipulate the reinforcement circuit and test whether reconsolidation occurs even if this circuit is blocked. A pharmacological approach similar to that applied for initial learning appears feasible [16] but an optophysiological manipulation of the reward system like in *Drosophila* [18] would be much more elegant, however an appropriate transgenic bee would be required. In that case it will be possible to ask whether the activity of the respective reinforcing neurons is required for reconsolidation to occur. Furthermore, imaging of the CS related memory trace will allow addressing the question whether it is weakened during extinction learning and/or consolidation, altered or erased, and whether a trace for reminder memory can be found.

Operant learning is the dominant form of learning under natural conditions in bees and other animals, but its neural basis is not well understood. A requirement of operant learning is spontaneous activity. Patterns of spontaneous activity differ in different contexts suggesting that the animal uses innate and/or learned searching mechanisms. In that sense the term spontaneous activity is an inappropriate term because it implies random fluctuations of behavioral acts, conditions that usually do not exist. Rather, spontaneous behavior is better conceptualized as searching, probing, inspecting behavior, and all these forms of behavior are created by some former knowledge (innate or learned) about a goal. The different forms of antennal search movement as documented so elegantly by the Erber group provide us with a unique opportunity to ask at which goals these movements are directed, which components are genetically fixed and which are modulated by experience, and whether there is a hierarchy of salience of stimuli for probing behaviors. Where are these goal directed behaviors initiated and controlled in the bee brain? How do internal and external conditions select the appropriate behavior? The further analysis of antennal movements will allow us to identify neural processes leading to the initiation and selection of goal directed probing movements in a paradigmatic situation.

Are there universal laws in biology and more specifically in behavioral biology? This question is raised in Chap. 6.5, and I would like to express a polite skepticism. Any science is an enterprise that searches for rules transcending the particular study case. For deriving universal laws it is necessary to define the conditions of all parameters having an impact on the test conditions. This is possible under ideal conditions in physics but not in biology. The complexity of the living organism and the close to unlimited and mostly unknown parameters make any kind of search for universal laws in biology impossible. However, some limited form of generality across conditions and animals in the form of a general rule is worthwhile searching for. What would be an indication (not to say a proof) for an observation in behavioral biology to follow a general rule? Comparison across animal species is an often used approach. Since sensory, motor and cognitive facilities differ between species one needs to categorize and normalize the measures of performance. These procedures are not neutral to the outcome of such a comparison. Take the two examples presented in Chap. 6.5. I bet, if a different measure of spatial separation would have been used no exponential function would have been found, or if one would normalize the data such that e.g. the circadian rhythm effect of retention scores would have been eliminated, some form of temporal weighting rule may have been found (which was not found in the example described by Ken Cheng). The justification for selecting particular procedures for data extraction and adjusting behavioral measures are often not so obvious. Furthermore, demonstrating the fitting of a curve (e.g. an exponential decay function) without formulating alternative functions may have little diagnostic value. Even more importantly the extremely large parameter space in behavioral research makes such comparison rather limited. Take the example on generalization reported in Chap. 6.5. If the animal would have been trained differently the outcome would have been very different [9]. These arguments of caution do not mean that comparison should be avoided or formal descriptions of behavioral measures are not useful. I only question that "universal laws" can reasonably be searched for by across animal comparison. For me as a neurobiologist the motivation of a "honeybee-too" approach is rather limited. We hope to track behavioral patterns to neural processes which reflect basic mechanisms that can be better studied in an animal with an accessible brain and an intermediate level of neural complexity such as the honeybee. Such basic mechanisms may reflect evolutionary history or common environmental adaptations, but detecting such basic mechanisms will not indicate a "universal law".

While we all read wikipedia.org it may be appropriate to cite the first sentences under the heading "cognition" since the points made are rather useful in our context:

> The term **cognition** (Latin: cognoscere, "to know", "to conceptualize" or "to recognize") refers to a faculty for the processing of information, applying knowledge, and changing preferences. Cognition, or cognitive processes, can be natural or artificial, conscious or unconscious. ... Within psychology or philosophy, the concept of cognition is closely

related to abstract concepts such as mind, intelligence; cognition is used to refer to the
mental functions, mental processes (thoughts) and states of intelligent entities (humans,
human organizations, highly autonomous machines).

Since this convincing definition of cognition includes intelligent entities like
autonomous machines it is certainly not premature to equip insects with intelligent
cognition. In this sense any battle about cognition in the honeybee appears to be
superfluous and irrelevant. Martin Giurfa's work has certainly contributed strongly
to the change of thinking in this respect. In my view, however, the cognitive dimensions of animal minds including that of the bee reach beyond the demonstration of
non-elemental forms of learning like categorization, context dependence, attentional processing, configural perception and learning, and learning by observation.
Cognition as I understand it stresses the intrinsic organizing capacity of the brain.
This capacity is expressed best in faculties of active or working memory including
directed attention, expectation, decision making, planning, rule extraction, spatial
cognition, and communication. Focusing on these capacities of the brain requires a
shift from a learning perspective to a memory processing perspective. The difference between these two perspectives can be nicely illustrated for navigation (see our
chapter on navigation in Part 2). The *stimulus-response* (route based) concepts favor
the single system perspective and root in the tradition of learning theory. Multiple
separate associations between objects and motor routines are assumed. New behavioral acts like a short cut to a goal without experience along the novel path are
explained as attempts to sequentially reduce the mismatch between the current
image and the image at the goal. Such an assumption does not require any internal
processing other than the retrieval of images stored in memory. The *cognitive map*
concept on the other side equips animals with the ability to internally plan novel
routes on spatial representations in active memory. The structure of this spatial memory allows for flexible integration of multiple information. Animals are thought to
store associations between stimuli (not only between CSs and USs) such that they
are cross referenced within a common reference (called a cognitive map irrespective
of whether it has a geometric structure or not) that promotes planning of actions on
the basis of expectations emanating from these associations. Planning requires
selection of actions not from a pool of stochastic movements but rather from multiple goal directed and value based sensory-motor programs. At the heart of these
concepts lies the assumption of neural (internal) search processes which predict and
evaluate potential future outcomes, and all these processes reflect the current working
of the active memory.

Addressing the properties and functioning of active working memory requires
for each paradigm a careful evaluation of whether elemental forms of learning and
memory retrieval are sufficient for an explanation. The tradition of the most parsimonious explanation provides a strong tool in science in general and is well observed
in behavioral studies particularly of those in insects. But let me point out a few
caveats. (1) The richness of behavior appears in studies that allow the animal to
behave close to normal in its natural environment. Rigid designs of experiments
might be necessary to quantify behavior and make it reproducible but they run into
the danger that the animal in its restriction can only do what the experimenter
allowed it to do. The conclusion from such experiments is often that, since the

animal did what was expected from it, this is the only thing an animal can do. (2) Although scientific progress is bound to search for the most parsimonious explanation it is not obvious what may be more or less demanding for the brain, particularly a small brain as that of an insect. For example, will it be more difficult for a small brain to follow a route based navigation strategy or a cognitive map based strategy, store many images or extract a rule connecting these images? Are neural processes derived from behavioristic learning theory less demanding than those derived from cognitive concepts? We must answer at this stage that we do not know, and that the only way to find out lies in the search for neural mechanisms within a broader conceptual frame. (3) We need to acknowledge that potential behavioral acts that are not performed by an animal are equally important as expressed behavior. Only by accepting that an attentive brain is constantly producing potential behaviors, most of which are not expressed, and that it is this capacity which shapes brain function in evolution we shall be able to search for the neural basis of the "inner doing" as a prerequisite of decision making processes.

So how do we find out about the "inner doing" of the brain as the essential preparatory processes for evaluation, decision making and planning? The optimal approach would be to watch the brain in action during a working memory task, and manipulate it in such a way that the neural options provided by working memory processes are changed. First attempts on a rather elementary level have been successful in *Drosophila* for the evaluating networks (e.g. [18]). The next step could be to manipulate these evaluating systems when a particular memory trace is active in working memory, and determine whether the animal behaves accordingly. Such experiments will not be possible in any close future for the bee, and we have to wait for methodological advances that are urgently needed. So what could we do in between? In my view the following questions (and many more) may be approached by behavioral analyses and will help to characterize the cognitive dimensions. Is the memory content changed during the process of memory consolidation? Generalization, reversal learning, overlearning and context dependence could be suitable paradigms. Do bees develop an expectation about reward strength? Greggers and Menzel [4] found for bees foraging in a patch of 4 or 8 feeders that delivered different flow rates of sucrose solution that bees store the reward properties of these feeders in a transient and active memory. The capacity of working memory was determined as containing at least eight items. The time range of these specific working memories could be estimated as lying around 6 min. Which forms of memory consolidation are sensitive to sleep deprivation? Consolidation after extinction learning [8], after navigation learning (Bogusch, personal comunication, [10]) are examples. Is there an on-line measure of attention? Such a measure has been claimed for *Drosophila* [3]. Are bees rejecting information transmitted during dance communication that is inconsistent with their prior experience?

Und so schließen wir mit einer offenen Frage wie deren viele in diesem Buch stehen. Das macht nichts. Die offenen Fragen sind es, an denen sich die Jungen für die Wissenschaft begeistern. Es wäre doch schade, wenn sie eines Tages sagen müßten: Nun wissen wir alles und haben nichts mehr zu tun. Habt keine Sorge! Dieser Tag wird nicht kommen. Denn der Menschengeist ist begrenzt, aber die Wunder der lebenden Natur sind ohne Ende.(Zitat aus dem Buch von Karl von Frisch (1965, p. 534)).

References

1. Bienenstock EL, Cooper LN, Munro PW (1982) Theory for the development of neuron selectivity: orientation specificity and binocular interaction in visual cortex. J Neurosci 2(1):32–48
2. Gervasi N, Tchenio P, Preat T (2010) PKA dynamics in a *Drosophila* learning center: coincidence detection by rutabaga adenylyl cyclase and spatial regulation by dunce phosphodiesterase. Neuron 65(4):516–529
3. Greenspan RJ, van Swinderen B (2004) Cognitive consonance: complex brain functions in the fruit fly and its relatives. Trends Neurosci 27(12):707–711
4. Greggers U, Menzel R (1993) Memory dynamics and foraging strategies of honeybees. Behav Ecol Sociobiol 32(1):17–29
5. Grünewald B (1999) Physiological properties and response modulations of mushroom body feedback neurons during olfactory learning in the honeybee, *Apis mellifera*. J Comp Phys A 185:565–576
6. Haehnel M, Menzel R (2010) Sensory representation and learning-related plasticity in mushroom body extrinsic feedback neurons of the protocerebral tract. Front Syst Neurosci 4:161
7. Hammer M (1993) An identified neuron mediates the unconditioned stimulus in associative olfactory learning in honeybees. Nature 366(6450):59–63
8. Hussaini SA, Bogusch L, Landgraf T, Menzel R (2009) Sleep deprivation affects extinction but not acquisition memory in honeybees. Learn Mem 16(11):698–705
9. Kehoe EJ (2009) Discrimination and generalization. In: Byrne JH (ed) Learning and memory: a comprehensive reference. Elsevier, Amsterdam, pp 123–149
10. Klein BA, Klein A, Wray MK, Mueller UG, Seeley TD (2010) Sleep deprivation impairs precision of waggle dance signaling in honey bees. Proc Natl Acad Sci USA 107(52):22705–22709
11. Kuwabara M (1952) Über die Funktion der Antenne der Honigbiene in bezug auf die Raumorientierung. Mem Fac Sci 1:13–64
12. Kuwabara M (1957) Bildung des bedingten Reflexes von Pavlovs Typus bei der Honigbiene, *Apis mellifica*. J Fac Sci Hokkaido Univ Ser VI Zool 13:458–464
13. Kuwabara M, Takeda K (1956) On the hygroreceptor of the honeybee, *Apis mellifica*. Physiol Ecol 7:1–6
14. Menzel R, Giurfa M (2001) Cognitive architecture of a mini-brain: the honeybee. Trends Cogn Sci 5(2):62–71
15. Menzel R, Erber J, Masuhr T (1974) Learning and memory in the honeybee. In: Barton-Browne L (ed) Experimental analysis of insect behaviour. Springer, Berlin, pp 195–217
16. Menzel R, Heyne A, Kinzel C, Gerber B, Fiala A (1999) Pharmacological dissociation between the reinforcing, sensitizing, and response-releasing functions of reward in honeybee classical conditioning. Behav Neurosci 113(4):744–754
17. Okada R, Rybak J, Manz G, Menzel R (2007) Learning-related plasticity in PE1 and other mushroom body-extrinsic neurons in the honeybee brain. J Neurosci 27(43):11736–11747
18. Schroll C, Riemensperger T, Bucher D, Ehmer J, Voller T et al (2006) Light-induced activation of distinct modulatory neurons triggers appetitive or aversive learning in *Drosophila* larvae. Curr Biol 16(17):1741–1747
19. Szyszka P, Galkin A, Menzel R (2008) Associative and non-associative plasticity in kenyon cells of the honeybee mushroom body. Front Syst Neurosci 2:3
20. Vareschi E (1971) Duftunterscheidung bei der Honigbiene – Einzelzell-Ableitungen und Verhaltensreaktionen. Z vergl Physiol 75:143–173
21. Wittstock S, Kaatz HH, Menzel R (1993) Inhibition of brain protein-synthesis by cycloheximide does not affect formation of long-term-memory in honeybees after olfactory conditioning. J Neurosci 13(4):1379–1386

Scientific Index

A
Aristotle, 4

B
Brenner, S., 487

C
Crick, F., 380

D
Darwin, C., 3–6
Dingman, W., 380

F
Fewell, J.H., 6

G
Golgi, C., 126, 146, 227
Gottlieb, G., 382
Griffin, D.R., 66

H
Heran, H., 57
Hölldobler, B., 3, 5
Holliday, R., 380
Hyden, H., 380

K
Kuwabara, M., 481, 485

L
Lindauer, M., 73, 98, 108, 117

M
Maeterlink, M., 3–7
Masuhr, T., 485, 486

R
Ramon y Cajal, S., 126, 227

S
Schneider, D., 244, 485, 486
Shepard, R.N., 461, 462, 468
Sporn, M.B., 380
Sweatt, J.D., 380

T
Tinbergen, N., 458

V
von Frisch, K., v, vii, 53–55, 57, 61, 62, 66, 77, 78, 93, 97, 108, 110, 117, 120, 237, 261, 270, 315, 317, 318, 325, 473, 485, 491

C.G. Galizia et al. (eds.), *Honeybee Neurobiology and Behavior: A Tribute to Randolf Menzel*, DOI 10.1007/978-94-007-2099-2,
© Springer Science+Business Media B.V. 2012

Subject Index

C.G. Galizia et al. (eds.), *Honeybee Neurobiology and Behavior: A Tribute
to Randolf Menzel*, DOI 10.1007/978-94-007-2099-2,
© Springer Science+Business Media B.V. 2012

Printed by Books on Demand, Germany